LONG DISTANCE TRANSPORT AND WELFARE OF FARM ANIMALS

LONG DISTANCE TRANSPORT AND WELFARE OF FARM ANIMALS

Edited by

**Appleby, M.C., Cussen, V.A., Garcés, L.,
Lambert, L.A. and Turner, J.**

www.cabi.org

CABI is a trading name of CAB International

CABI Head Office	CABI North American Office
Nosworthy Way	875 Massachusetts Avenue
Wallingford	7th Floor
Oxfordshire OX10 8DE	Cambridge, MA 02139
UK	USA

Tel: +44 (0)1491 832111
Fax: +44 (0)1491 833508
E-mail: cabi@cabi.org
Website: www.cabi.org

Tel: +1 617 395 4056
Fax: +1 617 354 6875
E-mail: cabi-nao@cabi.org

A catalogue record for this book is available from the British Library, London, UK.

Library of Congress Cataloging-in-Publication Data

Long distance transport and welfare of farm animals/editorial board:
Appleby, M.C. . . . [et al.].
   p. cm.
  Includes bibliographical references and index.

  ISBN 978-1-84593-403-3 (alk. paper)
1. Livestock--Transportation. 2. Animal welfare. I. Appleby, Michael C.

  SF89.L66 2008
  636.08'3--dc22

           2007041035

ISBN-13: 978 1 84593 403 3

Typeset by SPi, Pondicherry, India.
Printed and bound in the UK by Biddles, Kings Lynn.

Contents

Contributors

AgraCEAS Consulting Ltd, *Imperial College, University of London, Wye, Ashford, Kent, UK. E-mail: info@ceasc.com*

Libby Anderson, *ANNEX Consultancy, Edinburgh, UK. E-mail: libby.anderson@blueyonder.co.uk*

Michael C. Appleby, PhD, *Animal Welfare Policy Advisor, World Society for the Protection of Animals, 89 Albert Embankment, London, UK. Formerly Senior Lecturer, University of Edinburgh. E-mail: michaelappleby@wspa-international.org*

Philip Brooke, *Compassion in World Farming, River Court, Mill Lane, Godalming, Surrey, UK. E-mail: compassion@ciwf.org.uk*

Professor Donald Broom, *Cambridge University Animal Welfare Information Centre, University of Cambridge, Madingley Road, Cambridge, UK. E-mail: dmb16@hermes.cam.ac.uk*

Lissa Collins, PhD, *Compassion in World Farming, River Court, Mill Lane, Godalming, Surrey, UK. E-mail: compassion@ciwf.org.uk*

Sarah Corson, MSc, *E-mail: positivepetbehaviour@googlemail.com*

Victoria A. Cussen, MSc, BSc, DipCABT, *Research Manager, World Society for the Protection of Animals, 89 Albert Embankment, London, UK. E-mail: victoriacussen@wspa international.org*

Monica Engebretson, *Project Director, Animal Protection Institute, 1122 S Street, Sacramento, California, USA. E-mail: monica@api4animals.org*

Mark Fischer, PhD, *Kotare Bioethics, PO Box 2484 Stortford Lodge, Hastings, New Zealand. E-mail: kotare.bioethics@xtra.co.nz*

Carmen Gallo, PhD, *Instituto de Ciencia Animal y Tecnología de Carnes, Facultad de Ciencias Veterinarias, Universidad Austral de Chile, Chile. E-mail: cgallo@uach.cl*

Leah Garcés, Msc, *Director of Programmes, World Society for the Protection of Animals, 89 Albert Embankment, London, UK, E-mail: leahgarces@wspa-international.org*

Temple Grandin, PhD, *Assistant Professor, Department of Animal Sciences, Colorado State University, Fort Collins, Colorado, USA. E-mail: cheryl.miller@colostate.edu*

Bidda Jones, PhD, *Chief Scientist, RSPCA Australia, Australia. E-mail: bjones@rspca.org.au*

Lesley A. Lambert, DPhil, *Director of Research and Education, Compassion in World Farming, River Court, Mill Lane, Godalming, Surrey, UK. E-mail: lesley@ciwf.org.uk*

Peter, J. Li, PhD, *Assistant Professor, Social Sciences Department, University of Huston-Downtown, Office 682-S, One Main Street, Huston, USA. E-mail: LiPj@uhd.edu*

Xavier Manteca, PhD, *Facultad de Veterinaria, Universidad Autónoma de Barcelona, Spain. E-mail: Xavier.manteca@uab.es*

Gustavo A. Maria, PhD, *Faculty of Veterinary Medicine, University of Zaragoza, Spain. E-mail: levrino@unizar.es*

Karen Menczer, PhD, *Natural Resources Biodiversity Consultant, USA. E-mail: perros22002@yahoo.com*

Professor Clive Phillips, *Professor of Animal Welfare, Centre for Animal Welfare and Ethics, School of Veterinary Science, University of Queensland, Queensland, Australia. E-mail: c.phillips@uq.edu.au*

Abdul Rahman, PhD, FRVCS, *Secretary, Commonwealth Veterinary Association, #123, 7th 'B' Main Road, 4th Block (West), Jayanagar, Bangalore, India. E-mail: shireen@blr.vsnl.net.in*

Tamara A. Tadich, MV, MSc, *Programa Doctorado en Ciencas Veterinarias, Escuela de Graduados, Facultad de Ciencias Veterinarias, Universidad Austral de Chile, Chile. E-mail: cgallo@uach.cl*

Jacky Turner, PhD, *Research Consultant, UK. E-mail: turner-research@btinternet.com*

Foreword: Strategies to Improve Farm Animal Welfare and Reduce Long Distance Transport of Livestock Going to Slaughter

T. GRANDIN

Assistant Professor, Department of Animal Sciences, Colorado State University, Fort Collins, Colorado, USA

Most people interested in animal welfare would agree that transporting livestock destined for slaughter across either an ocean or a continent is a practice that should be discontinued. Shipping the chilled or frozen meat and processing the animals in the region of origin would improve welfare and reduce stress.

This book makes a valuable contribution to that aim by providing the first comprehensive coverage of the science, welfare problems and incidence of long-distance transport, with many authors contributing from many countries around the world.

In many situations, the time that an animal is on a vehicle and the condition of the roads are more important than the distance travelled. A 500km trip on a smooth highway will probably be less stressful than a 100km trip on a bumpy dirt road that takes the same length of time. This foreword will cover areas where welfare during transport is likely to have some of the greatest problems. I will then outline methods that are already being successfully used to improve animal welfare on farms, vehicles and slaughter plants, which could be used to eliminate long-distance transportation of slaughter animals. During a career spanning over 35 years, I have learned to understand more and more how economic forces can be used to improve animal welfare. In this foreword, I will discuss how economic incentives to treat animals better can be very effective. All of the things that I recommend are based on either first-hand experience implementing a programme, observations during extensive travel, research or interviews with other individuals who have implemented effective programmes.

Major Problem Areas

Customers require live animals

The Australian live sheep trade is a primary example. The welfare of the sheep would be greatly improved if the sheep were slaughtered in Australia and the meat was shipped to the Middle East. The religious requirements for halal slaughter could be met in Australia. The main barrier to eliminating this trade is that many customers want unchilled meat. The only ways to change this are to increase customers' awareness of animal welfare issues or to convince them that chilled or frozen meat is a good product.

Old cull breeding stock of little value

Some of the worst long-distance travels occur with old cull breeding stock. In the USA, these animals often travel greater distances than young animals. There is less economic incentive to treat these animals well. An effective way to reduce abuses is to increase the value of old breeding stock. Producers need to be educated that if they sell animals before they become emaciated, they will receive more money. In the USA and other parts of the developed world, programmes have been implemented in some areas to fatten old breeding stock so that they will become more valuable for meat.

Highly segmented marketing chains

In the developed world, such as Europe and North America, most high-quality young animals that are fattened for slaughter go directly from the feedlot or farm to a slaughter plant. The transport time is often less than 4h. Old breeding stock often passes through a series of auctions or dealers and the origin of the animals may not be able to be traced. In the developing world, all classes of livestock are often sold through middlemen and dealers. Sectors of the livestock market where animals go through a series of auctions, dealers or middlemen will be the most difficult to improve. Middlemen and dealers who do not own the animals have little economic incentive to reduce bruises, injuries and sickness because they are not held financially accountable for losses.

Transport requires less expertise and capital investment than alternatives

Importing animals from an unstable country to a slaughter plant in a stable economic zone requires much less expertise and financial risk compared to building and operating a slaughter plant in a foreign country. A large slaughter plant is a big investment and many merchants who are either exporting or importing livestock may be lacking either the expertise or the financial capital to build a plant. Transporting live animals usually requires less capital than building and operating a slaughter plant.

Feeding a hungry family has priority

Poor people will usually buy the cheapest meat they can get. There may be some situations where local sources of meat could be developed. Feeding the family is their first priority.

Well-intentioned legislation may have bad consequences

A prime example of this is the US law banning horse slaughter for human consumption, without appropriate and sufficient back-up scenarios for these animals. The closure of two of the three US horse slaughter plants has resulted in unwanted horses being transported even further distances to either Canada or Mexico for slaughter. Live horses are also being shipped to Japan. There are fates worse than slaughter: (i) longer transport times; (ii) transport under substandard conditions; (iii) neglect and starvation; and (iv) being ridden and worked until totally debilitated. Abolition of long-distance transport of slaughter animals must happen in a well thought out and realistic manner that considers the unintended as well as the intended consequences.

Stopping Long-distance Transport for Slaughter

Alliances between producers and meat companies

In these systems, ranchers and farmers produce animals which must meet specific requirements for animal welfare, food safety and other requirements (e.g. organic and natural meats). Producers are often eager to join these programmes in order to get higher prices. Most of these programmes emphasize local production of the animals. Promotion of alliances that include slaughtering animals as close to the point of origin as possible is important.

Welfare auditing by major meat buyers

The programmes that have been implemented by supermarkets and restaurants to inspect farms and slaughter facilities have resulted in great improvements in how animals are treated (Grandin, 2005, 2007a). These auditing programmes are currently operating in the USA, Canada, South America, Australia, Asia and Europe. Large meat buyers such as McDonalds and Tesco have brought about big animal welfare, environmental and labour improvements by using their tremendous purchasing power to enforce standards. Pressure from activist groups forced the upper management of many big companies to examine the substandard practices of their suppliers. Adoption of local sourcing by these companies as part of their purchasing standards would advance the welfare of animals transported for slaughter more than legislation or other avenues.

Educate consumers about the problems of long-distance transport

In the developed world, people are becoming more and more concerned about where their food is coming from. Many consumers may stop buying meat from animals transported long distances. When consumers are educated, they are willing to buy more socially responsible products. This method can be very effective with affluent consumers. Targeting tourist destinations in developing countries broadens the scope of influence that consumers have on animal welfare.

Develop local slaughter facilities with experienced management

In many countries there is a need for high-quality, small slaughter plants in areas where animals are raised. Efforts by the government to build local slaughter plants have failed in some countries due to a lack of experienced people to operate them. Some of these plants contained equipment that was too expensive and difficult for local people to maintain. It is important to build local slaughterhouses in a manner that takes into account local resources and knowledge, and augment this with strong experienced management. Efforts in building producer-owned cooperative plants have had mixed results.

Hire and train people who can implement practical solutions

There is a tremendous need for more experienced people to work on practical tasks. Policies and legislation are useless unless there are practical people on the ground who can implement them. Input from practical people will help create policy and legislation that will work. I urge governments, NGO animal activist groups and live-stock companies to support and educate skilled field workers and researchers.

Improving Animal Welfare

Make producers and transporters financially accountable for problems and losses

Bruises on slaughter cattle were greatly reduced when producers or transporters had to pay for them (Grandin, 1981). Parennas de Costa (2007, personal communication) reported that when supermarkets in Brazil audited bruises and made deductions from transporters' pay, bruising was reduced from 20% to 1% of cattle. Carmen Gallo has also reported that bruises were reduced when transporters in Chile were fined for damage to the animals (Grandin and Gallo, 2007). In another case, problems with weak pigs that were too fatigued to walk off the truck or move to the stunner were greatly reduced when producers were fined $20 for each fatigued pig. Producers reduced the dose of ractopamine, a feed additive that increases lean growth of pigs that in too high a dose may increase the percentage of non-ambulatory pigs.

Use objective methods for assessing losses

Vague guidelines which use terms such as *adequate space* or *proper handling* are impossible to implement because one person's interpretation of proper handling will be different from somebody else's. Loading and unloading of trucks should be measured with numerical scoring of variables such as the percentage of animals that fall, percentage that are electric prodded and the percentage that move faster than a trot (Grandin, 1998a, 2007b; Maria *et al.*, 2004). Alvaro Barros-Restano (2006, personal communication) reported that in auction markets in Uruguay, continuous monitoring has greatly improved handling. Handling practices need to be continuously measured to prevent them from gradually deteriorating. Measures of death losses, non-ambulatory animals, bruises, injuries, pale soft meat in pigs and dark cutting meat in beef should also be used either to provide bonuses or to make deductions from transporter or producer pay.

Use fitness-to-travel measures

Some of the worst abuses I have observed were of animals that were not fit for transport. Emaciated, weak, old cows, sows or ewes should be euthanized on the farm. Published materials for assessing body condition, lameness and injuries should be used. The use of pictures and videos to make assessment of the animal's condition more objective is strongly recommended.

Promote the use of livestock identification and traceback

In much of the developed world, animals are required to be identified with either an individual identification number or the identity of their farm of origin. Animal identification makes it possible to trace animals back to the farm of origin, which enables customers to determine where their meat comes from.

Implement simple practical ways to improve transport in developing countries

Fancy equipment such as hydraulic tailgate lifts or aluminium trailers are often not appropriate in developing countries. The people do not have the equipment or the money to maintain these items. In many countries I have observed that simple improvements can make a big difference. Non-slip truck flooring is essential in vehicles used to haul livestock. Floors can be made non-slip by welding readily available steel bars on the floor in a grid pattern. There is also a need to build ramps for loading and unloading. Many animals in developing countries are injured when they are forced to jump off a vehicle. Training people in animal behaviour and low stress handling methods is also essential (Grandin 1987, 1998b, 2007a,b; Smith, 1998; Ewbank and Parker, 2007).

Use economic incentives for loading and unloading

Give animal handlers who load animals extra pay for low levels of injuries and deaths. In the US and British poultry industry, broken wings were reduced from

5% to 1% by paying a bonus to the chicken loaders when broken wings were 1% or less. The same system has also worked well for people handling pigs and cattle. The worst way to pay animal handlers is based on how many they can move per hour.

Train and use professional handlers and transporters

In some developed countries, there is a shortage of qualified truck drivers. Hands-on jobs such as truck driver and animal handler need to receive more recognition and pay. I have observed effective programmes where handling improved when employees received training, better supervision, higher pay and recognition with a special humane-handler emblem for their hats. For these programmes to work, they must be backed by a firm commitment from upper managers. At the highly educated professional level there is also a shortage. In the USA there are not enough students who want to become large animal veterinarians (NIAA, 2007).

Conclusions

An understanding of how economic factors affect the way farm animals are treated will help policy makers to improve welfare. Holding people financially accountable for losses or providing incentives for low losses will greatly improve animal treatment. The wise use of the tremendous purchasing power of large meat buyers has already brought about some dramatic improvements. All these areas would be helped by avoiding unnecessary long-distance transport of livestock to slaughter.

References

Ewbank, R. and Parker, M. (2007) Handling cattle raised in close association with people. In: Grandin, T. (ed.) *Livestock Handling and Transport*. CAB International, Wallingford, UK, pp. 76–89.

Grandin, T. (1981) Bruises on Southwestern feedlot cattle. *Journal of Animal Science* 53 (Supl. 1), 213 (Abstract).

Grandin, T. (1987) Animal handling veterinary clinics of North America. *Food Animal Practice* 3, 323–338.

Grandin, T. (1998a) Objective scoring of animal handling and stunning practices in slaughter plants. *Journal of American Veterinary Medical Association* 212, 36–39.

Grandin, T. (1998b) Handling methods and facilities to reduce stress on cattle veterinary clinics of North America. *Food Animal Practice* 14, 325–341.

Grandin, T. (2005) Maintenance of good animal welfare standards in beef slaughter plants by use of auditing programs. *Journal of American Veterinary Medical Association* 226, 370–373.

Grandin, T. (2007a) Introduction: effect of customer requirements, international standards and marketing structure on the handling and transport of livestock and poultry. In: Grandin, T. (ed.) *Livestock Handling and Transport*. CAB International, Wallingford, UK, pp. 1–18.

Grandin, T. (2007b) Recommended Animal Handling Guidelines and Audit Guide, 2007 Edition, American Meat Institute Foundation, Washington, DC. Available at: www.animalhandling.org

Grandin, T. and Gallo, C. (2007) Cattle transport. In: Grandin, T. (ed.) *Livestock Handling and Transport*. CAB International, Wallingford, UK, pp. 134–154.

Maria, G.A., Villaroel, M., Chacon, C. and Gebresenbet, C. (2004) Scoring system for evaluating stress to cattle during commercial loading and unloading. *Veterinary Record* 154, 818–821.

NIAA (National Institute of Animal Agriculture) (2007) Shortage of food animal veterinarians: A call for action, Summer, p. 6.

Smith, G. (1998) *Moving 'Em, a Guide to Low Stress Animal Handling'*. The Graziers, Hui, Kamuela, Hawaii.

Overview

Introduction

Around 60 billion animals are reared for food each year worldwide. Most are transported for slaughter, often long distances, both within and between countries.

This massive movement of live animals – more than 1 billion every week – takes place against a background of:

- Increasing concern for animal welfare in many countries;
- Increasing understanding of animal welfare and the factors that affect it, including during transport;
- Increasing awareness that problems for animal welfare may be correlated with other problems, for example, in food safety and quality.

This book is a review of the effects of long-distance transport for slaughter on the welfare of farm animals, of its occurrence and variation worldwide and of ways in which problems for both animals and people may be addressed. It brings together authoritative views and experiences from 28 authors from 18 countries, all of them drawing from international sources and many drawing directly from an initial, even wider resource of contributors.

Science and Practice

The first seven chapters review the science and practice of long-distance transport of farm animals for slaughter, with particular emphasis on implications for animal welfare and prospects for reduction of welfare problems.

Concern for animal welfare has considerable scientific basis, as outlined in Chapter 1. Animal welfare has physical, mental and natural aspects, and science contributes to understanding and assessing all of these. Physical welfare problems caused by transport include injury, disease and stress, and in the worst cases animals die. Mental problems include hunger, thirst, discomfort, pain, frustration, fear and distress. All generally increase with time of deprivation or exposure, such as during long-distance transport. Whether transport is 'natural' has received less attention but consideration of animal biology is important in designing and managing husbandry systems including those for handling and transport. Integrative approaches, combining various welfare indicators, provide strong scientific evidence that long-distance transport causes many welfare problems for farm animals.

However, what we know about welfare and other problems of long-distance transport is often not put into practice. Chapter 2 explores the economic pressures that shape the animal transport industry, with particular reference to exports from Australasia to the Middle East; trade between the USA, Canada and Mexico; and exports from Poland to Italy. These trades are driven by factors such as supply and demand for particular types of livestock; demand and/or price premiums for freshly killed animals (for religious reasons or preference); availability of slaughterhouses and storage capacity; certification of slaughterhouses and of meat (which may be restricted partly for economic reasons); and historical development of the markets (e.g. availability of value-added processing at the destination). All of these factors can be addressed to at least some extent by economic mechanisms, but given local economic interests this often requires national or international action. This is more likely as concern for animal welfare increases. One promising approach to allow expression of that concern is labelling, to provide reliable information both about the welfare of the animals that produced the meat or other products, and about their place of origin.

In some respects, of course, safeguarding and promoting animal welfare will be beneficial to both the animals and the agricultural/processing industry – for example, by preventing disease and injuries such as bruising that reduce meat quality. Chapters 3 and 4 address those issues. Transporting animals causes stress and increases both their susceptibility to infection and their infectiousness. Furthermore, it augments the contacts between animals and hence the spread of diseases. These effects may be reduced by selecting only fit and healthy animals for transport, improving transport conditions and reducing contact between transported and non-transported animals. However, these measures can never be completely effective, so one of the main strategies to mitigate disease spread must be to minimize the numbers of live animals transported and the distances over which they are moved.

Chapter 4 makes similar arguments about meat quality, while making the important point that it takes a very strong stress to have visible effects on meat quality: absence of effects on meat quality does not indicate absence of suffering. Problems for meat quality are frequent, including injuries, morbidity, mortality and abnormal meat pH, causing partial or complete downgrading or condemnation of carcasses and major economic losses. Handling, loading, transport conditions and unloading can all be improved to reduce such problems, and payment systems should be implemented to reward improvements in welfare. However, journey time is the main variable that affects the biological cost of transport stress, and should therefore be minimized.

Together, these chapters suggest that as more information becomes available, the economic advantage of considering welfare becomes clearer. For example, short-term costs in slaughtering animals closer to the farm where they are produced may be covered by the long-term benefits of avoiding disease spread or reduced meat quality.

Nevertheless, the overlap between animal and human interests is not complete, so many countries are developing legislation to set minimum standards for treatment of animals during transport. Chapter 5 discusses the difficulties of enforcing such legislation, taking as a case study the European Union (EU), which has more comprehensive legislation for animal welfare during transport than anywhere else in the world. Compliance with this legislation varies within and between member states of the EU, depending on the investment in, and commitment to, monitoring implementation. Because of the difficulties in assuring the welfare of animals while in transit, it would be more logical to slaughter animals as close to the farm of origin as possible. This would also decrease resource pressure on the authorities concerned.

Chapters 6 and 7 review the principal issues involved in two of the main types of transport considered in the book: sea and road. Transport by train is also important, as seen in the second part of the book, but air transport is rarely used for slaughter animals and is not generally covered here.

Large numbers of livestock are reared for transport overseas, particularly cattle and sheep from Australia to South-east Asia and the Middle East. Chapter 6 describes the methods used, and emphasizes that multiple factors impacting on animal welfare are involved before and after the ship voyage; these include mustering, shearing of sheep, transport to feedlots and several changes of environment that can cause fear and anxiety. During the voyage, high stocking density and high ammonia levels contribute to problems that include failure to eat (the main cause of mortality in sheep), salmonellosis and heat stress. Other potential stressors include noise, motion sickness, changes in lighting patterns and novel environments. The long duration of sea journeys and the changes in the animals' environments provide special challenges compared to other short-distance transport.

Road transport is used worldwide, as discussed in Chapter 7. Some of the key factors affecting the welfare of animals are attitudes to animals and the need for training of staff; methods of payment of staff; laws and retailers' codes; genetics, especially selection for high productivity; rearing conditions and experience; mixing of animals from different social groups; handling procedures; driving methods; stocking density; susceptibility to disease; and journey length. Vehicles transporting animals should be driven more carefully than vehicles with human passengers, especially on roads with sharp bends or poor surfaces. Ventilation management and other methods to avoid harmful physical conditions are important, and transport should be managed to minimize disease susceptibility and spread. Better conditions are needed if journeys are long, but long journeys (the term long having different meanings for different species) should be avoided wherever possible.

The first part of the book sets the scene for a synopsis of the extent to which farm animals are transported long distances for slaughter around the world, the context within which that transport occurs, the procedures used and the measures that may be taken to address problems for welfare.

Global Survey

Given the vast scale of the livestock industry, the job of surveying the transport of animals to slaughter worldwide is an immense one and has not previously been attempted. The seven regions named in the titles of Chapters 8–14 were decided on to cover the world, with the exception of some island nations, particularly in Australasia. For each region, a team of colleagues and contacts was then established to carry out the survey either in entirety or by considering it on a subregional basis; in the latter case, one writer/editor then collated a regional report. In some cases, the survey involved visits and practical research in the region, while in others information was available in written form or on the Internet. Some of the regional authors found it necessary or appropriate to concentrate on particular subregions or countries within their region – usually because information on other countries proved impossible to obtain – so the survey is nearly comprehensive but not completely so.

Long-distance transport of animals for slaughter occurs in all regions of the world and also between regions – most notably from Australia to the Middle East, as covered in Chapters 12 and 14. Variation in practices is loosely associated with the degree of development. In developed countries (mostly covered in Chapters 9, 12 and 13) there is often more legislation protecting the welfare of transported animals; on the other hand, these countries tend to have good infrastructure, such as roads, which enables more systematic and often larger-scale transport of animals over long distances. Developing countries (which form the majority of those covered in Chapters 8, 10, 11 and 14) have fewer structures in place for legislation or for supervision of animal treatment. Their transport systems are generally less advanced, so animals are not often moved over such long distances; however, it is more common for unsatisfactory vehicles and other procedures to be used. All chapters consider approaches to reducing welfare problems in transport, appropriate to their regions.

Chapter 8 covers long-distance transport in North Africa, East Africa, Southern Africa and West Africa, particularly of cattle, goats and sheep, and discusses the cultural, religious and economic factors influencing livestock trade. Animal welfare issues include poor vehicles and infrastructure; lack of enforcement of legislation, where it exists; and inhumane handling of livestock throughout the production chain. Trucking and trekking are the most common means of long-distance transport in Africa, with road journeys of up to 6 days (or much longer including stops at markets) and some treks taking as long as 75 days. Livestock are also shipped round the coast, with journeys of up to 10 days, and imported to Egypt from Australia. One country making progress on animal welfare is South Africa, where non-governmental organizations (NGOs) are having an impact on livestock transport and slaughter, providing oversight of the livestock industry. There is potential for similar activity by NGOs elsewhere, particularly in East Africa.

Transport of animals in, and between, Canada, the USA and Mexico is reviewed in Chapter 9. Animals may be transported across multiple states, regions or provinces and across national borders for fattening and slaughter. Some may even be moved across national borders for slaughter only for the meat then to be shipped back to their countries of origin for consumption. This transport is influenced by the economic costs of transporting animals (which tend to be lower than the costs of transporting feed), geographical differences in feed and forage availability and prices, and the development and location of feedlots and slaughterhouses. These countries have varying laws, codes and regulations, and there are significant shortcomings in scope and enforcement which present challenges for ensuring animal welfare. For example, in the USA the '28-Hour Law' is supposed to limit transport to that period, but it does not apply across national borders and it has not been applied to road transport until recently. However, in North America civil society pressure can influence legislation, so public education and advocacy have the prospect of achieving improvements for transported farm animals, including reducing the number transported long distances. The most likely mechanisms for progress are the setting of travel time limits, rest periods and provisioning of food and water for livestock during road transport.

Chapter 10 points out that countries in the central and northern part of South America (Bolivia, Colombia, Ecuador, Perú, Venezuela) are less developed and give less priority to animal welfare than those in the south (Argentina, Brazil, Chile, Paraguay, Uruguay). There is also considerable variation in climate and geography. The continent includes some countries among the world's most important beef exporters (Brazil, Argentina) and others where this business is small but important (Chile, Uruguay). Livestock producers and veterinarians are aware that international commercial agreements require them to safeguard animal health, and they also consider consumer requirements, including ethical considerations. Export therefore encourages quality assurance schemes and good livestock practices that consider welfare, including during transport. Export of live animals is rare; transport within countries is mostly short distance but it also occurs over long distances and can occasionally reach up to 60h due to bad roads, bad weather and intermediate dealers. Bad practices during loading, transport and unloading of animals are common, as well as overstocking of trucks. Most countries are members of the World Organisation for Animal Health (OIE), and are starting to bring national regulations into line with OIE standards. Regional research and training are important at all stages of the meat production chain to achieve those standards.

Meat consumption and animal transport are increasing across Asia, and Chapter 11 reports a widespread preference for freshly killed meat, whether out of habit, beliefs about quality, religious preference or lack of refrigeration. In China (responsible for 29% of world meat production in 2005) there are long distances between some significant production areas and areas of consumption, and this problem is getting worse with urbanization as production is moved away from cities for environmental reasons. China also exports livestock to Muslim countries. In Taiwan distances are shorter, but similarly most production and consumption are in different regions. In India less meat is consumed but old cattle are transported to Kerala and West Bengal, the two states where cattle slaughter is permitted, or to Pakistan and Bangladesh. Across Asia, in addition to long journeys, animals are often subjected to poor truck design and driving, long periods without food or water, rough handling, overcrowding and extremes of weather. Many countries are developing welfare legislation, and a priority must be to reduce transport times. This requires the development of local slaughterhouses with high standards together with refrigerated transport. Consumers need to be educated in welfare issues and markets for refrigerated meats need to be developed. All those involved in animal transport and marketing need to be trained to a high professional standard and then treated and respected accordingly.

The majority of livestock within Australia and New Zealand (Chapter 12) are not transported to slaughter over long distances. Both countries have government-endorsed codes or guidelines and quality assurance programmes. In New Zealand codes of welfare have legal status. In Australia there are no enforceable standards for land transport, but reported instances of poor animal welfare in either country during such journeys are rare. However, in Australia some livestock and feral animals may be transported long distances, sometimes extremely long, from remote areas. These journeys may involve significant changes in environmental conditions. Animal responses to these journeys have not been adequately documented, but these animals are often unused to human contact, and may suffer from the stress of mustering, confinement and long-distance transport. Animals are also exported by sea from Australia, and to a lesser extent New Zealand, for slaughter overseas. In view of the conditions during transport, the mortality recorded, the equivocal benefits to the exporting country, and the alternative of supplying chilled product, this chapter concludes that the long-distance export of livestock for slaughter should not take place.

Chapter 13 reports that millions of animals are transported long distances in Europe each year. Routes causing particular problems for welfare include horses from Central and Eastern Europe to Southern Italy; pigs from the Netherlands to Southern Italy; and imports of sheep into Greece. Even where legislation is adhered to, animals may endure poor conditions, stress, regional variations including in climate, and infrastructure which can further compromise their welfare. Furthermore, inconsistencies in enforcement within the EU and/or ineffective penalties for non-compliance mean that legislation is often ignored. Reports of welfare infringements are not uncommon and include overstocking, illegal route plans, inadequate road vehicles, and sick, injured and dead animals. Many animal welfare organizations in Europe believe that transport of live animals over a set time limit should be ended completely, with trade over that limit in carcass form. However, the live export trade in Europe is complex, influenced by many factors including the economy and the religions and cultures prevalent in different regions. Nevertheless, legislative initiatives and action plans to be implemented within the next few years are generally supported by public opinion and will offer opportunities to reduce the welfare problems of livestock transported long distances to slaughter.

Most of the Middle East has an arid climate yet a large demand for meat. Chapter 14 describes how this, coupled with traditional patterns of slaughter and consumption, has led to imports of millions of live ruminants, particularly sheep, into the region. They are also traded locally. In addition to long sea journeys, animals are then kept for varying periods in feedlots, before being transported for considerable distances to cities, to markets and then to butchers for local trade; or some consumers buy straight from the feedlots and transport their purchases by road. Various vehicles are used including conventional trucks, open trucks, pickup vans and cars – with animals sometimes carried in car boots. Although cultural traditions on humane treatment of animals may exist, for example, in religious teachings, specific laws on animal welfare are mostly lacking in this region. Enforcement of legislation in countries where it exists and introduction in countries where there is none should be a priority. Furthermore, education of traders and butchers in animal welfare is important. For some communities, this could be done by highlighting religious teachings.

General Conclusion

Long-distance transport of animals for slaughter is common throughout the world, and there is considerable evidence that it causes many welfare problems for farm animals. An increasing number of authorities give due regard to such evidence, including intergovernmental organizations. For example, the European Food Safety Authority (see Chapter 1) says: 'Transport should therefore be avoided wherever possible and journeys should be as short as possible.' The data and interpretations reported in this book endorse that view. Long-distance transport for slaughter is unnecessary and should be replaced by trade in carcasses and meat. To apply that principle it is necessary to take cultural and religious sensitivities into account, but that can be done in a context where there are cultural traditions on humane treatment of animals, including in religious teachings – and against a background of increasing awareness worldwide about animal welfare issues. Avoiding long-distance transport for slaughter will require developments in infrastructure such as local slaughterhouses and refrigerated transport. Producers, processors, retailers and consumers need to be educated in welfare issues and markets for refrigerated meats need to be developed. All such matters can be addressed to at least some extent by economic mechanisms, but given local economic interests this often requires national or international action. It is time for countries and the international community to take this responsibility seriously.

Michael C. Appleby
Victoria Cussen
Leah Garcés
Lesley Lambert
Jacky Turner

Acknowledgement

The Editorial Board is grateful for the contribution made by all the organizations in the Global Long-Distance Transport Coalition: Bom Free USA united with API, CIWF, Dutch SPCA, Dyrenes Beskyttelse, HSUS, ILPH, Eurogroup, RSPCA, RSPCA AU, SSPCA and WSPA. We also wish to thank Paul Rainger, Sofia Parente, Virag Kaufer and Poonam Doshi.

1 Science of Animal Welfare

M.C. Appleby

World Society for the Protection of Animals, 89 Albert Embankment, London, UK

Abstract

Concern for animals has increased in many countries over a long time, particularly the last 50 years. Development of scientific disciplines relevant to animal welfare also has a long history and recent acceleration. These disciplines include animal husbandry, animal science, veterinary medicine and behavioural science, which contribute to our understanding of welfare and the welfare problems of long-distance transport of farm animals for slaughter. Welfare is affected by many factors, as expressed by the Five Freedoms, and by the fact that people vary in the emphasis that they place on physical aspects, mental aspects and natural-ness. Science copes with this complexity by applying a systematic approach to the asking of questions, and yielding answers that, as information accumulates from one or more compa-rable situations, are increasingly reliable and generalizable. As one example, science requires careful record keeping, so that factors affecting injury and mortality during transport can be proven rather than assumed, and therefore addressed.

Physical welfare problems caused by transport include injury, disease and stress – which may be detected from behaviour, from physical effects such as failure to grow or from physi-ological measurements. In the worst cases, animals die, and mortality is increased by high or low temperatures, by long journey times and by transporting very young animals. Evidence about mental aspects of welfare is mainly of two sorts: whether animals have what they want and whether they are suffering. Many preferences of animals may be frustrated by transport, both to avoid conditions such as vibration and noise, and to express normal behaviour. Forms of suffering caused by transport include hunger, thirst, discomfort, pain, frustration, fear and distress. All of these generally increase with time of deprivation or exposure, such as during long-distance transport.

The idea that welfare is related to naturalness has received less scientific attention, but is implicit in the scientific approach to using animal biology in understanding, designing and managing husbandry systems including those for handling and transport. It is therefore appropriate to point out that transport is clearly unnatural.

Decisions about treatment of animals will be most firmly based, and acceptable to the greatest possible number of people, if they take into account all three approaches to welfare: physical, mental and natural. There are various integrative approaches that do this, both with respect to the inputs affecting welfare and the outcomes hoped for. From such approaches, there is considerable scientific evidence that long-distance transport causes many welfare problems for farm animals. The European Food Safety Authority is one of an increasing number of organizations that take such evidence seriously, and says: 'Transport should therefore be avoided wherever possible and journeys should be as short as possible.'

Introduction

Concern for non-human animals (hereafter 'animals') has always been shown by many or most people: witness the many relevant stories in ancient religious texts. The story of Noah's Ark in the Bible could be described as the first published example of long-distance transport of animals, and the animals saved included those regarded as 'unclean' as well as 'clean', with little use as well as useful, wildlife as well as livestock. This concern for animals has increased in many countries over about the last two centuries and particularly the last 50 years. One factor in developed countries is industrialization, with a decline in the number of people earning their living from farming and pressure for animal protection mostly coming from urban populations. Affluence in these countries has increased, allowing people to diversify their interests beyond the question of whether they can afford the next meal. Perhaps most importantly, part of the industrialization process has been intensification of agriculture, and it is this that has provoked much of the concern for farm animal welfare. Development of scientific disciplines relevant to animal welfare has a similarly long history and recent acceleration. This chapter will outline the ways in which science contributes to our understanding of welfare with reference, where possible, to long-distance transport of farm animals for slaughter.

Development of Animal Welfare Science

Once, most people were involved in farming, and in some less-industrialized countries more than half of the population is still associated with agricultural employment. Rollin (1993, p. 6) has emphasized that in those conditions farmers practise enlightened self-interest: good husbandry can be summed up by the maxim 'I take care of the animals, the animals take care of me'. Much of that husbandry was based on enquiry and understanding that developed into what became known as animal science, and much animal science was relevant to welfare, such as nutrition, physiology (reproductive, respiratory, neural and so on) and control of behaviour. For example, more is known about the nutritional requirements of chickens than for any species other than humans, and arguably more than what is known for humans too. Similarly, diagnosis and treatment of injury and disease, and their prevention, have developed over centuries into veterinary science, which is fundamental to safeguarding some very important aspects of welfare.

However, as Rollin (1993, p. 7) puts it:

> Industrialized, high-technology agriculture has given us the ability to move beyond our implicit contract with the animal, to move beyond keeping square pegs in square holes. We can now . . . fit animals into environments which are good for us without necessarily being good for them.

Intensification of livestock agriculture proceeded apace in Europe and North America in the 1950s and 1960s, faster there than anywhere else, and was given widespread publicity when Ruth Harrison's book *Animal Machines* was published in 1964 and serialized in a British Sunday newspaper. Public reaction was vociferous on both sides of the Atlantic, and the UK government set up a committee chaired by Professor Rogers Brambell, which reported in 1965 (HMSO, 1965). Many of its recommendations were incorporated into the Agriculture (Miscellaneous Provisions) Act in 1968 and as a result an independent Farm Animal Welfare Council (FAWC) was established. The Brambell Report and FAWC have had international impact, including through their development of the concept of Five Freedoms (Table 1.1) and through FAWC's reports (e.g. 2006).

Professor Brambell was a veterinarian and the other members of his committee were veterinarians, agricultural scientists and the behavioural scientist Professor William Thorpe. FAWC and similar committees in other countries, such as the Danish Animal Ethical Council, have continued to have scientific input, while the European Commission appoints a Scientific Committee on Animal Health and Animal Welfare (now the Scientific Panel on Animal Health and Welfare of the European Food Safety Authority) that consists wholly of scientists.

The Brambell Report also stimulated interest among scientists in carrying out research relevant to welfare. One such project examined a contention of the Report that if hens are to be kept in cages the floors should be rigid. Interestingly, it rejected that contention, which had been based not on scientific evidence but on speculation. Hughes and Black (1973) tested this assumption scientifically, giving hens a choice between rigid cage floors and others of flexible mesh, and found that they chose the latter, probably because the small mesh gave better support for their feet.

Table 1.1. The concept of Five Freedoms originated from a phrase in the Brambell Report (HMSO, 1965) and was developed by the Farm Animal Welfare Council (FAWC, 2007). All of these freedoms are likely to be compromised during long-distance transport.

Brambell Report
Farm animals should have freedom to stand up, lie down, turn around, groom themselves and stretch their limbs

FAWC – Animals should have

Freedom from hunger and thirst	By ready access to fresh water and a diet to maintain full health and vigour
Freedom from discomfort	By providing an appropriate environment, including shelter and a comfortable resting area
Freedom from pain, injury and disease	By prevention or rapid diagnosis and treatment
Freedom to express normal behaviour	By providing sufficient space, proper facilities and company of the animal's own kind
Freedom from fear and distress	By ensuring conditions and treatment which avoid mental suffering

In 1976 the Council of Europe produced the Convention for the Protection of Animals Kept for Farming Purposes. In addition to the 21 member countries (there are now 46), the EU became a party to the Convention in 1978 and one consequence was the establishment of a 'farm animal welfare co-ordination programme' from 1979 (Tarrant, 1983). This financed background scientific work on welfare, and other research was funded by national governments in northern Europe. The main emphasis was research on husbandry systems, particularly those for egg-laying hens, and most of this work was by scientists, especially behavioural scientists. As this work developed, acceptance of ideas about animal welfare changed among all animal scientists, not just those who themselves were involved. Welfare is now taken seriously as an academic subject, is included in veterinary curricula and other courses, and is the subject of several scientific journals. There are also established positions in government and industry related to the field.

Research on farm animal transport has developed over the same period. As one illustration, the 20 chapters of the book *Livestock Handling and Transport* (Grandin, 2000) together cite 1375 references, mostly scientific. Of these, only 5% are from before 1970, while 13% are from the 1970s, 33% from the 1980s and 49% from the 1990s.

Concepts of Welfare

The fact that science contributes to our understanding of animal welfare should not be misunderstood to mean that welfare is a wholly scientific concept. This is well explained by Duncan and Fraser (1997, p. 20):

> 'Animal welfare' is not a term that arose in science to express a scientific concept. Rather it arose in society to express ethical concerns regarding the treatment of animals. The 'welfare' of an animal refers to its quality of life, and this involves many different elements such as health, happiness, and longevity, to which different people attach different degrees of importance. . . . However, because science plays an important role in interpreting and implementing social concerns over the quality of animal life, animal welfare was adopted as a subject of scientific research and discussion. This adoption has led to a remarkably protracted debate on how to conceptualize, in a scientific context, a concept that is fundamentally rooted in values.

Some scientists have formulated short, one-line explanations or definitions of animal welfare similar to those of purely scientific concepts, and these can help clarify the topic under discussion. For example, the Brambell Report (HMSO, 1965) said that welfare is 'a wide term that embraces both the physical and mental well-being of the animal'. Another widely quoted definition is by Broom (1996): 'The welfare of an individual is its state as regards its attempts to cope with its environment.'

However, such brief explications do not provide a framework for study or assessment. Perhaps more generally helpful in those respects have been broader approaches. One that is widely quoted, particularly in continental Europe, is from the Council of Europe's Convention (1976), which puts a strong emphasis on science:

> Article 3: Animals shall be housed and provided with food, water and care which –
> having regard for their species and to their degree of development, adaptation and

domestication – is appropriate to their physiological and ethological needs, in accordance with established experience and scientific knowledge.

It might be worth commenting at this point about the emphasis on behavioural science (particularly ethology, the naturalistic study of behaviour) in the field of animal welfare. This has arisen because behaviour is the interface between the animal and most aspects of its environment and may therefore be both the source of some problems (or an indicator of their absence) and a symptom of other problems such as disease (Appleby and Waran, 1997).

Probably the most widely used framework for expressing the complexity of animal welfare, though, is the Five Freedoms as formulated by FAWC (Table 1.1). These can also be considered in inverse, so to speak, to point out that there are different aspects of welfare, such as hunger, thirst, discomfort and pain, all of which are amenable to scientific study.

It is also interesting that FAWC (2006, p. 3) has recently produced a new one-liner that combines their own approach with that of Broom (1996): 'In our view, welfare encompasses the animal's health and general physical condition, its psychological state and its ability to cope with any adverse effects of the environment in which it is kept.'

The core of this chapter, however, will develop an even broader framework taking into account the fact that people do not all have the same concept of welfare (Tannenbaum, 1995). Indeed, one common response to this variation, implicit or explicit, is to avoid any direct consideration of welfare and settle instead for a list of factors regarded as important. Needless to say, such lists are quite variable. One complication is that they often include not only factors that affect welfare, but also the effects of those factors on the animals concerned. The Welfare Codes issued by the UK's Ministry of Agriculture, Fisheries and Food (MAFF, 1987), for example, provided such a list in the preface until recently (Table 1.2). Among the first few items on the list, we may note that shelter and fresh water are factors that should

Table 1.2. Factors relevant to welfare, listed in the preface to the Welfare Codes of the UK's Ministry of Agriculture, Fisheries and Food, in this case for turkeys. (From MAFF, 1987.)

Comfort and shelter
Readily accessible fresh water and a diet to maintain the birds in full health and vigour
Freedom of movement
The company of other birds, particularly of like kind
The opportunity to exercise most normal patterns of behaviour
Light during the hours of daylight, and lighting readily available to enable the birds to be inspected at any time
Floors/perches which neither harm the birds, nor cause undue strain
The prevention, or rapid diagnosis and treatment, of vice, injury, parasitic infection and disease
The avoidance of unnecessary mutilation
Emergency arrangements to cover outbreaks of fire, the breakdown of essential mechanical services and the disruption of supplies

be provided by people caring for animals, but comfort and freedom of movement are descriptions of the animals' responses to their environment. It is important to identify both how to safeguard welfare and what constitutes it (see section on 'Integrative approaches'), and revised Welfare Codes produced by the successor to MAFF, the Department for Environment, Food and Rural Affairs (DEFRA, 2002), explain this more clearly.

Three concepts are common among animal-centred approaches to the question of what welfare actually is, and people may believe one of these or a mixture of two or three (Duncan and Fraser, 1997; Fraser *et al.*, 1997). First, animal welfare may concern health and fitness, so that problems such as disease and injury are the most important challenges to welfare. Second, it may concern animals' preferences and feelings, such as pleasure and suffering. Third, welfare may concern the ability of animals to express their 'nature', for example by living in natural conditions. The concepts can be summarized as emphasizing animal bodies, minds and natures, respectively (Fig. 1.1).

Once recognized, these concepts can be identified in other treatments of animal welfare, including in lists of factors (Table 1.2). For example, the Brambell Report (HMSO, 1965) said, as we have seen, that welfare involves both physical and mental well-being. It also stated that '[i]n principle we disapprove of a degree of confinement which necessarily frustrates most of the major activities which make up the animal's natural behaviour'. In other words, it acknowledged the importance of the three elements of body, mind and nature. These three elements can also be identified in FAWC's Five Freedoms (Table 1.1), which include feelings such as hunger, physical aspects such as injury and aspects of naturalness such as expression of normal behaviour.

One other important issue is the differences between animal species. While different concepts of welfare vary in their conclusions on this issue, for example on the priority that should be given to the welfare of complex species compared to those less complex, it is perhaps generally true that more complex conditions are needed to safeguard the welfare of more complex species (Appleby, 1999).

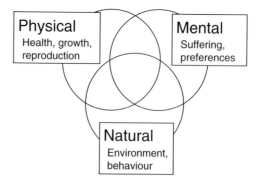

Fig. 1.1. Three concepts of animal welfare emphasizing physical, mental and natural aspects can be characterized as considering animal bodies, minds and natures, respectively. They overlap, but not completely. (After Appleby, 1999.)

Nevertheless, this issue can be set aside here, because this book is concerned only with a small range of species, those vertebrates predominant in farming. And while poultry are sometimes belittled in comparison to mammals, there is no scientific justification for this. On the basis of variety and complexity of behaviour shown, the mental processes of poultry are comparable in at least some ways both to parrots (Pepperberg, 1987) and many mammals (Rogers, 1995), and the welfare of birds is just as much a matter for concern as that of mammals.

In the following sections we will examine the scientific understanding of the three concepts of welfare illustrated in Fig. 1.1. For the purposes of discussion, we are considering each separately, although they are clearly connected, for example in mental aspects of physical injury and stress.

Physical aspects

Producers and owners of livestock tend to be most concerned with physical aspects of welfare, for reasons that include their own interests. Indeed, they sometimes claim that this overlap between owners' and animals' interests proves that concerns about welfare – in transport and in other stages of animal production – must be groundless, or at least that any problems that exist must be unavoidable. The National Pork Board (2005) of the USA, for example, says: 'Because the welfare of their animals directly affects their livelihood, pork producers work to ensure their animals are treated humanely. Anything less would be self-defeating.' However, their primary concern is with group performance. From the animals' perspective, it is the individuals that matter. Furthermore, the owner's decisions must be affected by financial considerations (see Chapter 2, this volume): for example, some modifications to transport methods may reduce weight loss in animals and therefore increase the price received from their sale, but nevertheless be considered too expensive to implement. So owners' and animals' interests do not overlap completely. As such, applying scientific approaches directly to animal welfare may produce different conclusions to those of traditional animal production science. In fact, the conventional approach, emphasizing financial efficiency, has not always identified the best methods even to achieve its own aims. Thus, it took the alternative approach, aimed at reducing problems for the animals concerned, to identify the fact that understanding animal behaviour can improve design of handling systems and hence efficient use of labour in handling livestock (Grandin, 2000).

This difference in approaches is perhaps most clearly demonstrated by the fact that records of welfare problems in handling and transport of farm animals have been sparse, even of unequivocal problems such as mortality. This reflects the commercial assumption that prevention of such problems must be impossible or financially prohibitive. Yet record keeping is a basic requirement for scientific understanding of a problem, and for conclusions on how to address it. This is now better understood, and record keeping is required for most on-farm assurance schemes, although still frequently underemphasized for transport.

Records of mortality in commercial or experimental conditions, both during and after transport, have now shown, for example, that it is increased by high or low temperatures (Knowles *et al.*, 1997; Knowles and Warriss, 2000), by long journey times (Warriss *et al.*, 1992) and by transporting very young animals (Staples and Haugse, 1974; Knowles, 1995).

Precursors to mortality are also, of course, important welfare problems in themselves. The incidence and severity of injury and disease are directly measurable, and aspects of transport that affect these are reviewed in other chapters (see Chapters 3 and 4, this volume). In addition, a considerable amount is known about the causes and effects of injury and disease, at both an anatomical and physiological level (Flecknell and Molony, 1997; Hughes and Curtis, 1997), together with their implications for welfare. For example: 'Injury is of concern both because of the consequent pain which is likely to arise from traumatized tissues, and also because of its incapacitating effect on the animal. This incapacity can lead to other problems such as hunger, thirst and inability to find shelter.' (Flecknell and Molony, 1997, p. 63)

More generally, the physiological causes and effects of injury and disease are part of the broader issue of stress. As early as 1914, Cannon showed that physical and emotional disturbances may trigger the same physiological responses, and Selye described this as the General Adaptation Syndrome in 1932. Nowadays the term stress is generally restricted to negative effects. Moberg (2000, p. 1) defines it thus: 'The biological response elicited when an individual perceives a threat to its homeostasis. The threat is the "stressor". When the stress response truly threatens the animal's well-being, then the animal experiences "distress".' Moberg (1999, p. 3) also makes clear both how this concept relates to biological functioning and the links between physical and mental aspects of welfare: 'A stress response begins with the central nervous system perceiving a potential threat to homeostasis. Whether or not the stimulus is actually a threat is not important; it is only the perception of a threat that is critical.'

It is widely agreed that '[t]ransport is generally an exceptionally stressful episode in the life of the animal and one which is sometimes far removed from an idealized picture of animal welfare' (Knowles and Warriss, 2000, p. 385). The European Food Safety Authority (2004) says: 'Transport should therefore be avoided wherever possible and journeys should be as short as possible.'

Transport is not a single, closely defined stressor. Transporting animals involves changes to their whole environment, or at least to many of its most important aspects: they will be moved from familiar areas and will encounter strange materials, smells, sights, sounds and vibrations, they may be handled and mixed with unfamiliar animals, subjected to changes in temperature and air movement, possibly hurt or injured and restricted in space, feed and water (see Chapters 6 and 7, this volume). As a result, it is common for welfare to be compromised in all the areas indicated by the Five Freedoms (Table 1.1). In this context, the fact that many different stressors have similar physiological effects means that physiological measurements can be used as integrative indicators of welfare during transport.

So while stress is sometimes deduced from behaviour or from physical effects such as failure to grow, it is more often assessed from physiological measurements, including heart rate, blood cell counts and compounds such as glucocorticoids (stress hormones released by the adrenal glands) (Terlouw *et al.*, 1997). Cell counts can indicate the state of an animal's immune system; stress such as that which can occur during transport sometimes causes immunosuppression, which may be a factor in disease susceptibility. Glucocorticoids can be detected in the blood or in body secretions such as saliva and faeces. Physiological measurements of body function can therefore be made using heart rate monitors, blood sampling and other non-invasive methods.

Such measurements, though, do not provide definitive indications of stress or its absence and hence of welfare. Increased heart rate or corticoids do not themselves show that a situation is stressful – indeed, animals show the same symptoms during mating. But there may be other reasons to assume that animals are stressed, including the 'threat to homeostasis' referred to by Moberg. For example, Mitchell and Kettlewell (1993) studied heat stress, which involves just such a threat. They used a number of physiological measurements to determine the effect of temperature and relative humidity on transported broilers. Further, in this particular case they were able to make absolute judgements of the severity of the stress, categorizing it as minimal, moderate or severe, and helping to prevent the worst possible effect of severe stress, i.e. death. In other circumstances, physiological evidence may be useful for comparisons between situations rather than for absolute judgements. Thus, when adult sows are mixed into a new group – a common practice prior to or during transport – they usually fight, and they may reasonably be assumed to be stressed. Barnett *et al.* (1996) compared different methods of mixing, and found that sows given stalls into which they could escape had lower cortisol concentrations than others. This helped them to conclude that this method was less stressful than the others.

If animals are functioning properly, they are not just healthy. They are eating, drinking and excreting, moving, breathing and responding to stimuli. All of these aspects of normal behaviour are measurable and liable to disruption during transport, and the physical effects of this are also measurable, such as by weight loss. Again, absolute judgements of the importance of such physical measurements are impossible. Nevertheless, understanding the ways in which the integrity of animals' physical bodies is threatened by handling and transport is at least an important element of welfare. It can and should also be combined with evidence of the animals' own perception of their treatment, with conclusions about mental aspects of welfare.

Mental aspects

The concept that welfare concerns the perceptions or feelings of animals is perhaps the most common of the three approaches to welfare shown in Fig. 1.1. Dawkins, a foremost authority on welfare, has suggested (1988, p. 209): 'To be concerned about animal welfare is to be concerned with the subjective feelings of animals, particularly the unpleasant subjective feelings of suffering and pain.' One criticism of this concept is that we cannot ever know about those feelings for certain. However, study of animal sentience has increased rapidly in recent years (Kirkwood *et al.*, 2001; Webster, 2006), so we do have evidence, which can be used to make conclusions just as a court uses evidence to produce a verdict even when there is no definite proof of guilt or innocence. This evidence is of two sorts: evidence about what animals want and evidence about whether they have positive feelings, whether they 'are happy'. These are not the same thing. If a person's or an animal's preferences are fulfilled, this may make them happier, but it may not, so these approaches are partly distinct, but they are often not distinguished in discussions of animal feelings and welfare (Appleby and Sandøe, 2002).

Animals' preferences for or against features of handling and transport may be indicated by variables such as ease of loading, and these can also be studied experimentally. Rushen (1986, p. 391) examined the aversiveness of electro-immobilization in sheep:

> Compared with allowing the sheep to run freely through the race, both physical restraint and electro-immobilization increased the time required to run through on subsequent occasions, suggesting that both treatments were aversive. The longest running times were recorded from the immobilized sheep, and considerably more time had to be spent pushing these animals up the race. These results suggest that electro-immobilization is a more aversive experience for sheep than simple mechanical restraint. The attachment of the electrodes and the necessary extra handling appeared to make no significant contribution to this aversiveness. Furthermore, sheep remember this experience for some 12 weeks.

Stephens and Perry (1990, pp. 50–51) used a different approach, training pigs to push a switch in a noisy, vibrating transport simulator. All the pigs learned to turn the simulator off, and soon kept the apparatus switched off for about 75% of the time. They concluded that young pigs found the vibration, typical of transport, to be aversive.

So animals may have preferences for certain changes in the environment, or for carrying out certain behaviour patterns. Frustrating those preferences may constitute a reduction in welfare, while fulfilling them may increase welfare, even if only temporarily.

The next consideration is feelings of pleasure and suffering. There has often been an emphasis on negative feelings, and their avoidance, as in the quotation from Dawkins (1988) above. Yet, just as '[h]ealth . . . is more than the absence of disease' (Hughes and Curtis, 1997, p. 110), welfare is not just avoidance of the negative, but includes promotion of the positive (Mench, 1998). Evidence about positive feelings is scant, but one possible source is play. It is reasonable that play should be pleasurable to animals, as it is to humans, because play is functional, allowing animals to learn or practise actions that will be useful later (Bekoff, 1998), and natural selection may be expected to have produced positive reinforcement for such behaviour. Similarly, it is possible to interpret much normal behaviour as evidence of positive feelings. Thus, the UK's Welfare Codes suggest that animals should be provided with comfort, and with 'the company of other [animals], particularly of like kind' (Table 1.2). Behaviour such as resting and affiliative behaviour (including grouping) is therefore relevant here. This is an area where there is a risk of uncritical anthropomorphism: it is easy to describe animals as looking contented, more difficult to obtain good evidence of such contentment. However, in recent years there has been considerable progress in qualitative assessment of behaviour, led by Wemelsfelder (2007), allowing systematic evaluation of both positive and negative aspects of quality of life.

More evidence is available on negative feelings. Probably the most useful general word for such feelings is suffering. Some people would not describe mildly negative feelings as suffering, so use of the word may concentrate attention on more important, stronger feelings. Phrases such as 'pain and suffering' are sometimes used, as in the Dawkins (1988) quotation, but this is probably to give particular emphasis to pain rather than to distinguish it from other forms of suffering. There is no definitive list of forms of suffering; we shall use here those included in the Five Freedoms (Table 1.1), and first those with physical causes that impact animals directly: hunger, thirst, discomfort and pain.

Hunger and thirst are major concerns in long-distance transport. In addition to the journey itself, it is common to withhold feed prior to transport to reduce defaecation and travel sickness (Bradshaw *et al.*, 1996) during the journey. There is good evidence to confirm that the effect on the animals increases with time of deprivation, both from experimental studies (Savory *et al.*, 1993) and from the response of transported animals if they are subsequently given feed and water (Hall *et al.*, 1997).

Discomfort frequently occurs in association with other problems or as a lesser form of such problems. Physical discomfort is probably common in transport, due to poor design of handling facilities and vehicles and crowding, but this has not been studied as a separate topic. Thermal discomfort was studied as part of the study by Mitchell and Kettlewell (1993), mentioned above, on heat stress in broilers during transport, as it could be considered equivalent to their category of minimal stress. Their method is now used in some countries in warning devices on transporters that indicate when stress will occur, to allow the driver to take preventative measures (Kettlewell and Mitchell, 2001).

Pain in mammals and birds is associated with receptor cells called nociceptors (Gentle, 1997). Farm animals suffer pain when they are subject to mutilations such as tail docking and beak trimming, to injury from others such as by aggressive attacks, to long-term problems such as foot and leg damage – which may be exacerbated during transport – to accidental injury such as bruising or bone breakage during handling or transport and to physical treatment such as electric goads and hanging poultry on shackles for slaughter (Flecknell and Molony, 1997). As just one example, laying hens frequently suffer broken bones when removed from battery cages and placed into transport crates (Gregory and Wilkins, 1989). The resultant pain will continue throughout the journey, however long (although increasing journey times will exacerbate this suffering), and increase during the subsequent handling and shackling prior to slaughter.

Other negative feelings may not have direct physical causes, but are also aversive and may thus constitute suffering, such as frustration, fear and distress.

The Five Freedoms (Table 1.1) include freedom to express normal behaviour, and animals being transported will be frustrated in this expression in many ways. Indeed, at typical stocking densities in vehicles, animals are not afforded even the earlier, much more modest recommendation of the Brambell Report (HMSO, 1965): 'An animal should at least have sufficient freedom of movement to be able without difficulty to turn round, groom itself, get up, lie down and stretch its limbs.'

Fear and distress are probably associated with many of the other problems in this and the previous section. They also occur independently, and are likely to be worse during handling and transport than at most other times in the animals' lives because of the intensity and variety of changes to their environment. Thus, many animals that have not had much human contact react adversely to approaching humans (Jones *et al.*, 1981), so the close contact involved in collecting them for transport will be highly stressful. Fear and distress are also caused by aggression, which is often increased following the mixing of animals for transport commonly practised to produce convenient group sizes.

We can also draw conclusions about fear and distress from physiological measurements. Duncan *et al.* (1986) measured the heart rate of broiler chickens that were being caught by hand or by machine. Heart rate went up in both groups, which suggested that the birds were frightened. It returned to normal more quickly in the latter group, so they concluded that it was less frightening for a chicken to be picked up by a machine than by a person.

Natural aspects

Transport is clearly unnatural. However, the third concept in Fig. 1.1, that welfare concerns the ability of animals to express their 'nature', has received less scientific attention than the other two. The idea of an animal's nature, comparable to human nature, has been emphasized by Rollin (1993), who also uses the Greek word *telos*, after Aristotle. There have been two main answers to the question of what is necessary for animals to express their natures. The first is that they must be kept in ways that allow them to perform natural behaviour and the second is that features of the natural environment such as sunshine and fresh air are important in themselves. It is difficult to translate these approaches into specific recommendations or requirements – particularly in relation to highly unnatural aspects of their treatment such as transport – but that is not the point. Human nature can be respected without saying what sort of house people should live in, and the same goes for animals. To consider animal nature is to emphasize that they are animals, not machines or mere economic units. One possible conclusion is that transport is so unnatural that it should be avoided altogether if at all possible, particularly long-distance transport.

Whether or not explicit, and whether or not explicitly scientific, this sort of thinking is relevant to an ethological or animal-centred approach to understanding, designing and managing husbandry systems, including those for handling and transport. This approach has been most publicized by Temple Grandin, in her own work and in her edited books (Grandin, 2000). For example, Hutson's chapter on 'Behavioural principles of sheep handling' (2000, p. 183) lists what he regards as important characteristics of the animals with respect to handling:

> These four characteristics of the sheep – its flocking behaviour, following behaviour, vision and intelligence – form the basis of all behavioural principles of sheep handling. I shall consider these principles in relation to the three key elements of an integrated sheep handling system – the design of the handling environment, the handling technique and the reason for being handled – the handling treatment. [Concerning design, I] recommended that the most crucial design criterion was to give sheep a clear, unobstructed view towards the exit, or towards where they are meant to move. This often becomes more evident by taking a sheep's eye view of the facility.

Similarly, when behaviour is studied under transport conditions or, say, in lairage – either a specific type of behaviour such as lying, or an overview of behaviour such as the time spent in different activities (the time budget) – it has to be interpreted by comparison with some baseline. This may be the behaviour in a home pen, but one other comparison commonly made is with behaviour in conditions that are as natural as possible, for example in extensive grazing. The implication is that departure from natural conditions and behaviour is associated with welfare problems, although this may then be interpreted in terms of its physical or mental impact on the animal. As one further example, mixing unfamiliar bulls in lairage increases the incidence of 'dark cutting beef', which is high in pH, dry in texture and worth less than normal beef. This effect was shown to be caused by the high levels of aggression and other activity among such newly mixed animals, including the fact that they sometimes spend the entire night moving round the pen rather than lying down (Bartoš *et al.*, 1993; Chapter 4, this volume). Such mixing is not part of the evolutionary, or natural, behavioural background of cattle. So it would be possible to regard this as a welfare problem because of the departure from natural conditions. In practice, most welfare scientists and other commentators would probably emphasize its physical or mental effects.

The very fact that transport is completely unnatural might suggest that this whole area of management requires more thought than it has previously been given. Such issues as whether pre-slaughter transport is necessary at all or whether animals could be slaughtered on the farm deserve much more attention. As just one example, end-of-lay hens are sometimes not used for food, but just killed and disposed of. If this is the case, the carcasses do not have to be inspected for food safety reasons, so the birds could be killed on farm rather than subjected to a long journey to a slaughterhouse.

Integrative approaches

Having considered physical, mental and natural approaches to welfare separately, it is now necessary to consider all of them in an integrated manner, since different aspects can impact on each other. The physical approach might suggest, for example, that animals should be handled and transported singly, to prevent any possibility of aggression, but this would be unnatural and might cause mental problems. However, there is also overlap between the approaches; this incomplete overlap is indicated schematically in Fig. 1.1. Decisions about welfare, including for legislation, will be most firmly based, and acceptable to the greatest possible number of people, if they take into account all three concepts of animal bodies, minds and natures. As already pointed out, some broad approaches to welfare such as the Five Freedoms do just that.

How can we translate the principles derived from welfare science into practice? Another important feature of the Five Freedoms as outlined by FAWC (Table 1.1) is that they also give guidance on how the Freedoms can be ensured. The two approaches – specifying what welfare is, and suggesting how it can be safeguarded – are, of course, complementary yet useful for different purposes. The Freedom Food programme of the UK's Royal Society for the Prevention of Cruelty to Animals (RSPCA) is intended to make food production more animal-centred, but animals' responses to the environment are variable. Farmers in the programme wanted to know what they were actually expected to do, for example, what stocking density they should use. So the programme's criteria issued by the RSPCA were developed to specify inputs as well as outcomes (RSPCA, 1997). Nevertheless, subsequent modification of the criteria has emphasized the importance of welfare outcome assessment, in addition to checks for compliance with input requirements (RSPCA, 2006).

So there is a tendency in development of standards for assessing animal treatment, including transport, and of guidelines for how such treatment should be managed, to emphasize outcomes (such as behaviour, frequency of injuries and mortality) rather than inputs – the actual effects on animal welfare rather than the ways in which those are achieved. This is often referred to as use of performance criteria rather than design criteria. One reason is to avoid 'micromanagement': if operators can find their own way of attaining desired outcomes they should be allowed to do so, especially as this may encourage innovation rather than conformity. In practice, most standards and guidelines include both design and performance criteria.

Conclusions

There is a considerable body of science relevant to transport of farm animals and its implications for welfare. That is not to say that all the conclusions of that scientific work are unanimous: that is not how science works. Different scientific disciplines are applied, and different individual projects are carried out, to address different types of questions about transport, and interpreted according to different concepts of animal welfare and other values (such as financial profitability and environmental sustainability).

Given this variability, the areas of consensus that do exist are remarkable. In particular, there is considerable scientific evidence that long-distance transport causes many welfare problems for farm animals, as described throughout this book.

Furthermore, what we know about welfare and other problems of long-distance transport is often not put into practice. As one major example, widespread movement of farm animals contributed to serious outbreaks of classical swine fever in the Netherlands in 1997 and foot-and-mouth disease in the UK in 2001 (Appleby, 2003), yet such movement continues in the livestock industry in most countries and has recently been implicated in the spread of avian influenza (Greger, 2006). Science suggests that for animal welfare, restriction of disease spread, sustainability and food safety there should be reduced transport of animal feed, animals and food from animals, within and between countries (Appleby, 2003).

References

Appleby, M.C. (1999) *What Should We Do About Animal Welfare?* Blackwell, Oxford, UK.
Appleby, M.C. (2003) Farm disease crises in the UK: lessons to be learned. In: Salem, D.J. and Rowan, A.N. (eds) *The State of the Animals II: 2003*. Humane Society Press, Washington, DC, pp. 149–158.
Appleby, M.C. and Sandøe, P. (2002) Philosophical debates relevant to animal welfare: the nature of well-being. *Animal Welfare* 11, 283–294.

Appleby, M.C. and Waran, N.K. (1997) Solutions: physical conditions. In: Appleby, M.C. and Hughes, B.O. (eds) *Animal Welfare*. CAB International, Wallingford, UK, pp. 177–190.

Barnett, J.L., Cronin, G.M., McCallum, T.H., Newman, E.A. and Hennessy, D.P. (1996) Effects of grouping unfamiliar adult pigs after dark, after treatment with amperozide and by using pens with stalls, on aggression, skin lesions and plasma cortisol concentrations. *Applied Animal Behaviour Science* 50, 121–133.

Bartoš, L., Franč, C., Rehak, D. and Stipkova M. (1993) A practical method to prevent dark-cutting (DFD) in beef. *Meat Science* 34, 275–282.

Bekoff, M. (1998) *Animal Play: Evolutionary, Comparative and Ecological Perspectives*. Cambridge University Press, Cambridge.

Bradshaw, R.H., Parrott, R.F., Forsling, M.L., Goode, J.A., Lloyd, D.M., Rodway, R.G. and Broom, D.M. (1996) Stress and travel sickness in pigs: effects of road transport on plasma concentrations of cortisol, beta-endorphin and lysine vasopressin. *Animal Science* 63, 507–516.

Broom, D.B. (1996) Animal welfare defined in terms of attempts to cope with the environment. *Acta Agriculturae Scandinavica, Section A, Animal Science Supplement* 27, 22–28.

Cannon, W.B. (1914) The emergency function of the adrenal medulla in pain and the major emotions. *American Journal of Physiology* 33, 356–372.

Council of Europe (1976) *Convention for the Protection of Animals Kept for Farming Purposes*. Available at: http://conventions.coe.int/treaty/en/Treaties/Html/087.htm

Dawkins, M.S. (1988) Behavioural deprivation: a central problem in animal welfare. *Applied Animal Behaviour Science* 20, 209–225.

DEFRA (Department for Environment, Food and Rural Affairs) (2002) *Code of Recommendations for the Welfare of Livestock: Laying Hens*. DEFRA, London.

Duncan, I.J.H. and Fraser, D. (1997) Understanding animal welfare. In: Appleby, M.C. and Hughes, B.O. (eds) *Animal Welfare*. CAB International, Wallingford, UK, pp. 19–31.

Duncan, I.J.H., Slee, G.S., Kettlewell, P., Berry, P. and Carlisle, A.J. (1986) Comparison of the stressfulness of harvesting broiler chickens by machine and by hand. *British Poultry Science* 27, 109–114.

European Food Safety Authority (2004) Opinion of the Scientific Panel on Animal Health and Welfare on a request from the Commission related to the welfare of animals during transport. *The European Food Safety Authority Journal* issue, The welfare of animals during transport.

FAWC (Farm Animal Welfare Council) (2006) *Report on Welfare Labelling*. FAWC, London.

FAWC (Farm Animal Welfare Council) (2007) *Five Freedoms*. Available at: www.fawc.org.uk

Flecknell, P.A. and Molony, V. (1997) Pain and injury. In: Appleby, M.C. and Hughes, B.O. (eds) *Animal Welfare*. CAB International, Wallingford, UK, pp. 63–73.

Fraser, D., Weary, D.M., Pajor, E.A. and Milligan, B.N. (1997) A scientific conception of animal welfare that reflects ethical concerns. *Animal Welfare* 6, 187–205.

Gentle, M.J. (1997) Acute and chronic pain in the chicken. In: Koene, P. and Blokhuis, H.J. (eds) *Proceedings, Fifth European Symposium on Poultry Welfare*. Wageningen Agricultural University, Wageningen, The Netherlands, pp. 5–11.

Grandin, T. (ed.) (2000) *Livestock Handling and Transport*, 2nd edn. CAB International, Wallingford, UK.

Greger, M. (2006) *Bird Flu: A Virus of Our Own Hatching*. Lantern, New York.

Gregory, N.G. and Wilkins, L.J. (1989) Broken bones in domestic fowl: handling and processing damage in end-of-lay battery hens. *British Poultry Science* 30, 555–562.

Hall, S.J.G., Schmidt, B. and Broom, D.M. (1997) Feeding behaviour and the intake of food and water by sheep after a period of deprivation lasting 14h. *Animal Science* 64, 105–110.

Harrison, R. (1964) *Animal Machines: The New Factory Farming Industry*. Vincent Stuart, London.

HMSO (Her Majesty's Stationery Office) (1965) *Report of the Technical Committee to Enquire into the Welfare of Animals Kept Under Intensive Livestock Husbandry Systems*. Command Paper 2836. HMSO, London.

Hughes, B.O. and Black, A.J. (1973) The preference of domestic hens for different types of battery cage floor. *British Poultry Science* 14, 615–619.

Hughes, B.O. and Curtis, P.E. (1997) Health and disease. In: Appleby, M.C. and Hughes, B.O. (eds) *Animal Welfare*. CAB International, Wallingford, UK, pp. 109–125.

Hutson, G.D. (2000) Behavioural principles of sheep handling. In: Grandin, T. (ed.) *Livestock Handling and Transport*. CAB International, Wallingford, UK, pp. 175–199.

Jones, R.B., Duncan, I.J.H. and Hughes, B.O. (1981) The assessment of fear in domestic hens exposed to a looming human stimulus. *Behavioural Processes* 6, 121–133.

Kettlewell, P.J. and Mitchell, M.A. (2001) Comfortable ride: concept 2000 provides climate control during poultry transport. *Resource Engineering and Technology for a Sustainable World* 8, 13–14.

Kirkwood, J.K., Hubrecht, R.C., Wickens, S., O'Leary, H. and Oakley, S. (eds) (2001) *Consciousness, Cognition and Animal Welfare*. Universities Federation for Animal Welfare, Wheathampstead, UK, supplement to volume 10 of *Animal Welfare*.

Knowles, T.G. (1995) A review of post transport mortality among younger calves. *Veterinary Record* 137, 406–407.

Knowles, T.G. and Warriss, P.D. (2000) Stress physiology of animals during transport. In: Grandin, T. (ed.) *Livestock Handling and Transport*, 2nd edn. CAB International, Wallingford, UK, pp. 385–407.

Knowles, T.G., Warriss, P.D., Brown, S.N., Edwards, J.E., Watkins, P.E. and Phillips, A.J. (1997) Effects on calves less than one month old of feeding or not feeding them during road transport of up to 24 hours. *Veterinary Record* 140, 116–124.

MAFF (Ministry of Agriculture, Fisheries and Food) (1987) *Codes of Recommendations for the Welfare of Livestock: Turkeys*. MAFF Publications Office, London.

Mench, J.A. (1998) Thirty years after Brambell: whither animal welfare science? *Journal of Applied Animal Welfare Science* 1, 91–102.

Mitchell, M.A. and Kettlewell, P.J. (1993) Catching and transport of broiler chickens. In: Savory, C.J. and Hughes, B.O. (eds) *Proceedings, Fourth European Symposium on Poultry Welfare*. Universities Federation for Animal Welfare, Hertfordshire, UK, pp. 219–229.

Moberg, G.P. (1999) When does stress become distress? *Lab Animal* 28, 22–26.

Moberg, G.P. (2000) Biological response to stress: implications for animal welfare. In: Moberg, G.P. and Mench, J.A. (eds) *The Biology of Animal Stress: Basic Principles and Implications for Animal Welfare*. CAB International, Wallingford, UK, pp. 1–21.

National Pork Board (2005) *Animal Care: Overview*. Available at: www.porkscience.org

Pepperberg, I. (1987) Evidence for conceptual quantitative abilities in the African grey parrot: labeling of cardinal sets. *Ethology* 75, 37–61.

Rogers, L.J. (1995) *The Development of Brain and Behaviour in the Chicken*. CAB International, Wallingford, UK.

Rollin, B.E. (1993) Animal production and the new social ethic for animals. In: Baumgardt, B. and Gray, H.G. (eds) *Food Animal Well-Being: Conference Proceedings and Deliberations*. USDA and Purdue University, West Lafayette, Indiana, pp. 3–13.

RSPCA (Royal Society for the Prevention of Cruelty to Animals) (1997) *Welfare Standards for Laying Hens*. RSPCA, Sussex, UK.

RSPCA (Royal Society for the Prevention of Cruelty to Animals) (2006) *Welfare Standards for Laying Hens and Pullets*. Available at: www.rspca.org.uk/farmanimals

Rushen, J. (1986) Aversion of sheep to electro-immobilization and physical restraint. *Applied Animal Behaviour Science* 15, 315–324.

Savory, C.J., Máros, K. and Rutter, S.M. (1993) Assessment of hunger in growing broiler breeders in relation to a commercial restricted feeding programme. *Animal Welfare* 2, 131–152.

Selye, H. (1932) The general adaptation syndrome and the diseases of adaptation. *Journal of Clinical Endocrinology* 6, 117–152.

Staples, G.E and Haugse, C.N. (1974) Losses in young calves after transportation. *British Veterinary Journal* 130, 374–378.

Stephens, D.B. and Perry, G.C. (1990) The effects of restraint, handling, simulated and real transport in the pig (with reference to man and other species). *Applied Animal Behaviour Science* 28, 41–55.

Tannenbaum, J. (1995) *Veterinary Ethics*. Mosby-Year Book, St Louis, Missouri.

Tarrant, P.V. (ed.) (1983) *CEC Farm Animal Welfare Programme Evaluation Report 1979–1983*. Commission of the European Communities, Luxembourg.

Terlouw, E.M.C., Schouten, W.G.P. and Ladewig, J. (1997) Physiology. In: Appleby, M.C. and Hughes, B.O. (eds) *Animal Welfare*. CAB International, Wallingford, UK, pp. 143–158.

Warriss, P.D., Bevis, E.A., Brown, S.N. and Edwards, J.E. (1992) Longer journeys to processing plants are associated with higher mortality in broiler chickens. *British Poultry Science* 33, 201–206.

Webster, A.J.F. (ed.) (2006) Sentience in animals. *Applied Animal Behaviour Science* 100(1–2), 1–151.

Wemelsfelder, F. (2007) How animals communicate quality of life: the qualitative assessment of behaviour. *Animal Welfare* 16(S), 25–31.

2   Economic Aspects

AGRA CEAS CONSULTING LTD

Imperial College, University of London, Wye, Ashford, Kent, UK

Abstract

After a preliminary analysis of global trade data, three main routes of global live animal transport, which were found significant globally, were considered in the report. The first part covers live animal exports from Australia and New Zealand to the Middle East, which concerns mainly sheep. This route represents 81% of Australian live sheep exports, and has been considered one of the most important in a global perspective. The second part of the report deals with the live animal trade in North America, on different routes between the USA, Canada and Mexico. Finally, the third part reviews a European case, which is live export from Poland to Italy.

Australia and New Zealand live exports for slaughter

While both Australia and New Zealand export significant volumes of live animals for a variety of breeding, fattening, etc. purposes the main trade for slaughter relates to castrated male sheep ('wethers') from Australia. This trade is directed primarily to the Middle East where due to climatic conditions domestic production is constrained. The animals shipped serve two main markets: the traditional 'wet markets' where consumers still buy their animals live and over 1 million sheep per year are sent for slaughter during the Hajj to Mecca. While the traditional wet market trade is expected to gradually decline as prosperity and hence consumer habits change, the pilgrimage results in the demand for the sacrifice of live animals with the meat then being donated to the poor. As is set out in detail in the report many actors are involved in the trade from Australia and under current circumstances there is a clear margin, i.e. economic incentive, for the trade both in live sheep and, to a lesser extent, cattle.

A likely way to discourage the live trade and facilitate change would be to take steps to encourage the use of vouchers for pilgrims based on slaughter in Australia and the distribution of the meat produced in this fashion to the needy in Islamic countries. Similarly, further support is given to educational efforts in consuming regions which highlight the quality, convenience, food safety and animal welfare benefits of purchasing meat at the retail level rather than live at a market.

North American live animal trade

In North America there are substantial cross-border trade flows for both cattle for fattening and for slaughter. The main trade flow of interest for the purposes of this study is the cross-border cattle trade from Canada to the USA for slaughter. This trade takes place in order to exploit the price differential which may exist in these two markets and in particular to attract the premium which is attributable to US Department of Agriculture (USDA)-graded beef in the USA. While there is no effective difference in the grading systems USDA-graded meat attracts a premium in the US retail and catering sectors while Canadian meat exports attract a lower price in that they can only be used where the origin of the meat is not specifically identified, i.e. some catering and meat processing. The US market in effect acts as a regulator of the prices on the Canadian market drawing in supplies when prices in Canada are lower.

In order to eliminate this flow of live animals and the associated potential welfare costs, action should be taken to ensure that meat produced for export to the USA in Canada be export-certified and graded in Canada by US inspectors. Prior to this such efforts could be reinforced by a campaign in the USA to highlight to consumers the fact that there are no effective differences in Canadian and US-produced beef but that the meat certification and labelling system is resulting in potentially high animal welfare losses.

Poland to Italy live animal trade

Historically shipments of cattle from Poland to Italy have been based on the fact that there is a limited market for beef in Poland and a strong market for this in Italy. This means there has always been a surplus of beef (as a result of the extra animals generated in the milk production process) in the Polish market and a strong price differential between the two markets. In addition to this main driver for trade in general, during the 1990s the trade in live animals specifically was encouraged by a number of other factors. After the collapse of the centrally planned system in Poland, there was a large surplus of cattle/meat as live-stock herds were drastically reduced. However, at that time there was a lack of recognized slaughter capacity, which would be licensed for EU exports. The trade in live animals was therefore an outlet for the surplus cattle. Apart from these factors, there was an additional incentive for live animals traded out of Poland as EU premiums were available for producers on animals fattened and slaughtered in the EU.

Many of the reasons for the historically high level of the live cattle for slaughter trade which were in place in the 1990s have now been removed and there has therefore been a corresponding decline in such trade. This having been said, a significant number of animals still are shipped for slaughter so as to obtain the premium derived from the fact that plants in Italy are more geared to adding value than their Polish counterparts. In principle, how-ever, over time most or all of the value-added attributes of slaughter in Italy could be repli-cated in Poland (e.g. if necessary hides could be shipped from Poland to Italy etc.). In order to achieve this there would need to be a concerted effort to highlight the potential animal welfare gains to be derived by a longer-term restructuring of the slaughter and processing process to favour slaughter in Poland.

As is the case for cattle, Poland has historically had high production and good pasturing conditions for horses and no demand for horse meat. Italy, on the other hand, has always had strong demand for horse meat and thus provides an attractive market. Apart from this basic feature in the past the trade has been driven by the fact that the slaughter capacity in Poland has historically not been EU licensed, and indeed, there was little horse slaughter capacity in the country. Additionally in the period before the EU entry there was also another incentive for live animal trade in the form of different tariff levels for meat and live animals.

As has been noted above many of the reasons for the historically high level of the live horse for slaughter trade which were in place in the 1990s have now been removed and there has therefore been a corresponding decline in such trade. This having been said, a sig-nificant number of animals, estimated at 20% of cattle shipped, are still shipped for slaughter so as to obtain the premium derived from the fact that plants in Italy are more geared to adding value than their Polish counterparts.

The trade in sheep meat and live sheep exists because of good sheep production conditions combined with a lack of demand for sheep meat in Poland. These factors combine with strong seasonal demand for sheep meat in Italy to create a strong incentive for trade. The incentive for the trade in live sheep destined for slaughter would appear to be primarily related to the fact that if the meat is slaughtered in Italy it can be marketed as being meat of Italian origin. This having been said, it is clear that the scale of this trade has been steadily diminishing since the latter half of the 1990s. To reduce this trade, Italian consumers need to be made aware of the potentially adverse animal welfare implications of this trade, and consideration should be given to changing the origin-labelling regulations so as to discourage such shipments.

Live Animal Export from Australia and New Zealand

Trade flows for meat and live animals

Livestock exported by Australia have included goats, buffaloes, horses, pigs, deer, camels, ostriches and honeybees, however, it is sheep and cattle which dominate the trade. Export stocks include breeding stock, stock destined for further finishing prior to slaughter (e.g. feedlotting) and stock for immediate slaughter. Historically stock for immediate slaughter has dominated the trade and continues to do so for sheep, but in the new century particularly in the dairy sector, there has been a rapid growth in stock for further value-adding prior to slaughter and stock for breeding purposes. For instance dairy cows for breeding and milk production now constitute AUS$100 million in a AUS$400 million live cattle export industry.

In the main, Australian live animals are transported to markets in Asia and the Middle East. There are approximately 55 licensed Australian exporters but most of the trade is concentrated in the hands of the ten largest operations (see Table 2.1.1A).

Cattle tend to be sourced from the northern Australian states (Queensland, Northern Territory and northern Western Australia) while sheep tend to be sourced from southern states (New South Wales, Victoria, South Australia and southern Western Australia). The vast majority of Australian live exports are shipped using sea freight with only a few high-value lines such as thoroughbred racehorses and stud animals travelling by air. In 2005, Australia exported livestock from 16 Australian seaports to 39 overseas destination ports in 19 different countries. In all, this involved 388 export consignments.

Australia has been able to develop an export trade in live animals as well as meat as a result of:

- A comparative advantage in cost of production, i.e. a low-cost product;
- Large supply volumes with potential for economies of scale;
- Low disease incidence and capacity to guarantee health status;
- Breeds that meet market requirements;
- Proximity to markets, e.g. Asia;
- Availability of land and sea transport facilities;

- Support infrastructure and expertise; and
- Feeding and veterinary infrastructure.

The dimensions of the industry's two largest sectors, live sheep and live cattle exports, are detailed separately in the sections below.

Australian live sheep exports

Australian live sheep exports from 1990 to 2005 are shown in Fig. 2.1. Broadly this shows that in volume and value terms exports rose sharply in the first half of the 1990s, then stabilized in the second half of the 1990s before reaching a peak in 2001–2002 and then falling back in 2004.

In 2005, live sheep export numbers increased by 18.4% (AUS$51.6 million) valued on a 'free on board' (fob) basis to 4.185 million head after reaching a 14 year low in 2004. The 2004 low was caused by a temporary cessation of the trade with Saudi Arabia from August 2003 following that country's rejection of an Australian vessel, the *Cormo Express*, at destination, which resulted in an official death count of over 5000 sheep (see Chapter 12, this volume, for more information on the incident). Resumption of the trade with Saudi Arabia in July 2005 boosted industry figures with Saudi Arabia again becoming the largest importer of live sheep for the year. Kuwait, Jordan and Bahrain were the next three largest importers of Australian sheep. Other markets of significance include Egypt, Lebanon, Oman, Qatar, United Arab Emirates (UAE) and Yemen. Some 98.8% of Australian markets for live sheep are located in the Middle East (see Table 2.1).

The most consistent market for Australian live sheep since the emergence of the volume trade in the late 1970s has been Kuwait. It is interesting to note that

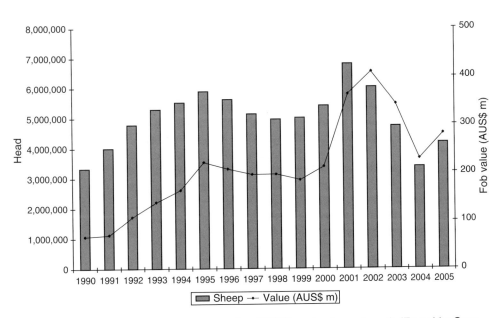

Fig. 2.1. Australian live sheep exports by value (AUS$) and volume (head). (From LiveCorp using Australian Bureau of Statistics (ABS).)

Table 2.1. Australia live sheep trade 2001–2005 by export destination (head). (From Australian Bureau of Statistics (ABS).)

	2001	2002	2003	2004	2005	Average 2000–2005
Kuwait	1,539,700	1,569,807	1,498,537	1,259,904	890,563	1,351,702
Saudi Arabia	2,147,950	1,873,041	1,411,195	0	1,072,089	1,300,855
Jordan	542,366	582,617	498,999	930,343	884,886	687,842
Bahrain	388,172	385,878	410,508	490,210	521,455	439,245
UAE	681,492	466,421	225,313	196,095	230,775	360,019
Oman	487,484	352,794	259,008	289,170	358,972	349,486
Total	4,787,164	5,230,558	4,303,560	3,165,722	3,958,740	4,489,149

volumes of live animals shipped into this market have been falling since the beginning of the decade, while at the same time there has been a very substantial growth in meat exports which reached a peak of 4145 t in 2005. Saudi Arabia has often imported larger numbers per year, however, trading disputes in the 1990s and again in 2003–2005 have resulted in lower average shipments to that country.

The Australian live sheep trade is generally considered by industry to be 'mature' or more or less stable at current levels at long-term average levels of around 5 million head per annum. These levels have been relatively stable now for over 20 years despite a significant decline in Australian sheep numbers from 170 million to around 100 million during that period.

Western Australia exported 83% of total live sheep exports in 2005 with Victoria and South Australia sharing the remainder. The 2005 result is typical of the trade over the last 20 years.

Live cattle exports

Australian live cattle exports from 1990 to 2005 are shown in Fig. 2.2. In the period since 1990 in volume and value terms exports rose almost continuously up to 1998 when they fell back sharply but recovered to 2002 since when they have been declining. The major factor that contributed to the decline of live cattle exports in 1998 was the economic downturn in the South-east Asian economies in 1997. However, the falls in these markets were partially offset by increases in live cattle exports to North Africa. This increase in demand from North Africa was due to the problems with bovine spongiform encephalopathy (BSE) and foot-and-mouth disease (FMD) in Europe. These disease problems in Europe meant that exports from this region, which had been the traditional supplier for North Africa, were either banned or constrained in terms of availability. Australian live exports of cattle for slaughter rose again in 1999/2000 and remained around this level in 2000/01. This was due to the recovery of the Indonesian economy, the continuing impact of BSE and FMD in European and South American countries and a weaker Australian dollar. In 2001/02 live exports for slaughter fell as cattle prices rose and economic recovery in Indonesia and the Philippines stalled. The record level in 2002 of 972,000 head (surpassing the previous high in 1997) was due to renewed demand from Indonesia along with increased demand from Malaysia,

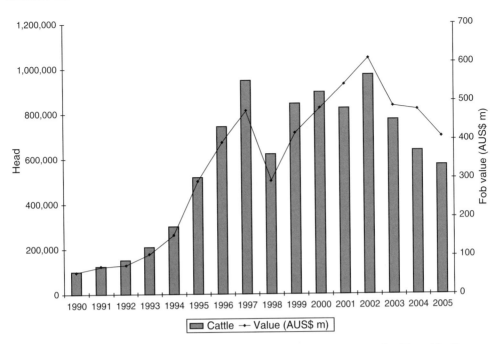

Fig. 2.2. Australian live cattle exports by value (AUS$) and volume (head). (From LiveCorp using ABS.)

Saudi Arabia and Brunei. However, the new level was not sustained in 2003/04 as currency fluctuations made exports of live slaughter cattle fall.

Live cattle exports for 2005 totalled 572,799 head, down a further 10.2% from 637,748 head at the end of 2004. The value also declined by AUS$70 million to AUS$407.5 million in the same period. Average price per head also decreased by AUS$38 to AUS$711 fob. The drop in values and volumes are associated with a higher Australian exchange rate relative to trading partners and less-favourable market conditions in importing countries. In the long term, the industry considers that trade volumes in live cattle are likely to be more or less stable. Indonesia's share of the market continues to increase and now accounts for 61%. In 2005, Malaysia followed by China and Israel were the next largest destinations at 6.2%, 5.7% and 5.6% respectively. Sales to the Middle East also rose during 2005 (see Table 2.1.2A).

The Australian live cattle trade has been evolving rapidly due to the growth and development in Indonesia with the addition of feedlot and other infrastructure to move the product out of wet markets. In addition the sector as a whole has benefited from increased sales of dairy breeding stock in China.

In 2004, Western Australia continued to export the largest number of cattle for all purposes with 43.3% of the market increasing by 5% on the previous year. The Northern Territory exported 30.9% in 2004 followed by Victoria at 12.4% and Queensland at 10.6%. Victoria's total doubled between 2003 and 2004 due to the growth in the dairy export trade to China. Victoria is Australia's premier dairy state.

In contrast to live cattle export, the main destinations for beef are the USA and Japan, which purchase 75% of Australia's beef exports. Since 2004, there has been a significant increase of export to Japan, which accounts for 36% of total Australian beef exports. Other important purchasers of beef are the Philippines, Indonesia and Malaysia, which are major destinations for Australian live cattle. Beef exports to Saudi Arabia have declined over the last 5 years and amounted to only 670 t in 2005 (see Table 2.1.3A).

Economic significance

In the 5 years to 30 June 2005, the Australian livestock export industry generated an average of AUS$750 million in direct export earnings (see Table 2.2). The significance of the economic activity provided by the export trade is estimated to provide direct employment for some 5600 people and indirect employment for a further 3400 people, a total of 9000 jobs in mainly rural and regional Australia.[1] While these figures are not disputed, it should be noted that other sources highlight the potential for further added value/employment if the trade were able to move from a live export to a meat-only basis. Thus, an economic report produced by Heilbron/Larkin published in the same year[2] states: '[I]f the sheep and cattle currently (1999/2000) exported live were instead processed in Australia, a further approximately $1.5 billion would be added to Australia's . . . GDP, around $250 million in household income and around 10,500 full-time jobs would be created.' This analysis is used to underscore the argument that by exporting live animals, Australia may be losing potential advantages in terms of employment and value added in the meat sector.

In numbers, export of the Australian beef and sheep meat industry compared to live exports amounts to the following:

- Australia exports around 60% of its beef and sheep meat production.
- Beef and veal exports totalled 977,363 t in 2005. Live cattle exports, not all of which were for meat production, totalled 572,799 head for the same period. Live cattle exports for slaughter and further fattening are therefore approximately 7% of the industry's export value.

Table 2.2. Australian live export revenues – average 2001–2005 (AUS$ million fob). (From ABARE Australian Commodity Statistics, 2005.)

	2001	2002	2003	2004	2005	Average 2000–2005
Sheep	258	392	408	266	207	306
Cattle	482	526	562	314	335	444
Total	740	918	970	580	542	750

[1] Source: LiveCorp and based on independent research by consultants Hassall and Associates, 2000.
[2] Source: Animals Australia citing a report by Dr Selwyn G Heilbron, Terry Larkin, 'Impact of the Live Animal Export Sector on the Australian Meat Processing Industry', April 2000. Further detailed information see under: http://www.animalsaustralia.org/default2.asp?idL1=1274&idL2=1670.

- Lamb and mutton exports totalled 268,000 t valued at AUS$1119 million fob in 2004/05. Live sheep exports totalled 3.233 million head valued at AUS$207 million fob for the same period. Live sheep exports are therefore approximately 18% of the industry's export value.

The live export industry thus contributes a relatively small but nevertheless significant proportion of the overall export value generated by the livestock sector. The importance of this trade is, however, particularly significant in certain production regions notably in Western Australia. The contribution of such exports for sheep and cattle producer income in Western Australia is outlined briefly in the following section.

Sheep prices received for live export versus next best alternative

Sheep producer returns are enhanced where a greater number of market opportunities exist. The live sheep export industry provides producers with access to market alternatives. The benefit of access to live export markets by Western Australian sheep producers is illustrated in Fig. 2.3.

Figure 2.3 shows that sheep producers in Western Australia achieve higher returns by supplying wethers to the live sheep trade. Similar stock (threescore ewes purchased for slaughter), consistently castrated male sheep receive lower prices than wethers purchased for live export. Ewes are unsuitable for the live export trade, markets prefer male stock.

On a regional level, e.g. in the north of Western Australia in the Broome-West Kimberley region, tourism, cattle production, social services provision and mining are the dominant industries. Cattle production is almost entirely dedicated to live export because of the limited number of alternative market opportunities. Road transport to Perth or Darwin would consume much of the cattle producers' profit

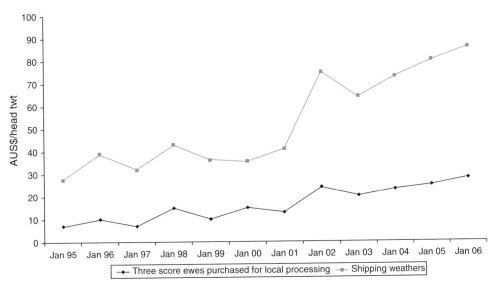

Fig. 2.3. Premiums for sheep destined for live export versus next best alternative. (From Western Australia Department of Agriculture and ABARE.)

margin and there is no local market of any size within this sparsely settled area. There are few alternative land uses for pasture suitable for cattle production in this remote region.

Economics of the Australian live export trade

See Appendix 2.2 for a list of main stakeholders.

Export value chain margins

The margins involved along the live export value chain for sheep and cattle are shown in Tables 2.3 and 2.4.

Table 2.3 values for live sheep exports are based on a cost insurance freight (CIF) value delivered to Kuwait for AUS$120.85. Value added (margin) per head is estimated at AUS$10.14. From this breakdown of the value chain it can be seen that the major revenue earners from live sheep exports are:

- Sheep producers;
- Livestock agents;
- Feedlot operators;
- Fodder suppliers; and
- Shippers.

Table 2.3. Typical value chain for live sheep 2004/05. (From Agra CEAS based on industry consultations.)

	AUS$ per head	Cost as %
Total gross value costs	120.85	
Livestock purchase (farm gate)	50.00	45
Additional veterinary on farm	0.08	0
Road transport to quarantine	1.50	1
Agents' fees	3.00	3
Transit insurance	0.58	1
Pre-export quarantine	1.75	2
Shearing	1.20	1
Freight to wharf	1.20	1
Wharf charges	0.02	0
Third-party vet	0.03	0
AQIS	0.01	0
Fodder for journey	10.00	9
Sea transport	40.00	36
On-board water	0.05	0
On-board veterinarian	0.35	0
Stockmen/women	0.14	0
Finance	0.03	0
Insurance	0.75	1
Total costs	110.70	100
Value added	10.14	

Table 2.4. Typical value chain for live cattle 2004/05. (From Agra CEAS based on industry consultations.)

	AUS$ per head	Cost as %
Total gross value costs	1,135.23	
Livestock purchase (farm gate)	650.00	60
Additional veterinary on farm	4.64	0
Road transport to quarantine	22.50	2
Agents' fees	39.00	4
Road transport to wharf	15.00	1
Pre-export quarantine	17.50	2
Wharf charges	2.00	0
Third-party vet		
AQIS	1.16	0
Stevedoring		
Fodder for journey	75.00	7
Sea transport	250.00	23
On-board water	1.20	0
On-board veterinarian	3.50	0
Stockmen/women	1.20	0
Total costs	1,090.23	100
Value added	45.00	

Table 2.4 values for live cattle exports are based on a CIF value delivered to Indonesia. Value added (margin) per head is estimated at AUS$45. From the value chain it can be seen that the major revenue earners from live cattle exports are:

- Cattle producers;
- Livestock agents;
- Road transport providers;
- Fodder suppliers; and
- Shippers.

Consumer preference for live animals versus meat

The determination of whether an animal goes to slaughter in Australia or to live export is a market decision and not usually a decision of the livestock producer. Livestock are available for purchase and the graziers' interest is in receiving the best return for the stock. There are also many Australian abattoirs capable of providing Halal-certified product to export customers, and this is not a factor leading to stock being exported live.

Furthermore, it is not simply a straightforward matter of stating that Australia's export customers for live sheep and cattle prefer this form of product to carcass meat since in practice most Australian live export markets purchase both carcass cuts and live animals.

Live animals are preferred in many markets when breeding stock is required. Breeding stock accounts for around 25% of Australia's live cattle exports. Very few live sheep are exported for breeding purposes. Over time live breeding animal requirements may decline as infrastructure for handling and distributing genetic material (such as semen straws and frozen embryos) improves in export destinations.

Live animals are also required when further value-adding can be achieved in destination countries at a lower cost than in Australia or when owners of value-adding infrastructure are foreign nationals who wish to capture this added value for themselves, i.e. by slaughtering, cutting and processing in the country. In these instances Australia has the greatest comparative advantage in specializing in live animal supply, i.e. provision of large numbers of low-cost and suitable livestock.

Live animals are preferred in some markets where there is inadequate infrastructure to handle large volumes of carcass meat. It is often stated that in these destinations refrigeration, storage and electrical infrastructure are inadequate and product would spoil prior to consumption. A live animal can be nourished on local feedstuffs until required and slaughtered and consumed on the same day without the need for refrigeration. Parts of the Middle East and Asia, Australia's main markets for live animals, are reliant on this form of traditional wet market retailing.

Consumers, especially poorer consumers, prefer to purchase fresh meat through traditional wet markets rather than purchase frozen product of indeterminate age and quality from wet market alternatives.

As live animal export markets mature or more or less stabilize at current levels and living standards in these countries increase, live export animals are increasingly being substituted for by packaged meat sales. Supply infrastructure improves and more sophisticated retailing, including supermarkets, are developed. Supply chains systems also improve and access for fresh chilled product becomes possible. There has, for example, been a strong relationship between an initial growth in live animal imports followed by growth in meat imports from Australia in South-east Asia since the early 1990s.

In a number of Australian live animal export markets demand exists simultaneously for live animals and meat imports. Live animals tend to be purchased through wet markets by poorer consumers while more wealthy consumers purchase imported meat through supermarkets and the hotel and restaurant sectors. Table 2.5 shows Australian live animal and meat exports to selected Middle Eastern and Asian markets that are simultaneously purchasing live animals and meat from the same species.

Table 2.5. Value of Australian cattle and sheep meat and live animals in selected markets, 2003 (AUS$ million fob). (From ABARE 2005.)

	Meat	Live animals
Cattle		
Indonesia	38.4	203.4
Japan	1,384.4	16.6
Malaysia	86.8	38.9
Philippines	23.0	39.1
Saudi Arabia	7.8	9.4
Sheep		
Saudi Arabia	65.0	104
United Arab Emirates	29.9	16

It is also difficult to establish a correlation between the absence of a trade in livestock and what effect this has on processed meat imports. By way of example: despite a reduction in sheep availability from Australia of over 2 million head, sheep meat imports declined in Saudi Arabia between 2003 and 2004. By contrast, the Heilbron Report cited earlier mentions that during the embargo on the live export trade to Saudi Arabia in the 1990s, frozen/chilled meat exports tripled.

Live animals may also be preferred to meat in some parts of export markets on religious grounds. In the Middle East some consumers may prefer that halal practices are carried out by their own nationals rather than appropriately qualified Muslims in Australia and it is a requirement of the Hajj to Saudi Arabia that a sheep is slaughtered by each pilgrim and the meat provided to the poor. The Hajj market accounts for approximately 1.2 million head per annum from Australia. Imported meat is not a suitable substitute in this cultural tradition. Saudi Arabia is Australia's single largest market for live sheep and there is an annual surge in exports to this market prior to the Hajj. When this market is closed to Australian sheep, Australian sheep are sourced for the Hajj through neighbouring countries such as Jordan and Kuwait.

The significance of the traditional wet markets and live sales for slaughter during the Hajj would, however, appear to be a matter of some dispute. Animals Australia estimated that out of the average 4.5 million Australian live sheep exported to the Middle East, 70% enter the wholesale market chain directly after being slaughtered. They are then reportedly distributed chilled to butchers for sale.[3]

A shortage of Australian infrastructure is not a reason for the trade. There are ample slaughtering facilities for both sheep and cattle. Human capital in the form of appropriately qualified Muslims to certify Halal slaughter may be a minor issue in periods of high demand.

Desire to label and market joint products such as leather, as a product of the importing country is not an issue of significance in the Australian live animal export trade.

Macroeconomic settings and discriminatory trade policy

The Australian live animal export industry does not receive any advantageous treatment from government in the form of subsidies that would favour live animal export over carcass cuts. There are no known discriminating tariffs or entry barriers that favour live exports over carcass cuts.

Reasons for New Zealand's decision to switch to carcass-only exports

Live sheep exports

There are fundamental differences between the sheep industries in Australia and New Zealand. The Australian industry was developed as a wool industry with meat production as a secondary product. With the advent of lower wool prices and

---

[3] Set out by Animals Australia: http://www.animalsaustralia.org/default2.asp?idL1=1274&idL2=1670
Also The Heilbron Report, as set out by AACT http://www.aact.org.au/liveexport.htm

higher lamb and mutton prices in more recent times, this has altered the industry's structure to some extent, but wool production remains the principal driver.

In New Zealand the sheep industry has been based on meat production with a secondary carpet wool industry. The New Zealand sheep meat industry has to a large extent been responsive to a preferential level of access to the higher-value EU market of 227,854 t, compared to Australia's 18,650 t. This has provided the New Zealand industry with a significantly different focus to that of the Australian industry over many years.

The New Zealand live sheep trade is an opportunistic industry that grew through the 1990s to supply Saudi Arabia when Australia withdrew from this trade (New Zealand Ministry for Agriculture and Forestry, February 2004). The trade reached a peak in the mid-1990s when over 1 million sheep were exported annually to the Middle East. Since 2000, the New Zealand live sheep trade has dwindled to a single ship per year, i.e. approximately 40,000 head per annum. The reasons for this decline include:

- Falling New Zealand sheep numbers;
- Higher costs of supply from New Zealand than Australia; and
- Age restrictions on sheep that can be exported from New Zealand.

Reasons for the decline include higher returns from deer production, forestry and dairy together with prevailing drought conditions throughout much of the early 2000s. Another determining factor is that Australia is more competitive for live sheep export than New Zealand. Only a slow rebuild in New Zealand sheep numbers is forecast and current estimates are for a national flock of less than 40 million head.

In supplying the Middle East, the New Zealand live sheep export industry faced higher costs of supply than Australia. Australia, especially Western Australia, is up to a week closer to major Middle East ports and on a more direct and frequent shipping route than New Zealand. Shipping from Western Australia can take as little as 14 days. When Australia is in the market, New Zealand finds it difficult to supply on a cost-effective basis.

In 2000, in response to Saudi Arabian requirements for younger sheep to supply the Hajj festival and resulting higher than acceptable in-transit deaths, minimum age limits were placed on New Zealand sheep by the Ministry for Agriculture and Forestry. This made it increasingly difficult for New Zealand to supply the type of stock required by Saudi Arabia. Lambs and hoggets (1–8 months old) are required for the New Zealand lamb trade into the EU and would not enter the live export trade. Sheep older than 3 years are not required by Saudi Arabia. Live export sheep originating from both Australia and New Zealand fall within this age bracket.

New Zealand continues to export frozen meat to the Middle East (about 15% of the total amount of frozen lamb exported from New Zealand), however, for religious and cultural reasons, Saudi Arabia chooses to sacrificially slaughter live animals, particularly for the observance of religious occasions such as the Hajj. Saudi Arabia does not produce enough sheep on its own and therefore has to import sheep from other countries, not only for sacrificial purposes but also to satisfy domestic consumption (New Zealand Ministry for Agriculture and Forestry, February 2004). But still, in 2004 and 2005 alone, Saudi Arabia (13,667 t and 13,924 t, respectively) was the biggest importer of sheep and lamb meat in the

Middle East, followed by Jordan (2861 t and 1837 t) and Oman (1375 t and 1098 t). Further export destinations for sheep and lamb meat in the Middle East are Kuwait, United Arab Emirates and Bahrain and to a minor degree Egypt, Qatar and Lebanon.[4]

In the absence of Australia as a supplier to the Saudi market in the 1990s, sheep were sourced from many countries including South America and China. Increased numbers were also sourced from North African countries such as Sudan, Somalia and Ethiopia despite disease concerns.

In February 2004, the New Zealand Ministry for Agriculture and Forestry commented on whether the New Zealand live sheep trade will increase from its current levels. Factors considered to determine the industry's future included:

- The economics of alternative enterprises in New Zealand;
- Ongoing Australian involvement in live exports;
- International and domestic pressure on the trade including Saudi Arabian rejections of apparently suitable and disease-free shipments; and
- Public debate on the issue of live animal exports.

Live cattle exports

The New Zealand live cattle trade mainly consists of dairy heifers for breeding purposes. Prior to the emergence of China as a market for dairy cattle, export levels averaged about 10,000 head per year. In recent years this has increased to over 30,000 head per annum with particularly strong demand for dairy cattle in 2003–2004.

Nature of the current animal welfare debate on Australian live exports

The debate is acute and long running in the Australian press.

Both major political parties (Liberal and Labour) have maintained support for the trade although there continues to be attack and reference to government failings by the opposition when publicity about the trade occurs. This support reflects an acceptance of the contribution of live animal trade to regional areas of Australia. However, the Democrat and Green parties publicly advocate the banning of livestock exports and the replacing of them with carcass trade. The idea behind it is based on economic as well as animal welfare grounds. On the one hand, animal cruelty could be greatly reduced and on the other hand, the processing of high-quality meat on a local level could provide to regain regional meat workers' jobs that were lost when the live export trade began.[5]

The Australian government has introduced regulatory changes following the *Cormo Express* rejection by Saudi Arabia. Changes require a move from industry self-regulation based on risk assessment and quality assurance principles to mandatory

---

[4] Source: UN Comtrade.

[5] For further information see: http://www.vic.greens.org.au/media/media-releases-2006/the-greens-lead-on-humane-treatment-for-animals; and http://www.democrats.org.au/speeches/index.htm?speech_id=1794

government-prescribed standards. The change is in who enforces the standards (government not industry) rather than their content. The Australian Government has also been actively negotiating Memoranda of Understanding (MOUs) with governments in principal market countries to gain an assurance that stock consigned to the country will be able to be discharged on arrival. Both of these initiatives have provided a basis for a government response to community concerns about the trade following the *Cormo* rejection.

The government also suspended the trade to Egypt following recent negative publicity on the 60 min television programme about cattle mishandling in that country. At the time of writing, it is understood that arrangements to enable a reopening of that trade are nearing completion.

Conclusions on the economics of the Australian trade

Consumer preferences in importing countries and market forces drive the Australian live animal export trade in sheep and cattle. In the absence of significant production capacity in the region, the trade in sheep for immediate slaughter is considered to be stable and driven, to a significant extent by Saudi Arabian requirements for live sheep for cultural and religious events, principally the Hajj as well as by demand, albeit declining, for live sheep in traditional markets in the region. A further driver for a live as opposed to a meat-only trade would appear to be the limited alternative outlets for sheep in the significant north Western Australia production region.

By contrast there would appear to be no particularly strong reasons why live exports for slaughter at destination rather than meat exports take place for cattle going to South and South-east Asia. This having been said, the main growth taking place in the live cattle export industry to this region is in the supply of breeding stock for improvements of domestic herds. In the longer-term further growth in this form of live cattle trade can be expected.

Supply chains that support the trade are long and relatively complex. The trade brings significant advantages to rural Australia although it is also pointed out that a shift to a meat-only trade would in theory have the potential to provide significant economic benefits to the farming sector and associated downstream industries, at least assuming that such a trade were to absorb a similar number of animals.

New Zealand is relatively inactive in the sheep trade at the current time because of lack of supply and its inability to compete with Australia's low-cost advantage.

Recommendations

As is evident from the analysis undertaken a significant proportion of the trade in sheep for slaughter into the Middle East takes place to address the religious requirement that a live sheep be slaughtered and distributed as a charitable gift as part of the Hajj pilgrimage to the Islamic holy sites in Saudi Arabia. The balance will go into traditional wet markets in the region where modern refrigeration and distribution systems have not yet penetrated.

With respect to the animals sacrificed as part of the Hajj[6] it is our understanding that traditionally the pilgrim slaughtered the animal himself or oversaw the slaughtering, while today many pilgrims buy a sacrifice voucher in Mecca before the greater Hajj begins. This allows an animal to be slaughtered in their name without the pilgrim being physically present.[7] Given that this already occurs in relation to animals which are then presumably slaughtered locally, it would seem entirely logical that in order to avoid the animal welfare risks associated with shipping large numbers of animals such great distances, to enable such a voucher system to be extended to enable the sheep to be slaughtered in Australia with the meat then distributed to those in need in poorer Islamic countries. Clearly there would need to be safeguards and inspections acceptable to all parties to ensure that the animals were of the right type and slaughtered fully in accordance with Islamic requirements. There would also need to be an agreed system to ensure that the meat produced in this fashion could be distributed to those in need[8] and a suitable certification and control system would need to be put in place.

With respect to the animals shipped into traditional markets it is clear that these are gradually being displaced by meat produced and slaughtered in Australia itself or other supplying countries, and that therefore this trade will tend to disappear over time as consumer habits change. This trend could be encouraged by raising awareness and highlighting the food safety, quality, environmental and, clearly above all, welfare benefits of such change to consumers in the markets concerned.

Live Animal Export from Canada/Mexico to the USA

Trade flows for meat and live animals

The North American market comprising Canada, the USA and Mexico for cattle and beef can be considered as one market with the production of calves, feedlot cattle and beef spread over this geographic area.[9] This would mean that there is no specific specialization of breeding, fattening or slaughter in any of these areas. The borders between these countries are relatively open, and as long as this is the case, the various products being raised will seek out the highest market return to their operations. As a result if, for example, the demand for calves in some regions

---

[6] During the Hajj or pilgrimage to Mecca and the other holy places on the day of Eid ul-Adha (Arabic: عيدالاضحى *ʿĪd al-ʾAḍḥā*), the religious festival celebrated by Muslims worldwide as a commemoration of Prophet Ibrahim's (Abraham's) willingness to sacrifice his son Ismael for God, an animal is sacrificed. The sacrificed animals, called 'udhiya' Arabic: اضحية, have to meet certain age and quality standards or else the animal is considered an unacceptable sacrifice. The meat is then distributed among the sacrificer's family, friends, relatives, and the poor and hungry.

[7] It is our understanding, although this could not be verified, that the meat produced in this way cannot necessarily all be distributed as there is a substantial surplus only at this particular time.

[8] In this context it would need to be ensured that any such distribution did not interfere with the livelihood of producers in those markets where such meat would be distributed for free.

[9] For the purposes of this report, the term 'cattle' will refer to live animals and the term 'beef' will refer to post-slaughter product, i.e. meat (carcasses, cuts, etc.).

of the USA exceeds the local demand and prices are relatively high, Canadian producers will sell calves into those premium markets. This is also the case for yearling feeder cattle, fed cattle and carcass/boxed beef.

In effect, since there are currently few trade restrictions at the border, movement is mostly dictated by price-driven opportunities. These again change on a regular basis depending on relative supply/demand conditions in various markets across North America. This creates a rather fluid product flow situation that is dictated by the markets and subject to the actual costs of arbitrage (transportation, border inspection fees, etc.).

Trade directions – North America

Within this relatively open border environment, several dominant movements of live animals and beef products between the countries can be identified. These are:

- Live 'fed cattle', i.e. fully fattened cattle for immediate slaughter, exports from Canadian feedlots into US slaughter operations;
- Live 'feeder cattle', i.e. cattle which are further fattened, movements from Canada into US feedlots;
- Live feeder cattle movements from Mexico into the USA; and
- Beef exports from Canada to the USA.

Because the focus of this chapter is on live cattle movements, specifically fed cattle movements for slaughter, the emphasis of the analysis presented will be on movements of live cattle from Canada to the USA. The USA is the most important outlet for Canada's live cattle export accounting for 99% of the Canadian live bovine exports.[10]

Trade volumes in North America

The USA is the biggest importer of beef and live cattle (for all purposes, i.e. immediate slaughter, further fattening and breeding[11]) out of the three North American countries. Canada has historically played an important role for live cattle export to the USA but this flow was abruptly interrupted in May 2003 when the USA closed the border after BSE was discovered in a Canadian beef animal. Therefore, the amount of Canadian live cattle export in 2003 (505,000 heads and a value of US$365 million; see also Table 2.6) went down to less than one-third of the amount of 2002 (1.7 million head and a value of US$1 billion). After the USA reopened the border for Canadian cattle in mid-2005, imports from Canada started to grow again and indeed exceeded the pre-BSE level of 2003. It should be noted that from 2003 onwards, Mexico became more important as an exporter of live cattle to the USA than Canada but this trade is not our particular concern here since it almost exclusively consists of animals for further fattening rather than slaughter. While, in 2005, Canada exported 575,000 head to the USA, Mexico exported 1,256,000 head. Apart from live cattle, during the period 2002–2005, Canada exported an average of 450 million kg of beef to the USA, despite the existence of what is effectively a barrier in the form of the US grading system,

---

[10] Source: UN COMTRADE data.
[11] A breakdown by type of import is provided in the section on 'Canadian beef exports'.

Table 2.6. US imports and exports of cattle and beef from Canada and Mexico (values in $'000). (From USTIC.)

	2001	2002	2003	2004	2005
US cattle imports ($'000)					
Canada	978,362	1,067,576	365,854	1,189	525,494
Mexico	408,349	300,500	470,379	542,872	515,519
US cattle exports ($'000)					
Canada	152,468	46,601	23,225	4,190	3,159
Mexico	91,829	62,027	11,363	594	20
US beef imports ($'000)					
Canada	1,083,876	1,096,237	842,634	1,184,198	1,246,823
Mexico	15,211	15,928	25,580	33,209	47,106
US beef exports ($'000)					
Canada	230,414	230,197	256,236	58,701	138,911
Mexico	663,424	678,645	687,722	431,783	653,962

which does not apply to Canadian beef. During 2003, beef imports from Canada were only banned from May to September 2003 because of the BSE crisis.

Canadian exports of live and feeder cattle to the USA

Prior to the BSE crisis, Canada experienced significant growth in live cattle exports to the USA. Exports are predominantly (generally over 90%) cattle for slaughter, along with some feeder cattle and minor exports of cattle for breeding purposes. Total exports have exceeded 1 million head in most years from 1992 through 2002. During 2002, a record 1.7 million head (total cattle) were exported with nearly 1.1 million head being slaughter cattle shipped to the USA. However, the number of feeder cattle exports to the USA jumped from 85,000 head in 2000 to 160,000 head in 2001 and more than 575,000 head in 2002. These large increases in the number of animals shipped for further fattening were due to severe drought on the Canadian prairies, particularly in 2002, which limited feed supplies and created an incentive to export more cattle. As is indicated above, Canada's live cattle exports to the USA were banned in May 2003 following the BSE incident in Alberta. The USA reopened the border to trade of live cattle under 30 months of age in July 2005, but cattle over 30 months are still banned from import. By the end of 2005 trade began to recover as Canada shipped nearly 310,000 head of steers and heifers for immediate slaughter to the USA and about 240,000 head of feeder cattle. However, further evolution of the trade depends heavily on the relative values of live cattle in Canada and the USA.

CANADIAN CATTLE EXPORTS BY DESTINATION. Virtually all cattle shipped to the USA from Canada are moved by truck with the majority of the feedlot to slaughter animals originating in the major Canadian cattle feeding centres in southern Alberta. Not surprisingly, the majority of cattle move into border states (Washington and Utah) close to southern Alberta or move to states with large concentrations of slaughter operations (Nebraska and Colorado). Feeder cattle on the other hand tend to move into US states that have major feedlot operations (Nebraska, Colorado and Kansas) or grain surplus areas close to Alberta such as Iowa and Minnesota. The destination largely depends on the business relationships that the sellers have with various packers and the relative sales prices in the various end (slaughter) markets.

NON-TARIFF BARRIERS. The largest issue facing the Canadian cattle feeder is one of getting fair and equal value for the cattle being sold. Due to strong dependence on the US export market (either cattle or beef), pricing in Canada will almost always be at least equivalent to the price on the US market taking into account the freight cost. There is a grading issue, which creates a non-tariff barrier that results in lower prices paid in the US market for beef imported from Canada. Beef produced in Canada but sold in the USA cannot carry a USDA grade. However, carcass beef produced from live cattle originating in Canada but slaughtered/boned in the USA can carry a USDA grade. The USA and Canada have reciprocity in the inspection process but due to Canadian-processed beef not being able to gain USDA grading certification, the Canadian beef is discounted. The inability to get the US grading on Canadian beef creates a discount which means there may not be a margin over and above the cost of preparation and delivery. Many retail and food service operations advertise they sell or serve only USDA choice beef. If they sell or serve equivalent Canadian grade, they will be charged with false advertising.

FREIGHT COSTS. The freight rates out of southern Alberta into the various US export destinations are as follows:

- Pasco, Washington: AUS$4.50/cwt;
- Hyrum, Utah: AUS$6.00/cwt;
- Greeley, Colorado: AUS$7.50/cwt.

Freight cost comprises roughly 80% of total costs of exporting a fed steer. The other export costs of veterinary inspection, border inspection and import duty are estimated to be AUS$1.50/cwt [12] (see Fig. 2.4).

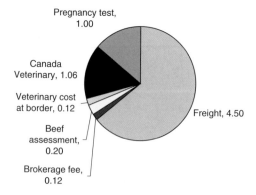

Costs of exporting a feed steer ($/cwt)

Pregnancy test, 1.00

Canada Veterinary, 1.06

Veterinary cost at border, 0.12

Beef assessment, 0.20

Brokerage fee, 0.12

Freight, 4.50

Fig. 2.4. Breakdown of exporting costs to Pasco, Washington, from southern Alberta. (From Agra CEAS.)

[12] Cwt is an abbreviation for hundredweight and equals 100 pounds or 45.359 kg.

Canadian beef exports

Beef trade has become an increasingly important aspect of the Canadian industry over the last 20 years. It shifted into high gear during the 1990s as Canadian cattle herds and the slaughter industry in general grew. From the 1980s onwards both Canadian production and slaughter capacity grew at a faster pace than in the USA due to the fact that there were large investments in the slaughter industry in Canada to take advantage of favourable feeding conditions. Some of the investors were American companies creating the Canadian export capacity to the USA. From levels around 80,000–100,000 t through most of the 1980s, beef exports surpassed 500,000 t by 2000 and reached a record in excess of 612,000 t in 2002. The trade restrictions brought on by the BSE incident caused a dramatic drop in beef exports for 2003, but the reopening of the US market to boneless beef from cattle under 30 months of age in September 2003 helped exports to rebound in 2004. There was a modest decline in 2005, as trade in live cattle under 30 months of age was renewed in July 2005, leading to significant numbers of fed cattle being shipped to the USA for slaughter.

At the beginning of the 1990s, beef exports accounted for about 12% of total beef production in Canada. This proportion grew rapidly as beef exports expanded, reaching just under half of production in 2002. The proportion fell to less than 35% in 2003, but rebounded to approximately 40% in 2004 and 2005.

The USA has traditionally been the primary market for Canada's beef exports, usually accounting for 75–85% of total exports. Japan had accounted for 5–6% of Canadian beef exports, but that market was closed following the BSE incident in May 2003 and has just recently reopened to boneless beef from cattle not over 21 months of age. Mexico surpassed Japan in 1999 as the second leading market for Canadian beef and has remained an important market behind the USA. Exports to Korea have been up and down over the years and Canadian beef has been locked out of that market since May 2003, although Korea is expected shortly to resume beef trade with Canada and the USA.

Economics and drivers of trade

The most important economic driver for the export of Canadian live cattle and also of Canadian beef to the USA is the price difference between the Canadian and US markets. Other drivers are geographical proximity of the cattle feeding centres of Alberta/Canada to the US border states as well as an oversupply of feedlot cattle in Alberta which outgrew the local slaughter capacity prior to 2003.

Export incentive through price differences

Figure 2.5 shows the price differential a cattle feeder in Alberta, Canada, would receive from a domestic (Canadian) packer versus prices available in the export (USA) market. A negative value on this chart indicates that there is at least some incentive to move cattle for immediate slaughter into the USA from Canada. The deep discount for Canadian cattle in the post-BSE period created a tremendous incentive for live cattle movements to take place from Canada to the USA. As a result, fed cattle exports for immediate slaughter from Canada to the USA rebounded sharply in 2005.

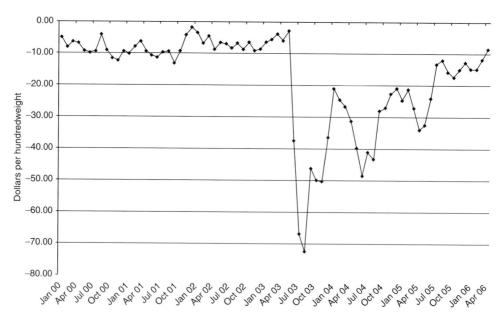

Fig. 2.5. Fed cattle price difference paid by packers (Canada minus USA). Note: UTM indicates the age of cattle and means 'under 30 months'.

Typically, the Canadian cattle feeder will demand a slight premium to export cattle to the USA for slaughter. If the producer cannot obtain a premium, he will sell to a local Canadian slaughter operation. A premium is needed for the producer to compensate for additional time and expense related to export inspections and to compensate for the risk of additional weight shrinkage en route to a US slaughter operation.[13] The volumes of cattle that regularly move to the USA are not large relative to USA and Canadian cattle inventory but the potential of usually being able to export the cattle does keep the prices Alberta slaughter operations can obtain in line with live cattle market prices in the USA.

Export incentive through limited slaughter capacity
In the pre-BSE period (before May 2003), fed cattle moved to the USA partially because the feedlot supply in Alberta outgrew the local slaughter capacity available to Alberta cattle feeders. Thus, 15,000–20,000 head per week would be exported to the USA to find available slaughter capacity. Without the capacity in Alberta to slaughter all the cattle, there was little alternative except to move live fed cattle to the USA. In the post-BSE period, slaughter capacity has been added in Canada and the industry is probably in near balance in terms of kill capacity relative to sustainable slaughter volumes. This should reduce the weekly movement of cattle

---

[13] Once cattle are taken away from the feedbunk, the cattle begin to lose weight. The stress of handling and shipping, long time off feed and long distance of transport will cause extra weight loss. The amount of weight loss is apparently highly variable, and is often in a range of 3–5% of body weight over a short or medium distance (and time). It is reported that some companies will let the cattle rest and have access to water and feed following a very long haul, before taking them into the slaughterhouse.

to the USA for slaughter (relative to the pre-BSE period). However, relative pricing between the two markets will be the determining factor. The stronger Canadian dollar is an impediment to fed cattle movement as long as currency-adjusted prices in Alberta are in line with the US market and not severely discounted.

Export incentive for beef through price differences

Because the USA plays such a prominent role in Canadian beef exports, not surprisingly the price difference between the Canadian and US markets is a primary driver in the beef trade between the two countries. Figure 2.6 shows the monthly price difference between Canadian AAA and USDA choice beef. Large disparities will cause significant increases in exports from Canada with the opposite impact when prices differences are low.

In future there will be a tendency for more beef to transit the border as packers in Canada will seek to utilize their capacity to the fullest extent possible. However, as is indicated above this development will also depend on the price differences between both markets. Once the Canadian beef has entered the US market, it mainly moves through the same channels of distribution as does the beef from US slaughter operations. In terms of ultimate consumption point, it is estimated that Canadian beef is consumed as follows:

- Retail: 49%
- Foodservice: 47%
- Exported Products: 4%.

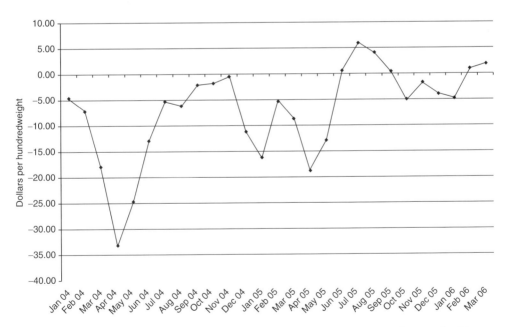

Fig. 2.6. Cattle cutout value difference (Canada AAA minus USDA choice). Source: CanFax and USDA. Note: the cutout value is an aggregation of individual meat cut prices designed to estimate an average carcass-related price. The value is driven by seven key elements of the carcass generally referred to as 'primals': rib, chuck, round, loin, brisket, plate and flank.

Trade with Mexico

Mexico exports of live cattle and feeder cattle to the USA

No live cattle from Canada are shipped to Mexico for slaughter because the transportation is too expensive. The same goes for live cattle trade from Mexico to Canada. The Mexican market prefers leaner beef, which tends to be the lower grades of meat. For example, in the Mexican market a Canada AA piece of meat is usually preferred to a Canada AAA[14] piece of meat. Many products in Mexico utilize shredded or thinly cut beef, which minimizes the need for the beef to be tender. However, Mexico is becoming a wealthier country and is eating more beef and in the future is expected to demand higher grades of beef.

Cattle export forms 99% of all live animal trade in Mexico. Out of this, 99.7% goes to the USA. The Mexican/USA trade is predominately calves and feeders exported from Mexico to the USA; virtually no fed cattle are exported. Mexico is a grain-deficit area and lacks enough pasture to feed all the cattle. Some Mexican operations specialize in producing higher-quality calves with the intention of shipping the cattle to the USA after the animals are weaned. The calves will be placed into a feedlot in the USA. When Mexico ships cattle to the USA, the movement depends on the location of the cattle and the pasture conditions.

Background on production chain

As noted earlier, because of the level of integration and the relatively open borders (the BSE-related trade disruptions notwithstanding) the cattle and beef industries in Canada, Mexico and the USA can be viewed as a North American industry. The key segments of the Canadian, USA and Mexican cattle and beef supply chains are described in the next sections of this chapter. From this exercise, points of difference in the respective supply chains are identified whether of a structural, environmental or political origin.

Production at farm level

CANADIAN INVENTORIES RELATIVE TO USA.   Each sector of the Canadian beef industry has many similarities as well as some differences with the comparable sectors in the USA. Total inventory of cattle and calves on Canadian farms on 1 January 2006 was estimated by Statistics Canada at 14.8 million head. This was down 1.5% from the record level of nearly 15.1 million head in 2005, the first decline in 3 years. But the inventory was still more than 1.3 million head larger than in 2003 or before the ban on Canadian cattle exports following the first case of BSE in a domestic cow.

The 2006 inventory number includes an estimated 704,000 head of cull cows that were being held on farms. While exports to the USA of cattle under 30

---

[14] Canada AA and Canada AAA are Canadian beef grades. AAA grade would mean better quality beef than AA grade. Canada A/AA/AAA/Prime are the highest-quality grades. See Appendix 2.1 for explanation.

months of age resumed in mid-July 2005, Canadian cattle over 30 months of age are still banned in the USA and other countries. Despite an increase in plant capacity in Canada over the last 2 years for the slaughter of cows, there has been a hold back on the number of cows that would have been culled under more normal circumstances because of the disruptions caused by BSE.

Cattle inventories in Canada and the USA followed similar cyclical trends into the late 1990s. However, the Canadian industry grew at a faster rate from the late 1980s onward. While the US cattle herd-size numbers were falling from 1997 to 2004, the Canadian herd showed a more modest downturn in the late 1990s followed by renewed growth in 2000 and beyond. A temporary decline in the inventory occurred in 2003 as a result of a severe drought in the Canadian prairies in 2001/02. The US industry only started to show a turn from liquidation to expansion since 2004. As of 1 January 2006, the Canadian cattle herd is 32% larger than in 1990, while the US cattle herd is only 1% larger. The ratio of US cattle inventory to Canadian cattle inventory has declined from 10:1 in the mid-1980s to 6.5:1 in recent years.

Production Structure

THE US CATTLE AND BEEF PRODUCTION AND MARKETING CHAIN.   The participants in the US cattle and beef marketing chain are all specialized in one of the following key sequences of the marketing chain: inputs, production, slaughter/rendering/packing, processing, distribution and retail/food service.

Conclusions on the live trade in North America

As has been indicated in the above analysis, in North America there are substantial cross-border trade flows for both cattle for fattening and for slaughter. The main trade flow of interest for the purposes of this study is the cross-border cattle trade from Canada to the USA for slaughter.

Under normal market conditions, Canada is a net exporter to the USA of both cattle and beef. Canada has large land assets that do not have many alternative economic uses other than cattle and grain farming and can therefore usually produce cattle at very competitive prices. Historically, Canada did not have enough packing capacity to slaughter all the cattle raised in Canada, but post-BSE Canada has ample slaughter capacity. The reason cattle are therefore not slaughtered in Canada but rather shipped live into the USA is to exploit the price differential which may exist between these two markets and in particular to attract the premium which is attributable to USDA-graded beef in the USA. Cattle slaughtered in the USA can be advertised as USDA-inspected. Since Canadian packing plants cannot label product as USDA-inspected, the American retailer will not pay the same amount for Canadian beef or will not buy the beef. Lower-price restaurants and processors in the USA where the origin of the meat is not specifically identified would rather have the cheaper Canadian beef because it improves profit margins and the customers will not generally be aware of the difference. The US market in effect acts as a regulator of the prices on the Canadian market regularly drawing in supplies when prices in Canada are lower. Cattle could theoretically be killed in

Canada and the carcass then shipped to the USA to be graded by US inspectors, however, the transportation costs and handling issues minimize this practice.

Recommendations

As has been indicated above the trade in live cattle for slaughter from Canada to the USA is largely driven by the premium obtained for having USDA-graded meat in the retail and higher-price[15] catering sectors. As has been pointed out there are no real differences of significance in the Canadian and US meat grading systems, and therefore this trade is being largely driven by an erroneous consumer perception that USDA-graded meat has different quality attributes from that of Canadian-produced and Canadian-inspected beef. There would appear to be two potential solutions to this issue. These are:

- To highlight to US retailers and consumers that their non-acceptance of equivalent-quality Canadian-inspected beef has potentially adverse implications for animal welfare, and may result in their paying more for their product and does not produce any other tangible consumer benefit[16]; and
- To seek to have specific Canadian slaughterhouses designated as export-accredited and the meat certified as conforming to USDA standards by US inspectors thus allowing the meat to obtain a label equivalent to US-inspected meat; such a move would obviously be resisted by US slaughterhouses and producers as it would tend to remove the throughput and premium obtaining for 'US-produced and US-inspected beef' but against this it could be argued that there would be potential animal welfare and price gains for consumers.

Live Animal Export from Poland to Italy

This part of the chapter analyses the economics of live animal transport from Poland to Italy focusing on cattle, horses and sheep.

Cattle

Background of the production chain
The largest cattle herds in Europe are to be found in France, Germany and the UK. France and Germany together contribute 37% to the cattle stock in Europe. The four leading countries have a share of more than 50% of European cattle stocks. Poland has the seventh largest herd and in 2004 it accounted for 6% of EU cattle

---

[15] These need not necessarily be higher price although if they are sourcing USDA-graded beef it would generally be the case that they are.
[16] Apart from the possible benefit associated with buying 'US' produce which would in any case still be a matter of choice.

Table 2.7. Consumption of meat in Poland and Italy, 2003 (kg/person). (From Polish Statistical Office (GUS) and Eurostat.)

	Total meat	Beef/veal	Pig meat	Sheep meat	Equidae[a]
Italy	93	24	39	1.5	1.1
Poland	79	6.5	48	0.05	0

[a]'Equidae' is the term used by Eurostat to name animals belonging to the family of horses, which apart from horses also includes wild asses, zebras, wild horses.

stocks. Italy, which is the main recipient of live cattle from Poland, is also an important EU cattle producer, accounting for 7.6% total cattle stocks in 2004.

The structure of the cattle herd is closely related to Polish eating habits as far as meat is concerned with the bulk of animals held for milk rather than meat production. Poles eat mainly pork, as can be seen in Table 2.7. Furthermore, within the 79 kg of meat per person consumed on average in Poland in 2003, beef only accounts for some 8%. By contrast in Italy beef accounts for some 25% of total meat consumption. Due to a growth in beef/cattle exports to the EU-15 countries (see Table 2.8) retail beef prices in Poland rose significantly after Polish entry to the EU. In 2005, prices were 16.5% higher than in 2004. At the same time the price of pork and poultry fell, leading to a further 25% fall in beef consumption in 2005 to 4 kg per person. Currently, retail beef prices are 40% higher than the level at the end of 2003.

It is important to note that Polish cattle production structures are highly fragmented with the majority (62.8%) of Polish farms having less than 5 cattle. This having been said, the average size of holding is 5–9 head and 43.8% of the herd is held in farms with 20 or more animals. Farm structures in Poland are changing and the trend towards an increasing number of larger farms will continue.

Since accession to the EU, the position in the Polish cattle market has changed in that demand, and to a lesser extent supply, has increased. Also, due to the adoption of the Common Agricultural Policy (CAP), producer support prices for beef have risen. The combined effect has been that beef/cattle prices in Poland rose sharply and in 2005 the average ex-slaughterhouse price in Poland was 63% higher than the price in 2003. The price of young slaughter cattle in 2005 was 45% higher than in 2003. Figure 2.7 shows how cattle prices at slaughter have increased within the last 3 years with the sharpest rise taking place in the 17th week of 2004, when Poland joined the EU. In 2005, prices remained at the new higher level.

Table 2.8. Export of bovine meat and live cattle from Poland, 2003–2005 ('000 t). (From Instytut Ekonomiki Rolnictwa i Gospodarki Zywnosciowej (Institute of Agricultural and Food Economics – Polish National Research Institute).)

	2003	2004	2005
Total exports	92.5	118.3	178.1
Meat	54.8	73.2	128.1
Live	37.7	45.1	50

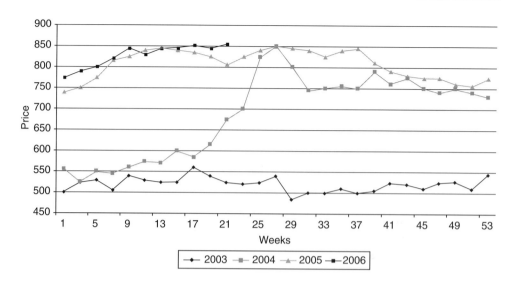

Fig. 2.7. Development of prices at slaughterhouse for cattle in Poland, 2003–May 2006. (From Polish Ministry of Agriculture and Rural Development, 2006.) Note: Prices are expressed weekly in zl/100 kg of cooled slaughtered meat.

The increasing profitability of cattle breeding, especially slaughter cattle, has encouraged farmers to keep young cattle on farm, which has resulted in growing stocks.

Table 2.9 presents balance sheets for bovine meat. This shows clearly that there is an exportable surplus of cattle in Poland. In 2004, 14% of Polish bovine production was exported, either live or in meat form.

Trade flows

As described in more detail in the section above, beef/cattle production in Poland has become more profitable in recent years, which is related to the increase in prices received by beef/cattle producers, and growing demand from EU-15 countries. In 2003, beef production in the EU-15 fell, and from then onwards the Union started importing more meat from third countries, including Poland particularly after the removal of trade barriers on accession. This trend is illustrated in Fig. 2.8 which shows the development of Polish exports of bovine meat and live animals over the last 25 years. It should be noted that while meat exports have risen sharply live cattle exports have remained more or less stable after accession.

Table 2.9. Polish bovine meat balance sheet, 2002, 2004 ('000 t carcass weight equivalent). (From Polish Statistical Office (GUS).)

	National gross production	Import of live animals	Export of live animals	Meat import	Meat export	Domestic use
2002	311	3	37	0	71	245
2004	340	4	45	2	75	228

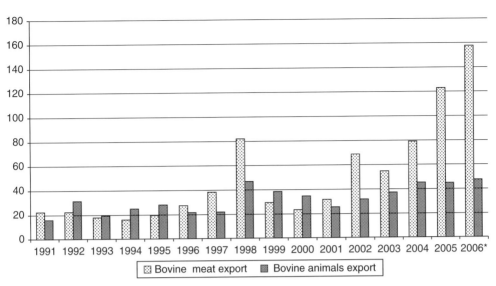

Fig. 2.8. Bovine meat and live animal exports from Poland, 1991–2006. (From IERiGZ (Institute of Agricultural and Food Economics – Polish National Research Institute).) Note: expressed in '000 t of meat equivalent.
*estimated

At the beginning of the 1990s the difference between volumes of live export and meat export was very small. This is primarily related to the fact that Polish slaughter-houses and the processing industry had to undergo a long process of adaptation to European standards in order to be able to export meat. This is confirmed by the fact that a review of the main trade destinations for exports in 1991 shows that live animals were exported to Western Europe while meat (fresh, chilled or frozen) went to Eastern Europe. Throughout the period examined, an average of 80% of total Polish live bovine exports was sent to EU-15 countries. The main driver for the live trade in the early 1990s was therefore the fact that the sector had to restructure and the processing sector was not able to export westwards. After the increase in live bovine animals export fol-lowing the system change in 1989, there was a drop in the numbers of animals exported in 1993. This fall corresponds to the timing of the signing of the Association Agreement between Poland and the EU, which limited live bovine trade by introducing quotas.

Table 2.10 presents live animal exports from Poland by main recipient in the period 1996–2004. The data come from the UN Comtrade database, and are presented for every year, as well as for the whole period analysed. As far as numbers of exported animals are concerned, Italy is the major recipient receiving more than 2.6 million head of cattle from Poland or an average of just under 300,000 animals per year in the period 1996–2004. This represents 53% of total cattle exported from Poland in this period.

One feature of the trade which has changed over the period reviewed is the fact that in terms of the composition of the trade to Italy there is a growing share in terms of young animals exported. This can be seen by the fact that the number of animals exported has been growing while the average weight of animals has been falling. In 2004, the average weight was only 90 kg. It therefore can be inferred that the majority of cattle exported are calves. In 2005, 70% of cattle

Table 2.10. Cattle exports from Poland by main receivers, 1996–2004 (head). (From UN COMTRADE and FAOSTAT 2005.)

Partner	1996	1997	1998	1999	2000	2001	2002	2003	2004	Overall 1996–2004	%
Italy	262.8	293.5	250.9	267.7	285.2	239	289	310	411.6	2,631	53
Bosnia Herzegovina	6.2	9.6	74.6	76	58.4	52.8	179	177	121.5	729	15
The Netherlands	28.7	16.9	15.6	19.2	20.4	8.7	19.5	43	111	301.5	6
Spain	3	11.4	10.5	6.4	8.3	10	52.6	67	96.5	269	5
Croatia	6.9	5.2	32.3	31.2	68.7	44	37.7	0	24.6	237	5
Germany	20.2	22.8	16.8	14.3	21.7	8.6	12.5	21	56.7	204	4
France	3.2	0.340	0	0.678	0	2.3	0	0	7.3	15.8	0
Hungary	0	0	3	1.3	1.2	0	5.9	0.483	2.3	13.7	0
Belgium	0	0	0	0.699	1.7	0.651	0.	2.5	7.2	13.6	0
World	347.8	405.3	513.6	492.1	591.6	448.7	654	653	867.6	4,968	100

exported were calves for further fattening. Figure 2.9 shows how the share of calves in total cattle exports from Poland has developed.

Structure of the marketing chain

The marketing chain for live cattle shipments from Poland and Italy is characterized by the fact that there are not many participants beyond the farm gate. Traders either operating on their own or on behalf of a number of transport enterprises usually

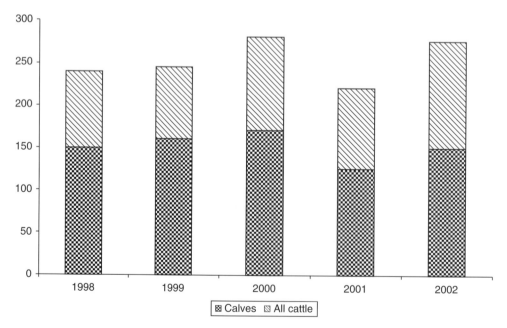

Fig. 2.9. Export of cattle and calves from Poland, 1998–2002 ('000 head). (From IERiGZ (Institute of Agricultural and Food Economics – Polish National Research Institute).)

purchase direct from producers. Cattle are bought directly on farm rather than via markets. The purchasers in Italy are the companies operating feedlots primarily in northern Italy. As has been indicated above they are mainly interested in young cattle, which are either fattened to a weight of some 200 kg in order to provide veal or they are fed up to a slaughter weight of approximately 500–600 kg when they are slaughtered as beef animals. Approximately 20% of the exports (or approximately 70,000–80,000 animals based on 2004/05 numbers) are, however, estimated to consist of older animals with a weight of 450–650 kg destined for immediate slaughter. This provides the Italian market with lower-quality beef. It is important to mention that the feedlot sector in Italy, especially in the north of the country is a well-established sector with a large capacity. It is estimated that there are some 50,000 feedlots in Italy.

Economics and drivers of trade

Transport of live cattle from Poland to Italy has been taking place for many years. As the trade flow data presented show commercial relations between Polish farmers or traders and Italian feedlots/slaughterhouses have a long history. In this section, the economic incentives behind the live cattle trade are analysed by first looking at the situation before the 1991 Europe Agreement which associated Poland to the EU, then reviewing the period after this Agreement and before Polish accession to the EU. Finally, we will analyse the most recent period since May 2004 when Poland has been a member of the EU.

POLISH BEEF/CATTLE MARKET CHARACTERISTICS. As indicated in the section on cattle above, Poland has a cattle production capacity in excess of domestic production requirements. Given the low demand for beef in Poland, Italian demand creates a market for beef/cattle from Poland. Italian demand for beef or cattle from Poland provides incentives for Polish farmers to breed more cattle, and indeed to start to introduce more meat breeds into the Polish cattle herd. As shown in Fig. 2.9 there has been a huge upsurge in beef exports from Poland since 2004. It is estimated that the volumes of exported beef will double this year relative to 2004. This is related to growing standards of Polish meat processing industry and the inclusion of Poland in the EU's internal market. It is most likely that beef exports will continue to grow.

MARKETING CHAIN. The structure of the marketing chain for live cattle in Poland acts to the advantage of the Italian buyers. As has already been indicated, the buyers in Italy create the growing market for beef and cattle in Poland and also set the prices on this market. Production in Poland is fragmented and this disadvantages the sellers in Poland. Recently the first attempts have been made to organize cattle producers in Poland in order to provide them with more negotiating power, and to provide the trade with greater transparency. It seems likely that in the near future this market will therefore become more structured.

PRICE LEVEL DIFFERENTIALS BETWEEN POLAND AND ITALY. The major driver for trade in both meat and live cattle between Poland and Italy is the differential in prices between the two markets. It is very clear that prices for calves and heifers in Italy are the highest in the current EU, well above the EU average of approximately

€320/100 kg carcass equivalent while prices in Poland are among the lowest in the EU. In May 2006, the differential per kg for calves amounted to €125/100 kg. This difference less transport costs represents the margin available on the trade and clearly provides a strong economic incentive to engage in the trade.

On the question of why live exports rather than meat exports have taken place, Fig. 2.10 shows how export prices for cattle and beef in Poland have developed over the last 11 years. As is evident for most of the period covered and particularly in the mid-1990s before Polish slaughtering facilities would have been approved for exports there was a marked price advantage in favour of live exports. However, since 2001 this price advantage in favour of live exports has for the most part disappeared and this is also reflected in the fact that these exports have largely been reduced to those animals (calves) which are destined for further fattening in Italy rather than immediate slaughter.

The driver for the trade in calves for further fattening is explained by Fig. 2.11, which presents a comparison of prices of calves in Poland and in the EU. The clear gap between EU-15 prices and the prices on the Polish market provides a major incentive for Polish cattle producers to export to Italy. Italy has the highest prices for cattle within the EU. Therefore, the gap in Fig. 2.11 is the gap between the EU maximum price and the Polish price. In May this year, 100 kg of calf carcass meat equivalent was sold for €200 more in Italy than in Poland.

ROLE OF EU POLICY.   Before Poland entered the EU, trade was governed by the 'Europe Agreement' between Poland and the EU.[17] The agreement specified

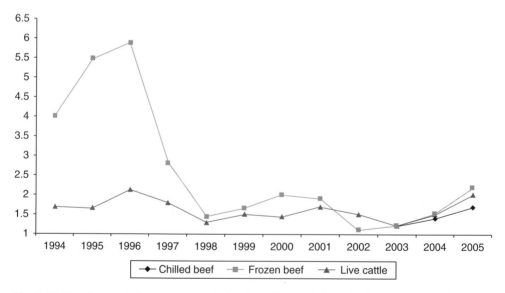

Fig. 2.10. Development of export prices for beef and live cattle in Poland, 1994–2005. US$/kg of carcass weight equivalent. (From IERiGZ (Institute of Agricultural and Food Economics – Polish National Research Institute).)

[17] The Association Agreement (called the Europe Agreement) entered into force on 1 January 1994. The Agreements were concluded with Poland, Hungary, Slovakia and the Czech Republic.

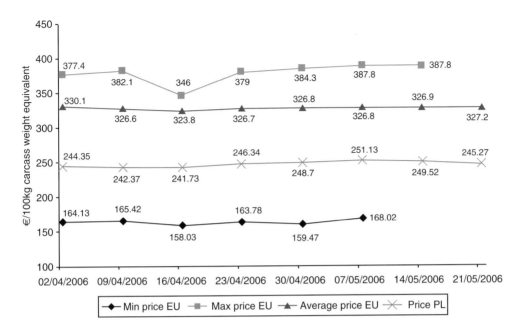

Fig. 2.11. Comparison of prices of male calves between Poland and the EU. April–May 2006. €/100 kg carcass weight equivalent. (From Polish Ministry of Agriculture and Rural Development.)

annual import quotas for live cattle between the EU and Poland. The volumes introduced under the import quota were subject to a 25% levy reduction. For the first half of 1995, the quota for male calves was 99,000, and for cattle of the weight of 160–300 kg it was 39,600. For these volumes, therefore there was an incentive to enter the trade.

Another driver for the live trade in the pre-accession period was the fact that the common organization of markets in the beef and veal sector, provided income support measures to farmers in the form of the 'beef special premium', and 'slaughter premium'. As is explained below these two measures, combined with the price differentials, were also important drivers of live cattle trade from Poland to Italy.

BEEF SPECIAL PREMIUM.   Prior to the 2003 Common Agricultural Policy (CAP) reforms, and to a much more limited extent up until now, EU producers can claim a special premium on male bovine animals. The premium for young bulls is €210 per head, while the premium for steers is €150 per head. The beef special premium on young bulls is payable only once during the lifetime of an animal from the age of 9 months. For steers the premium can be paid twice in the lifetime of the animal. Steers are eligible for their first payment once they have reached the age of 9 months. The second payment is made after the animal has reached the age of 21 months. To qualify for the premium, the animal must be held by the producer for fattening for a 2-month period.

Up until 2005 the possibility of obtaining such a premium provided an additional incentive for Italian producers/fatteners to import live calves from Poland.

SLAUGHTER PREMIUM.   Prior to the 2003 Common Agricultural Policy (CAP) reforms, and to a much more limited extent up until now, EU producers are also eligible for a slaughter premium on bulls, steers, cows and heifers from the age of 8 months and on calves between 1 and 7 months and weighing less than 160 kg. The adult cattle slaughter premium is €80 per head and the veal slaughter premium is €50 per head. To be eligible to claim the slaughter premium the animal must have been retained by the producer for a minimum period of 2 months, ending less than 1 month before slaughter. In case of calves slaughtered at the age of 3 months, the retention period is 1 month.

It is important to highlight that the premia system as described above was for the most part in place only until the end of December 2004. As a result of the most recent CAP reform,[18] there have been changes in the payment modalities which entered into force in 2005. Member states were provided with the possibility of partial 'decoupling', i.e. delinking of payments from production, allowing them to retain some of the existing headage payments while decoupling others. Italy is among the group of countries which have decided to proceed with full decoupling, which in practice means that it is no longer paying headage-based premia as described above. It is too early to judge what effect this will have on the volumes of live cattle trade to Italy. It is very probable that as a result of this CAP reform, as well as recent attempts to limit live animal transport in the EU in general, less cattle will be traded between Poland and Italy.

Conclusions on live cattle trade

Historically shipments of live animals from Poland to Italy have been based on the fact that there is a limited market for beef in Poland and a strong market for this in Italy. This means there has always been a surplus of beef (as a result of the extra animals generated in the milk production process) on the Polish market and a strong price differential between the two markets. In addition to this main driver, during the 1990s specifically the trade in live animals (as opposed to meat) was encouraged by a number of factors including:

- The fact that following the collapse of the centrally planned system a large surplus of cattle/meat emerged as the livestock herds were drastically reduced;
- The lack of recognized slaughter capacity in Poland licensed for export to the EU; and
- The availability of EU premiums for producers on animals fattened and slaughtered in the EU.

---

[18] The latest reform of the CAP took place in 2003. It introduced a single farm payment scheme (SFP), which is based on historic payment levels and replaces most of the directly production-linked aid paid to farmers for their crops and livestock. In other words the payment is no longer linked to what the farmer produces, it is decoupled. The amount of payment is calculated on the basis of the direct aid paid to a farmer in the reference period 2000–2002. It is also linked to the farmer's respect for environmental, food safety, animal and plant health and welfare standards, as well as the requirement to keep all farmland in good agricultural and environmental condition. This is known as cross-compliance, and it is compulsory.

Since accession to the EU, none of the factors listed above have applied but the price differential has remained highly significant making it attractive to ship animals for fattening to Italy where for many years there has been a large and intensive feedlot sector geared to taking such animals for the veal and beef trade.

Beyond this it is estimated that approximately 20% of cattle shipped would consist of older animals for immediate slaughter. The main reason for the live (as opposed to meat) trade would appear to be the fact that it creates employment and added value in the slaughter and processing sector in Italy and that an industry generating these effects has been built up historically (e.g. in meat processing and notably in the form of leather manufacture).

Recommendations

As has been noted above many of the reasons for the historically high level of the live cattle for slaughter trade which were in place in the 1990s have now been removed and there has therefore been a corresponding decline in such trade. This having been said, a significant number of animals still are shipped for slaughter so as to obtain the premium derived from the fact that plants in Italy are more geared to adding value than their Polish counterparts. In principle, however, over time most or all of the value-added attributes of slaughter in Italy should be able to be replicated in Poland (e.g. if necessary hides could be shipped from Poland to Italy etc.). In order to achieve this there would need to be a concerted effort to highlight the potential animal welfare gains to be derived by a longer-term restructuring of the slaughter and processing process to favour slaughter in Poland.

Horses

Background on production chain

Horses have always had an important role on farms in Poland, particularly as draught animals. There were huge numbers of horses in Poland throughout the 1950s and well into the 1980s. Due to the fact that Poland retained a substantial number of very small holdings (over 2 million) and these were mechanized only to a limited degree, horses were very commonly used on farms. Later, with the changes to the Polish agriculture sector which were accelerated by the shift to a market economy, horses became less needed on the farms and a significant decline in horse numbers was initiated in the late 1960s which continued up to about 2000.

As far as can be ascertained horses are generally sold on markets where they are purchased by agents for the Italian slaughterhouses. The seller's market is highly fragmented with Polish farmers usually selling one or two horses at local markets. The price will therefore almost exclusively be determined by the buyers.

Very recently a first attempt has been made to structure horse trading in Poland. A regional association of horse breeders in the north-east of Poland has for the first time organized a market. If they prove successful, such efforts will enable more control over the live horse trade which has tended to be conducted outside official supervision and is often seen by Polish farmers as a 'business of the

Italian mafia' although this may also reflect the fact that the trade and price formation within the market are highly untransparent.

Since there is no market for horse meat in Poland, the country has for many decades been a supplier of both horse meat and live horses for slaughter to those countries in Western Europe where horse meat is consumed. As a result of the large number of horses in Poland and the continuing decline in the number of horse needed in the agricultural sector, Poland has long been in a strong position to supply these markets. The conditions for horse grazing in Poland are very good as pasture is widely available, and the relatively lower feed prices make production cheaper than in Western Europe. In the early period of the trade in horses out of Poland this was a marginal activity and it was in effect the surplus of animals being exported. Currently, however, we understand that some farms in Poland specialize in the production of horses specifically for slaughter. This has meant that live horse exports have been maintained at a stable level. Furthermore, in recent years, a new type of 'slaughter horse' has been created.

Looking in more detail at the types of horses transported out of Poland, it is apparent that the Italian market demands exactly the type of horse which is commonly bred in Poland. In Poland there is a tradition of breeding draught 'cold-blooded' horses. These horses are considered relatively easy to breed, have low feeding costs and are reported to have particularly good disease resistance. Currently, there are approximately 180,000 such horses in Poland. The main regions in Poland where these horses are bred are Warminsko-Mazurskie voivod-ship in the north-east and Podlaskie voivodship in the east which are both regions with extensive natural pasture and are therefore also the regions from which the greatest numbers of live horses are transported.

We understand that Italian consumers prefer light, non-fat meat and for this reason young horses up to the age of 7 years are mainly shipped. As is indicated above, Poles do not eat horse meat and horses have therefore never been seen as meat-producing animals. The growth of the live horse trade and the profitability of this activity have therefore attracted some media attention in recent years as organizations protecting the rights of animals have organized widely supported awareness raising campaigns.

Trade flows

As Fig. 2.13 shows, live horse shipments from Poland date back to the 1960s. It has only been very recently that the volumes have started falling. Over the last 40 years, the numbers of horses traded live has in some years exceeded 100,000 animals per year but currently, the number appears to have stabilized at approximately 40,000 head per year. Figure 2.13 shows that the largest number of animals was transported out of Poland during the 1990s, and meat trade was very limited. By contrast in recent years increasingly more meat is being traded, and the live horse trade has been decreasing. As for cattle this trend can also be explained by the improving standards within the Polish slaughtering and meat processing industry.

Horses are generally transported in lorries by transport companies, which are mostly Polish-owned but there are apparently also Italian transport companies operating in the Polish market.

During the communist period, trade in horses was centralized and only one public body was authorized to conduct trade in horses.[19] After the collapse of the system, the structure became much more disparate and unstructured. In essence, it would appear that producers generally sell their horses at local markets where they will be purchased by a broad range of traders.[20] These markets have been functioning for many years and it has therefore created mechanisms of its own still using the market linkages which in many cases were established some decades ago. Overall, however, the trade has declined and the regulatory framework has also been improved. Figure 2.12 presents the structure of the marketing chain in live horse trade in Poland.

In 1998, a report from the Polish Audit Commission[21] highlighted a number of irregularities related to the transport of live animals. The report draws attention to the lack of proper infrastructure at collection points, markets and border crossings; it also mentions insufficient veterinary controls and operator ignorance of regulations relating to transport of live animals. More generally, it highlights the existence of unlicensed markets where horses are traded as well as the use of fraudulent documentation.

This chapter, as well as the increasing interest of animal welfare organizations in issues relating to the transport of horses out of Poland and the introduction of a Polish animal welfare policy due to EU accession, have attracted attention to the transport of live horses. However, as far as we can ascertain the profitability of the trade remains high and it would appear that the trade still remains inadequately controlled.

The horses traded out of Poland are generally slaughtered within 4–5 days of arrival in Italy. The horse meat is then on the market within 48 h.

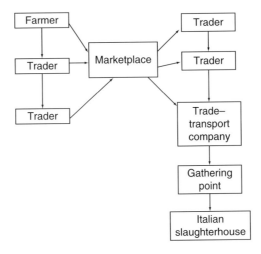

Fig. 2.12. Marketing chain in the live horse trade in Poland. (From Agra CEAS.)

---

[19] A company called 'Animex'.
[20] There is virtually no infomation on the structure of the trade as it is highly 'informal'.
[21] Najwyzsza Izba Kontroli (NIK) Report 87/1998/P97121.

Table 2.11 shows clearly that Italy is the most important recipient of slaughter horses from Poland. According to UN Comtrade data, in the period 1996–2004, Italian buyers accounted for 90% of live horses exported from Poland. In 2005, out of the total of 35,154 head exported out of Poland, 28,268 horses were traded to Italy. By comparison, the number of breeding horses exported out of Poland is relatively small, and in 2005 the number exported came to 1852 head.

The recipients of the live horses in Italy are slaughterhouses. In the north of the country there are well-organized complexes made up of slaughterhouses and processing plants. In the south it seems that horses are sometimes sold on more local small-scale markets. Horse meat is sold as meat, and also in the form of different processed products. It appears to be most in demand in the south of Italy, where most horse meat in Italy is consumed, often in raw form.

An increasing proportion of horses are slaughtered in Poland. It is our understanding that there are three abattoirs where horses are slaughtered, and the meat is entirely exported to Western Europe. These abattoirs are entirely owned by Italians. It is understood that in part the horses which are slaughtered in Poland are those which are considered as being unfit for transport to Italy although we have not been able to establish what 'criteria' are used to determine this.

Economics and drivers of live horse trade

GEOGRAPHICAL LOCATION OF POLAND. Due to its geographical location, Poland acts as a transit point for horses coming from neighbouring countries. These countries which include Lithuania, Belarus and Romania are also important exporters of live horses to Italy, and transports from these countries are often combined with Polish transports.

CHARACTERISTICS OF THE POLISH MARKET. Since horse meat is not consumed in Poland, Italian demand creates a market for slaughter horses/horse meat produced in Poland. Without this demand the market would not exist.

BREEDING COST DIFFERENCE BETWEEN POLAND AND ITALY. It has already been mentioned that the cold-blooded horses which the Italian market demands have traditionally been bred in Poland. Polish stocks of horses, though they have decreased over the last 20 years, are sufficient to cover Italian demand. The availability of cheap pasture and labour in Poland as well as the tradition of breeding such horses mean that production costs in Poland are lower than in Italy thus generating a further incentive for trade.

POLICY FACTORS. In the 1990s, when Poland was not yet a member of the EU, import quotas were introduced which allowed duty-free import of slaughter horses from Poland to the EU. In the framework of this quota system, licences were also introduced, and they were necessary for Polish producers in order to be able to trade. However, licences could only be granted to producers who had already traded with the EU: The bulk of the import licences were granted to those who had traditionally traded in the market, i.e. the Italian slaughterhouse owners. In addition to this there was also an incentive to engage in live trade as a result of differences in customs duties on live horses and horse meat which were introduced in the 1990s. The 1990 trade agreement governing between Poland and the EU

Table 2.11. Live horses exported from Poland by main destination, 1996–2004 (head). (From UN COMTRADE and FAOSTAT.)

	1996	1997	1998	1999	2000	2001	2002	2003	2004	Total 1996–2004	%
Italy	77,744	82,693	80,010	67,192	22,437	35,770	27,794	36,399	34,659	464,439	89
France	9,481	10,006	7,547	4,260	1,803	2,158	1,319	2,305	2,501	40,485	8
Belgium	–	–	–	73	116	128	–	–	7	403	–
Total	89,904	96,346	90,815	74,066	25,206	39,052	29,748	39,780	38,283	521,808	100

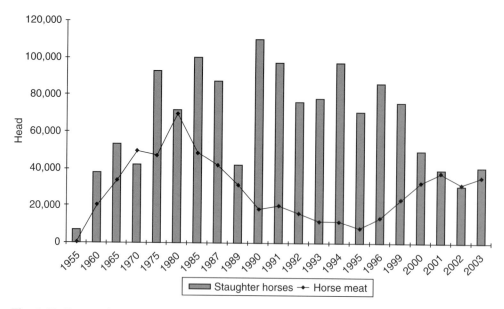

Fig. 2.13. Export of live slaughter horses and horse meat from Poland, 1955–2003 (head). (From Polski Zwiazek Hodowcow Koni (Polish Horse Breeders Association).)

removed the duty on live horses entering the EU from Poland, while the duty on horse meat from Poland was maintained at a level of 8%. Annual quotas were also agreed for the number of horses entering the EU with 0% duty. The likelihood that this had an influence is confirmed by the trade data presented in Fig. 2.13, which shows a significant increase in live horse exports from Poland starting from 1990, and a parallel fall in meat exports. An increase in meat exports from 1995 most probably marks the period when horse slaughter capacity appeared in Poland, and the abattoirs received accreditation for meat export to Western Europe.

PRICE DIFFERENTIALS FOR LIVE HORSES IN POLAND AND ITALY. Figure 2.14 provides an analysis of the differences in price levels for live horses which can be obtained by producers for sale in Poland, and the prices obtained on horses for export. It is not clear from the data provided whether these are prices obtained for the same quality of horses and in fact we would assume that the price paid for horses which are not transported is lower since as has been indicated above they are generally considered not to be transportable. What is clear is that export prices are very advantageous for Polish producers.

For Italian slaughterhouses prices for live horses in Poland are very attractive. Industry investigation has revealed that currently, i.e. early 2007, the average cost per kg of horse bought in Poland, including transport costs, is some €2.00. Maximum prices for high-quality horses can reach €3.2/kg. Assuming an average horse weighs some 600 kg, it can be estimated that after slaughter and removal of all by-products, some 300 kg of horse meat are produced. The ex-slaughterhouse price in the north of Italy can be estimated at €6/kg of horse meat,[22] which would generate an

---

[22] The retail price for horse meat in the north of Italy is currently at €12–13/kg. If we assume transport costs and retailer margins of approximately 50%, this gives an estimate of an ex-slaughterhouse price of 6/kg.

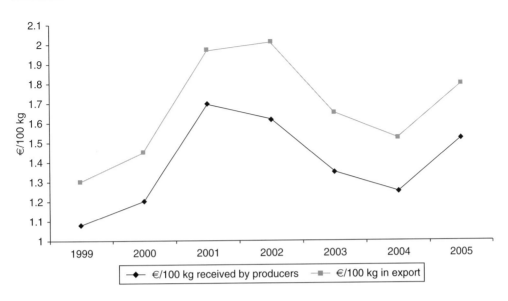

Fig. 2.14. Producer prices of live horses for export and for sale in Poland, 1999–2005. €/100 kg. (From IERiGZ (Institute of Agricultural and Food Economics – Polish National Research Institute).)

income of €1800 per horse. If we then assume that a live horse bought in Poland costs the Italian slaughterhouse €2/kg including the costs of transport, we can estimate the slaughterhouse revenue before costs at €600 per horse. As stated this estimation does not take into account any costs of the slaughterhouse, but it also does not include revenue from the sale of by-products which may also be significant. Due to strong regional fluctuations this cannot be taken as an Italian average. It should therefore only be treated as an indication of the possible scope for margin.

IMPORTANCE TO THE ITALIAN ECONOMY. While, as has been highlighted above, there are some Italian-owned and Italian-operated slaughterhouses in Poland, there is a clear incentive for Italian slaughterhouses to maintain throughput/capacity utilization in their own plants in Italy. The slaughter of the animals in-country maintains employment in relatively poorer regions of Italy and by slaughtering in-country they are able take advantage of the sale of by-products (hide, bones, hooves, etc.) into an established processing sector. The sale of horse by-products constitutes an additional income for a slaughterhouse, and may be seen as one of the incentives for live horse transport. Industry sources in the north of Italy have confirmed that hides of animals are further sold to the leather industry at a price of approximately €1.00/kg of horse hide. Also, horse tendons are further used in other European countries to produce materials for brain surgery. A North Italian slaughterhouse can sell these tendons to its users in Germany for €25/kg. Other parts of horses go into fertilizer production. It is understood that the offals are also used for pet food. It is not possible within the scope of this study to precisely calculate the profit the slaughterhouse makes from selling by-products but these sales clearly constitute a significant additional revenue flow.

An argument which is also applied to rationalize the live trade is that Italian slaughterhouses cut the meat in line with the regional preferences of Italian consumers.

While clearly there are specific characteristics which are required in the production of horse meat for consumption[23] we do not consider there are any reasons, apart from possible short- to medium-term capacity and skill constraints, why the meat could not be cut in line with Italian requirements in Poland.

Conclusions on live horse trade from Poland

As is the case for cattle, Poland has historically had high production and good pasturing conditions for horses and no demand for horse meat. Italy, on the other hand, has always had strong demand for horse meat and thus provides an attractive market. Apart from this basic feature in the past the trade specifically in live animals (as opposed to meat) has been driven by the fact that:

- Many animals from the whole region were transhipped via Poland.
- Slaughter capacity in Poland has historically not been EU licensed.
- During the period up to EU accession there was a slight incentive in favour of live animals provided by the tariff structure prevailing on meat and live animals.

In this context it should be noted that in recent years there has been a significant shift away from live export and towards export of meat. While between 1995 and 1999 live exports averaged at some 80,000 head per year this had fallen to half this number in the period up to 2003. At the same time the volume of horse meat exported rose from the carcass equivalent of some 10,000 head in 1995 to an average of carcass equivalent of almost 40,000 head in the period 2000–2003. This suggests Polish slaughter capacity has improved and it is therefore expected that this general trend towards a decline in live exports will have been maintained post accession.

Recommendations

As has been noted above many of the reasons for the historically high level of live horses for slaughter trade which were in place in the 1990s have now been removed and there has therefore been a corresponding decline in such trade. This said, a significant number of animals still are shipped for slaughter so as to obtain the premium derived from the fact that plants in Italy are more geared to adding value than their Polish counterparts. In principle, however, over time most or all of the value-added attributes of slaughter in Italy should be able to be replicated in Poland (e.g. if necessary hides could be shipped from Poland to Italy etc.). In order to achieve this there would need to be a concerted effort to highlight the potential animal welfare gains to be derived by a longer-term restructuring of the slaughter and processing process to favour slaughter in Poland. In this context it would also be necessary to persuade consumers that horse meat slaughtered in Poland was of equivalent quality (and freshness) as that slaughtered in Italy.

---

[23] It has to be chilled to a temperature of approximately $-18°C$ to $-26°C$ in order to be further processed and has to be mature.

Sheep

Sheep are also transported live from Poland to Italy. The numbers of animals are, however, much smaller than in the cases of cattle and horses, and the trade is very seasonal.

Trade flows

Table 2.12 and Fig. 2.15 show the volume of Polish exports of live sheep and meat. The number of animals transported out of Poland at the beginning of the 1990s during

Table 2.12. Live sheep and meat exported from Poland, 1990, 1992, 1994, 1996, 1998, 2000–2005. (From IERiGZ (Institute of Agricultural and Food Economics – Polish National Research Institute).)

	1990	1992	1994	1996	1998	2000	2001	2002	2003	2004	2005
Sheep '000 head	753.1	972.4	388.4	295.0	206.3	155.9	154.9	115.2	118.2	100	90
Including: to the EU	638.6	655.0	338.0	257.0	198.2	154.7	154.4	112.6	113.6	100	90
Live sheep '000 t	25.0	31.7	10.2	6.7	4.5	3.3	3.2	–	–		–
Including: to the EU	21.7	15.4	10.0	5.7	4.1	3.3	3.1	–	–		–
Meat '000 t	12.0	15.1	4.8	3.1	0.2	0.1	0.0	0.0	0.0	0.3	0.0
Including: to the EU	10.0	7.0	4.7	2.4	0.2	0.1	0.0	0.0	0.0	0.3	0.0

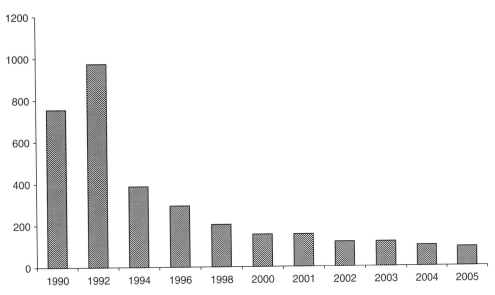

Fig. 2.15. Live sheep exports from Poland, 1990–2005. (From IERiGZ (Institute of Agricultural and Food Economics – Polish National Research Institute).)

the period of restructuring was very significant but since then there has been a sharp fall in live exports from Poland. By 2004 the numbers to Italy had fallen to some 75,000 head but we do not have data indicating the trade flows since accession. As Table 2.13 shows, the transport of live sheep, as well as sheep meat, has been almost entirely destined for the EU. Again, as for the other species of animals analysed, Italy is the major receiver as can be seen in Table 2.13. Over 90% of sheep transported to Italy consist of lambs up to 1 year of age destined for immediate slaughter.

Almost the entire live sheep transport from Poland to Italy takes place in the spring when the Italian market requires fresh sheep meat for Easter.

Economics and drivers of live sheep trade

ITALIAN MARKET PREFERENCES AND POLISH MARKET CHARACTERISTICS. It is very clear that sheep are transported out of Poland at only one specific time of the year. The demand of Italians for live sheep for Easter creates the market for sheep trade out of Poland. The market for sheep meat in Poland is very small. In 2003, 2000 t of sheep meat were consumed in Poland, which can be compared with 86,000 t consumed in Italy in the same year. As for beef and horse meat, there is therefore limited demand for sheep meat in Poland and a potential exportable surplus. In addition we understand that by slaughtering the animals in Italy the meat can be designated as being of Italian origin thus rendering it more attractive to consumers.

PRICE DIFFERENTIALS BETWEEN POLAND AND ITALY. Figure 2.16 shows the difference in prices for live sheep received on the Polish market, and for exports to the EU. Italian prices for live sheep are on average 40% higher than the price for the same product in Poland providing a strong incentive for trade.

Conclusions on live sheep trade from Poland

The trade in sheep meat and live sheep exists because of good sheep production conditions combined with a lack of demand for sheep meat in Poland. These factors combine with strong seasonal demand for sheep meat in Italy to create a strong incentive for trade. The incentive for the trade in live sheep destined for slaughter would appear to be primarily related to the fact that if the meat is slaughtered in Italy it can be marketed as being meat of Italian origin. This having been said, it is clear that the scale of this trade has been steadily diminishing since the latter half of the 1990s.

Recommendations

As has been noted in the analysis the bulk of shipments of live sheep to Italy from Poland take place at Easter and are thought to largely take place so that the meat can be labelled as being of Italian origin. To address this issue it is recommended that:

- Italian consumers be made aware of the potentially adverse animal welfare implications of this trade; and
- That consideration be given to changing the origin-labelling regulations so as to discourage such shipments.

Table 2.13. Live sheep exports from Poland by recipient country, 1996–2004 (head). (From UN COMTRADE and FAOSTAT.)

	1996	1997	1998	1999	2000	2001	2002	2003	2004	Overall 1996–2004	%
France	1,911	0	0	0	10,688	22,027	25,671	23,123	10,105	88,882	6
Germany	6,810	5,775	9,665	2,301	0	6,227	7,119	3,915	6,867	52,005	3
Greece	0	0	0	29,159	3,655	0	0	0	0	28,225	2
Italy	217,773	184,571	172,645	132,530	178,083	114,107	82,400	80,561	74,347	1,240,942	78
The Netherlands	0	0	11,300	16,028	8,821	0	14,512	12,506	9,541	69,691	4
Spain	16,649	5,793	2,197	4,197	10,966	2,502	0	3,731	916	46,816	3

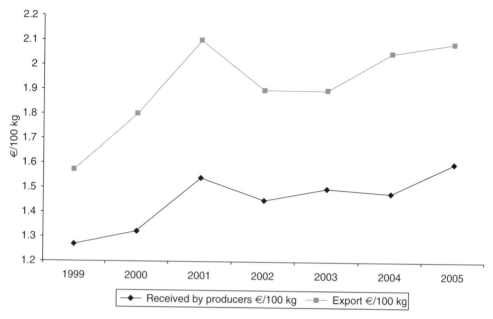

Fig. 2.16. Producer prices of live sheep for export and for sale in Poland, 1999–2005. €/100 kg. (From IERiGZ (Institute of Agricultural and Food Economics – Polish National Research Institute).)

Appendix 2.1

Table 2.1.1A. Major Australian livestock shipping companies. (From Industry consultations.)

Name of the company	Shipping port	Location	Type of export
Emanuel	Australia-wide/ International	Perth Western Australia	Sheep, cattle, goats
Wellard	Australia-wide/ International	Perth Western Australia	Sheep, cattle, goats
Elders	Australia-wide/ International	Melbourne	Sheep, cattle, goats
Livestock S.S.	Mainly from Western Australia	Perth Western Australia	Sheep, cattle
Australian Rural Exports (Austrex)	Mainly Northern Australia	Brisbane	Cattle
Seals	Darwin	Brisbane/Darwin	Cattle
Aust Asia	Northern Australia	Brisbane	Cattle
Halleen	Western Australia	Perth	Cattle
AAA Livestock	Northern Australia	Brisbane	Cattle, goats, camels
Austock	Eastern Australia	Brisbane	Cattle

Table 2.1.2A. Australia live cattle trade 2001–2005 by export destination (head). (From ABS.)

Partners	2001	2002	2003	2004	2005	Average 2000–2005
Indonesia	289,525	426,458	387,160	359,560	347,967	362,134
The Philippines	97,411	115,522	96,016	46,918	20,941	75,362
Malaysia	77,925	92,009	87,955	47,541	38,067	68,699
Saudi Arabia	20,800	54,277	15,969		17,522	21,714
Israel	34,966	47,777	43,213	20,947	32,027	35,786
Mexico	20,541	17,706	2,552	5,633	17,481	12,783
Japan	17,957	14,028	22,034	18,098	25,269	19,477
China	1,985	9,372	44,138	73,911	32,512	32,384
Jordan	13,186	4,765	23,065	34,154	16,980	18,430
Other	284,178	189,966	52,146	30,986	24,033	116,262
Total	858,474	971,880	774,248	637,748	572,799	763,030

Table 2.1.3A. Australia bovine meat trade 2001–2005 by export destination (t). (From UN Comtrade and Agra CEAS.)

Partners	2001	2002	2003	2004	2005	Average 2000–2005
Indonesia	13,037	19,457	16,937	8,555	10,234	10,644
The Philippines	19,950	14,760	10,202	5,817	6,172	11,380.2
Malaysia	8,153	11,079	9,664	7,395	4,461	8,150.4
Saudi Arabia	5,522	3,233	2,837	1,161	670	2,684.6
Japan	320,157	238,654	280,046	406,240	427,617	334,542.8
Subtotal	366,819	287,183	319,686	429,168	449,154	378,140.4
Other	603,823	663,631	552,531	534,643	528,209	568,829
Total	970,642	950,814	872,217	963,811	977,363	946,969.4

Appendix 2.2

Export chain structure, key players and beneficiaries

Livestock exporting is an industry with a series of businesses, livestock owners and goods and service inputs and outputs up and down a value chain. In many respects the links in the value chains for live sheep and cattle exports are the same. A generic description of each link is provided below.

Importers

A livestock importer may be either a commercial company or an entity associated with the government of an importing country that contracts with an Australian exporter to supply sheep or cattle. Typically a number of companies will operate independently in larger countries and in smaller countries or centrally planned economies, requests will come directly from a government-related entity. By way of example, the Kuwait Livestock Transport and Trading Company is a publicly listed company and is majority government-owned. While it is the major importer into Kuwait, other private companies emerge as competitors on occasion. Livestock requests may be made simultaneously to one or more Australian exporters, as well as to competing suppliers such as China, South America or countries in the Horn of Africa.

Exporters

Exporters are responsible for pricing, purchase, assembly and preparation of an importer's order. This role may involve specialist suppliers (farmers) who simply source livestock for exporters.

Supplier/Procurement

Where these are involved a supplier buys livestock for export, to a specific order for an exporter. The supplier provides a procurement service. The supplier does not own the livestock and is paid on the basis of a contract rate.

Agents

Agents are large pastoral supply companies such as Elders and Landmark in Australia that work on behalf of the grazier vendor to sell to a supplier or directly to an exporter. Their role includes initial inspection of livestock to ensure they meet the required specification. Specification may include age, weight, sex, breed and disease status. An agent will negotiate the sale price between the grazier vendor and exporter/supplier purchaser. After purchase the agent will arrange transportation to the pre-export quarantine facility. Agents serving this market tend to be the larger operators in each area from which livestock are sourced. Agents generally attain a percentage of the farm gate sale price, typically this figure is set at 3% for sheep and between 3–5% for cattle.

Graziers

Graziers produce sheep or cattle for sale to an exporter. A cattle breeding or growing-out operation may be specifically geared towards supply of livestock suitable for the live export trade such as in the Northern Territory. Alternatively, sale to an exporter/supplier may be opportunistic and this is more often the case in southern Australia. Sheep graziers may include wool producers selling older sheep or specialist producers selling a younger animal for higher-value markets. The market is split roughly between 45% young sheep and 55% older stock. In some cases, where economies of scale permit, graziers may supply directly to an exporter, without an agent. Alternatively, graziers may act as suppliers and buy trade stock to compile an appropriate lot for an exporter.

Preparation specialist

The preparation specialist or backgrounder works with the exporter or agent to ensure consistency of livestock supply and the conditioning of stock for sea travel. Opportunities for specialist preparation usually occur when specified requirements are outside a grazier's normal management programme. Some markets (e.g. China dairy cattle) require an extended period in a registered quarantine facility with nominated disease testing and treatments during the period. Preparation specialists tend to be more active in live cattle exports than with live sheep. Preparation specialists were more active in the sheep industry when returns were higher for wool and profits could be made by buying, shearing and on selling to exporters. Preparation specialists were also active when the trade included suppliers of fat-tailed sheep such as the Awassi. Approximately 10% of export livestock pass through the preparation/backgrounding process. Independent preparation specialists are active in Victoria, Western Australia and Queensland.

Vaccinators

Arrangements for the reopening of the sheep trade to Saudi Arabia in 2005 included a requirement that sheep are vaccinated twice for the disease scabby mouth prior to export. The first of these vaccinations is carried out by the grazier and occurs on the property at lamb marking. The second occurs following selection for live export to Saudi Arabia and is conducted by teams of accredited vaccinators arranged by the buyer. The Australian Livestock Export Corporation (LiveCorp) accredits these vaccinators following training. The vaccinators also have the responsibility for ensuring age specifications are met by sheep selected for export.

Land transport

Road transport is required at a number of stages throughout the livestock export value chain. Road transport of livestock takes place between the grazier producer, preparation specialist, the quarantine area and the ship. Further road transport is also required for transportation of fodder from fodder mills and feed producers to quarantine areas, fodder mills and on-board ship. Specialist transporters are present in each area where livestock and fodder are sourced as well as within close proximity of quarantine area feedlots. Rail transport, particularly in Queensland, is important in the live cattle export industry.

Veterinary services

Traditionally, commercial veterinary services to the industry were required for health compliance for diseases such as brucellosis and tuberculosis. In recent years, the role of commercial vets has expanded and now includes district health status certification, animal welfare, tests and treatments associated with quarantine compliance and final inspection. Under revised regulatory arrangements introduced in 2004/05, all veterinarians undertaking this role are accredited by the Australian Quarantine Inspection Service (AQIS). Commercial vets who have built dedicated practices in the industry are found in Darwin, Katherine, Townsville, Kununurra, Perth and Western Victoria. An accredited veterinarian is also required to travel aboard the vessel to Saudi Arabia or to other destinations required by AQIS. These veterinarians also provide a reporting function back to AQIS.

Pre-export Quarantine Facility (PEQ)

Formerly known as the export feedlot or assembly depot, the PEQ acts as an assembly and holding facility for stock prior to export. These facilities must be accredited by AQIS. AQIS also inspect stock held in the PEQ prior to granting approval to load. A livestock quarantine area can be located at port or inland. Quarantine areas are established for shipments prior to boarding and as disease controls between States. For example, Cloncurry in north-western Queensland has quarantine facilities for cattle entering and leaving the Northern Territory. Large feedlots exist as marshalling areas for live sheep exports to the west or Perth. The Northern Territory government, in association with Northern Territory exporters, has developed quarantine facilities for cattle near the port of Darwin.

Quality control specialists

The industry currently supports a limited number of third-party independent quality control specialists who inspect livestock against specification prior to vessel loading. Quality control specialists are not employed in every value chain.

Fodder manufacturers

During the sea journey, live sheep are fed on a pellet typically consisting of 50% hay or straw, 30% grain barley, 10% lupins and the balance of bulk roughage, urea and possibly antibiotics. This pellet is also fed to cattle on southern export routes. On northern routes, lucerne (legume-based) cubes are fed to export cattle. Lucerne cubes are used in cases where on-journey weight gain is desirable.

Fodder growers

Fodder manufacturers may source fodder for production in their local region or from further afield. Pellet producers in southern Victoria, for example, are known to source fodder from growers in the New South Wales Riverina region, as are live cattle exporters from Queensland and the Northern Territory. In the north growers produce fodder specifically for pellet production.

Chemical manufacturers and retailers

Chemical suppliers to the livestock export industry include local stock and station agents, manufacturers of veterinary drugs and suppliers to vets such as the manufacturers of antibiotics and electrolytes and vaccine companies. Chemicals are also required onboard during the journey and are supplied in bulk to the ship.

Shearing contractors

Sheep are normally shorn prior to live export, if wool growth is greater than 2.5 cm or approximately 10-weeks growth. This requirement provides additional work for shearing contractors either in regional areas from where the sheep are sourced, or in the pre-export quarantine facility.

Port authority

The port authority provides port and marine services, it owns and maintains the port facility and wharves. It holds a long-term lease on the channels of the harbour. The port authority leases space for storage and wharves, provides pilotage, towage and so on. The port authority is responsible for all aspects of inbound and outbound shipping.

Stevedores

Stevedores are responsible for ship loading of stock into pens on board the vessel. Stevedores employ livestock handlers.

Providores

Providores include suppliers of food for the voyage, marine engineers for ship repairs, yards and fencing materials, protective clothing and boots, semi-trailers for stock management at destination and any other materials required during the sea journey.

Ships agents

Ships agents notify port authorities of boat arrival time, organize quarantine clearance on arrival, manage customs documentation and provide services to the ship (repairs and maintenance) and ship's crew (doctors, dentists, etc).

Ships owners

Over 90% of vessels plying the live export trade are foreign-owned. There are a small number of Australian-owned vessels operating the northern live cattle trade, although this number has declined in recent times.

Stockmen

Stockmen accredited by LiveCorp are involved in the process of load planning and loading the vessel. Accredited Australian stockmen are required to accompany all vessels to the destination port, completing onboard stock work and assisting with disembarkation. Reporting requirements are fulfilled by onboard stockmen.

Government agency services and industry representation

In addition to the activities ascribed in AQIS vets, government services to the industry, and hence employment, include customs and immigration officers, Australian Marine Safety Authority (AMSA) employees, who check ship's engines, survey works, ventilation, general ship safety, and Australian Government Department of Agriculture Forestry and Fisheries (DAFF). DAFF responsibilities include animal welfare, policy, guideline and protocol formulation and trade delegations. Meat and Livestock Australia (MLA) represents the industry from a producer perspective and works in close association with the export industry service company LiveCorp on market access, policy, promotion and research and development.

Insurance

Insurance services purchased by the industry include stock transit insurance from the graziers' property to the quarantine area and mortality and cargo insurance for stock and fodder. In addition the Export Finance Insurance Corporation (EFIC) provides underwriting of credit risk in destination countries. Around 70% of industry insurance is provided by foreign-owned companies.

Banking

Banking services required include exporter finance and letter of credit guarantees. A significant proportion of banking activity is undertaken by Australian banks, *inter alia* due to the fact that (interest rates and premiums are lower, than e.g. at Indonesian financial institutions).

Other services

Other services purchased by the industry include auditors, accountants, solicitors and interpreters. Auditors, for example, are regularly employed by importers to review the Australian arms of their operations. Travel and accommodation and various other services for visiting buyers and inspectors from destination countries are also required. Similarly, a number of hotels in towns were livestock facilities, such as quarantine yards and the ports, are located rely on the regular patronage of agents, stockmen and transport operators.

Meat processing

The meat processing sector purchase cull stock from live exporters up until the point of ship departure, this activity is a by-product of the trade. For example, some 3% of sheep are rejected prior to loading and these are sold to the processing sector.

3 Physiology and Disease

X. Manteca

Facultad de Veterinaria, Universidad Autónoma de Barcelona, Spain

Abstract

During transportation animals are exposed simultaneously to a variety of stressors. Stress may have a negative effect on the immune system and this may result in increased susceptibility to infection and increased infectiousness. The intensity of the stress response during transport can be measured using a combination of physiological and behavioural measures. Transport augments the intensity and frequency of contacts between animals and this may result in diseases being spread. Some of the diseases that may be spread due to transport are very infectious and economically damaging and are of considerable importance in their effect on the welfare of animals. Transport may also result in tissue damage and malfunction in transported animals. Some of the strategies that may help reduce the effects of transport on disease susceptibility and spread are inspection of the animals that are to be transported so that only fit and healthy animals are transported, improvement of transport conditions and reduction of contact between transported and non-transported animals. However, these measures cannot completely prevent the negative effects of transport on disease susceptibility and spread. Clearly one of the main strategies to mitigate disease spread is to minimize the distances live animals are transported.

Introduction

During long-distance transportation animals are exposed simultaneously to a variety of stressors in a relatively short period of time (Grandin, 2000). Such stressors include fasting and water deprivation, mixing of unacquainted individuals, handling by humans, exposure to a novel environment, noise and vibration, forced physical exercise and extremes of temperature and humidity (Sainsbury and Sainsbury, 1988). All these factors contribute to activate the stress response through different physiological pathways. The stress response is known to be additive, i.e. the higher the number of simultaneous stress factors, the bigger the response. It is important to take into account the biology of the stress response in order to understand its relationship with disease (pathology) as well as its importance as a pre-pathological process. Health is an important part of welfare and any increase in disease or in pre-pathological states means poorer welfare. Disease can result from different factors in relation to transport, including: (i) tissue damage; (ii) tissue malfunction; (iii) increased susceptibility to infection and disease; (iv) increased infectivity; and (v) increased contact between animals. Both increased susceptibility to infection and disease and increased infectivity are a consequence of the effects of the stress response on the immune system.

Biology of Stress

Animals live in a changing environment, but life needs relatively constant conditions. The tendency of a living being to keep internal conditions constant is called homeostasis. In 1929, Cannon described stress as the response of the organism (involving mainly the sympatho-adrenomedullary system) to attempt to regulate homeostasis when threatened by environmental challenges. Later on, Selye (1936) conducted some of his classic studies on the response of the hypothalamic-pituitary-adrenal (HPA) axis to noxious stimuli. Selye suggested that organisms react in a non-specific manner to a wide variety of aversive stimuli, and this stress reaction was termed 'General Adaptation Syndrome'. In 1971, Mason suggested that the psycho-logical component of the aversive stimuli is the main determinant of the stress response. For example, animals that could control and/or predict the occurrence of an electric shock showed less-pronounced stress responses than counterparts with no control or warning signals (Weiss, 1970, 1971). Most researchers agree that the animal's appraisal of the situation is a major determinant of the stress response (Terlouw et al., 1997). This is important when looking at transport of animals, as very often transport is a new experience for animals that may perceive it as unpredictable.

In the late 1980s, research on stress biology addressed the role of the brain (Chrousos et al., 1988). Several areas of the brain are involved in the organization of responses to aversive or threatening stimuli, and these areas interact extensively. Neurons in the hypothalamus, the main area in the brain related to stress response, are sensitive to internal and external (both physical and psychosocial) stimuli (Laborit, 1991). To a great extent the stress response is mediated by corticotropin releasing factor (CRF), a hormone that is secreted mainly by the paraventricular nucleus of the hypothalamus (Dunn and Berridge, 1990).

How to measure stress: an overview

Stress can be measured using behavioural and physiological parameters. The proportion of animals that lie down during transport and the frequency of aggressive interactions between animals are some of the most important behavioural indicators. Plasma levels of glucocorticoids, catecholamines, prolactin and endorphins as well as heart rate are among the most frequently used physiological parameters to study the effects of transport (Broom and Johnson, 1993). Levels of neurotransmitters in the brain – although less frequently used – are also of interest in an experimental setting (Broom and Johnson, 1993). Acute phase proteins are also useful to assess the extent of tissue damage caused by transport (Saco *et al.*, 2003) (see Chapter 7, this volume, for details of behavioural and physiological indicators of stress).

Effects of stress on the immune system

The immune system is the main defence mechanism against infection and stress impairs immune function. However, the mechanisms responsible for this effect are highly specific, and only some types of defence against disease are affected. As seen before, the stress response involves the release of glucocorticoids and/or catecholamines, which can reduce immunity (Elenkor *et al.*, 2000). A clear example of this is shipping fever, a transport-related disease of cattle. It is caused by the interaction of the reduced capacity of the immune system caused by the stress response, and the increased contact between animals and pathogens caused by transport (Quinn *et al.*, 2000). Stress may have pronounced effects on some diseases, including respiratory infectious diseases and *Salmonella* infection (Gregory, 2004), the latter being a disease that can be transmitted from animals to humans.

Increased susceptibility to infection

The susceptibility of individuals to disease is affected by the efficacy of body defence systems, especially the immune system. Poor welfare, mainly through the effect of the stress response on the immune system, can make these systems less efficient and result in predisposition to infection (Broom and Kirkden, 2003). Transport has been shown to increase pneumonia caused by bovine herpes virus-1 in calves (Filion *et al.*, 1984), pneumonia caused by pasteurellosis and mortality in calves and sheep (Brodgen *et al.*, 1998; Radostits *et al.*, 2000), and salmonellosis in sheep (Higgs *et al.*, 1993) and horses (Owen *et al.*, 1983). In general, a combination of different stressors has a greater effect than a single stressor.

Tissue malfunction in transported animals (pigs)

Tissue malfunction refers to alterations in biological functions that may lead to transport-related diseases. The most important is malign hyperthermia or Porcine Stress Syndrome (PSS), which is a serious welfare problem in pigs. This involves a cascade of physiological changes that may cause death. Stress and forced physical exercise may lead to an increase in body temperature, cardiac arrest and death. Death rates are higher when conditions are hot and humid (Abbott *et al.*, 1995). Colleu and Chevillon (1999) found that death rate increased with the outside temperature from 0.07% for outside temperatures lower than 5°C to 0.11% for temperatures higher than 15°C. Death rate also significantly increases with the length of the journey from 0.08% for journeys shorter than 75 km to 0.12% for journeys longer than 150 km, which means a substantial increase of 50%.

According to Warriss (1996) losses in pigs during transports shorter than 8 h vary between different European Union countries from 0.03% to 0.5%. About

70% of the deaths occur on the lorry and the rest occur during lairage (Christensen and Barton-Gade, 1997).

Apart from the environmental conditions and the length of the journey, the genotype of the pigs also has an effect on the death rate. Scientific evidence has indicated that reduction of the frequency of the halothane gene (*Hal*) in commercial pigs would lead to a major reduction in pre-slaughter death rate (Murray and Johnson, 1998). The *Hal* is now considered to be equivalent to the *ryr-1* gene, which encodes a muscle protein called ryanodine receptor or calcium release channel (Fujii *et al.*, 1991). Pigs with an altered form of the protein perform a prolonged muscle contraction (forced physical exercise) that induces increase in body temperature (hyperthermia). These changes add to those occurring during transport and make these animals more susceptible to stress.

Several studies have suggested that any stressful situation such as transport can trigger the onset of PSS, increasing mortality rates in both genotypes nn and Nn, i.e. in pigs that have one or two copies of the gene that increases stress susceptibility (Murray and Johnson, 1998). These authors found that frequencies of death during lairage and transport were 0.05%, 0.27% and 9.2% for the NN, Nn and nn genotypes, respectively. These data are in agreement with findings of other authors obtained from different studies and environmental conditions (Webb *et al.*, 1982; McPhee *et al.*, 1994). One study carried out in Spain found that the frequencies of pre-slaughter deaths within each genotype were estimated in 0.02%, 0.09% and 2.29% for NN, Nn and nn genotypes, respectively. According to these results, the removal of both nn and Nn genotypes would imply an 11 times reduction in the pre-slaughter mortality rate (from 0.22% to 0.02%) (Fàbrega *et al.*, 2002).

Tissue damage in transported animals

Mixing of unacquainted animals leads to an increase in agonistic activity in several species, such as cattle and pigs (McBride *et al.*, 1964), due to the animals establishing a new hierarchy in the group. Fighting involves increased physical exercise and may cause injuries. Both social mixing and fighting act as stress factors (Arnone and Dantzer, 1980; Blecha *et al.*, 1985).

Mixing of unacquainted animals is a common practice in transport and lairage, both to homogenize the weight of the animals and to increase profit. This may lead to stressed and injured animals, a combination that facilitates infections and decreases welfare.

Increased infectivity

Transmission of pathogenic agents begins with shedding from the infected host through oronasal fluids, respiratory aerosols, faeces, or other secretions or excretions. Stress related to transport can enhance the level and duration of pathogen shedding in subclinically infected animals and thereby enhance their infectiousness. This effect is likely to be greater when mixed animals remain together for a prolonged period of time, as it may happen, for example, in long-distance transport by ship (see Chapter 6, this volume). For example, calves subclinically infected with a vaccine strain of bovine herpes virus BHV1 started to shed virus after having been transported (Thiry *et al.*, 1987). A disease that deserves particular attention is salmonellosis in pigs, as it is a major public health problem. It has been shown that transport stress increases the proportion of animals positive for *Salmonella* sp. Although feed withdrawal before transport may considerably reduce or even prevent the effect of transport (Isaacson *et al.*, 1999), it can have serious welfare implications (see Chapter 1, this volume).

The way animals are transported is also important. For example, if poultry crates are stocked one of top of the other during transport, faeces may pass from one cage to another, increasing the risk of diseases being spread. This is likely to be a problem in multi-tiered vehicles.

Increased contact

It is clear that transport increases the intensity and frequency of contacts between animals and this may result in diseases being spread. This applies not only to transport from farm to slaughterhouse, but also from farm to farm or farm to market. For example, if animals with foot-and-mouth disease or classical swine fever are transported there is a major risk of spreading the disease. This was well documented in 1997 with classical swine fever in the Netherlands and in 2001 with foot-and-mouth disease in the UK. Schlüter and Kramer (2001) reviewed the outbreaks of classical swine fever and foot-and-mouth disease in the EU and concluded that once the diseases were in the farm animal population, at least 9% of further spread was caused by animal transport and the diseases were transmitted a considerable distance – around 110 km – when transport for slaughter occurred.

Other important diseases which may be transmitted by animal transportation include bovine viral diarrhoea, African swine fever, swine dysentery, swine vesicular disease, porcine reproductive and respiratory syndrome, post weaning multi-system weaning syndrome, porcine dermatitis and nephropathy syndrome, enzootic pneumonia, bovine rhinotracheitis, rinderpest, glanders, sheep scab, Newcastle disease and avian influenza (SCAHAW, 2002).

Discussion

In summary, the effects of transport on disease result from three main factors:

- Stress caused by transport increases susceptibility to disease by reducing the effectiveness of the immune system.
- Stress caused by transport increases the infectivity of subclinically infected animals.
- Transport increases the frequency of contacts between animals and can therefore contribute to diseases being spread.

The first and second factors above may be controlled by improving the welfare of animals during transport, so that stress is less intense. In short, the main points to be considered to improve welfare during transport are inspection of animals before transport, loading and unloading facilities, stocking density in the vehicle, journey duration and quality of driving. Reduction in journey duration could play a signifi- cant role in reducing welfare problems. It is also important – at least in pigs and cattle – not to mix unfamiliar animals. In addition, the second factor could also be controlled by inspection of animals prior to transport, so that unhealthy animals are not transported. However, considering that the increased infectivity is an issue for subclinically infected animals which, by definition, would not be detected by veterinary inspection, this strategy is unlikely to be very useful, other than to reduce the negative effects that transport may have on the welfare of unfit animals.

Finally, the increase in the frequency of contacts between animals could theoretically be reduced if animals do not leave the transport vehicle during the journey for periods of rest and to allow the animals to be fed and watered. This, however, has the potential to increase stress, fatigue and dehydration if animals are not adequately rested. Also, unless the vehicle is very well equipped (and they are often not, see Chapter 5, this volume), the fact that animals stay in the vehicle for a long period of time may have very negative consequences on their welfare.

The risk of diseases being spread by transport is likely to be greater in longer- distance transportation and also when animals are transported between places with a different livestock health status.

In view of all the evidence reviewed above, it seems clear that the disease risk caused by transport of animals needs to be taken seriously and would be reduced by minimizing the number of animals being transported and the distances that they cover.

References

Abbott, T.A., Guise, H.J., Hunter, E.J., Penny, R.H., Bayen, P.J. and Easby, C. (1995) Factors influ- encing pig death during transit: an analysis of drivers' reports. *Animal Welfare* 4, 29–40.

Arnone, M. and Dantzer, R. (1980) Does frustration induce aggression in pigs? *Applied Animal Ethology* 6, 351–362.

Blecha, F., Pollman, D.S. and Nichols, D.A. (1985) Immunological reactions of pigs regrouped after near weaning. *American Journal of Veterinary Research* 46, 1934–1937.

Brodgen, K.A., Lehmkuhl, H.D. and Cutlip, R.C. (1998) *Pastereulla haemolytica* complicated respiratory infections in sheep and goats. *Veterinary Research* 29, 233–254.

Broom, D.M. and Johnson, K.G. (1993) *Stress and Animal Welfare.* Chapman & Hall, London.

Broom, D.M. and Kirkden, R.D. (2003) Welfare, stress, behaviour and pathophysiology. In: Dunlop, R.H. (ed.) *Veterinary Pathophysiology.* Iowa State University Press, Ames, Iowa, pp. 337–369.

Cannon, W.B. (1929) *Bodily Changes in Pain, Hunger, Fear and Rage: An Account of Recent Researches into the Function of Emotional Excitement.* Appleton, New York.

Christensen, L. and Barton-Gade, P. (1997) *Heart Rate and Environmental Measurements During Transport and Experience from the Routine Transports with the Experimental Vehicle.* Report number 02.647 AIR project. Danish Research Institute, Frederiksberg, Denmark.

Chrousos, G.P., Loriaux, D.L. and Gold, P.W. (1988) The concept of stress and its historical develop- ment. In: Chrousos, G.P., Loriaux, D.L. and Gold, P.W. (eds) *Mechanisms of Physical and Emotional Stress.* Plenum Press, New York, pp. 3–7.

Colleu, T. and Chevillon, P. (1999) Incidence des paramètres climatiques et des distances sur la mor- talité des porcs au cours de transport. *Techni-Porc* 22, 31–36.

Dunn, A.J. and Berridge, C.W. (1990) Physiological and behavioral responses to corticotropin-releasing factor administration: is CRF a mediator of anxiety or stress responses? *Brain Research Review* 15, 71–100.

Elenkor, I.J., Chrousos, G.P. and Wilder, R.L. (2000) Neuroendocrine regulation of IL-12 and TNF-alfa/IL-10 balance. *Annals of the New York Academy of Sciences* 917, 94–105.

Fàbrega, E., Diestre, A., Carrión, D., Font, J. and Manteca, X. (2002) Effect of the halothane gene on pre-slaughter mortality in two Spanish commercial pig abattoirs. *Animal Welfare* 11, 449–452.

Filion, L.G., Wilson, P.J., Bielefeldt-Ohmann, L.A. and Thomson, R.G. (1984) The possible role of stress in the induction of pneumonic pasteurellosis. *Canadian Journal Comparative Medicine* 48, 268–274.

Fujii, J., Otsu, K., Zorzato, F., De Leon, S., Khanna, V.K., Weilter, J.E., O'Brien, P.J. and MaClennan, D.H. (1991) Identification of a mutation in porcine ryanodine receptor associated with malignant hyperthermia. *Science* 253, 448–451.

Grandin, T. (2000) Introduction: management and economic factors of handling and transport. In: Grandin, T. (ed.) *Livestock Handling and Transport*, 2nd edn. CAB International, Wallingford, UK, pp. 1–9.

Gregory, N.G. (2004) *Physiology and Behaviour of Animal Suffering*. Blackwell, Oxford, UK.

Higgs, A.R.B., Norris, R.T. and Richards, R.B. (1993) Epidemiology of salmonellosis in the live sheep export industry. *Australian Veterinary Journal* 70, 330–335.

Isaacson, R.E., Firkins, L.D., Weigel, R.M., Zuckerman, F.A. and Di Pietro, J.A. (1999) Effect of transportation and feed withdrawal on shedding of *Salmonella typhimurium* among experimentally infected pigs. *American Journal Veterinary Research* 60, 1155–1158.

Laborit, H. (1991) The major mechanisms of stress. In: Jasmin, G. and Proschek, L. (eds) *Stress Revisited II: Systemic Effects of Stress*. Karger, New York, 1–26.

Mason, J.W. (1971) A re-evaluation of the concept of "non-specificity" in stress theory. *Journal Psychiatric Research* 8, 323–333.

McBride, G., James, J.W. and Hodgens, N.W. (1964) Social behaviour of domestic animals IV: growing pigs. *Animal Production* 6, 129–139.

McPhee, C.P., Daniels, L.J., Kramer, H.L., Macbeth, G.M. and Noble, J.W. (1994) The effects of selection for lean growth and halothane allele on growth performance and mortality of pigs in a tropical environment. *Livestock Production Science* 38, 117–123.

Murray, A.C. and Johnson, C.P. (1998) Impact of halothane gene on muscle quality and pre-slaughter deaths in Western Canadian pigs. *Canadian Journal Animal Science* 78, 543–548.

Owen, R.A., Fullerton, J. and Barnum, D.A. (1983) Effect of transportation, surgery, and antibiotic therapy in ponies infected with *Salmonella*. *American Journal Veterinary Research* 44, 46–50.

Quinn, P.J., Markey, B.K., Carter, M.E., Donnelley, W.J. and Leonard, F.C. (2000) *Veterinary Microbiology and Microbial Diseases*. Blackwell, Oxford, UK.

Radostits, O.M., Gay, C.C., Blood, D.C. and Hinchcliff, K.W. (2000) *Veterinary Medicine: A Textbook of Cattle, Sheep, Pigs, Goats and Horses*, 9th edn. Saunders, London.

Saco, Y., Docampo, M.J., Fàbrega, E., Manteca, X., Diestre, A., Lampreave, F. and Bassols, A. (2003) Effect of stress transport on serum haptoglobin and pig-MAP in pigs. *Animal Welfare* 12, 403–409.

Sainsbury, D. and Sainsbury, P. (1988) *Livestock Health and Housing*. Baillière Tindall, London.

SCAHAW (Scientific Committee on Animal Health and Animal Welfare) (2002) *Report on the Welfare of Animals During Transport*. Health and Consumer Directorate-General, Brussels, Belgium.

Schlüter, H. and Kramer, M. (2001) Epidemiologische Beispiele zur Seuchenausbreitung. *Deutsche Tierarztliche Wochenschrifte* 108, 338–343.

Selye, H. (1936) A syndrome produced by diverse nocuous agents. *Nature* 138, 32–33.

Terlouw, E.M.C., Schouten, W.G.P. and Ladewig, J. (1997) Physiology. In: Appleby, M.C. and Hughes, B.O. (eds) *Animal Welfare*. CAB International, Wallingford, UK, pp. 143–158.

Thiry, E., Saliki, J., Bublot, M. and Pastoret, P.-P. (1987) Reactivation of infectious bovine rhinotracheitis by transport. *Comparative Immunology Microbiology and Infectious Diseases* 10, 59–63.

Warriss, P.D. (1996) Guidelines for the handling of pigs antemortem – Interim conclusions from EC-AIR3-Project CT920262. *Lanbauforschung Völkenrode* 166, 217–225.

Webb, A.J., Cardin, A.E., Smith, C. and Imlah, P. (1982) Porcine stress syndrome in pig breeding. In: *Proceedings of the Second World Congress on Genetics Applied to Livestock Production.* Madrid, Spain, pp. 588–608.

Weiss, J.M. (1970) Somatic effect of predictable and unpredictable shock. *Psychosomatic Medicine* 32, 397–408.

Weiss, J.M. (1971) Effects of coping behaviour with and without a feedback signal on stress pathology in rats. *Journal of Comparative and Physiological Psychology* 77, 22–30.

4 Meat Quality

G.A. María

Faculty of Veterinary Medicine, University of Zaragoza, Spain

Abstract

The stress undergone by domestic animals as they are transported to the abattoir is one of the most severe they suffer during the process of meat production. Transport of live domestic animals is also very visible to the consumer, so the industry should consider ways to improve it and the ethical quality of the final product. There is a wide range of potential markets available for meat producers, each with defined specifications for animals and pricing criteria. Injury or damage to carcasses due to poor transportation practices, including long-distance transport, will downgrade the meat. In order to meet market requirements, animals should be transported in such a way as to avoid downgraded carcasses. Transport is stressful for animals, even under optimal conditions, and – as with any other negative or damaging stressor – should be minimized. Journey time is the main variable to control, to reduce the biological cost of transport stress. The physical condition of the animal before transport, and food and water deprivation, play an important role in the way that animals should be managed during the transport process. Domestic species will vary in their reaction to transport stress. Handling, loading, transport conditions and unloading are critical points that should be performed properly. The quality of driving, road conditions and weather constraints vary widely and transport limits and recommendations should be adapted to those variable situations. The meat industry and the authorities in charge of enforcing the legislation should inform personnel involved in the transport/slaughter chain that improvements can be made – by investing in training and by improving the design of facilities. It is important to improve training and education on methods of animal handling and transport to minimize the risk to welfare and to avoid animal suffering. A payment system should be implemented in terms of transport quality. That payment should be directly related to welfare. Journey time limits and rest periods must be enforced. Journey time should be short enough to avoid the requirement to unload the animals for resting. There need to be strict controls at the loading point to prevent poor transport practice and logistics and minimize stress to the animals. Along those lines, an approved route plan should be compulsory based on a proper decision support system. Research results should be used to help propose modifications to improve animal welfare. Meat quality indicators can reveal major animal welfare problems during transport, but otherwise have limited usefulness in assessing welfare. A very strong stress is required to have a visible effect on meat quality. When there are even small effects on the meat, it is clear that the animals have suffered, because other welfare criteria such as behavioural changes, physiological constants or plasmatic indicators are normally greatly affected. An absence of an effect on meat quality is not a clear sign of absence of suffering due to poor welfare. We need to evaluate animal welfare from a multiple perspective approach to have a clear idea about the real extent of suffering of animals during transport (see Chapter 1, this volume). Transportation represents a combination of several stressors, which can have additive and deleterious effects on the animal. All attempts to decrease the amount of stress that livestock undergo should consider the complexity of stressors involved in the process.

Introduction

The stress undergone by domestic farm animals as they are transported to the abattoir is one of the most severe they suffer during the process of meat production. Transport of live domestic farm animals is also very visible to the consumer, so the industry should consider ways to improve it and the ethical quality of the final product (Warriss, 2003; María *et al.*, 2004). It is necessary to bridge the gap between the objective quality approach (product characteristics) pursued in food sciences and the subjectively perceived quality approach (ethical quality) (Becker, 2000).

A successful meat industry makes a profit by growing and converting pasture and concentrated feed into meat products that consistently satisfy market require-ments. Most producers strive to gain a reputation as a reliable supplier of quality meat – to maintain access to as many markets as possible and to be economically rewarded when a high proportion of the meat produced consistently meets customers' expectations. There is a wide range of potential markets available for meat producers in the European Union (EU), for example, each with defined specifications for animals and pricing criteria (see Chapter 2, this volume). Injury or damage to carcasses due to poor transportation practices, including long-distance transport, will downgrade the meat. In order to meet market require-ments, animals should be transported in such a way as to avoid downgraded carcasses. In addition, producers should also aim to meet customer preferences in the target markets, which involves a regular evaluation of quality, including wel-fare. Those preferences should include systems that avoid low animal welfare standards and reduce stress and its impact on meat quality (María, 2006). Hughes (1995) explored the key trends explaining the consumer concern and action on animal welfare. He concluded that animal welfare concerns are not a fad, but are deep-seated and here to stay. For example, the member countries of the World Organisation for Animal Health (OIE) – recognized by the World Trade Organization (WTO) as the reference organization for guaranteeing the sanitary safety of world trade in animals and animal products (Vallat, 2004) – decided that the link between animal health and animal welfare was so clear-cut that the OIE should also become the international reference organization in the field of animal protection.

Transport and the Meat Market

According to a recent Eurobarometer poll, a large majority of European consumers would prefer more attention to be given to the welfare and protection of animals (European Commission, 2005). Intensive livestock production in confinement conditions has resulted in considerable criticism by various segments of society in the USA (Jamison, 1992; Cheeke, 1993). Pettitt (2001) makes a similar observation in the UK. A particular area of concern is the protection of animals during long-distance transport, where the international dimension has proven to be very problematic.

While animal welfare improvements are often expected to increase costs, it should be noted that in some cases, a trade in meat instead of live animals could result in lower production costs and higher efficiency of the meat production system (Hobbs, 1997).

One of the main causes of poor welfare may be the cheap food policy that prevails in society. Cheaper food for humans sometimes involves greater pain and suffering for food animals (Appleby, 2005); and the proportion of income spent on food has been declining (Hodges, 2005). The consumer has become used to the idea of cheap food, and has not generally been expected to pay a premium for animal produce reared in a more traditional manner (Pettitt, 2001). Americans spend a smaller percentage of their income on food than anyone else (Schlosser, 2006), but there are costs associated with cheap food produced from animals that reach well beyond the dollars paid by citizens at the supermarket or the fast-food restaurants (Tegtmeier and Duffy, 2004). A morally significant effect of pressure for cheap food has been modifications to production methods that may have effects on animal welfare (e.g. by decreasing space allowance). The introduction of new technologies and the creation of specialized genetic lines have greatly increased the production per animal and made the animals' productive life shorter. As an example, a broiler chicken is produced in half the time it took four decades ago. The lifespan of dairy cattle is now typically less than five lactations. A laying hen produces more than 12 times its weight in eggs in less than 80 weeks. Animals lack enough time or space to eat more and their genes demand more than the animal can consume. Nevertheless, these specialized animals need controlled housing systems that produce crowded environments with poor welfare, especially regarding their freedom to express natural behavioural needs. In the frame of these low-cost production systems, transport costs become a relatively more costly step in the chain. In consequence, any improvement on transport conditions will be reflected on the final price in a higher proportion. Some production systems do not get enough income from the carcasses to pay transport costs. This is the case of laying hens (Newberry *et al.*, 1999; HSUS, 2006). Hails (1978) mentioned that the care of animals during transportation is directly proportional to the individual price of the animal transported. The intensification of production systems has consequences on the number of times that animals suffer transportation during their productive life. In many cases, the animals are born in one place, fattened in another place and slaughtered in a third place (i.e. beef production). However, improvements in welfare could be achieved with affordable increases of price to the consumer

(Appleby, 2005). The main obstacle to change is the economic inertia of producers who resist change because buyers expect low prices. Economic incentives for welfare improvements will be necessary to break this inertia. The public's demand for cheap food has contributed to the consolidation of the food processing industry and as a result, live animals are regularly transported long distances from their home farms. These animals are loaded into commercial semi-trailer trucks and transported for hours on end in the dead of winter and the sweltering heat of summer. There can be no question that the majority of these animals suffer en route (BCSPCA, 2006).

Many production systems have been modified to comply with welfare regulations and that has had an impact on production costs that should be borne by the market. Indeed, some consumers (e.g. in the EU) already pay more by buying products produced under legislated, minimum welfare standards (Carlsson et al., 2004; Blokhuis, H. Brussels, 2005, personal communication). European consumers are not well informed about this aspect of the quality of European animal products versus animal products produced where there is little or no welfare legislation. In the USA, some companies (i.e. McDonalds) are using their welfare standards as publicity to attract more customers. Blandford et al. (2002) analysed the potential implications of animal welfare concerns and public policies in industrialized countries for international trade. In a submission to the WTO on animal welfare and trade agriculture, the EU has proposed that multilateral agreement be developed to deal with the protection of animal welfare; labelling of products be used to inform consumers' choice; and producers' compensation be permitted to offset the higher costs of meeting welfare standards.

Among the welfare issues related to meat production, in many countries the transport component is one of the most visible to consumers (Hobbs, 1997). In general, quality of handling and care during transport is proportional to the value of the individual animals (Hails, 1978). Poor handling during a few hours of transport can potentially ruin months of careful work by farmers during the fattening period (Grandin, 2000). In addition, negative press can have a strong impact on the public perception of animal welfare (María et al., 2004). The treatment of downed, injured, or immobile animals has sometimes been a major issue in the public eye in the EU, for example. The transport of animals to and from markets, to slaughter and particularly, on export journeys, is a matter of increasing public concern. The EU has recognized this concern and places a high priority on ensuring that the welfare of animals is protected during transport. Many EU countries have been urged to ban the export of live animals for slaughter or fattening. EU consumers apparently prefer the export of meat (carcasses) to the long-distance transport of live animals for slaughter (Special Eurobarometer, 2005). There is a strong resistance among European residents to animals being transported long distances for slaughter (Moynagh, 2000). According to Eurostat (2006), at least 26 million cattle, 6 million calves, 202 million pigs, 72 million sheep, 9 million goats and more than 4200 million chickens are transported per year in the EU alone. The total value of these animals on the market is approximately €50,000 million.

Little is known about the proportion of downgraded carcasses in Europe, or elsewhere. The proportion of pigs that die during transport in the EU has been estimated to range from 0.033% to 0.5% (Warriss, 1996). More than 70% of the deaths occur on the truck and 30% after unloading at the abattoir. An average mortality rate of 0.25% would represent more than half a million animals per year (worth approximately €60 million). Animals can die during transport for several reasons. In pigs, the main reason is hyperthermia (overheating), especially during summer. In poultry, the main causes of death during transport are heat stress and congestive heart failure and, in sheep, it is asphyxia (Warriss, 1996). According to EU regulations, the carcasses of dead animals should be considered a total loss. The duration of the journey can also affect mortality rates. The longer the journey before slaughter, the greater is the likelihood that there will be some deaths (Gregory, 1998). In broilers, a large-scale study found that mortality ranged from 0.16% to 0.28%, depending on the duration of the journey (Freeman *et al.*, 1984). Similarly, Weeks and Nicol (2000) reported an average broiler mortality during transport of approximately 0.33%. In laying hens, mortality was 0.5% but some studies have found 25% mortality (Swarbrick, 1986; Knowles and Broom, 1990b; Warriss *et al.*, 1992). The main cause of death was heat stress. In Europe, mortality during transport of broilers averages between 0.2% and 0.3% and the mortality rate increases with journey length (European Food Safety Authority, 2004).

One way that transport affects carcass quality is by bruising (Knowles, 1999). Bruising provides an indication of the number and severity of physical insults sustained during transportation. In the past, studies on bruising have concentrated on economic consequences. When bruising is severe, some meat has to be trimmed away from the carcass. Bruising in broilers in the UK affects about 2.6% of the carcasses (Mayes, 1980). Knowles and Broom (1990b) pointed out that according to several studies in this field in Europe, the average incidence of bruising varies widely between 2.63% and 20%. That wide range probably reflects the subjectivity of carcass grading and differences in inspection procedures. Carcass damage that may be caused during processing must be differentiated from damage to the live bird.

In cattle, the incidence of bruising is estimated to be 7.8% (McNally and Warriss, 1996). More recent studies (Weeks *et al.*, 2002) have found that 75% of cattle that passed through a UK market had some bruising. In pigs, a study in five EU countries found skin blemish damage after transport in 63% of the animals (Warriss *et al.*, 1998b). In cattle, carcass bruising and stick-marking in cattle from different auction markets in the UK were estimated to be 8.1% and 2.2%, respectively (McNally and Warriss, 1997).

Broken bones is one of the most common injuries during bird transportation, along with dislocated bones, bruising or skin damage (Gregory and Wilkins, 1990). These authors found 3% of broilers with broken bones and more than 4.5% with dislocated bones. All of these problems degrade the value of the meat (e.g. Scottish Executive, 2006). There is some discrepancy as to whether bones are broken when the animals are caught in preparation for transport, or during transport itself (Gregory and Wilkins, 1989).

Transport Stress and Meat Quality

An animal's stress response has three general stages: the perception of the stressor, the physiological changes triggered to defend against the stressor and the consequences of the stress response (Moberg, 1985). In long-distance transport, the animal suffers chronic fatigue states that lead to exhaustion and suffering. There is an accumulation of acute stressors and the system is overloaded, leading to a chronic stress situation with different metabolic costs than a unique acute stress response. Exhaustion during long journeys depletes muscle glycogen and increases the incidence of dark, firm and dry (DFD) meat; animals (particularly cattle) yielding such meat are known as 'dark cutters'. This is one of the most important meat quality problems due to transport, together with bruising (Knowles, 1999).

Immediately after a stressful event, catecholamines, a class of hormones secreted by the adrenal glands, are produced and help to convert glycogen to readily utilizable energy. Once the stressor is mitigated, the glycogen store is replenished to pre-stress levels. When the ability of the animal to restore its normal state is exceeded, the biological cost is too high and the stress response becomes distress, with negative biological consequences. If the stressor or new stressors continue to threaten homeostasis, energy reserves are depleted and resources must be taken from other functions, including muscle growth, food conversion efficiency, immune response or reproductive functions. In the short term, the quality of the meat is also affected. Enough glycogen must be present to produce the post-mortem metabolic changes in the muscle, which result in a high-quality meat. The effects of dealing with stressors can thus have an important impact on meat quality and, from an economic perspective, the output of the meat production system. The system becomes less efficient because the cost of production is higher and the quantity and quality of production are lower. Producers will suffer economic losses since the lesser-quality meat will be penalized by the market.

Transport is considered stressful for farm animals because their normal routine of feeding and drinking and resting is altered, they are exposed to novel environments, sometimes mixed with unfamiliar animals, closely confined and subjected to noise and vibration and possibly extreme temperatures (Warriss, 2004). Distress can result from both acute and chronic stress. The physiological mechanisms involved in these two types of stressors are similar, but they differ in duration and intensity (Moberg and Mench, 2000). Transport is a source of both acute and chronic stress. Short journeys can produce distress because of a cumulative effect of consecutive stressors derived from loading, short transport and unloading, without time for recovery. During long journeys, acute and chronic stressors can act together to exhaust the animals' ability to cope with stress. There is a direct relationship between transport time and the possibility of increased biological cost during transportation (Grandin, 2000) which will produce carcass damage or low-quality meat (Gregory, 1998).

A demonstration that better welfare means better meat quality would be welcome to those seeking to improve animal welfare. Indicators of welfare during handling and transport can include the quantification of injuries, bruises, mortality, morbidity and carcass and meat quality (Broom, 2000; Smith et al., 2004). Bruises, scratches, blemishes, broken bones and the incidence of abnormal meat pH provide information about the welfare of the animals during handling, transport and lairage (Grandin, 2000). However, it is difficult to establish a cut-off point for journey time based on the criteria used to evaluate the consequences of adaptation during transport. As in all the definitions of cut-off points for any welfare criteria adopted, an ethical decision should be made (Mendl, 1991). This is the most difficult part in the development of the welfare decision support systems. In this decision, all the parts involved should have an agreement, based on scientific, ethical and economic aspects. The threshold above which a significant effect is observed depends on the type of criteria considered. The most sensitive criteria are physiological variables associated with the 'fight-or-flight' response (e.g. heart rate, respiration rate), followed by animal behaviour.

Stronger stressors are required to affect physiological traits associated with the hypothalamus–pituitary–adrenal (HPA) axis, and, finally, changes in meat quality are due to very strong stressors, such as very long-distance transport. Importantly, even small negative effects on meat quality suggest that animal welfare has been seriously compromised (María *et al.*, 2003; Villarroel *et al.*, 2003a). Conversely, an absence of an effect on meat quality does not guarantee the absence of suffering. We need to evaluate animal welfare using a multiple perspective approach to provide a clear idea about the real degree of suffering during transport (Blokhuis *et al.*, 2003; Chapter 1, this volume).

Poor welfare during transport can decrease meat quality (Gregory, 1994). In the fresh meat market, that can lead to a loss of yield and loss of sales due to rejection or downgrading of a poor-quality product. Transportation can adversely affect meat quality by altering the ultimate pH and producing DFD meat. DFD meat is more frequent in cattle and pale, soft, exudative (PSE) meat in pigs and turkeys. Heat shortening (sarcomere shortening due to pre-slaughter heat stress) and broken bones are a common problem after poor transportation in poultry. Many studies have provided evidence that transport stress before slaughter can affect several aspects of meat quality in domestic animals (Tarrant, 1980; Mitchell *et al.*, 1988; Tarrant *et al.*, 1992; Smith *et al.*, 2004). Meat pH is the most commonly measured parameter in studies that consider ante-mortem (pre-slaughter) effects on meat quality. Good-quality meat has a final pH (ultimate pH, 24 h post-mortem: pH_{ult}) of approximately 5.5. At pH values above 5.8, both the tenderness and shelf life of fresh meat decrease (Gregory, 1998; Grandin, 2000). Tarrant (1980) estimated that about 10% of carcasses are downgraded due to high pH_{ult}. Those types of carcasses are commonly known as 'dark cutters' due to their dark coloration; long-distance transport increases dark-cutters, as described by Wythes *et al.* (1981), Tarrant *et al.* (1992) and Honkavaara and Kortesniemi (1994). Carcass bruising and skin damage are other important effects of transport, and cause important financial losses (Grandin, 1980; Wythes and Shorthose, 1984). In the USA, economic losses per carcass due to bruising are estimated to be US$12 (Smith *et al.*, 1994). The proportion of bruising in the carcass increases with journey duration (Hoffman *et al.*, 1998). The skills of the driver and the type of road are other important factors affecting the incidence of bruising of the carcass (Harris, 1996).

PSE and DFD meats

After an animal has been slaughtered its muscle continues to metabolize energy, contract and produce heat. Part of the energy produces convulsions and rigour contractions and another part is used in non-contractile biochemical changes. PSE meat is associated with increased stress reactions to pre-slaughter handling and, in particular, stress reactions to transport (Lister, 1969). After slaughter, acid forms in the muscle when glycogen is converted to lactic acid. The PSE condition occurs when glycolysis (the metabolic breakdown of glucose that releases energy) is rapid and the pH of the muscle, 45 min post-mortem (pH_{45}), is less than 6.0. The interaction between acidity and temperature reduces the capacity of the proteins in the muscle to hold water, making the meat exudative, with higher drip release. As a consequence of the acid, some proteins become insoluble and are deposited on the structural components of the muscle. This makes the meat paler than normal meat. The PSE meat is caused by stress before slaughter when there is a shortage of glycogen and a rapid acidification of the muscle due to a genetic low capacity to metabolize lactic acid. Pre-slaughter handling can contribute to PSE, inducing a low starting pH and a high muscle temperature at the moment of slaughter. The process is also influenced by genotype. Some animals are more sensitive to stressors, due to their genetic makeup. However, genetic predisposition is now well controlled in new selected lines of commercial pigs (Webb, 2003). Thus, the appearance of PSE meat in modern production systems is mostly due to pre-slaughter stress, including transport (especially long-distance transport).

Meat becomes DFD if muscle glycogen reserves are depleted before slaughter due to prolonged exposure to stressors like long transport, which impose a high biological cost on the animal (Tarrant et al., 1992; Batista de Deus et al., 1999; Grandin, 2000). Glycogen depleted muscle from stressed animals does not have sufficient glycolytic substrates to allow the muscle to acidify properly after slaughter. Lowered acidification levels can be observed by measuring the pH of the meat 24 h after slaughter. If the pH_{ult} is above 6, the meat is classified as DFD. Nestorov et al. (1970) were among the first to describe an increase in meat pH_{ult} in beef with increasing transport time, later corroborated by other studies, including Shorthose et al. (1972), Wythes et al. (1980) and Tarrant (1989).

The current opinion is that short journeys (less than 4 h) have little effect on meat pH_{ult} as long as conditions are good and there is a low level of trauma (Tarrant et al., 1992; Grandin, 2000). While it is clear that even travelling short distances can reduce live weight (shrinkage), decrease glycogen reserves and increase meat temperature (Agness et al., 1990), this is not always reflected in pH_{ult}. For example, Fernandez et al. (1996) found a pH of 5.43 after 1 h of transport and 5.42 after 11 h transport. Similar results were reported by Lensink et al. (2001) who compared one short transport of approximately 1 h (pH_{ult} 5.42) with two consecutive transports (pH_{ult} 5.45). There were no significant differences between treatments. The lack of an effect on pH could occur when transport is only a slight stress and the animals are in good condition. The relationship between initial muscle glycogen content and ultimate pH is only linear at very low levels of glycogen. Thus, glycogen stores do not decrease enough to have a substantial effect on the pH_{ult}, especially when the animals are allowed to recover during lairage (Purchas and Aungsupakorn, 1993). Thus, meat pH is not a definitive indicator of the real extent of suffering experienced by the animal.

Other authors (Jones et al., 1988; Mohan Raj et al., 1992; Klont et al., 1999) have considered changes in meat texture and colour with respect to ante-mortem stress, but normally the effect of transport time is mixed with other confounding factors (e.g. breed, production system, nutrition, etc.). Several studies have also shown that ageing (i.e. storage of the meat) improves the tenderness of most muscles (Campo *et al.*, 2000). Thus, ageing may potentially attenuate any negative effects of transport. Moreover, sensory characteristics of the meat are not well correlated to differences in pH_{ult} (Fernandez *et al.*, 1996; Villarroel *et al.*, 2003a). Recently, instrumental and sensory meat qualities have been considered in relation to ethical quality, that is, aspects of animal welfare that may be compromised during the production process. It remains unclear to what extent these three qualities are correlated or how to obtain optimal levels of one without compromising another. In this chapter, we will review this problem during the transport process from the farm to the abattoir using examples relating to particular species.

Pig transport and meat quality

For pigs, transportation is particularly stressful since they can suffer motion sickness (Randall and Bradshaw, 1998). The main problems associated with transport and meat quality include the Porcine Stress Syndrome (a rapid onset of extremely high fever with muscle rigidity occurring during slaughter, precipitated in genetically susceptible animals), live weight loss (4–6%) and mortality (0.1–0.4%) in addition to injuries, bruises, skin damage, abnormal colour, PSE or DFD meat and contamination by *Salmonella* (Grandin, 2000). Long journeys are more likely to cause stress in pigs than in ruminants (Gregory, 1998). Important sources of stress include vibration of the vehicle, jolts, shocks and sudden impacts caused by bad road conditions and driving style, together with the temperature and mixing of animals (Hall and Bradshaw, 1998). The EFSA (2004) report emphasizes the importance of bad driving on animal welfare during transport. The effect of fasting and transportation on meat quality of slaughter pigs was analysed by Becker *et al.* (1989). In pigs transported 700 km (for slaughter), homeostasis was disrupted but there was no apparent negative effect on meat quality. Studies have shown that meat tenderness can be increased by limited transportation (Becker *et al.*, 1989). Mayes *et al.* (1998) reported that transportation (700 km) and associated handling imposed an acute demand on energy metabolism and fluid regulation of pigs for slaughter. The consequences of this demand decreased carcass yield (−4.64 kg) but did not have a detrimental effect on meat quality, in contrast to previous studies (Hails, 1978; Moss and Rob, 1978; Moss, 1980; Warriss and Brown, 1985).

Lundström *et al.* (1987) found that transport and lairage time significantly increase the incidence of PSE and DFD meat, which was made worse by mixing animals before slaughter. Hambrecht *et al.* (2005) found a tendency towards more rapid acidification of muscle and increased electrical conductivity after long-distance transport, suggesting that long journeys had a greater biological impact than short journeys. Warriss and Brown (1983) and Brown *et al.* (1999) found that transport had no effect on muscle glycogen. During stress from low-intensity exercise, fatty acids are the preferred source of energy (Klein *et al.*, 1994), which could partially explain the absence of an effect on glycogen stores. However, Leheska *et al.* (2003) found strong glycogen depletion after transportation, but in that case, it could also be explained by the influence of temperature.

Warriss and Brown (1994) found 1769 pig deaths (0.061% mortality) during transport after a survey of 2.9 million pigs taken to seven different slaughterhouses, with some seasonal variation. McPhee and Trout (1995) analysed the effect of transport time in pig lines selected for lean growth. The pH_{ult} was significantly effected by transport time in both control and selected pigs. Meat from animals that were transported for a longer duration had higher pH_{ult} values than short journeys, with a higher proportion of DFD meat. However, PSE meat was more frequent after short journeys. Lundström *et al.* (1987) analysed the effect of transport and lairage before slaughter; they concluded that a lairage time of 2 h decreases the incidence of PSE, but that the amount of DFD increases. Even after short transport, meat quality was affected in both groups (with and without lairage). When lairage conditions are good, a reasonable lairage period is necessary to permit the recovery of the animal; however, if lairage conditions are not good (i.e. absence of anti-mounting devices), the lairage period can increase the incidence of DFD. Pre-slaughter stress also influences microbiological contamination in live animals because microorganisms grow better in PSE or DFD meat, resulting in more contaminated carcasses (Eikelenboom *et al.*, 1990). Shedding of *Salmonella* spp. has been reported in pigs during transport by Berends *et al.* (1996) and Isaacson *et al.* (1999) (and see Chapter 3, this volume).

Martoccia *et al.* (1995) considered the effect of long (650 km) or short (180 km) transport on some metabolic parameters and meat quality in pigs. Journey duration significantly increased pH_{45} and pH_{ult} and had an effect on meat colour. Even though the differences were significant, the authors concluded that they would have little practical impact on the marketing of the meat in question. However, they do suggest that improvements can be made in the pre-slaughter treatment of animals. It is important to note that, although the meat could be marketed, the effects show that the journeys caused stress and that the farmer is potentially affected by the reduced value of the meat and in terms of his reputation as a reliable supplier of quality meat.

Warriss (1995) presented *Guidelines for the Handling of Pigs*, suggesting that transport followed by fighting during lairage significantly affects plasma cortisol, plasma CK, pH_{45} and pH_{ult}. The whole period is important, from the farm to stunning, since poor handling at any stage can compromise the benefits of earlier care. Larger groups, where unfamiliar animals have been mixed, will exacerbate the negative effects of transport stress and the problem can be made even worse by increasing the duration of lairage. The implementation of practical guidelines should be adapted to the wide range of conditions under which pigs are transported, especially for long-distance transport.

Hall and Bradshaw (1998) analysed the welfare aspects of road transport in pigs. Most journeys in Europe last less than 3 h (Warriss, 1996). However, relatively few studies consider the direct effect of journey duration on plasmatic indicators, partly due to the difficulty of obtaining blood samples from pigs. Bradshaw *et al.* (1996) studied catheterized pigs and found cortisol increased after longer journeys (8 h). Becker *et al.* (1989) observed a 3.19% loss in body weight in slaughter pigs after 11 h transport.

Barton-Gade and Christensen (1998) analysed the interaction between stocking density and transport on both welfare and meat quality in pigs. Varying the space allowance from 0.35 to 0.50 m²/100 kg pig (typical slaughter weight) had little effect on blood parameters and meat quality after a 3h journey. Skin damage increased at densities above 0.35 m². The effect of stocking density on meat and blood parameters after 3h journeys was also analysed by Warriss *et al.* (1998a) studying more than 2400 pigs. Higher stocking densities (0.31 m²/100 kg) caused more physical stress but did not affect meat quality. The authors recommended 0.36 m²/100 kg for short journeys but lower densities for long-distance transport so that the animals can lie down. A similar recommendation has been proposed by the Scientific Committee on Animal Health and Animal Welfare (SCAHAW, 2002) and the OIE (2004).

In an analysis of carcass quality and stress indicators in more than 5000 pigs slaughtered in the EU (Warriss *et al.*, 1998b), the incidence of PSE varied considerably, suggesting a wide variability in stress during transport. There was no relation between plasma stress parameters and meat quality traits. Those data confirm the difficulties of interpreting physiological data and productive data to evaluate animal welfare (Rushen, 1991; Mason and Mendl, 1993). Animal welfare indicators should be valid, accurate and reliable. It is possible that we need to review the validity of some productive traits as welfare indicators. Probably, the effects on meat quality may only pick up the worst welfare insults. Very high stress tended to be associated with more DFD meat (Scanga *et al.*, 1998). The stress threshold necessary to cause an observable effect on meat quality is higher than other conventional welfare indicators (i.e. plasmatic measurements, physiological constants or behavioural data). Skin damage was associated with higher pH_{ult}, indicating a possible effect of fighting on meat quality due to mixing animals during transport and lairage. Nanni Costa *et al.* (1999) studied the effect of loading method and loading density during 1h journeys in heavy Italian pigs; there was no effect of either factor on meat quality (the main difference in meat pH among samples was related to genotype). The effect of transport time on welfare and meat quality in pigs was also analysed by Pérez *et al.* (2002) after commercial journeys in Spain. Transport affected both meat quality and blood welfare indicators; pigs subjected to short transport showed a more intense stress response and poorer meat quality than pigs subjected to longer transport.

Hambrecht *et al.* (2004, 2005) analysed the critical factors affecting pork meat quality with special attention to pre-slaughter stress, including transportation. Long journeys tended to increase muscle glycolytic potential, increasing electrical conductivity and significantly reducing redness, yellowness and, to a lesser degree, lightness. Plasmatic indicators of stress were also affected, but not the pH_{ult}. However, the authors conclude that the physiological stress that transport imposes on pigs cannot be completely described by measuring plasma lactate, cortisol and muscle glycolytic potential; and stress produced by transport includes both a psychological and physical component, influencing welfare in ways that are not reflected by physiological changes. This finding is in agreement with Warriss *et al.* (1993) and Brown *et al.* (1999), but disagrees with Leheska *et al.* (2003) who showed that transportation dramatically reduced ante-mortem glycogen reserves prior to slaughter. The reasons for the disagreement between those studies could be the low temperature and loading density used in the latter study. Hambrecht *et al.* (2005) also analysed the effect of stress produced by suboptimal transport and lairage on pork quality.

The effect of transport in hot climates on pork meat quality was analysed by Mota-Rojas *et al.* (2006). Long journeys of 8, 16 and 24h were analysed in Mexico in hot weather. Live weight loss was 2.7%, 4.3% and 6.8% for each transport duration, respectively. The weight loss was not compensated for during the lairage period. Animals recovered weight during lairage but only between 0.05% and 1.15%. In general, transport affected meat quality in different ways, depending on the duration of transport. The 8h transport significantly affected pH_{45}, producing more PSE meat. On the contrary, 24h transport affected pH_{ult}, producing an increase in DFD meat. Transport duration significantly influenced bruising score. As transport time increased, the percentage of animals with bruises also increased. Tremor in hind limbs increased with transport duration. Hyperventilation and rectal temperature were affected by transport time with higher values in the 16h journeys. The position at arrival was affected by the duration of transport. In shorter journeys, more pigs arrived standing than in longer journeys, where animals were more fatigued and arrived lying down (which may explain increased bruising and highlights the link between meat quality effects and welfare). This is in agreement with Grandin (1994) and Knowles (1999) who have both stated that after long journeys animals tend to lie down for the last 2h of their transportation. During transportation, animals may lose their balance or fall. Both those events can be associated with the truck structure and the loading density (Tarrant *et al.*, 1988, 1992; Gallo *et al.*, 2001). Weight losses after transport were higher than those losses observed by Brown *et al.* (1999), who found values ranging from 2.2% to 4.3% for the same journey durations. Casas (1990) reports weight losses of about 3.48–5.4% in journeys that lasted 24–36h. For longer duration (36–72h), weight loss increased from 3.88% to 6.37%.

Smith and Grandin (1998) estimated that the cost of quality defects in pigs could be US$12.40 per animal, partly due to penalties, bruises, PSE and DFD pork caused by transport and pre-slaughter handling. In a study performed by Scanga *et al.* (1998), the proportions of PSE and DFD meat in pigs in the USA were 15.5% and 1.9%, respectively. Bidner (2003) concluded that PSE meat causes a loss of US$0.90 per carcass, mainly due to the decreasing industrial fitness of meat for processing. To convince the industry to invest in the improvement of animal welfare, it would be useful to demonstrate that better welfare means better meat quality. For pigs, links between aspects of meat quality and transport are either known or strongly suspected, as detailed in this chapter. It seems clear that pigs should have enough space to lie down, especially in long journeys, as is mandatory in the new EU normative about protection of animals during transport. More information is needed about travel sickness and dehydration in pigs (Hall and Bradshaw, 1998).

Implications

Transport is an important source of stress for pigs, so physiological and behavioural indicators of welfare are always affected. The effect is directly proportional to transport duration – the longer the transport, the greater is the effect on welfare indicators. As a result, travel time should be limited in some way. Short transport can also be very stressful due to the cumulative effect of loading, travelling and unloading. Journeys between 3 and 6h seem to have the least impact on animal welfare. From the welfare point of view, the aim of many studies was to identify the maximum acceptable journey time, based on behaviour and physiological responses. This can be a difficult task due to the absence of clear cut-off values for the welfare criteria used and the amount of acceptable deviation from these values. Pigs are transported and slaughtered under a wide range of conditions. Many countries like Canada and Australia will soon have a new animal welfare code about animal transportation that applies across the board. It is very difficult to produce simple guidelines to establish optimal handling under all those conditions. However, it is important to minimize the different aspects of handling and transport that may contribute to welfare insult. More often than not, the recommendation should be adapted to specific situations.

The link between aspects of meat quality and transport stress has been proven. Studies have demonstrated a reduction in meat quality after both short-distance and long-distance transports. Less of an effort has been made to study long-distance transport compared to short-duration or medium-duration transports. Changes in meat quality after long-distance transport are different after short distances. It is clear that transport tends to increase carcass bruising and skin damage in direct proportion to transport distance. The relationship between transport duration and pH is less consistent. In general, short journeys tend to increase PSE meat, while long journeys tend to increase DFD meat. Long-distance transport may exhaust pigs due to excessively long feed withdrawal times (fasting) in combination with physical exercise and psychological stress (which all tend to produce DFD meat). The metabolic consequences of long-distance transport are not reflected in the typical physiological indicators of stress. Other criteria like haematocrit or immune response variables could be more important than cortisol or glucose. The activity of the enzyme creatine kinase (CK) is also an important parameter indicating muscle damage in long-distance transport. Meat quality traits are not the best indicators of poor welfare during transport. In Table 4.1, we summarize the effects of transport on meat quality variables and other welfare variables in 22 different studies on pigs. All the studies found an effect on the welfare variables and 87% found an effect on meat quality. The effect of transport was stronger on other welfare variables than on meat quality, this is evidence that small effects on meat quality could indicate a large amount of suffering. In addition, the absence of an effect on meat quality does not imply that welfare is optimal. The approach for assessing welfare during transport should be multifactorial, including blood parameters associated with the stress response, behavioural parameters associated with psychological stress, indicators of physical damage as a consequence of poor welfare and, of course, carcass and meat quality criteria. The long-distance transport problem is not only an economic problem related to the instrumental and sensorial quality of the meat. It is also an ethical problem, which is more difficult (but not less important) to tie in with the economic aspects of the production system. It is necessary to develop a new concept of quality including instrumental and sensory meat quality and the ethical quality of the product, leading to optimal levels of one without compromising the others.

Pigs should have enough space to lie down on the moving vehicle, particularly during long journeys. Loading density and avoiding mixing unfamiliar animals are crucial for acceptable welfare during long journeys. In hot climates, this aspect is more important. Future research in pigs could consider analysing the effect of travel sickness and dehydration. All of these aspects can affect carcass and meat

Table 4.1. Summary of results of several studies on the effect of transport on carcass/meat quality and other measures of welfare.

Study	Carcass/meat quality (CMQV)	Other welfare measures/ non-meat quality (NMQV)
Hails, 1978	**	**
Moss and Robb, 1978	**	**
Warriss and Brown, 1985	**	**
Warriss and Brown, 1994	NA	***
Becker *et al.*, 1989	*	***
Lundström *et al.*, 1987	*	NA
Mayes *et al.*, 1998	NS	*
McPhee and Trout, 1995	**	NA
Lundström *et al.*, 1987	**	NA
Martoccia *et al.*, 1995	**	NA
Warriss, 1995	***	***
Bradshaw *et al.*, 1996	NA	***
Barton-Gade and Christensen, 1998	(*)	(*)
Warriss *et al.*, 1998a	NS	**
Warriss *et al.*, 1998b	*	*
Nanni Costa *et al.*, 1999	NS	NA
Perez *et al.*, 2002	*	*
Hambrecht *et al.*, 2004	(*)	**
Mota-Rojas *et al.*, 2006	**	**
Leheska *et al.*, 2003	**	**
Hambrecht *et al.*, 2005	(*)	**
Hambrecht *et al.*, 2005	(*)	NA

CMQV (i.e. bruising, pH, water-holding capacity, meat colour, meat texture, etc.)
NMQV (i.e. plasmatic measurement of stress, behaviour variables, physiological constants)
*** $p \leq 0.001$; ** $p \leq 0.01$; * $p \leq 0.05$; (*) $p < 0.1$ or tendency; NS non-significant; NA not available.

quality (i.e. bruising, PSE, shrinkage). Ensuring rapid access to water is essential to good welfare during transport. Research shows that dehydration in pigs can occur after only 6 h transit (Warriss *et al.*, 1988; Becker *et al.*, 1989). The association between thirst and dehydration is a serious welfare problem in pigs. This aspect should be considered to help decide time limits for journeys.

One way to reduce welfare problems and low carcass and meat quality would be to pay for the service of transport and pre-slaughter handling in terms of welfare quality. When people receive economic benefits for loading or driving quickly, welfare will be worse. Paying bonuses to drivers and handlers for low or decreased carcass bruising or skin damage and appropriate pH levels will improve animal welfare.

Cattle transport and meat quality

The literature covering cattle transportation is quite extensive. However, there is less information available on long road journeys compared to short journeys. The adverse effects of transport on animal welfare, meat quality and yield are expected to increase with the duration of journey, because the adverse effects increase with journey length (Tarrant *et al.*, 1992). There is a concern about the economic consequences of long-distance transportation to slaughter. Excellent reviews are available by Leach (1981), Tarrant (1990), Warriss (1990), Tarrant and Grandin (1993), Morris (1994), Knowles (1999) and Smith *et al.* (2004). The welfare of transported livestock is directly proportional to the economic value of the animals transported (Hails, 1978). From that point of view, in general, cattle welfare during transport is normally under low potential risk. However, the high individual value of the carcass means that even a low incidence of problems has a strong negative impact on the final income. The situation is complex on a moving vehicle since the environmental conditions are constantly changing, and that places constant demands on the adaptive physiology and behaviour of the animal. There is also a limited time frame for the animal to cope with the new environment. If it fails to adapt, the consequences could be important in terms of welfare and for the farmer's economy.

The most common problem with cattle transport is overloading, which greatly increases the risk of carcass damage (Tarrant, 1990). Driver skill and truck design are other important causes of poor welfare. Transport is a source of stress, even under optimal conditions (Grandin, 2000). Observing appropriate rest periods and limiting the duration of the journey are the best way to avoid stress turning into distress. Sound animal handling practices, smooth driving, correctly designed pens and loading ramps are also crucial to assure good welfare during the journey (Tarrant, 1990; Mantecca and Ruiz de la Torre, 1996).

In cattle, transport may lead to weight loss, shipping fever (fibrinous pneumonia) in calves, altered behaviour and carcass and meat quality damage. All these aspects affect the welfare status of the animals (Fraser and Broom, 1997). Environmental variables could be summarized as journey time and distance, stocking density, handling and the microclimate within the truck. The main subject of this review is the effect on carcass and meat quality. In Table 4.2, we summarize the studies or reviews analysing the effect on carcass or meat traits and other welfare variables. The main meat quality problem associated with cattle transport, especially long-distance journeys, is carcass bruising (Knowles, 1999). When bruising is severe, it decreases the amount of marketable meat. If bruising is not so severe, meat is downgraded, producing a partial loss. Bruising in cattle increases with transport distance and stocking density (Tarrant *et al.*, 1992). Transporting horned animals also increases the amount of bruising (Wythes, 1979). Cattle arriving at slaughterhouses from auction markets have more bruises than those sent directly to slaughter from the farm (Jarvis *et al.*, 1995; McNally and Warriss, 1996, 1997). The effect of breed temperament (a particular reaction pattern of a breed of animal in a stressful situation) (Tyler *et al.*, 1982; Keeling and Gonyou, 2001) on welfare during transport has been assessed, but not found to influence welfare.

The effect of breed on the incidence of DFD meat in cattle was analysed by Shackelford *et al.* (1994). The study demonstrated that the incidence of DFD meat varied with genetic background, based on the reactivity of the breed under stressful situations. However, the low heritability of the traits would impede rapid genetic progress through selection. Bartos *et al.* (1993) proposed a practical method to prevent DFD in beef. The method is based on social stability in groups with or

Table 4.2. Summary of results of several studies on the effect of transport on carcass/meat quality variables (CMQV) and non-meat quality (NMQV) variables. Cattle.

Study	CMQV	NMQV
Shorthose *et al.*, 1972	**	NA
Staples and Haugse, 1974	(*)	**
Hails, 1978	**	**
Wythes *et al.*, 1980	**	**
Wythes *et al.*, 1981	**	NA
Leach, 1981	**	**
Tarrant, 1990	**	**
Warriss, 1990	**	**
Tarrant and Grandin, 1993	**	**
Morris, 1994	**	**
Augustini and Fischer, 1981	*	NA
Tyler *et al.*, 1982	**	NA
Jones *et al.*, 1988	*	NA
Tarrant *et al.*, 1988	**	**
Tarrant, 1989	**	(*)
Schaefer *et al.*, 1990	**	**
Jarvis *et al.*, 1995	**	NA
Warriss *et al.*, 1995	**	**
Tarrant *et al.*, 1992	**	**
Honkavaara, 1993	**	**
Honkavaara and Kortesniemi, 1994	**	**
Shackelford *et al.*, 1994	**	NA
Fernandez *et al.*, 1996	*	NA
McNally and Warriss, 1996	**	(*)
Den Hertog-Meischke *et al.*, 1997	(*)	NA
Schaefer *et al.*, 1997	**	**
Batista de Deus *et al.*, 1999	**	**
Tarrant and Grandin, 2000 (review)	**	**
Smith and Gandin, 1999	**	NA
Knowles *et al.*, 1999	(*)	**
Nani Costa *et al.*, 2002	NS	NA
Gallo *et al.*, 2003	**	NA
Van de Water *et al.*, 2003	*	*
María *et al.*, 2003	*	NA
Villarroel *et al.*, 2003b	*	**
María *et al.*, 2004	*	*
Grigor *et al.*, 2004	NS	*

CMQV (i.e. bruising, pH, water-holding capacity, meat colour, meat texture, etc.)
NMQV (i.e. plasmatic measurement of stress, behaviour variables, physiological constants)
*** $p \le 0.001$; ** $p \le 0.01$; * $p \le 0.05$; (*) $p < 0.1$ or tendency; NS non-significant; NA not available.

without lairage before slaughter. In the socially unstable groups, pH_{ult} values increased substantially after overnight lairage at the abattoir. The authors recommended loose housing within groups with stable social relationships after long transportation. In long transportation of animals with unstable social relationships (i.e. feedlots), it is very important to avoid mixing social groups. This practice significantly reduces the incidence of DFD after long transports (Grandin, 2000).

The effect of breed on DFD incidence in stressed young bulls was analysed by Sanz *et al.* (1996). The authors concluded that there was no difference between Spanish Brown Swiss and Pirenaica in the incidence of DFD meats. The possible reason for this contradictory result (compared to other studies) could be that these two breeds are not very different in their reactivity to stressful situations. Other breeds (i.e. Spanish Lidia breed), which have been selected for their behaviour in the bullring, are subjected to both physical and emotional stress before their death (Hernández *et al.*, 2006). Practically all animals analysed presented pH_{ult} clearly above the cut-off value for DFD (6.37 ± 0.28).

The effect of long-distance transport was studied in Finland by Honkavaara and Kortesniemi (1994). They concluded that there is a great variation in transport distances and external temperatures in cattle transportation in that country and that transport should be limited to distances less than 400 km. Surprisingly, those authors do not give special importance to transport time. However, in this study, the increase of bruising and DFD meat is significant and could be a cause of important economic losses.

Cattle carcasses were assessed in the UK by McNally and Warriss (1996) to determine the incidence of bruising. The study found a strong relationship between animal welfare, stress, and bruising and meat quality. More than 180,000 cattle were downgraded due to bruising and more than 49,000 for stick-marking. In addition, carcasses rejected by the consumers due to abnormal pH or dark cutting appearance should also be considered. The study demonstrated that the amount of meat with a pH_{ult} above 5.8 increased with the bruising score. The appearance of meat from the carcasses was inferior and packing quality was affected. Obviously, bruising or stick marks indicate incidence of painful injury and poor welfare during transport and lairage. The meat industry and the authorities in charge of enforcing the legislation should inform personnel involved in the transport/slaughter chain that improvements can be made by investing in training and improving the design of facilities to reduce the substantial losses incurred from bruising and the impropriate use of sticks.

Simulated transport and its effect on meat quality were evaluated by Den Hertog-Meischke *et al.* (1997). The study demonstrated that meat water-holding capacity was affected by transport conditions, especially vibration, but it depended on the intrinsic water-holding capacity of the meat itself. The effect of position (in the truck) on welfare and meat quality traits was analysed by Van de Water *et al.* (2003), who concluded that transport affects welfare and meat quality indicators and that the effect depends on the animal's position within the truck – animals travelling in the back had higher pH_{ult} values and darker meat than those travelling in the front. However, contradictory results were found for plasma cortisol, where the animals that travelled in the front had higher values of cortisol than those in the back, which is difficult to explain.

Dehydration and fatigue after 24 h journeys by road demonstrate that longer-transport times or deterioration in transport conditions are detrimental to the welfare of animals and to carcass and meat quality (Tarrant et al., 1992). Those authors found that high stocking density (i.e. 1.05 m²/600 kg animal) reduced cattle welfare during transport and increased the stress response, the number of falls and the amount of bruising.

The effects of animal behaviour and environmental parameters on the incidence of DFD meat in beef was reviewed by Tarrant (1989). Glycogen breakdown is rapidly activated by increasing circulating adrenaline (fight-or-flight response) or by exhausting muscular activity. Glycogen is slowly depleted during long fasting periods (Jones et al., 1988). Any behaviour and environmental circumstances that trigger glycogen breakdown will increase the amount of dark-cutters. Sexual behaviour (i.e. mounting) or agonistic behaviour during transport and lairage is costly and will deplete glycogen reserves and increase the incidence of DFD meat. Those behaviours are stimulated by mixing groups at loading or by including females in oestrus. The detrimental effect of glycogen depletion can be exacerbated by low temperatures or long fasting periods (i.e. >24 h). The use of beta-agonists as growth promoters may also increase the incidence of dark meats. Another important aspect, mentioned in Tarrant (1989), is that the rate of glycogen re-synthesis is slower in ruminants than in monogastric animals, probably because glucose is less available in the former. However, there is less evidence to demonstrate that glycogen depletion is slower in ruminants.

Important factors that affect bruising include the slope of the ramp (which should be no more than 20°) and the height of the truck (Lapworth, 2004). The slope of the ramp is an important aspect when loading and unloading animals (SCAHAW, 2002). The heart rate of the animals increased by 165% when they were made to climb a ramp (Chacón et al., 2005). Steeper ramps caused greater increases in heart rate, up to maximum level (Van Putten, 1982). Overloading is one of the most common problems affecting bruising (Tarrant, 1990). According to this author, loading density during transport should be 0.77 m²/250 kg animal, 1.13 m²/450 kg animal and 1.63 m²/650 kg animal.

The interaction between stocking density during transport and carcass bruising was analysed by Tarrant et al. (1988, 1992). High stocking densities adversely affected the animals' welfare (assessed by physiological and behavioural indicators). Carcass bruising increased with stocking density. Severe bruising was associated with high stocking densities (591 kg/m²). Eldridge and Winfield (1998) found that space allowance during transport affected the bruising score, with more bruising when less space was available.

The significant increase in creatine kinase activity found by Warriss et al. (1995) and Tarrant et al. (1992) after 1-day journeys, and the increase in free fatty acids and haematocrit, also suggests that animals undergo extreme physical fatigue (Shorthose et al., 1972; Staples and Haugse, 1974).

The effect of rehydration on carcass and meat quality after long journeys was analysed by Wythes et al. (1980) and Schaefer et al. (1997). The experiments demonstrate that cattle must have access to water after long journeys to ensure that body tissues are not dehydrated at slaughter. To prevent dehydration, a rest period of 4–5 h is necessary. However, to prevent abnormal pH_{ult}, a 24 h rest is needed.

Long-transport stress altered the fundamental acid-base and electrolyte physiology of the animals resulting in degraded meat and lower carcass weight (Schaeffer *et al.*, 1988). Transportation was a factor of great importance affecting meat quality, because of its influence on meat pH (Wythes *et al.*, 1981). Journeys longer than 8 h can seriously affect meat quality and animal welfare and also jeopardize the farmers' incomes (Tarrant and Grandin, 2000). The European rules on the protection of animals during transport that entered into force on 5 January 2007, recommend a maximum of 14 h transport followed by at least 1 h rest with watering, to continue another 14 h. After this second phase, animals must be unloaded in a proper place, watered, fed and rested for 24 h (http://ec.europa.eu/food/animal/welfare/transport). According to the author's experience, in practice, vehicles are not well prepared for long journeys and often there are inadequate facilities to unload and house the animals so that they can recover properly (Villarroel *et al.*, 2003b). Even if the trucks are well equipped, the personnel in charge may not use the facilities properly (fans, watering devices, etc.). The same conclusions were reached by Batista de Deus *et al.* (1999). Tarrant and Grandin (2000) and Knowles *et al.* (1997) believe that, in practice, unless resting facilities are adequate and the animals are unloaded with care, rest stops may be negative for animals and serve only to prolong the overall journey time. The author of this chapter believes that the most advisable decision is to transport animals for no longer than 8 h and recommend slaughtering animals as close to the farm as possible. Premiums could be developed for farmers to promote this recommendation and avoid long-distance journeys. Economic incentives can greatly reduce bruising and DFD incidence (Tarrant and Grandin, 2000). The beef producer has the most influence on handling and transport strategies that affect meat quality. They can select the proper animals to travel, decide how to sort them, prepare them prior to transport, define the handling standards and, more importantly, decide their final destination (i.e. journey duration). The driver, in agreement with the farmer, will decide the loading density.

It has been estimated that 80% of the aspects that contribute to dark cutting beef occur before the cattle reach the slaughterhouse (Smith and Grandin, 1999). As a result, limiting transport time will help to avoid damaging the meat. Other effects of transport on meat quality include increased toughness and decreased palatability (Jones *et al.*, 1988). The proportion of DFD meat in Europe varies widely among countries and depends on the structure of the production system. In Scandinavian countries, cattle farms are often quite small and animals are frequently mixed since several loadings are required to fill the truck, which increases transport time. This situation dramatically increases the incidence of DFD meat, representing more than 25% of the carcasses in some cases (Honkavaara, M. Slovenia, 2001, personal communication).

Economic incentives are likely to be one of the most effective ways of raising animal welfare standards. The situation in other countries (e.g. Spain) is different. In general, cattle feedlots have a very standard production system with fattening groups of 40 animals (based on EU subsidies), with two fattening cycles per year. In those feedlots, normally enough animals are loaded at once to fill one truck. Animal groups are rarely mixed and there are almost no subsequent loading stops (once or maximum twice). Thus, transport time is shorter and there are fewer loading stops. As a result, the incidence of DFD meat is lower, approximately 5% of the carcasses (Brown *et al.*, 1990; Villarroel *et al.*, 2001). Similar results are found for carcass bruising.

Augustini and Fischer (1981) analysed the effect of transport and fasting on some quality characteristics of bulls. Transport affected the incidence of dark cutting beef. The authors suggest that biochemical reactions in dark cutting beef differ from DFD in pork. The possible influence of these differences between animal species should be taken into consideration when considering the effect of very fast chilling on meat quality characteristics.

Knowles *et al.* (1999) analyse the effect of long-distance transport on welfare and meat quality. Transport significantly affected the bruising score and glycogen depletion. The pH_{ult} was not affected by transport. In addition, a 1-h rest stop during transport (with access to water) did not appear to serve its intended purpose as many animals refused to drink during the stop. One important advantage of the rest stop was the opportunity to inspect the animals during the journey. The authors suggested that 24 h lairage provides an adequate period of recovery for animals after 14–31 h journeys (with no mixing of animals). Gallo *et al.* (2003) found that long journeys were associated with more weight loss, increased pH_{ult}, decreased muscle luminosity and more DFD meat. There is good evidence that fasting reduces glycogen stores in the body. Jones *et al.* (1988) suggest that fasting and transportation had a slight but significant effect on reducing tenderness and making the muscle darker.

Villarroel *et al.* (2001) analysed the critical points for welfare and meat quality in the cattle transport chain in Spain. The study concluded that the transport chain for young bulls seems to work well, which is partly demonstrated by the low incidence of DFD meat (5–8%). The probable reasons are that many feedlots are relatively new and have a standard design and management, animals are not regrouped, regulations about densities are respected and retailers and consumers will penalize products from stressed animals. Although the situation has improved in recent years, efforts should be made to improve training, education and – more importantly – to enforce transport regulations.

The effect of transport on instrumental and sensory meat quality was analysed by Fernandez *et al.* (1996) for journeys up to 11 h. Long journeys produced significant decreases in weight loss, glycolitic potential and meat pH_{4h}. Water-holding capacity, pH_{ult} and cooking loss or sarcomere (the basic unit of a cross-striated muscle's myofibril) length were unaffected. However, long transport decreased tenderness and increased myofibrillar resistance. Long transport had negative effects on sensory attributes of the meat. In general, increasing transport time will decrease instrumental and sensory meat quality. The sensory aspects of meat from young bulls that were transported for varying lengths of time were analysed by Villarroel *et al.* (2003a). The study demonstrates that transport time affects tenderness and overall liking or acceptability. Thus, sensory meat quality can be compromised by transport time, even under optimal transport conditions. Under suboptimal conditions, the effect of transport could be much higher. María *et al.* (2003) studied the effect of transport on instrumental aspects of young bull meat. The results indicate that transport time has a moderate negative effect on instrumental meat quality in terms of compression and colour. There were no significant changes in pH and toughness. Similar results were found by Villarroel *et al.* (2003a) who report a significant effect of transport time on physiological welfare indicators. A scoring system for evaluating stress in cattle at loading and unloading was developed by María *et al.* (2004) and the results of that study were that loading was more stressful than unloading, and that higher scores implied significantly higher stress levels. Batista de Deus *et al.* (1999) analysed the effect of transport time on welfare and meat quality, concluding that transport significantly affected meat pH_{ult} and physiological welfare indicators. Their recommendations are to slaughter animals as close as possible to the farm of origin to minimize the detrimental effects of long-distance transport.

Grigor *et al.* (2004) compare cattle slaughtered at the farm without transport with those subjected to transport and lairage and slaughtered at the abattoir in the UK. There was no effect of transport on either carcass bruising or muscle pH_{ult}. The transport and handling associated with the journey to the slaughterhouse was stressful for the animals, resulting in less resting behaviour and greater disturbance. Those authors conclude that transport does not have a significant effect on meat quality if journey time is short (less than 3 h) and the transport conditions are good in space and environment.

Implications

The main problems caused by transport on cattle carcass/meat quality are bruising and DFD meat. These two problems are directly associated with transport time. The longer the transport, the higher is the probability of bruising and DFD meat. The effect of transport on carcass/meat quality and welfare has been analysed for the last 40 years. The detrimental effect of transport stress on meat and carcass quality has been demonstrated in several studies. Only 5% of the studies reviewed failed to find a significant effect of transport on carcass or meat quality. All the studies that analysed carcass/meat quality and welfare indicators simultaneously found a significant effect of transport on animal welfare traits. Transport is an important stressor for cattle and is directly associated with glycogen depletion in the muscle. The major influence of transport and pre-slaughter handling on lean meat quality is through the potential effect on muscle glycogen stores. Long-distance transport in cattle leads to a higher probability of carcass bruising and DFD meat. Glycogen breakdown is rapidly activated by the fight-or-flight response or by exhausting muscular activity. Any behavioural or environmental circumstance that triggers glycogen breakdown will increase the amount of dark-cutters. Bruising tends to be more of a problem in cattle than in pigs. Another important difference with pigs is that PSE meats are rare in cattle. The market consequences in beef caused by damage from transport, especially long-distance transport, are important. The high individual value of the carcass means that even a low incidence of problems like bruising or abnormal colour of the meat have a strong impact on the final income to the farmers. Retailers place importance on the appearance of the carcass and meat colour, because it is one of the most important criteria for customers. Although the problem is more important in bulls, dark cutting can occur in all types of cattle. The high cost of rearing and the increasing importance of meat quality in the market justify taking the maximum amount of care of the live animal to prevent loss of yield and to ensure the best quality of meat, preserving animal welfare during the whole process. Good handling and decreasing journey time to avoid very long transports are likely to promote welfare as well as optimize the economic income to the farmers. The concern for quality and welfare, with a wide concept of quality including the ethical quality, will be profitable for the industry and for society. However, it is important to recognize that improving these conditions and including a new concept of quality (which includes the ethical aspects of the process) could involve an additional cost that would be borne by the market.

With the development of integrated systems and with a reduction in the number of increasingly larger-sized slaughterhouses, transport distances and times will increase, as will the number of times animals are loaded or unloaded. The author is of the view that the most advisable decision is to ban transport longer than 8 h and recommend slaughtering animals as close to the farm as possible.

Poultry transport and meat quality

Poultry transportation represents the largest movement of commercial livestock in the world (EFSA, 2004). More than 40 billion chickens were transported globally in the year 2000, according to the FAO. More than 4 billion were transported in the EU alone. The average transport time, estimated by Warriss *et al.* (1990), is 3.6 h, with a maximum of 13 h. This long transport influences both animal welfare and meat quality. Welfare impacts occur through heat stress, vibration, noise, sudden accelerations, turns, breaks, fasting, lack of access to water and social disruption (Freeman *et al.*, 1984; Nicol and Scott, 1990; Mitchell and Kettlewell, 1993, 1998). Among those factors, heat stress constitutes the major threat to poultry welfare and carcass/meat quality (Mitchell *et al.*, 2001).

During transportation and lairage, birds may become heat stressed, particularly on hot and humid days. If there is no difference between body temperature and environmental temperature, the coping mechanism fails and the main way to lose heat is by panting. When the truck is moving, air is also moving around the birds, helping them to cope with heat stress. When the truck is at a stand still, the situation can be very risky for the animals' welfare. To maintain thermal comfort, the animals have to adapt their behaviour to be able to lose or gain heat effectively. Housing in crowded crates and limited space, due to high animal density during poultry transport, strongly limits the time budget behaviour of the animals. The combination of crowding with heat stress could be fatal for the birds. It is essential to provide an adequate airflow around the animals with fresh, dry air. The distribution of dead chickens within the truck at arrival at the abattoir was analysed by Hunter *et al.* (1997) and Mitchell *et al.* (1998a). The distribution of dead birds was closely associated with the thermal loads within the truck. In the future, vehicles must have proper ventilation systems to make the internal microclimate more homogeneous, dissipate the heat and moisture loads resulting from the bird's metabolism and prevent cold stress in winter.

Heat and moisture production in poultry has been analysed by Mitchell *et al.* (1998b), suggesting that proper ventilation rates to maintain temperature in the optimum zone should be 0.6 m³/s/t. The interaction between air temperature, air humidity and air quality is a complex equation to solve in a moving vehicle full of chickens. During the hot season, ventilation rates should be high to maintain the temperature within the thermo-neutral zone of the animals, and air quality is not a big problem. During the cold season, ventilation rates are lower (to keep the temperature elevated) and air quality deteriorates. When vehicles are stationary, there is no external force to ventilate properly and heat and moisture removal depend on the convection rate. The solution to these problems must involve modification and improvement of the ventilation regime, developing active systems of environmental control. Thermal stress in transit was analysed by Webster *et al.* (1993) using a model chicken. The authors conclude that it is possible to ensure thermal comfort over a wide range of ambient air temperatures by appropriate control of air movement within the vehicle, whether at rest or in motion. These aspects should also be considered during lairage. Analyses of air movement during lairage by Quinn *et al.* (1998) and Hunter *et al.* (1998) indicate that environments for broiler lairage facilities should be controlled more closely to increase air movement at the bird level and ensure a proper microclimate within the optimal thermal zone of the chicken.

During transport from the production unit to the processing plant, birds are subjected to the stress of transport and deprived of food and water. These factors can lead to a weight loss, reduced carcass yield and changes in the post-mortem acidification of the meat (Warriss *et al.*, 1988). Excessive exercise during handling and transport can also make feather removal more difficult. Broilers are transported in crates and cannot be effectively fed, watered or inspected. Thus, journeys must be considerably shorter than for other species (i.e. red meat animals). Stress is a cumulative response of an animal to its surroundings and may be increased when birds are subjected to major changes. During transportation, birds are subjected to several stressors, including catching and handling, deprivation of food, water and freedom of normal movement, changes in climatic conditions and unfamiliar surroundings, noises, vibration and other sensations. Any transport should be carried out safely and in a manner that minimizes stress, pain and suffering (Grandin, 2000). Particular care needs to be taken with end-of-lay hens as they may be vulnerable to injury and their bones may be weak (Gregory *et al.*, 1990). The degree of animal welfare during transport is often directly proportional to the financial value of the individual animal, which is particularly low in poultry (Hails, 1978). Indeed, the unit price of a hen is often lower than what it costs for transportation (Perry, 2004).

Stress and its consequences (or biological cost) increase in all types of poultry as transportation time increases (Weeks and Nicol, 2000). Warriss *et al.* (1993) investigated the effects of 2, 4 and 6h journeys with 1–10h of food withdrawal. They found that transport reduces liver weight but not carcass weight. In 1999, Warriss *et al.* analysed the effect of lairage on glycogen depletion in poultry. According to their results, 1–3h lairage decreases glycogen levels significantly, compared to those slaughtered immediately after arriving at the meat plant. These results agree with Freeman *et al.* (1984). Fasting affects glycogen content in the muscles of broilers (Warriss *et al.*, 1988). Liver glycogen was reduced to negligible concentrations within 6h of food deprivation and the initial muscle pH was elevated. Glycogen depletion occurs in the muscle after 12h of food deprivation, producing an increase in the ultimate pH and decreasing meat tenderness. When birds are transported for 6h or longer, the pH_{ult} can be raised (Warriss *et al.*, 1993), although another study found that pH_{ult} is normal, but pH_{15min} is lower (Gregory, 1998). These data underline the difficulties of interpreting meat quality data as a welfare indicator.

Birds are one of the most easily frightened domestic farm animals. Chickens perceive contact with humans as an alarming predatory encounter. Two of the most common frightening events for domestic fowl are sudden changes in their social or physical environment and exposure to people (Weeks and Butterworth, 2004). Capturing, crating and transportation of broilers are among the most traumatic events in their lives. One of the most critical is transport (Elrom, 2000). Cashman *et al.* (1989) found that journey duration was a major determinant of fear levels in transported broilers. The transportation of poultry from the farm to the slaughterhouse is a multifactorial process, which includes a number of potentially stressful events, beginning with loading, disruption of social environment, motion, sudden acceleration and deceleration, exposure to unfamiliar environments, noise, temperature shifts and pre-slaughter processing. Other factors to consider are the social, housing and genetic backgrounds, the loading density and the shape, size and position of the loading cages.

One of the problems with bird transport is the lack of information regarding real transport time. Warriss *et al.* (1990) analysed the time spent by broiler chickens in transit to slaughterhouses. The average journey lasted 3–5 h, including loading time, and it normally took place during the night or early morning. Capturing in the dark is beneficial for broiler welfare, reducing the fear response (Nicol and Scott, 1990). Six per cent of the birds travelled for 7 h. Longer transport (more than 3 h) is associated with increased mortality, bruising and reduced carcass yield (Warriss *et al.*, 1992). These authors report that more than 600,000 chickens died during transport (0.2% of 3.2 million birds), but mortality varied with journey duration. Journeys less than 4 h long had a mortality level of around 0.15%, while above 4 h, mortality rose to 0.28% (more than 800,000 birds).

Birds may suffer physical damage during the transport process, which includes capturing and loading. Broken bones, haemorrhaging and bruising are the most important problems (Knowles and Broom, 1990a). Bruising is a great concern for the industry and, presumably, for consumers. It has been estimated that 6.7% of the carcasses are downgraded due to bruising or black spots (Taylor and Helbacka, 1968). There are two main forms of ante-mortem bruising in poultry: breast bruising and red wingtips. The former can occur during transport because of loading in transport crates. Red wing tips can occur during electrical stunning or after catching for loading.

The catching crew and catching method have a significant effect on the frequency of bruising. Capturing animals by hand produces more bruising than using properly designed machines (Duncan *et al.*, 1986). Elrom (2000) reports that the main problems associated with poultry transportation are metabolic exhaustion, bruising, dehydration, broken bones, torn skin and thermal stress. The time of handling and transportation are the main factors affecting these problems. Extreme injuries during transport include broken bones, mainly caused by poor handling.

Laying hens are more susceptible to suffering broken bones during transport (Gregory and Wilkins, 1989). The fact that hens are restricted from exercising to such an extent that they are unable to maintain the strength of their bones is probably the greatest single indictment of the battery cage. The increased incidence of bone breakage results is a serious welfare insult (Baxter, 1994). The amount of trauma incurred by birds at depopulation can be reduced by improving bone strength to resist injuries during handling. More importantly, efforts should be made to improve handling to reduce insults inflicted during the depopulation process (Knowles and Wilkins, 1998). The effect of the husbandry system on the incidence of broken bones at slaughter was analysed by Gregory *et al.* (1990) and Knowles and Broom (1990b) and concluded that 'battery' (cage) housed birds had a higher incidence of recent bone fractures in comparison with free range or perchery systems. The incidence of broken bones in broilers was estimated by Gregory and Wilkins (1990) to be around 95%, with three or four breaks per carcass, due to the whole transport process, including catching, loading, unloading and slaughter. The incidence of haemorrhages during slaughter was analysed by Wilson and Brunson (1968), concluding that handling and stunning had a significant effect. The method of harvesting broiler chickens could also affect broken bones and bruising scores (Duncan *et al.*, 1986) and, therefore, the number of birds transported with broken bones. The results indicate that the potential for injury associated with catching can be considerably reduced using a well-designed machine instead of catching by hand. It is difficult to distinguish whether broken bones were produced during transport or during the catching/loading process as birds are transferred immediately after catching to transport crates and assessment of damage often takes place following transportation. However, birds with painful traumatic injuries (e.g. broken bones and dislocations) will suffer more as the journey duration increases. Even when broken bones are mainly caused during catching, the animal suffers intense pain after the bone is broken and it is impossible to identify and treat or humanely destroy injured birds while in transport crates (Gregory, 2004). All efforts to prevent bone breakage should be taken during catching and animals with broken bones are not suitable for transport. Unfortunately, broken bones are often only detected after slaughter, if they are noticed at all.

The amount of downgrading of broiler chicken carcasses was analysed by Griffiths and Nair (1984). The average proportion of birds downgraded per day varied from 9.6% to 15.7% of the production and the most important cause was bruising. The bruises were located mainly in the breast and the legs.

Broilers may be transported considerable distances to processing plants. The minimum distance transported, the journey time and the total marketing time were estimated by Warriss *et al.* (1990) for 5819 journeys in which 19.3 million broilers were transported to four processing plants. The average distance travelled was 33.5 km, the average journey including unloading time was 2.7 h and the average total time from the start of loading to the completion of unloading was 3.6 h. The maximum recorded journey including unloading time was 12.1 h and the maximum recorded total time was 12.8 h. There was considerable variation between plants, those with smaller annual throughputs tending to have longer average times. In 46% of journeys, the birds were unloaded within 3 h of the start of loading, 78% within 5 h and 94% within 7 h. Long journeys may cause fatigue, specifically, depleting glycogen stores in the body, especially when birds have been subjected to long periods of food deprivation (Warriss *et al.*, 1993). Depletion of glycogen may explain the exhaustion observed in some birds after long transport as there is evidence that fatigue is associated with a depletion of muscle glycogen (Warriss *et al.*, 1993).

Long transport can cause dehydration, particularly when birds are exposed to heat stress. Periods without food and in transit lead to losses in liver and muscle glycogen, associated with apparent fatigue. Longer transport produces dehydration in the birds. The progressive changes observed with longer journeys imply that transport stress is cumulative and should be minimized. It is necessary to establish a cut-off value for poultry transport time, minimizing the risk of poor welfare and carcass/meat quality damages. Progress is urgently needed to raise standards for poultry reared for meat. Maximum transport times for poultry travelling to slaughter need to be reduced further (Burgess and Pickett, 2006).

There is evidence that PSE meats can occur in turkeys because of pre-slaughter stress, including the stress due to transportation (Owens and Sams, 2000). Like in pigs, halothane gas can be used to identify PSE prone turkeys (Owens *et al.*, 2000). The results indicate that HAL+ turkeys have a higher incidence of PSE meat than those with the HAL− genotype. One genetic cause of PSE is associated with the presence of the halothane (HAL) gene, so named because animals that carried two copies of the gene (called homozygous carriers or nn) were discovered to undergo physiological stress and die when exposed to halothane anaesthesia. However, it has been shown that animals carrying one copy of the HAL gene (called heterozygous carriers or Nn) tend to produce leaner carcasses but more PSE than HAL gene-free animals (called homozygous negative or NN).

Babji *et al.* (1982) analysed the administration of electrolyte solutions to turkeys to reduce pre-slaughter stress. Those authors conclude that the administration of electrolytes does not prevent the changes in meat quality produced by pre-slaughter stress. Maintaining a cool pre-slaughter environment (within the thermal comfort zone for this animal) seems to produce meat with better carcass quality. Other problems associated with transportation and pre-slaughter handling that degrade the carcass include bruised drumsticks, breast buttons or blisters, back and leg scratches and leg oedema (McEwen and Barbut, 1992).

Implications

Poultry transport is a multifactorial process, which includes a number of potentially traumatic events. Temperature and humidity, air quality, ventilation rate, vibration and noise are some of the physical stressors that challenge the adaptation mechanisms of the animals. Food and water deprivation are other important stressors during transport which produce fear and pain during the journey, including catching, loading, transportation, unloading and pre-slaughter handling. The biological cost of exposure to those stressors includes injuries, bruising, painful bone fractures and dislocations, glycogen depletion and abnormal pH in the meat. In extreme cases, the result of the process is death. As chickens are fearful and flighty, journeys should be limited to less than 4 h. After that time, glycogen depletion will occur because of food deprivation and a failed stress response. It has been demonstrated that, as journey time increases, bruising, injuries and mortality increase. As a result, meat quality is directly affected by journey time. Long journeys can produce important economic losses and the impact of transport effects on the economics of the poultry sector is important since the profit margin per animal is quite low. The method of catching, crating and transporting should be adapted to climatic conditions and be well controlled. The payment system should be proportional to the condition of the animal upon arrival at the slaughterhouse and the incidence of bruising, injury, broken bones and mortality should be considered when judging or placing economic penalties for transport quality. The payment system should also avoid payment by piecework, in which the worker is paid for the number of tasks completed rather than being paid for the time worked. Placing emphasis on speed during capture, transport and processing is detrimental for welfare and meat quality. A specific welfare problem in poultry production is end-of-lay hens since the transport costs may exceed their economic value. Depopulation on the farm by euthanasia with specific gases, according to regulations, can be a solution to help to maintain acceptable standards of welfare.

Discussion

Transport is stressful for animals, even under optimal conditions, and, as any other stressor, should be minimized. The main variable to be controlled to reduce the biological cost of transport stress is journey time. The physical condition of the animal before transport, and food and water deprivation, play important roles in the way that animals should be managed during the transport process and in their fitness to travel. Domestic species vary in their reaction to transport stress. Stress in pigs more often produces PSE meat than DFD meat, which is more frequent in cattle. Chickens are more sensitive to heat stress, producing heat shortening of muscles. All species can suffer bruising and skin damage as a consequence of transportation, but poultry have the highest incidence of broken or dislocated bones. The loading process is especially stressful in poultry since the animals have to be caught before loading, most often by hand.

Handling, loading, transport conditions and unloading are critical processes that should be performed properly (see Chapters 6 and 7, this volume). The quality of driving, road conditions and weather constraints vary widely and transport limits and recommendations should be adapted to these variable situations. The meat industry and the authorities in charge of enforcing legislation should inform personnel involved in the transport/slaughter chain that improvements can be made by investing in training and by improving the design of facilities to reduce the substantial losses incurred from bruising and the inappropriate use of sticks. It is important to improve training and education on methods of animal handling and transport to minimize the risk to welfare and to avoid animal suffering. The training programmes should be supervised by competent authorities and must be a practical learning process. Drivers, farmers and slaughter associations should participate in these training programmes. A payment system should be implemented in terms of transport quality. That payment should be directly related to welfare; and criteria like bruising score and meat pH should also be considered in that payment system. Long-distance travel should be penalized when other abattoirs are closer to the loading point. The goal should be to slaughter the animals as close to the farm as possible and provide a bonus to the farmer or driver for complying with that objective.

In order for transport regulations to be correctly implemented, journey time limits and rest periods must be enforced (in proper rest areas with food and water). If the resting areas are not adequate, as is often the case in many European countries, transport time should be limited, possibly to less than 6 h. Off-loading stock at regular intervals during a long journey may decrease the stress associated with the lack of water or food and rest, but can increase the stress associated with handling, loading and unloading and possibly social changes, exposing the animals to another novel environment. Thus, journey time should be short enough to avoid the requirement to unload the animals for resting. The net effect of unloading animals for rest periods could be an increase of the biological cost of the stress response.

The author would like to comment on the interpretation of using meat quality indicators to assess animal welfare during transport. A very strong stress is required to have a visible effect on meat quality. When there are even small effects on the meat, it is clear that the animals have suffered, because other welfare criteria such as behavioural changes, physiological constants or plasmatic indicators are normally greatly affected. An absence of an effect on meat quality (i.e. no effect on pH) is not a clear sign of absence of suffering due to poor welfare. We need to evaluate animal welfare from a multiple perspective approach to have a clear idea about the real extent of suffering of animals during transport. All attempts to decrease the amount of stress that livestock undergo should consider the complexity of stressors involved in the process. Transportation represents a combination of several stressors that can have additive and deleterious effects on the animal.

References

Agness, F., Sartorelli, P., Abdi, B.H. and Locatelli, A. (1990) Effect of transport loading or noise on blood biochemical variables in calves. *American Journal of Veterinary Research* 51, 1679–1681.

Appleby, M.C. (2005) The relationship between food prices and animal welfare. *Journal Animal Science* 83, E9–E12.

Augustini, C. and Fischer, K. (1981) Effect of transport and pre-slaughter holding of bulls on some meat quality characteristics and the post-mortem glycolysis in Germany. *Fleischwirtsch* 61, 775–784.

Babji, A.S., Froning, G.W. and Ngonka, D.A. (1982) The effect of preslaughter environmental temperature in the presence of electrolyte treatment on turkey meat quality. *Poultry Science* 61, 2385–2389.

Barton-Gade, P. and Christensen, L. (1998) Effect of different stocking densities during transport on welfare and meat quality in Danish slaughter pigs. *Meat Science* 48, 237–247.

Bartos, L., Franc, C., Rehák, D. and Stípková, M. (1993) A practical method to prevent dark-cutting (DFD) in beef. *Meat Science* 34, 275–282.

Batista de Deus, J.C., Silva, W.P. and Soares, G.J.D. (1999) Efeito da distância de transporte de bovinos no metabolismo post-mortem. *Rev Bras Agrocienc* 5, 152–156.

Baxter, M.R. (1994) The welfare problems of laying hens in battery cages. *The Veterinary Record* 134, 614–619.

BCSPCA (2006) Transportation of farm animals. BC Society for the Prevention of Cruelty to Animals. Available at: www.spca.bc.ca/farminfo/transportation/

Becker, B.A., Mayes, H.F., Hahn, G.L., Nienaber, J.A., Jesse, G.W., Anderson, M.E., Heymann, H. and Hedrick, H.B. (1989) Effect of fasting and transportation on various physiological parameters and meat quality of slaughter hogs. *Journal Animal Science* 67, 334–341.

Becker, T. (2000) Consumer perception of fresh meat quality: a framework for analysis. *British Food Journal* 102, 158–176.

Berends, B.R., Urlings, H.A., Snijders, J.M. and Van Knapen, F. (1996) Identification and quantification of risk factors in animal management and transport regarding *Salmonella* spp. in pigs. *International Journal of Food Microbiology* 30(1–2), 37–53.

Bidner, B.S. (2003) Factors impacting pork quality and their relationship to ultimate pH. PhD thesis, University of Illinois at Urbana-Champaign, Illinois.

Blandford, D., Bureau, J.C., Fulponi, L. and Henson, S. (2002) Potential implications of animal welfare concerns and public policies in industrialized countries for international trade. In: Krissoff, B., Bohman, M. and Caswell, J. (eds) *Global Food Trade and Consumer Demand for Quality*. Kluwer, New York.

Bradshaw, R.H., Parrot, R.F., Goode, J.A., Lloyd, D.M., Rodway, R.G. and Broom, D.M. (1996) Stress and travel sickness in pigs: effects of road transport on plasma concentrations of cortisol, beta-endorphins and lysine vasopressin. *Animal Science* 63, 507–516.

Blokhuis, H., Jones, R.B., Geers, R., Miele, M. and Veissier, I. (2003) Measuring and monitoring animal welfare: transparency in the food product quality chain. *Animal Welfare* 12, 445

Broom, D. (2000) Welfare assessment and problem areas during handling and transport. In: Grandin, T. (ed.) *Livestock Handling and Transport*, 2nd edn. CAB International, Wallingford, UK, pp. 43–63.

Brown, S.N., Bevis, E.A. and Warriss, P.D. (1990) An estimate of the incidence of dark cutting beef in the United Kingdom. *Meat Science* 27, 249–258.

Brown, S.N., Knowles, T.G., Edwards, J.E. and Warriss, P.D. (1999) Behavioural and physiological responses of pigs to being transported for up to 24 hours followed by six hours recovery in lairage. *Veterinary Record* 145, 421–426.

Burgess, K. and Pickett, H. (2006) *Supermarkets and Farm Animal Welfare 'Raising the Standard'*. Compassion in World Farming Trust, Petersfield, UK.

Campo, M.M., Santolaria, P., Sañudo, C., Lepetit, J., Olleta, J.L., Panea, B. and Alberti, P. (2000) Assessment of breed type and ageing time effects on beef quality using two different texture devices. *Meat Science*, 55, 371–378.

Carlsson, F., Fryblom, P. and Lagerkvist, C.J. (2004) Consumer willingness to pay for farm animal welfare transportation of farm animals to slaughter versus the use of mobile abattoirs. *Working Paper in Economics* no 149, Department of Economics, Götegorg University, Sweden.

Casas, S.A. (1990) Merma de peso en canales. *Tesis de Licenciatura, Facultad de Medicina Veterinaria y Zootecnia*, Universidad Nacional Autónoma de México, México.

Cashman, C.J., Nicole, C.J. and Jones, R.B. (1989) Effect of transportation on the tonic immobility fear reaction of broilers. *British Poultry Science* 30, 211–222.

Chacón, G., Garcia-Belenguer, S., Villarroel, M. and Maria, G.A. (2005) Effect of transport stress on physiological responses of male bovines. *Deutsche tierärztliche Wochenschrift – German Veterinary Journal* 112, 465–469.

Cheeke, P. (1993) *Impacts of Livestock Production on Society, Diet/Health and the Environment*. Interstate Publishers, Danville, Illinois.

Den Hertog-Meischke, M., Vada-Kovacs, M., Kasella, M. and Smulders, F.J.M. (1997) Increased water holding capacity of PSE pork during post-mortem storage. In: Den Hertog-Meischke, M. (ed.) The water holding capacity of fresh meat – with special reference to the influence of processing and distribution. PhD thesis, University of Utrecht, Utrecht, The Netherlands, pp 133–144.

Duncan, I.J.H., Slee, G.S., Kettlewell, P.J., Berry, P.S. and Carlisle, A.J. (1986) Comparison of the stressfulness of catching broiler chickens by machine and hand. *British Poultry Science* 27, 109–114.

Eikelenboom, G., Bolink, A.H. and Sybesma, W. (1990) Effects of feed withdrawal before delivery on pork quality and carcass yield. *Meat Science* 29, 25–30.

Eldridge, G.A. and Winfield, C.G. (1998) The behaviour and bruising of cattle during transport at different space allowances. *Australian Journal of Experimental Agriculture* 28, 695–698.

Elrom, K. (2000) Handling and transportation of broilers-welfare, stress, fear and meat quality. Part III: fear; definitions, its relation to stress, causes of fear, responses of fear and measurement of fear. *Israel Journal of Veterinary Medicine* 55(3), 1–7.

European Commission (2005) Attitudes of consumers towards the welfare of farmed animals. Special Eurobarometer 229/Wave 63.2 – TNS Opinion & Social. Available at: http://ec.europa.eu/food/animal/welfare/euro_barometer25_en.pdf

European Food Safety Authority (EFSA) (2004) The welfare of animals during transport. Scientific Report of the Scientific Panel on Animal Health and Welfare on a request from Commission related to the welfare of animals during transport. EFSA-Q-2003-094.

Eurostat (2006) European Commission. Available at: http://epp.eurostat.cec.eu.int/portal/page?_pageid=1090,1 &_dad=portal&_schema=PORTAL

Fernandez, X., Monin, G., Culioli, J., Legrand, I. and Quilichini, Y. (1996) Effect of duration of feed withdrawal and transportation time on muscle characteristics and quality in Friesian–Holstein Calves. *Journal Animal Science* 74, 1576–1583.

Fraser, A.F. and Broom, D.M. (1997) *Farm Animal Behaviour and Welfare*. CAB International, Wallingford, UK.

Freeman, B.M., Kettlewell, P.J., Manning, A.G.C. and Berry, P.S. (1984) The stress of transportation for broilers. *Veterinary Record* 114, 286–287.

Gallo, C., Warris, P., Knowles, T., Negrón, R., Valdés, A. and Mencarini, I. (2001) Densidades de carga utilizadas para el transporte de matadero en Chile. *Archioos de Medicina Veterinaria* 37(2), 155–159.

Gallo, C., Lizondo, G. and Knowles, T.G. (2003) Effects of journey and lairage time on steers transported to slaughter in Chile. *Veterinary Record* 152, 361–364.

Grandin, T. (1980) Designs and specifications for livestock handling equipment in slaughter plants. *International Journal of the Study of Animals Problems* 1, 178–200.

Grandin, T. (1994) Farm animal welfare during handling, transport and slaughter. *Journal of the American Veterinary Medical Association* 204, 372–377.

Grandin, T. (2000) *Livestock Handling and Transport*, 2nd edn. CAB International, Wallingford, UK.

Gregory, N.G. (1994) Preslaughter handling, stunning and slaughter. *Meat Science* 36, 45–56.

Gregory, N.G. (1998) *Animal Welfare and Meat Science*. CAB International, Wallingford, UK.

Gregory, N.G. (2004) *Physiology and Behaviour of Animal Suffering*. Blackwell, Oxford, UK.

Gregory, N.G. and Wilkins, L.J. (1989) Handling and processing damage in end-of-lay hens. *British Poultry Science* 30, 555–562.

Gregory, N.G. and Wilkins, L.J. (1990) Broken bones in chicken: effect of stunning and processing in broilers. *British Poultry Science* 31, 53–58.

Gregory, N.G., Wilkins, L.J., Elepheruma, S.D., Ballantyne, A.J. and Overfield, N.D. (1990) Broken bones in domestic fowl: effect of husbandry system and stunning method on end-of-lay hens. *British Poultry Science* 31, 59–69.

Griffiths, G.L. and Nair, M.E. (1984) Carcass downgrading of broiler chickens. *British Poultry Science* 25, 441–446.

Grigor, P.N., Cockram, M.S., Steele, W.B., McIntyre, J.M., Williams, C.L., Leushuis, I.E. and van Reenen, C.G. (2004) A comparison of the welfare and meat quality of veal calves slaughtered on the farm with those subjected to transportation and lairage. *Livestock Production Science* 91, 219–228.

Hails, M.R. (1978) Transport stress in animals: a review. *Animal Regulation Studies* 1, 289–343.

Hall, S.J.G. and Bradshaw, H.R. (1998) Welfare aspects of the transport by road of sheep and pigs. *Journal of Applied Animal Welfare Science* 1, 235–254.

Hambrecht, E., Eissen, J.J., Nooijen, R.I.J., Duero, B.J., Smiths, C.H.M., den Hartog, L.A. and Verstegen, M.W.A. (2004) Pre-slaughter stress and muscle energy largely determine pork quality at two commercial processing plants. *Journal of Animal Science* 82, 1401–1409.

Hambrecht, E., Eissen, J.J., Newman, D.J., Smits, C.H.M., den Hartog, L.A. and Verstegen, M.W.A. (2005) Negative effects of stress immediately before slaughter on pork quality are aggravated by suboptimal transport and lairage conditions. *Journal of Animal Science* 83, 440–448.

Harris, T. (1996) *AATA Manual for the Transportation of Live Animals by Road*. Animal Transportation Association, Surrey, UK.

Hernández, B., Lizaso, G., Horcada, A., Beriain, M.J. and Purroy, A. (2006) Meat colour of fighting bulls. *Archivos Latinoamericanos de Producción Animal* 14, 6.

Hobbs, J.E. (1997) Measuring the importance of transaction costs in cattle marketing. *American Journal of Agricultural Economics* 79, 1083–1095.

Hodges, J. (2005) Cheap food and feeding the world sustainably. *Livestock Production Science* 92, 1–16.

Hoffman, D.E., Spire, M.F., Schwenke, J.R. and Vnrah, J.A. (1998) Effect of source of cattle and distance transported to a commercial slaughter facility on carcase bruising in mature beef cow. *Journal of the American Veterinary Medical Association* 212, 668–672.

Honkavaara, M. (1993) Effect of a controlled-ventilation stockcrate on stress and meat quality. *Meat Focus International* 2, 545–547.

Honkavaara, M. and Kortesniemi, P. (1994) Effect of long distance transport on cattle stress and meat quality. *Meat Focus International* 3, 405–409.

Hughes, D. (1995) Animal welfare: the consumer and the food industry. *British Food Journal* 97, 3–7.

HSUS (2006) *An HSUS Report: The Welfare of Animals in the Egg Industry*. The Humane Society of the United States, Washington, DC.

Hunter, R.R., Mitchell, M.A. and Matheu, C. (1997) Distribution of 'dead on arrivals' within the bioload on commercial broiler transporters: correlation with climate conditions and ventilation regimen. *British Poultry Science* 38, S7–S9.

Hunter, R.R., Mitchell, M.A., Carlisle, A.J., Quinn, A.D., Kettlewell, P.J., Knowles, T.G. and Warriss, P.D. (1998) Physiological responses of broilers to pre-slaughter lairage: effects of the thermal micro-environment? *British Poultry Science* 39, S53–S54.

Isaacson, R.E., Rirkins, L.D., Weigel, R.M., Zuckermann, F.A. and Di Pietro, J.A. (1999) Effect of transportation and feed withdrawal on shedding of *Salmonella typhimurium* among experimentally infected pigs. *American Journal of Veterinary Research* 60, 1155–1158.

Jamison, W. (1992) The rights of animals, political activism, and the feed industry. In: Lyons, T.P. (ed.) *Proceedings of the Alltech 8th Annual Symposium*. Alltech Technical Publications, Nicholasville, Kentucky, pp. 121–138.

Jarvis, A.M., Selkirk, L. and Cockram, M.A. (1995) The influence of source, sex class and pre-slaughter handling on the bruising of cattle at two slaughterhouses. *Livestock Production Science* 43, 215–224.

Jones, S.D.M., Schaefer, A.L., Tong, A.K.W. and Vincent, B.C. (1988) The effects of fasting and transportation on beef cattle. 2. Body component changes, carcass composition and meat quality. *Livestock Production Science* 20, 25–35.

Keeling, L.J. and Gonyou, H.W. (2001) *Social Behaviour in Farm Animals*. CAB International, Wallingford, UK.

Klein, S., Coyle, E.F. and Wolfe, R.R. (1994) Fat metabolism during low-intensity exercise in endurance trained and untrained men. *American Journal of Physiology* 267, E934–E940.

Klont, R., Barnier, V., Smulders, F., Van Dijk, A., Hoving-Bolink, A. and Eikelenboom, G. (1999) Post-mortem variation in pH, temperature, and colour profiles of veal carcassess in relation to breed, blood haemoglobin content, and carcass characteristics. *Meat Science* 53, 195–202.

Knowles, T.G. (1999) A review of the road transport of cattle. *Veterinary Record* 144, 197–201.

Knowles, T.G. and Broom, D.M. (1990a) The handling and transport of broilers and spent hens. *Applied Animal Behaviour Science* 28, 75–91.

Knowles, T.G. and Broom, D.M. (1990b) Limb bone strength and movement in laying hens from different housing systems. *Veterinary Record* 126, 354–356.

Knowles, T.G. and Wilkins, L.J. (1998) The problem of broken bones during the handling of laying hens: a review. *Poultry Science* 77, 1798–1802.

Knowles, T.G., Warriss, P.D., Brown, S.N., Edwards, J.E., Watkins, P.E. and Phillips, A.J. (1997) Effects on calves less than one month old of feeding or not feeding them during road transport of up to 24 hours. *Veterinary Record* 140, 116–124.

Knowles, T.G., Brown, S.N., Edwards, J.E., Phillips, A.J. and Warriss, P.D. (1999) Effect on young calves of a one-hour feeding stop during a 19-hour road journey. *Veterinary Record* 144, 687–692.

Lapworth, J.W. (2004) *Cattle Transport: Loading Strategies for Road Transport.* Department of Primary Industries and Fisheries, Brisbane, Australia.

Leach, T.M. (1981) Physiology of the transport of cattle. In: Moss, R. (ed.) *Transport of Animals Intended for Breeding Production and Slaughter.* Martinus Nijhoff, The Hague, The Netherlands, pp. 57–72.

Leheska, J.M., Wulf, D.M. and Maddock, R.J. (2003) Effects of fasting and transportation of pork quality development and extent of post-mortem metabolism. *Journal of Animal Science* 81, 3194–3202.

Lensink, B.J., Fernandez, X., Cozzi, G., Florand, L. and Veissier, I. (2001) The influence of farmers' behaviour on calves' reactions to transport and quality of veal meat. *Journal of Animal Science* 79, 642–652.

Lister, D. (1969) Some aspects of physiology of pale, soft and exudative muscle. In: Sybesma, W., van der Wal, P.G. and Walstra, P. (eds) *Recent Points of View on the Condition and Meat Quality of Pigs for Slaughter.* Research Institute for Animal Husbandry, Zeist, The Netherlands, pp. 123–131.

Lundström, K., Malmfors, B., Malmfors, G. and Stern, S. (1987) Meat quality in boars and gilts after immediate slaughter or lairage for two hours. *Swedish Journal of Agricultural Research* 17, 51–56.

Mantecca, X. and Ruiz de la Torre, J.L. (1996) Transport of extensively farmed animals. *Meat Science* 49, 89–94.

María, G.A. (2006) Public perception of farm animal welfare in Spain. *Livestock Science* 103, 250–256.

María, G.A., Villarroel, M., Sañudo, C., Olleta, J.L. and Gebresenbet, G. (2003) Effect of transport time and ageing on aspects of beef quality. *Meat Science* 65, 1335–1340.

María, G.A., Villarroel, M., Chacón, G. and Gebresenbet, G. (2004) Scoring system for evaluating the stress to cattle of commercial loading and unloading. *Veterinary Record* 154, 818–821.

Martoccia, L., Brambilla, G., Macrì, A., Moccia, G. and Cosentino, E. (1995) The effect of transport on some metabolic parameters and meat quality in pigs. *Meat Science* 40, 271–277.

Mason, G. and Mendl, M. (1993) Why is there no simple way of measuring animal welfare? *Animal Welfare* 2(4), 301–319.

Mayes, F.J. (1980) The incidence of bruising in broilers flocks. *British Poultry Science* 21, 505–509.

Mayes, H.F., Hahn, G.L., Becker, B.A., Anderson, M.E., Nienaber, J.A., Hedrick, H.B. and Jesse, G.W. (1998) A report on the effect of fasting and transportation on liveweight losses, carcass weight losses and heat production measures of slaughter hogs. *Applied Engineering in Agriculture* 4, 254–258.

McEwen, S.A. and Barbut, S. (1992) Survey of turkey downgrading at slaughter: carcass defects and associations with transport, toenail trimming, and type of bird. *Poultry Science* 71, 1107–1115.

McNally, P.W. and Warriss, P.D. (1996) Recent bruising in cattle at abattoirs. *Veterinary Record* 138, 126–128.

McNally, P.W. and Warriss, P.D. (1997) Prevalence of carcass bruising and stick-marking in cattle bought from different auction markets. *Veterinary Record* 140, 231–232.

McPhee, C.P. and Trout, G.R. (1995) The effects of selection for lean growth and the halothane allele on carcass and meat quality of pigs transported long and short distances to slaughter. *Livestock Production Science* 42, 55–62.

Mendl, M. (1991) Some problems with the concept of a cut-off point for determining when an animal's welfare is at risk. *Applied Animal Behaviour Science* 31, 139–146.

Mitchell, M.A. and Kettlewell, P.J. (1993) Catching and transport of broiler chickens. In: Savory, C.J. and Hughes, B.O. (eds) *Proceedings of the IVth European Symposium on Poultry Welfare.* Universities Federation for Animal Welfare, Wheathampstead, UK, pp. 219–229.

Mitchell, M.A. and Kettlewell, P.J. (1998) Physiological stress and welfare of broiler chickens in transit: solutions not problems. *Poultry Science* 77, 1803–1814.

Mitchell, G., Hattingh, J. and Ganhao, M. (1988) Stress in cattle assessed after handling, after transport and after slaughter. *Veterinary Record* 123, 201–205.

Mitchell, M.A., Kettlewell, P.J., Carlisle, A.J. and Matheu, C. (1998a) The use of apparent equivalent temperature (AET) to define the optimum thermal environment for broilers in transit. *Poultry Science* 75, 18.

Mitchell, M.A., Kettlewell, P.J., Hoxey, R.P. and Macleod, M.G. (1998b) Heat and moisture production of broilers during transportation: a whole vehicle direct calorimeter. *Poultry Science* 77, 4.

Mitchell, M.A., Kettlewell, P.J., Hunter, R.R. and Carlisle, A.J. (2001) Physiological stress response modelling-application to the broiler transport thermal environment. *Livestock Environment IV: Proceedings of the 6th International Symposium.* American Society of Agricultural and Biological Engineers (ASABE), St Joseph, Michigan.

Moberg, G.P. (1985) Biological response to stress: a key to assessment of animal well-being? In: Moberg, G.P. (ed.) *Animal Stress.* American Physiological Society, Bethesda, Maryland, pp. 27–29.

Moberg, G.P. and Mench, J.A. (2000) *The Biology of Animal Stress. Basic Principles and Implications for Animal Welfare.* CAB International, Wallingford, UK.

Mohan Raj, A.B., Moss, B.W., Rice, D.A., Kilpatrick, D.J., McCaughey, W.J. and McLauchlan, W. (1992) Effect of mixing male sex type of cattle on their meat quality and stress-related parameters. *Meat Science* 32, 367–386.

Morris, D.G. (1994) Literature review of welfare aspects and carcass quality effects in the transport of cattle, sheep and goats (LMAQ.011). Queensland Livestock and Meat Authority for the Meat Research Corporation.

Moss, B.W. (1980) The effect of mixing transport and duration of lairage on carcass characteristics of commercial bacon weight pigs. *Journal of the Science of Food and Agriculture* 31, 308–315.

Moss, B.W. and Rob, J.D. (1978) The effect of pre-slaughter lairage on serum thyroxine and cortisol levels at slaughter and meat quality of boars, hogs and gilts. *Journal of the Science of Food and Agriculture* 29, 689–696.

Mota-Rojas, D., Becerril, M., Lemus, C., Sánchez, P., González, M., Olmos, S.A., Ramírez, R. and Alonso-Spilsbury, M. (2006) Effects of mid-summer transport duration on pre- and post-slaughter performance and pork quality in Mexico. *Meat Science* 73, 404–412.

Moynagh, J. (2000) EU Regulations and consumer demand for animal welfare. *AgBioForum* 3, 107–114.

Nanni Costa, L., Lo Fiego, D.P., Dall'Olio, S., Davoli, R. and Russo, V. (1999) Influence of loading method and stocking density during transport on meat and dry-cured ham quality in pigs with different halothane genotypes. *Meat Science* 51, 391–399.

Nanni Costa, L., Lo Fiego, D.P., Dall'ollio, S. and Russo, V. (2002) Combined effect of pre-slaughter treatments and lairage time on carcass and meta quality in pigs of different halothane genotype. *Meat Science* 61, 41–47.

Nestorov, N., Tomov, T. and Krestev, A. (1970) A study on transport stress in cattle and conditions for its manifestation. *Proceedings of the 16th Meeting of the European Meat Research Workers* (Vol. 1 Paper A24) Sofia, Bulgaria.

Newberry, R.C., Webster, A.B., Lewis, N.J. and Van Arnam, C. (1999) Management of spent hens. *Journal of Applied Animal Welfare Science* 2, 13–29.

Nicol, C.J. and Scott, G.B. (1990) Preslaughter handling and transport of broiler chickens. *Applied Animal Behaviour Science* 28, 57–73.

OIE (2004) *The Proceedings of a Global Conference on Animal Welfare: An OIE Initiative.* World Organisation for Animal Health, Paris, France.

Owens, C.M. and Sams, A.R. (2000) The influence of transportation on turkey meat quality. *Poultry Science* 79, 1204–1207.

Owens, C.M., Matthews, N.S. and Sams, A.R. (2000) The use of halothane gas to identify turkeys prone to developing pale, exudative meat when transported before slaughter. *Poultry Science* 79, 789–795.

Pérez, M.P., Palacio, J., Santolaria, M.P., Aceña, M.C., Chacón, G., Gascón, M., Calvo, J.H., Zaragoza, P., Beltrán, J.A. and García-Belenguer, S. (2002) Effect of transport time on welfare and meat quality in pigs. *Meat Science* 61, 425–433.

Perry, G.C. (2004) *Welfare of the Laying Hen.* CAB International, Wallingford, UK.

Pettitt, R.G. (2001) Traceability in the food animal industry and the supermarket chains. *Revue Scientifique et Technique Office International des Epizooties* 20, 584–597.

Purchas, R.W. and Aungsupakorn, R. (1993) Further investigations into the relationship between ultimate pH and the tenderness for beef samples from bulls and steers. *Meat Science* 34, 163–178.

Quinn, A.D., Kettlewell, P.J., Mitchell, M.A. and Knowles, T.G. (1998) Air movement and the thermal microclimates observed in poultry lairages. *Poultry Science* 39, 469–476.

Randall, J.M. and Bradshaw, R.H. (1998) Vehicle motion and motion sickness in pigs. *Animal Science* 66, 239–245.

Rushen, J. (1991) Problems associated with the interpretation of physiological data in the assessment of animal welfare. *Applied Animal Behaviour Science* 28, 381–386.

Sanz, M.C., Verde, M.T., Sáez, T. and Sañudo, C. (1996) Effect of breed on the muscle glycogen content and dark cutting incidence in stressed young bulls. *Meat Science* 43, 37–42.

SCAHAW (Scientific committee on animal health and animal welfare) (2002) *The Welfare of Animals During Transport (Detail for Horses, Pigs, Sheep and Cattle)*. European Commission, Brussels, Belgium. Available at: http://ec.europa.eu/comm/food/fs/sc/scah/out71_en.pdf

Scanga, J., Belk, K., Tatum, J., Grandin, T. and Smith, G. (1998) Factors contributing to the incidence of dark cutting beef. *Journal of Animal Science* 76, 2040–2047.

Schaefer, A.L., Jones, S.D.M., Tong, A.K.W. and Vincent, B.C. (1988) The effect of fasting and transportation on beef cattle. 1. Acid base electrolyte balance and infrared heat loss of beef cattle. *Livestock Production Science* 20, 15–24.

Schaefer, A.L., Jones, S.D.M., Tong, A.K.W. and Young, B.A. (1990) Effects of transport and electrolyte supplementation on ion concentrations, carcass yield and quality in bulls. *Canadian Journal of Animal Science* 70, 107–119.

Schaefer, A.L., Jones, S.D.M. and Stanley, R.W. (1997) The use of electrolyte solutions for reducing transport stress. *Journal of Animal Science* 75, 258–265.

Schlosser, E. (2006) Cheap food nation. *Sierra Magazine*, November/December 2006. Available at: http://www.sierraclub.org/sierra/200611/cheapfood.asp

Scottish Executive (2006) The welfare of animals during transport: consultation on the implementation of EU regulation. 1/2005 May 2006. Available at: http://www.scotland.gov.uk/Publications/2006/05/25101804/18

Shackelford, S.D., Koohmaraie, M., Wheeler, T.L., Cundiff, L.V. and Dikeman, M.E. (1994) Effect of biological type of cattle on the incidence of the dark, firm, and dry condition in the *Longissimus* muscle. *Journal of Animal Science* 72, 337–343.

Shorthose, W.R., Harris, P.V. and Bouton, P.E. (1972) The effects on some properties of beef of resting and feeding cattle after a long journey to slaughter. *Proceedings of the Australian Society for Animal Production* 9, 387–391.

Smith, G.C. and Grandin, T. (1998) Animal handling for productivity, quality and profitability. pp. 1–12. Presented at the Annual Convention of the American Meat Institute, Philadelphia, Pennsylvania.

Smith, G.C. and Grandin, T. (1999) The relationship between good handling/stunning and meat quality. Presented at the American Meat Institute Foundation, Animal Handling and Stunning Conference, Kansas City, Missouri. pp. 1–22.

Smith, G.C., Morgan, J.B., Tatum, J.D., Kukay, C.C., Smith, M.T., Schnell, T.D. and Hilton, G.G. (1994) *National Non-fed Beef Quality Audit*. National Cattlemen's Beef Association, Centennial, Colorado.

Smith, G.C., Grandin, T., Friend, T.H., Lay, D. and Swanson, J.C. (2004) Effect of transport on meat quality and animal welfare of cattle, pigs, sheep, horses, deer, and poultry. Available at: http://www.grandin.com/behaviour/effect.of.transport.html

Special Eurobarometer (2005) Available at: ftp://ftp.cordis.lu/pub/innovation/docs/innovation_readiness_final_2005.pdf

Staples, G.E. and Haugse, C.N. (1974) Losses in young calves after transportation. *British Veterinary Journal* 130, 374–379.

Swarbrick, O. (1986) The welfare during transport of broilers, old hens and replacement pullets. In: Gibson, T.E. (ed.) *The Welfare of Animals in Transit*. British Veterinary Association, London, pp. 82–97.

Tarrant, P.V. (1980) An investigation of ultimate pH in the muscles of commercial beef. *Meat Science* 4, 287–297.

Tarrant, P.V. (1989) Animal behaviour and environment in the dark-cutting condition in beef: a review. *Irish Journal of Food Science and Technology* 13, 1–21.

Tarrant, P.V. (1990) Transportation of cattle by road. *Applied Animal Behaviour Science* 28, 153–170.

Tarrant, P.V. and Grandin, T. (1993) Cattle transport. In: Grandin, T. (ed.) *Livestock Handling and Transport*. CAB International, Wallingford, UK, pp. 109–126.

Tarrant, P.V. and Grandin, T. (2000) Cattle transport. In: Grandin, T. (ed.) *Livestock Handling and Transport*. CAB International, Wallingford, UK, pp. 151–173.

Tarrant, P.V., Kenny, F.J. and Harrington, D. (1988) The effect of stocking density during four hour transport to slaughter on behaviour, blood constituents and carcass bruising in Friesian steers. *Meat Science* 24, 209–222.

Tarrant, P.V., Kenny, F.J., Harrington, D. and Murphy, M. (1992) Long distance transportation of steers to slaughter: effect of stocking density on physiology, behaviour and carcass quality. *Livestock Production Science* 30, 223–238.

Taylor, M.H. and Helbacka, N.V.L. (1968) Field studies of bruised poultry. *Poultry Science* 47, 1166–1169.

Tegtmeier, E.M. and Duffy, M.D. (2004) External costs of agricultural production in the United States. *International Journal of Agricultural Sustainability* 2, 1–20.

Tyler, R., Taylor, D.J., Cheffins, R.C. and Rickard, M.W. (1982) Bruising and muscle pH. In Zebu crossbred and British breed cattle. *Veterinary Record* 110, 444–445.

Vallat, B. (2004) The OIE: historical and scientific background and prospects for the future. *Proceedings of the Global Conference on Animal Welfare: An OIE Inititative*. World Organisation for Animal Health, Paris, France, pp. 5–6.

Van Putten, G. (1982) Handling of slaughter pigs prior to loading and during loading on a lorry. In: Moss, P. (ed.) *Transport of Animals Intended for Breeding Production and Slaughter*. Martinus Nijhoff Publishers, Boston, Massachusetts.

Van de Water, G., Verjans, F. and Geers, R. (2003) The effect of short distance transport under commercial conditions on the physiology of slaughter calves; pH, colour profiles of veal. *Livestock Production Science* 82, 171–179.

Villarroel, M., María, G.A., Sierra, I., Sañudo, C., García-Belenguer, S. and Gebresenbet, G. (2001) Critical points in the transport of cattle to slaughter in Spain that may compromise the animals' welfare. *Veterinary Record* 149, 173–176.

Villarroel, M., María, G.A., Sañudo, C., Olleta, J.L. and Gebresenbet, G. (2003a) Effect of transport time on sensorial aspects of beef meat quality. *Meat Science* 63, 353–357.

Villarroel, M., María, G., Sañudo, C., García-Belenguer, S., Chacón, G. and Gebresenbet, G. (2003b) Effect of commercial transport in Spain on cattle welfare and meat quality. *Dtsch Tierärztl Wschr* 110, 105–107.

Warriss, P.D. (1990) The handling of cattle pre-slaughter and its effects on carcass and meat quality. *Applied Animal Behaviour Science* 28, 171–186.

Warriss, P.D. (1995) Pig handling: guidelines for the handling of pigs ante-mortem. *Meat Focus International* 4, 491–494.

Warriss, P.D. (1996) Guidelines for the handling of pig ante-mortem: interim conclusion from EC-AIR-Project CT920262. *Landbauhandforschung Volkenrode* S166, 217–225.

Warriss, P.D. (2003) Modern meat production and animal welfare. *Proceedings of 49th International Congress on Meat Science and Technology* (ICoMST). Brazil, pp. 39–40.

Warriss, P.D. (2004) The transport of animals: a long way to go. *Veterinary Journal* 168, 213–214.

Warriss, P.D. and Brown, S.N. (1983) The influence of pre-slaughter fasting on carcass and liver yield in pigs. *Livestock Production Science* 10, 273–282.

Warriss, P.D. and Brown, S.N. (1985) The physiological responses to fighting in pigs and the consequences for meat quality. *Journal of the Science of Food and Agriculture* 36, 87–92.

Warriss, P.D. and Brown, S.N. (1994) A survey of mortality in slaughter pigs during transport and lairage. *Veterinary Record* 134, 513–515.

Warriss, P.D., Kestin, S.C., Brown, S.N. and Bevis, E.A. (1988) Depletion of glycogen reverses in fasting broiler chickens. *British Poultry Science* 29, 149–154.

Warriss, P.D., Bevis, E.A. and Brown, S.N. (1990) Time spent by broiler chickens in transit to processing plants. *Veterinary Record* 127, 617–619.

Warriss, P.D., Bevis, E.A., Brown, S.N. and Edwards, J.E. (1992) Longer journeys to processing plants are associated with higher mortality in broiler chickens. *British Poultry Science* 33, 201–206.

Warriss, P.D., Kestin, S.C., Brown, S.N., Knowles, T.G., Wilkins, L.J., Edwards, J.E., Austin, S.D. and Nicol, C.J. (1993) The depletion of glycogen stores and indices of dehydration in transported broilers. *British Veterinary Journal* 149, 391–398.

Warriss, P.D., Brown, S.N., Knowles, T.G., Kestin, S.C., Edwards, J.E., Dolan, S.K. and Phillips, A.J. (1995) Effects on cattle of transport by road for up to 15 hours. *Veterinary Record* 136, 319–323.

Warriss, P.D., Brown, S.N., Knowles, T.G., Edwards, J.E., Kettlewell, P.J. and Guise, H.J. (1998a) The effect of stocking density in transit on the carcass quality and welfare of slaughter pigs: 2. Results from the analysis of blood and meat samples. *Meat Science* 50, 447–456.

Warriss, P.D., Brown, S.N., Barton-Gade, P., Santos, C., Nanni Costa, L., Lambooij, E. and Geers, R. (1998b) An analysis of data relating to pig carcass quality and indices of stress collected in the European Union. *Meat Science* 49, 137–144.

Warriss, P.D., Knowles, T.G., Brown, S.N., Edwards, J.E., Kettlewell, P.J., Mitchell, M.A. and Baxter, C.A. (1999) Effects of lairage time on body temperature and glycogen reserves of broiler chickens held in transport modules. *Veterinary Record* 145, 218–222.

Webb, J. (2003) How we produce a uniform high quality market pig. *Proceedings of the London Swine Conference – Maintaining Your Competitive Edge.* London, pp. 105–111.

Webster, A.J.F., Tuddenham, A., Saville, C.A. and Scott, G.B. (1993) Thermal stress on chickens in transit. *British Poultry Science* 34, 267–277.

Weeks, C. and Nicol, C. (2000) Poultry handling and transport. In: Grandin, T. (ed.) *Livestock Handling and Transport.* CAB International, Wallingford, UK, pp. 363–384.

Weeks, C.A. and Butterworth, A. (2004) *Measuring and Auditing Broiler Welfare.* CAB International, Wallingford, UK.

Weeks, C.A., McNally, P.W. and Warriss, P.D. (2002) Influence of the design of facilities at auction markets and the animal handling procedures on bruising in cattle. *Veterinary Record* 150, 743–748.

Wilson, J.G. and Brunson, C.C. (1968) The effects of handling and slaughter method on the incidence of haemorrhagic thighs in broilers. *Poultry Science* 47, 1315–1318.

Wythes, J.R. (1979) Effects of tipped horns on cattle bruising. *Veterinary Record* 104, 390–392.

Wythes, J.R. and Shorthose, W.R. (1984) *Marketing Cattle: Its Effect on Live Weight Carcass and Meat Quality.* Australian Meat Research Committee Review No. 46, AMRC, Sydney, Australia.

Wythes, J.R., Shorthose, W.R., Schmidt, P.J. and Davis, C.B. (1980) Effects of various rehydration procedures after a long journey on liveweight, carcasses and muscle properties of cattle. *Australian Journal of Agricultural Research* 31, 849–855.

Wythes, J.R., Arthur, R.J., Thompson, P.J.M., Williams, G.E. and Bond, J.H. (1981) Effect of transporting cows various distances on liveweight, carcase traits and muscle pH. *Australian Journal of Experimental Agriculture and Animal Husbandry* 21, 557–561.

5

Enforcement of Transport Regulations: the EU as Case Study

V.A. CUSSEN

World Society for the Protection of Animals, 89 Albert Embankment, London, UK

Abstract

The European Union (EU) has a more comprehensive legislation for animal welfare during transport than anywhere else in the world. However, legislation must be adequately enforced to ensure compliance and, thereby, satisfactory levels of animal welfare. Enforcement of EU legislation is resource-intensive, in both working hours and money, and the level of compliance varies greatly depending on both the investment in, and commitment to, effectively monitoring implementation of legislation. A large degree of variability in enforcement exists both within and between member states of the EU, leading to persistent areas of non-compliance with legislation and diminished animal welfare – especially for those animals transported on long-distance journeys. Standards of enforcement activity have not improved dramatically over the last 8 years, as assessed by Food and Veterinary Office missions to member states. Because of the difficulties in assuring the welfare of slaughter animals while in transit, it is more logical to slaughter them as close to the farm of origin as possible. This would also decrease resource pressure on already overextended authorities, and complement the current drive to refine the regulatory landscape in general.

Introduction

Many countries have, to varying degrees, legislation in place to govern conditions for the transportation of live animals. The EU has more comprehensive legislation for animal welfare during transport than anywhere else in the world. Therefore, evaluating the landscape of animal welfare during long-distance transportation in the EU is important for understanding how effective legislation can be in protecting live animals transported long distances. The conclusion is that legislation must be adequately monitored and enforced to ensure compliance and, thereby, satisfactory levels of animal welfare. Further, it is very difficult to provide adequate resources on the ground to achieve acceptable levels of enforcement and compliance.

Background

Enforcement of legislation regulating the welfare of transported animals is a complex and difficult undertaking, not the least component of which is remaining up to date on a complicated and evolving legislative framework. Until recently, animal welfare was regulated mainly by a directive (91/628/EEC) on the protection of animals during transport. That piece of legislation was amended in 1995 by a directive (95/29/EC) requiring reporting of enforcement activity, in 1997 by a regulation ([EC] No 1255/99) setting criteria for staging points, and again in 1998 by another regulation ([EC] No 411/98) that outlined additional requirements for journeys that exceeded 8 h. Some of these new laws were themselves (subsequently) amended by additional legislation. Overall, the EU body responsible for farm animal welfare (the Directorate for Consumer Health and Safety) lists ten pieces of 'main' legislation related to welfare during transport (European Commission, 2006a); however, that list grows if older pieces of legislation these works update or amend are considered. This web of legislation is positive, in that it represents a continuing process to update the framework in light of past experience, but it is negative in that it makes it very difficult, for both transporters and authorities, to be thoroughly familiar with all the relevant legislation. Partly because of problems with enforcement (see section on 'Enforcement – Current Situation and Future Prospects'), the European legislative framework for the protection of animals during transport has recently been changed and is now governed by Regulation (EC No 1/2005) on the protection of animals during transport and related operations, which entered into force in January 2007.

Regulations and directives differ in how they are implemented in the member states. Regulations, such as the new Transport Regulation, are directly applicable (European Commission, 2006b). That is they are transposed word for word into each member state's national regulations. If any of a member state's national legislation contradicts the regulation it is the national law that must be changed, not the regulation. For directives, the end result must be achieved but each member state can decide how best to accomplish the required result (European Commission, 2006b) – in theory, each country should achieve the same end result, but in reality there will be variability in the effectiveness of the different methods (Appleby, 2003). Until January 2007, animal welfare during transport was governed mainly by the Directive (91/628/EEC) mentioned above; as such, each member state had unique national implementing legislation. In the UK, for example, it was the Welfare of Animals (Transport) Order 1997 also referred to as WATO (DEFRA, 2006a).

This approach allowed for the spirit of the legislation to be achieved while accommodating the individual member state differences; however, it also gave rise to inadequacies in the implementation and enforcement of European legislation. In their animal welfare mission reports, the Food and Veterinary Office (FVO) of the European Commission has cited numerous occasions where legislation is improperly or incompletely transposed into national legislation resulting in poor animal welfare (see sections below). Partly because of the difficulties in uniformly enforcing the former directive, the new legislation which came into force on 5th January 2007 is a regulation (EC 1/2005) (European Commission, 2003). As such, it is intended to mitigate some animal suffering, because it must be written directly into each member state's national legislation. However, even with a regulation and uniform national legislation, vague terms or phrases within the legislation are open to different interpretations. It is the attitude of the competent authorities towards animal welfare that will ultimately determine the quality of enforcement in a given member state (DG SANCO, 2004a) (see section on 'Enforcement – Current Situation and Future Prospects' for further detail).

International agreements

In Europe, in addition to the EU legislation, there is a limited amount of international regulation for animal welfare during transportation. The EU is party to the Revised European Convention for the Protection of Animals During International Transport (Council of Europe, 2003), which was adopted by the Council of Europe on 11 June 2003 and opened for signatures on 6 November of the same year (Council of Europe, undated a). It entered into force on 14 March 2006 and is open to EU member states, as well as European and non-European non-member states (Council of Europe, undated b). As a convention, it is an 'expression of will' on the part of the signatories (Council of Europe, undated c). The original convention was proposed because it was recognized that: 'Considering that the humane treatment of animals . . . even in member states of the Council of Europe, the necessary standards are not always observed' (Council of Europe, undated d). Currently the convention has been signed and ratified by seven countries and signed, but not ratified, by a further nine countries (including the UK) (Council of Europe, undated e). Parties that have ratified the convention agree to be bound by its terms and disputes between parties are brought before a tribunal for resolution (Council of Europe, 2003). Multilateral consultations, at least every 5 years, are used to monitor implementation of the convention, and the convention can be revised based on experiences in adhering to it. Work is currently under way to draft specifications for transport vehicles and maximum journey time limits (European Commission, 2006c).

The World Organisation for Animal Health (OIE) is an intergovernmental organization that is recognized as the international reference for issues relating to animal disease (OIE, undated a). Since its inception in 1924 the OIE standards and guidance have focused on animal disease and zoonoses, and these standards are binding to members of the World Trade Organization. In 2002, the OIE established a permanent Working Group on Animal Welfare whose purpose is to provide the OIE with recommendations on animal welfare, particularly agricultural animals (OIE, undated b). At the 2005 General Session, the International Committee adopted welfare standards applicable during transport by road and by sea as part of the *Terrestrial Animal Health Code* (OIE, 2006). However, the animal welfare standards are non-binding.

Enforcement Infrastructure

The EU enforcement and auditing landscape is a complex of European government bodies and national member state government agencies and local authorities. Because animal welfare is a 'horizontal issue' (European Commission, 2006c) (i.e. different aspects may fall under the remit of several different bodies, including those responsible for agriculture, trade, the environment, etc.). This section is not exhaustive, but will provide a brief background on the structure and function of the main regulatory bodies as concerns welfare during transport.

European Commission role

Of the three central European institutions (the Council of Ministers, Commission and Parliament), it is the Commission that is mainly responsible for the initiation and enforcement of European legislation. If a member state fails to comply with EU law it is the Commission that issues instructions and/or refers the state to the European Court of Justice for further action (see section on the 'European Court of Justice'). The Commission is also responsible for drafting and submitting the EU budget (EUR-Lex, undated).

The Commission is composed of various Directorate Generals with specific areas of focus (European Commission, 2006d). Animal welfare falls under the remit of the Health and Consumer Protection Directorate General of the European Commission (DG SANCO), whose vision statement includes the aim to: 'Promote the humane treatment of animals' (European Commission, undated a). DG SANCO comprises six directorates, two of which have explicit welfare remits; the Animal Health and Welfare Directorate and the Food and Veterinary Office Directorate (European Commission, undated a). It is the FVO's mission to '. . . monitor, report on and assist in the enforcement of Community legislation on food safety, animal health, plant health and animal welfare systems' (FVO, 2000). It does this by sending inspectors on 'missions' to member states to perform inspections and audits of the domestic implementation of EU legislation, and to third countries (i.e. non-EU members) to check on compliance with Community law with particular reference to the export of animals, meat and meat products. While the FVO does not directly enforce the legislation, it does draft reports that include recommendations for the competent authorities of the member states. The inspection teams report to a standing committee within the FVO where mission reports are discussed, and the FVO has the authority to approve the action plans received from state governments in response to their mission reports (DG SANCO, 2004b). In theory, if the government fails to follow up on these mission report actions, they can be used in the Commission infringement proceedings (European Commission, undated b). However, in practice it can take years of violations before infringement proceedings are initiated (see section on 'Enforcement – Current Situation and Future Prospects').

European Court of Justice
If a member state fails to carry out its action plan to correct shortcomings found in the FVO reports, the Commission can instigate infringement proceedings for failure to fulfil obligations (e.g. not implementing or enforcing Directives and Regulations) before the European Court of Justice, described as 'the final arbiter of European law' (EUR-Lex, undated). For 2005, the most recent year with data available, the Court heard 179 cases for direct action; of these 95% (170 cases) were actions for failure to fulfil obligations. It is very difficult, however, to determine the nature of each case because, as their own report points out 'a judgment, classified under a

given subject, may broach issues of great interest in relation to another subject' (Court of Justice, 2006, p. 11). For example, cases with implications for animal welfare may be heard under the subject of 'environment and consumer' or 'agriculture' but some may also fall under the less-intuitive subject 'free movement of goods'. Even if the Court finds the member state to be at fault they are allowed a further period of time to correct the issue and the Commission must bring another follow-up action before the Court in order for the member state to be penalized (CURIA, undated). This procedure is quite time-consuming given the nature of the Court procedure and the number of cases before it at any time. In 2005, the average time for direct actions and appeals was just over 21 months and the Court had 474 new cases brought before it, in addition to the old cases still pending at the close of the previous year (Court of Justice, 2006) (see section on 'Enforcement – Current Situation and Future Prospects' for implications for animal welfare).

Member states

Competent authorities

At the member state level, the body responsible for monitoring and enforcing the transport regulations is called the 'competent authority'. This is a generic term as the structure and remit of the competent authority will vary from state to state; this structure can range from the fairly simple to the quite complex. For example, Austria is divided into six *Länder* (provinces) that are autonomous for animal welfare in general, but report to a central Ministry for issues of animal welfare during transport (DG SANCO, 2002a), while France is divided into 100 *departements*, each of which has a local veterinary service, which ultimately reports to the central competent authority in the Directorate for Food (DG SANCO, 2002b). In the UK, enforcement falls to the local authorities, with the Animal Health executive agency of the Department for Environment Food and Rural Affairs, as the central competent authority (DEFRA, 2006b, 2007).

Auditing

1997–1998 – Mission series

Over a year from 1997 to 1998, the FVO carried out a series of missions to inspect the level of compliance with the new Directive (95/29/EC) that amended the existing Directive (91/628/EEC) on the protection of animals during transport. Nine missions were carried out in eight member states (one was a follow-up mission). The results of these missions were summarized in a report to the Commission (European Commission, 2000) that used these missions and other information to evaluate the implementation of the relatively new transport directive from 1995; the following is derived from that communication.

The report summarized the findings in the various member states by comparing the ten most prevalent points of non-conformity; that is, areas where member states were in violation of the EU legislation. The eight member states visited ranged from being cited in 2–8 of the 10 summarized points of non-conformity, with a median citation level of six. Every member state visited had at least one area in which it breached EU legislation.

The infringements of community legislation identified by the Commission report had the potential to drastically reduce the welfare of transported animals. Six of the member states visited were cited for deficiencies in the vehicles used to transport live animals. The FVO found that vehicles were 'often identified as potentially dangerous to transported animals' (European Commission, 2000, p. 22). Five of the member states were cited for breaching legislation concerning the care and handling of animals. The FVO identifies failure to provide food and water for the animals as the most common infringement found in this mission series, but also notes that beatings and 'excessive' electric prod use were found to occur.

Furthermore, six member states were found to have shortcomings in the competent authorities themselves, the very bodies responsible for enforcing compliance with EU legislation. The identified problems ranged from a lack of resources to competent authorities which did not impose penalties when infringements of the transport Directive were identified, a deficiency which the report characterized as 'commonly mentioned' (European Commission, 2000, p. 22). Where these deficiencies had been noted on previous missions, the FVO found that little had been done to rectify the problems. These persisting deficiencies were the impetus for drafting the newly ascendant Regulation (1/2005) on the protection of animals during transport and related operations.

2003 – Mission series

In 2004, the FVO released another summary report (DG SANCO, 2004a) on a series of missions, carried out in nine member states during 2003 (5 years after the initial overview mission series) that evaluated animal welfare during transport, including assembly centres and at the time of slaughter. Six of the nine member states visited were follow-up missions to assess how well the states had implemented previous mission recommendations, missions also occurred in the intervening years between the two overview series (DG SANCO, 2004a). The impetus for the mission series was again to evaluate the implementation of Directive (91/682/EEC) on the protection of animals during transport, as well as Directive (93/119/EC) on the protection of animals at the time of slaughter and killing; the results as presented in the overview report are discussed below.

The summary report does not specify individual member states when listing infringements – the FVO says this is due to the general, widespread nature of the deficiencies across the member states visited. The main conclusions of the report included specific breaches of the regulations in the way animals were transported. The most common infringement found in the 2003 mission series was a lack of 'monitoring and enforcement' (Overview of a Series of Missions (DG SANCO, 2004a, p. 4)) of adherence to the legal limits for journey times. This is supported by investigations conducted by non-governmental organizations (NGOs) who monitor the conditions of animal transport (Eurogroup, 2003). Especially in the case of long-distance transport, this can result in major welfare problems (JRC, 2006a). It is to address this problem that article 6 (requiring the use of navigation systems) of the new Regulation (EC 1/2005) was introduced, but there are questions about the effectiveness of the proposed solution (see section on 'Enforcement – Current Situation and Future Prospects').

Other widespread infringements were identified. For example, at assembly centres animals often had no food or water – either because none was provided or because it was provided just prior to departure and the animals did not have sufficient time to access it. This was despite the fact that animals were held for over 20 h. Once on board the vehicles, stocking densities were such that animals could not access drinkers, and ventilation systems were inadequate. Furthermore, the vehicles were designed in such a way that inspectors did not have direct access to the animals on board, thereby reducing the possibility of identifying problems during transport. Taken together, the picture that emerges from the FVO mission series reports is of food and water-deprived animals loaded on to vehicles lacking proper ventilation, without access to water during journeys that probably often exceeded the legally stipulated maximum time.

The problem of access occurred despite the fact that, according to the report, properly designed vehicles have been available for years and such a design is required by the legislation. It is indicative of deficiencies in the enforcement of EU transport legislation at national and local levels; the report indicates that the competent authorities accept this breach of law because it is prevalent – as they also found with unacceptably dirty vehicles and unfit animals on a previous mission (DG SANCO, 2002b). In addition to a lack of enforcement stemming from overwhelming infringements, the report identified competent authorities that simply ignored aspects of the regulations with which they did not agree, and member states that failed to transcribe the EU legislation in its entirety.

As far as inspections themselves are concerned, the report highlights systematic deficiencies in the structure and function of various competent authorities. Article 8 of the transport directive at the time of this mission (Directive 91/628/ EEC) stipulated that inspections be carried out on 'an adequate sample of the animals transported each year' (EUR-Lex, 1991). However, it did not define a specific target sample size. The FVO missions found that authorities were not inspecting an adequate sample and attributed this deficiency to a lack of information regarding the number of animals transported, in addition to the lack of a benchmark for adequacy (see section on 'Enforcement – Current Situation and Future Prospects' for implications). For example, competent authorities in this report based their estimates on slaughter statistics and information from transport associations. The authorities themselves held no records on the number of animals transported in their area, making an accurate and adequate enforcement regime impossible. Even at a central member state level it is difficult for the authorities to obtain and monitor accurate statistics on the number of animals transported through the state. Furthermore, based on the above estimates, the authorities reported that, with current staffing levels, they could inspect only an estimated 2–4% of eligible shipments.

Of the inspections that were carried out, few were inspections in transit. For the most part, inspections occurred at the point of departure or at the destination – for example, previous missions showed that the majority of checks occurred at slaughterhouses or ports (DG SANCO, 2002b,c). For example, in one member state these checks comprised only a small proportion of all inspections for both years, 13% and 8%, respectively, while staging points represented less than 1% of checks for both years (DG SANCO, 2002b). Roadside checks are difficult to orchestrate, they require cooperation between competent authorities and the police, as they are necessary to stop vehicles on the motorway (DG SANCO, 2004a, p. 9). These detection rates are probably far below the actual infringement level for several reasons. For one thing, there are invariably no facilities for unloading and inspecting the animals at the inspection site. According to this report, ports are the most frequent sites of checks in transit, and in addition to a lack of facilities, other FVO missions have identified issues of cooperation with other port authorities that limit the usefulness of these inspections (see sections on 'Staging points and Border inspection posts'). As we have seen above, the stocking densities and vehicle designs are such that access to animals is limited while on the vehicle, so any physical check of the animals would be severely hampered in the absence of facilities for unloading them first (see Chapter 7, this volume). In spite of the many limitations, the report indicates that in-transit inspections do detect serious infringements, and are especially successful when police services are involved, as is the case with roadside checks. However, the competent authorities consider in-transit checks to be 'costly in terms of planning and resources' (DG SANCO, 2004a, p. 6). Given the necessity for coordinating roadside checks with other agencies such as the police, that have many responsibilities and resource pressures it is perhaps unrealistic to expect more checks in transit to take place. However, these resource constraints mean there is, overall, an inability to check the welfare of animals during transit, as opposed to at the destination when it is too late to offset or prevent animal suffering.

A lack of communication within and between member states and between member states and the central European administration further hampers enforcement (again, this was found in previous missions, for example, a lack of formal reporting between veterinary inspectors and the head of veterinary services (DG SANCO, 2001, 2002a)). This FVO review cites inspection reporting as ineffective both because inspections are not organized against a preset target (as discussed above) and because limited feedback is received from the central level on the inspections that have occurred. It further says that this causes disenfranchisement at the competent authority level because inspections come to be viewed as 'a bureaucratic exercise' (DG SANCO, 2004a, p. 6), leading to an under-reporting of identified deficiencies. A similar disenfranchisement also hampers enforcement on international shipments of animals. If the competent authority in a particular member state discovers an infringement, the competent authority at the site of destination may not be informed, due to poor reporting. If they are made aware, they often fail to update the reporting authority on action taken (if any) against the infringer. This lack of communication at the various levels is a disincentive for competent authorities to report infringements (found too, in previous missions (DG SANCO, 2002c)). According to this report, even serious infringements are merely cautioned because of the resources, both time and staff, necessary to prepare sanction documents – for example, the report states penalties were 'rarely imposed' for exceeding journey time limits. This reluctance to pursue sanctions may also stem from the lack of success in such endeavours. Previous mission reports have highlighted the difficulty in successfully prosecuting infringers, leading the FVO to characterize the penalty process as too slow to act as an effective deterrent (DG SANCO, 2001).

The FVO concludes by suggesting that the (then) forthcoming Regulation (EC 1/2005) will address many of the issues cited above. However, as their own reports demonstrate, good legislation does not equate with enforcement or compliance and it is, therefore, too early to assess the impact of the new legislation (see section on 'Enforcement – Current Situation and Future Prospects' for implications).

1999–2001 – Staging point mission series

Staging points, defined as areas where the required rest periods for animals transported long distances occur, were evaluated by the FVO during a series of missions carried out in seven member states between October 1999 and February 2001 (DG SANCO, 2004c). The requirement for staging points had entered into force in January 1999 (Council Regulation (EC) No. 1255/97 (EUR-Lex, 1997)) and the missions were conducted to assess implementation of the regulation requirement.

Staging points were required so as to improve the welfare of animals being transported long distances by affording them a period of rest and an opportunity for rehydrating themselves during their journey (EUR-Lex, 1997). As such, and in light of previous discussion, these staging points offer an opportunity to evaluate the fitness of transported animals mid-journey because they are offloaded from the vehicle during the prescribed rest period. It is, therefore, worrisome that the FVO was unable actually to audit the animal inspections that occur at staging points at 'many' of the staging points visited because no animals were present (this would also be the case as regards ports visited in 2002 (DG SANCO, 2002c)). Because of this, the mission predominantly inspected facilities and procedures rather than evaluating the competent authority performance on physical inspections of animals (something that would be repeatedly found lacking, see section on 'Enforcement – Current Situation and Future Prospects' for implications).

The staging point procedures were decided by each member state central competent authority; the quality of these instructions was found to range greatly between member states, from 'detailed guidance' to little or no instruction for veterinarians working at the staging points. A high degree of variability was also found in the level of control exerted by the veterinarian in charge at the staging point, and this variability existed between staging points within a single member state as well as between different states. Furthermore, this had serious implications for animal welfare because staging point veterinarians with little control did not enforce provisions 'with serious consequences for animal health and welfare' (DG SANCO, 2004c, p. 8). Record keeping was also found to be inconsistent across staging points visited, with records characterized as 'inadequate' at several staging points in the majority of member states visited such that traceability of animals that travelled through them was diminished (DG SANCO, 2004c, p. 7). Poor record keeping may encourage transporters to breach the staging point requirement; at one facility visited the FVO found that 40% of the vehicles they observed failed to stop at the staging point, despite route plans that indicated they would.

This would be echoed in later missions where port veterinarians reported sending suspicious consignments to a staging point facility for closer inspection and the FVO later finding no records of those consignments arriving at the staging point – with a result that the port veterinarian of referral receives no feedback and the central competent authority cannot properly audit the outcomes of inspections (or even be certain the inspections actually occurred) (DG SANCO, 2002c). This report recommended better communication regarding adherence to route plans, despite 'mutual assistance' already being required by the Directive (91/628/EEC) (EUR-Lex, 1991) and the FVO further recommended the directive be amended to specify the requirements more exactly.

Facilities at inspected staging points were found consistently lacking. For example, many competent authorities had approved facilities for ill and injured animals that the FVO team found to be inadequate. Furthermore, many of the staging points visited lacked such basic features as ramps for unloading animals, or had ramps characterized as potentially dangerous to the animals – a major short-coming given the importance of loading/unloading animals in the transport process (see Chapter 7, this volume). Multiple competent authorities responded that, as trucks are equipped with ramps, it was unnecessary for them to provide them. However, this is another example of authority's attitudes impacting animal welfare, as this is not a point legally open to their interpretation (see section on 'Enforcement – Current Situation and Future Prospects' for implications). Other shortcomings had more immediate welfare impacts. For example, one staging point was found to have empty water troughs in a corral where horses had been rested for 24 h. When this was noted in the FVO report, the authority said they would carry out 'assessments' for how best to provide water, but did not mention any immediate action to prevent dehydration of transported horses in the interim.

In their report conclusions, the FVO notes the major animal welfare problems observed fell into the categories of provision of care, records, structure and animal health; it could be argued that the staging points are failing to fulfil their purpose of improving animal welfare during transport. The member states responded to the FVO and gave written assurance that the deficiencies would be addressed. However, a later mission identified staging point inspectors not using a checklist and performing 'superficial' animal welfare inspections (DG SANCO, 2002c, 2004b, p. 9), so these assurances are not a guarantee of future improvements (see section on 'Enforcement – Current Situation and Future Prospects' for implications).

2004 – Border inspection posts mission series

Border inspection posts (BIPs) were established to carry out veterinary checks on shipments from third countries (including live animal shipments) as they enter the EU. The Commission approves inspection points, but the member states are responsible for implementing the proper procedures, with the FVO responsible for assessing the performance of individual states. In 2004, the FVO published a review report on a series of missions carried out in 13 member states during 2002–2003 (DG SANCO, 2004b). The purpose of the review report was to give a general overview of the situation in the member states as a whole. As they note at the beginning of the review, non-compliance at BIPs can 'seriously undermine the effectiveness of the control system' (DG SANCO, 2004b, p. 1). As with other control points, such as staging points, procedures and facilities were found to be lacking at the BIPs visited by the FVO.

During the mission series a total of 47 active inspection points (previously approved by the Commission) and nine proposed BIPs (not yet approved by the Commission) were visited. Deficiencies were identified in all member states visited, including shortcomings in animal inspections. Training for veterinarians and other staff was characterized as non-comprehensive in most facilities visited. Non-compliance in staff training was identified at 96% of active BIPs visited, with 46% found to have inadequate staff. Major or minor veterinary procedure non-compliances were identified at 75% of active BIPs inspected; the review cited a lack of supervision and a failure to follow up on problems identified during inspections as the two main deficiencies in veterinary procedure. The review further cites lack of resources, training and information as the cause of some veterinary inspections not adhering to EU legislation. These problems with inspections were further compounded by some BIPs lacking facilities to offload the animals. An absence of offloading facilities was also cited by port veterinarians in another FVO mission (DG SANCO, 2002c) as a major impediment to carrying out comprehensive inspections of the animals during transport. The review found this problem made proper inspection of animals impossible at some of the BIPs visited. Furthermore, the problem with facilities at certain inspection points had been previously identified by the FVO but continued to persist, even though the member state had submitted an accepted action plan for meeting prior mission recommendations. For example, in one instance a border inspection point was submitted by the member state for re-approval without any of the FVO-identified deficiencies being rectified as promised in the action plan. These persistent deficiencies were found even in the case of major non-compliances – defined as those that can have a clear impact on the efficiency of the control system.

In addition to the problems with facilities and inspections, communication and record keeping were problematic at BIPs (as at other control points – see section on '1999–2001 staging point mission series'). Although communication was described as 'adequate' between competent authorities, the communication between the competent authority for animal welfare inspections and other authorities at the BIPs (e.g. customs officials) was limited. For example, the veterinarians in most of the member states visited were unaware of what consignments were arriving or of the consignments' contents; in some cases notification was not received until after the consignment arrived. Other BIPs lacked written records of consignments altogether. Often, it was customs officials, who decided which consignments to inspect, with the veterinarians limited to their selection. And, while veterinary authorities should inspect all consignments of live animals, the report says the customs authorities systems did not always identify consignments 'of veterinary interest' (DG SANCO, 2004b, p. 14). Perhaps, because of a lack of information, incomplete consignment information was uploaded by inspection point authorities; previous FVO missions had identified this shortcoming but it persisted. Furthermore, information reported to the Commission by the member state central competent authority was often found to conflict with the statistics held by the inspection points themselves – with some BIPs failing to report to a central authority. This situation makes it difficult for the central authorities of member states to ensure adequate checks are being performed, and the FVO report notes that for some states it is impossible to be confident that checks are being performed in compliance with the legislation.

The FVO mission series review notes that authorities in the member states were generally supportive, open and constructive regarding the FVO inspectors' findings. In some cases, FVO inspections and recommendations did result in an improvement in member states' facilities. However, as stated in the staging point report (DG SANCO, 2004c), uptake of suggestions is indicative of lack of proactive monitoring on the part of the state authorities, especially in the face of persistent deficiencies in almost all member states visited and a failure of several member states to fully address the findings of this mission series. Despite the persistence of previously identified shortcomings (e.g. transport conditions from third countries and inadequate inspection point facilities (European Commission, 2000)), the FVO cites the difficulty in inspecting all facilities each year and recommends the Commission shift to a risk-based approach in the future.

Member state reports – 1997–1998

Article 8 of the transport Directive (91/628/EEC) required each member state to submit an annual report on enforcement activity (EUR-Lex, 1991). These reports were summarized as part of the Report to the Council (European Commission, 2000) discussed above. The member state reports should address shortcomings in enforcement of animal welfare during transport, by enabling better auditing of member state activity at the central level. However, despite the legal requirement (which came into force on 31 December, 1996) to submit a summary of enforcement activity each year, as of 2000, the FVO found member state reports of 'limited' use. The problems predominantly arose for two reasons. First, member states did not submit reports as legally required. Of the 15 member states, only 12 reports were received each year and these were not submitted by the same member state each year; only six of the 15 member states submitted a report for both years. This lack of reporting occurred, according to the report, despite repeated reminders to member states from the Commission. Of the submitted member state reports, the Commission reported many were submitted late or with incomplete information. It could be argued that the above was due to the recent implementation of the requirement; however, as late as 2002, 6 years after the reporting requirement came into force, the Commission was still noting the difficulty it experienced in obtaining member state reports (DG SANCO, 2002c).

Because individual member states rarely submit reports for consecutive years the Commission (and, subsequently, the Council and the Parliament) is unable to accurately audit enforcement activities between years to see if improvements occur. It is possibly because of this that the Commission report summarizes data by year for all reports (unlike the FVO reports discussed above, which are summarized by individual member state), therefore, we cannot analyse the six consecutive member state reports the Commission did receive.

With that caveat in mind, it is still possible to assess the general enforcement picture across the 2 years evaluated. As with the FVO mission reports, the Communication summarized the member reports on ten categories of non-conformity. Between the 2 years, the number of citations per category remained the same or increased across nine of the ten categories. Only one category decreased in the number of points cited (loading and unloading), which fell from 2 to 1 point cited. Fitness of animals, loading densities and ventilation all remained at the same level of citation for the 2 years (despite the FVO position that fitness to travel is the 'most critical factor' concerning welfare in transport (DG SANCO, 2004a)). Vehicle design and maintenance increased from three to six points of infringement cited, a 100% increase, while infringements of duration and transport and rest increased from one point of non-conformity cited to three. The points of non-conformity for care and handling of animals nearly doubled, from five citations to eight.

Again, the nature of the member state data as presented in the Commission summary is such that interpretations are difficult. For instance, the Commission states that the reports should include the number of inspections carried out, any infringements noted and what was done as a result. The Commission summary table merely lists categories and 'points of non-conformity'. It fails to describe what a point represents, that is if a point of non-conformity equates to the findings of an inspection, of a single infringement, or if each point of non-conformity references the number of member state reports they are listed in, regardless of frequency of infringement? Further, we do not know from the presented summary the sample size of live animal consignments evaluated. Therefore, it is difficult to determine if the state reports provide an accurate representation of the welfare of animals during transportation in a given member state; however, the noted discrepancy between the FVO's own mission findings and the enforcement reports submitted by member states makes it unlikely. Finally, the increased points of citation could represent an improvement in surveillance and detection of welfare infringements; again, FVO missions over the course of these years tend not to support this interpretation because, despite some general improvements, they identify persistent problems with enforcement infrastructure in member states (European Commission, 2000). Because of difficulties with record keeping and central reporting at the member state level, and inconsistent report submission from the states to the Commission, enforcement reporting is (while potentially quite valuable) in reality of limited use.

Enforcement – Current Situation and Future Prospects

FVO reports are used when developing government policy on animal welfare (DG SANCO, 2004b) and communicating with the public and the European Parliament and Council (European Commission, 2000; EUR-Lex, 2006a). As such, they are an important source of the official assessment of the EU enforcement landscape. As summaries, the FVO mission overview reports produced by the Commission are not as extensive in detail as the individual mission reports and fail to document the welfare insults endured by animals during transport, such as those uncovered by NGO investigations (Eurogroup, 2003). Despite this, the FVO reports and Commission communications reviewed above have exposed problems with implementation (including transposition of legislation), interpretation, inspections and auditing and penalty systems in various member states.

Infrastructure

European Commission

Ensuring animal welfare during transport requires complex infrastructure and resources at all levels from the administration of the EU to the local authorities within member states. At the European level, in addition to drafting legislation, this includes the necessary infrastructure for auditing the implementation of legislation in member states (as discussed above). In the most recent EU draft budget, under the heading Health and Consumer Protection, €15 million were appropriated for 2007 under the heading that includes providing assistance 'for the monitoring of compliance with animal-protection provisions during the transport of animals for slaughter' (European Commission, 2006e). This sum is proportionately quite small; for example, the total sum budgeted for DG SANCO was €544 million, with €209 million appropriated for animal disease eradication and monitoring and over €97 million appropriated for DG SANCO's administrative costs (EUR-Lex, 2006a). Nevertheless, a substantial sum of money is already spent on enforcement of animal welfare standards and in addition to the money appropriated for DG SANCO, the true cost of auditing and enforcement at the EU level would include a portion of the Court of Justice budget – over €282 million (EUR-Lex, 2006b) – and TAIEX (Technical Assistance Information Exchange Office) budget – over €11 million appropriated for enlargement of 'Information and Communication' (EUR-Lex, 2006c) – in addition to research budgets and other horizontally related bodies.

According to their annual report, in 2004 the FVO conducted 232 inspections (FVO, 2005a). Of those, 6% evaluated animal welfare generally and 2% focused on animal welfare during transport (although no mission title for 2004 reflects a specific transport remit so it may be incorporated within a larger mission). The projected programme of inspections for 2005–2006 foresaw animal welfare accounting for 5% of inspections conducted in 2005, with missions split between multiple categories of which welfare during transport and slaughter was only one (FVO, 2004a). Likewise, in 2006 animal welfare during transport was projected to account for a mere 3% of total FVO missions (FVO, 2005b). According to their own web site, FVO inspection topics are decided by perceived risk and 'key policy issues' (FVO, 2001, p. 4); the overall number of missions carried out is no doubt limited by resources. For example, the FVO is unable to meet its legally required inspection levels for BIPs (DG SANCO, 2004b), despite the Commission's view of the importance of BIPs as a control point (European Commission, 2000), 'due to other priorities and resource constraints' (FVO, 2004b, p. 7). In 2000, the FVO spent 1524 person-days training staff for the 92 inspectors on staff in the directorate. New inspectors have to be trained on procedures and legislation as well as undergo external auditing training courses and on-the-job training (FVO, 2001). Animal welfare is a small component of the overall mission of the FVO and so would represent a small component of overall training for new inspectors – especially in an enforcement landscape that is increasingly risk-based. In addition to training its own staff the Commission, via TAIEX, offers training for new and ascending member state official authorities (European Commission, undated c). Current resource constraints on the FVO, together with increasing pressures from enlargement of the EU and the resource-intensive nature of increasing inspectorate staff, are likely to exacerbate the short-comings of animal welfare auditing and enforcement in the future.

Member states

Given resource shortfalls at EU level, it is, perhaps, unsurprising that at the member state level competent authorities were described as lacking training, and in some cases staff, in most of the member states visited by the FVO. The inspectors ascribed this to a lack of sufficient resources from the central member state government level. Where inadequately resourced, there was an inability to apply controls in compliance with EU legislation (DG SANCO, 2004b) leading to multiple infringements, including the transportation of unfit animals (DG SANCO, 2002b). Over one mission series, the FVO found six of eight member states visited had deficiencies in their competent authorities (European Commission, 2000). Furthermore, a lack of central support leads to inadequate inspections and persistent deficiencies in monitoring EU legislation. Just as the FVO cannot meet its border inspection point auditing requirements, competent authorities, too, can inspect only a small fraction of facilities (DG SANCO, 2004a). For example, multiple FVO missions have found inadequate training, instruction or other guidance from a central level for inspectors on the ground (DG SANCO, 2004a) and the Commission has stated that member states have difficulty in implementing and enforcing EU transport legislation (European Commission, 2000; FVO, 2001). This difficulty may be a result of the low priority placed on enforcement by member states, which is, according to that Commission communication, the most commonly cited complaint of both the FVO and NGO's reports. The fragmented responsibility for animal welfare across several departments may also play a role in the inadequate resources devoted to enforcement. In Austria, for example, animal welfare during transport falls to the Federal Ministry for Traffic, Innovation and Technology, whose staff reported dedicating only 20% of their time to animal welfare, and which has no veterinary expertise. However, in the same country, certification of animals for export and inspection of BIPs falls under the Ministry for Social Security and Generations. There is little communication between the agencies (DG SANCO, 2002a). Even in the UK, where animal welfare standards are cited as best practice (DEFRA, 2006c), the central competent authority has acknowledged the problems associated with multiple agencies responsible for implementing different aspects of the animal welfare measures (DEFRA, 2006b). Furthermore, according to the UK State Veterinary Service (now Animal Health), 5.5% of their budget in 2006–2007 is allocated to animal welfare (SVS, 2006), while the report of the Chief Veterinary Officer indicated only 2% would be spent on animal welfare during transport (DEFRA, 2006d).

Auditing

European Commission

The FVO missions are necessary tools to evaluate implementation because of the variability in enforcement that exists between member states. This was highlighted in the Commission report of 2000 (European Commission, 2000). Many of the inconsistencies in enforcement stem from differences in interpretation of the EU legislation (e.g. what constitutes a rest period (DG SANCO, 2002d)). Interpretation differences range from inspectors approving unfit animals for transportation because they feel they are fit for transport, to approving unfit animals for transport because they feel it is acceptable under the legislation (such as one state response that the transport of ill or injured animals is compatible with the law (DG SANCO, 2004c)). Furthermore, member states have failed to accurately assess the level of infringements that occur in their country (European Commission, 2000). For example, member state reports indicated that route plans and journey times were not problem areas but the FVO found these among the most commonly occurring infringements. The FVO also found them to be commonly overlooked and, therefore, under-reported by member state competent authorities.

Member states

Many member states do place importance on animal welfare, but this in itself does not ensure adequate welfare is enforced. In the executive summary of their review report the FVO states: '[W]here the CA does not set targets, analyse the results of inspections or establish a clear enforcement policy there are persistent areas of non-compliance. Even where animal welfare is given a higher priority, weaknesses in the operation of controls arise.' (DG SANCO, 2004a). The FVO reports discussed above have shown that inspection targets are not set or adhered to in many member states, likely rendering non-compliances. Over a 3-year period (2002–2004) in the UK, the average annual inspection rate was just 0.59% of the more than 1 billion total animals transported on average each year (unpublished annual enforcement reports for years 2000, 2001, 2002, available from DEFRA). However, FVO missions have found that even member states which have a benchmark for inspections experience inconsistent implementation and auditing of inspections rendering them less effective. Essentially, they have an arbitrarily set number of inspections to conduct, but do not conduct them thoroughly enough, or there is no follow-up on infringements detected during the inspections that are carried out. According to the Commission, it is the attitude of competent authority inspectors and others on the ground that determines whether some deficiencies are corrected (European Commission, 2000).

Sanctions and infringement proceedings

When the FVO audited member state competent authorities and found 75% with deficiencies (see above), these shortcomings were described as including 'insufficient involvement' and a common 'unwillingness to penalize breaches' (European Commission, 2000, p. 22). Multiple individual mission reports mention a failure of the competent authorities to impose penalties on offenders. On one mission, the FVO team identified approximately 40 pigs in a casualty pen that had arrived in an unfit condition in a single slaughterhouse on a single day. Yet, the report notes that in the first 3 months of the year (2002) only 66 written warnings were issued regarding unfit animals transported to slaughter. Unsurprisingly, the report suggests inspectors were failing to report all unfit animals (DG SANCO, 2002b). In the UK, over the 3-year period of 2002–2004, inspections resulted in an average of 4881 enforcement actions being taken annually. Of these, an average of 70% of actions were oral warnings, with only 1.5% being prosecutions and less than 1% resulting in a conviction (an average of 34 convictions of 4881 actions annually) (unpublished annual enforcement reports for years 2000, 2001, 2002, available from DEFRA). In a recent consultation commissioned by the UK government, two of the six listed principles of effective penalties were 'Aim to eliminate any financial gain or benefit from a non-compliance' and 'Aim to deter future non-compliance' (McRory, 2006). It is unlikely that oral or written warnings, which make up the bulk of enforcement action, fulfil either of these criteria for an effective penalty.

The matter is further complicated by the international nature of much long-distance transport. Communication between competent authorities in different member states is frequently cited as deficient with the result that even when authorities in the member state of transit or destination notify the competent authority in the member state of origin (where the transporter is registered and can be sanctioned), no action is ultimately taken. This could be due to a number of factors, from lack of resources to follow up on reports, to a lack of evidence for a successful prosecution. Moreover, when penalties are applied, they are of limited use. One authority told the mission team that they had never received payment of fines by foreign transporters and that member state's Ministry of Justice informed them there was no way to force payment (DG SANCO, 2002c). Further, member state penalties are subject to appeal, one example was a state that revoked a transporter's approval, after repeated infringements, only to have a court rule this punishment to be excessive (DG SANCO, 2004a).

When enforcement of animal welfare legislation is relegated to a low priority (as reported by the FVO and NGOs), it is the responsibility of the member state to correct any shortcomings based on the FVO recommendations, but problems are not always rectified. One mission concluded: 'Limited progress had been made' and 'An effective level of enforcement of the provisions of Council Directive 91/628/EEC (as amended by 95/29/EC) had not yet been achieved' (DG SANCO, 2002c, p. 5). The central competent authority had sought to address the previous missions' identified shortcomings by circulating information circulars, which the subsequent mission described as lacking in information that would enable appropriate inspections on the ground. It is usual for a time lag to exist between an FVO mission and the correction of noted deficiencies (FVO, 2004). However, multiple reports have described limited or no progress on deficiencies identified in previous missions (DG SANCO, 2002a,b,c, 2004a; FVO, 2005a) including instances where the member state government had submitted 'guarantees' that the deficiencies had been fixed (DG SANCO, 2004a). It is perhaps unsurprising, then, that deficiencies and recommendations identified and suggested in both the 2000 and 2003 FVO overview reports are strikingly similar, notwithstanding the assertion that implementation was improving (European Commission, 2000; DG SANCO, 2004a). Even in these cases, where deficiencies persist in the face of FVO recommendations and member state action plans, infringement proceedings are rarely initiated. In the rare case that they are, it is only after years of persistent infringement – in the most recent case the FVO '. . . between 1998 and 2006 reported consistent and serious shortcomings' (European Commission, 2007). That represents 8 years of serious animal welfare infringements before the member state was even referred to the Court, given the average Court proceeding time those infringements will have persisted for over a decade. The Commission's reluctance to bring infringement proceedings against member states may be explained by the lengthy nature of the proceedings, as discussed in the section 'Introduction'. If heard consecutively, it would take approximately 829 years to resolve only the new cases introduced before the Court of Justice in 2005, based on the number of cases and an average proceeding time of 21 months (estimate from Court of Justice, 2006). While there are provisions for expediting Court procedures, an application must by made and exceptional urgency demonstrated (CURIA, 2005); all such applications for expedited proceedings were denied in the six cases (subject matter not specified) in which they were lodged in 2005 (Court of Justice, 2006). The Court procedure is so resource-intensive and so drawn out that it is difficult for the Commission to use often and of little value in dissuading a persistently infringing member state to improve their animal welfare oversight.

Economics

Economics is another driver in the inconsistent enforcement of legislation. Despite being identified by the Commission as, 'the most critical factor to protect against major abuses of animal welfare' it was found that 'where economic considerations take precedence, unfit animals continue to be transported' (DG SANCO, 2004a p. 5). These 'economic considerations' include transporting unfit animals to a destination where a higher price will be received for them (DG SANCO, 2004a), or transporting via member states with lax enforcement or penalties to avoid the costs associated with higher standards (DG SANCO, 2004b). For example, a staging point in Italy was allowed to remodel gradually to adhere to the regulations because 'Otherwise, it was stated it would not be financially viable to set up a staging point' (DG SANCO, 2004c, p. 5). Furthermore, it has been repeatedly noted by the Commission that the variability in the implementation and enforcement of transportation regulations could lead to trade distortions whereby hauliers avoid member states with strict enforcement procedures (DG SANCO, 2004a,b), effectively punishing those member states that adhere to the EU legislation because, 'efforts to improve animal welfare can impose extra costs on either the keepers of such animals or administrations responsible for enforcing such controls' (European Commission, 2006c). This has been anecdotally reported as already occurring with transporters driving around certain states with stringent controls.

New EU Transport Regulation (1/2005)

It was to address many of the persistent shortcomings associated with the transport directives that the Commission and Council drafted and adopted the new Regulation (1/2005) on the welfare of animals during transport (EUR-Lex, 2005). The new regulation seeks to improve the harmonization of implementation and enforcement in several ways, one of which is by virtue of its being a regulation

(European Commission, 2003). Although the Commission sees the new regulation, 'radically improving the enforcement of animal transport rules in Europe' (European Commission, 2004), it is unlikely the new regulation will be an enforcement panacea for several reasons.

When first introduced, the old transport directive was seen as affording 'significant innovations in the way the welfare of animals in transport is controlled' (European Commission, 2000, p. 6), yet the Commission's own reports have shown these promised innovations did not translate into harmonious and effective implementation across member states. The same can be anticipated for the new transport regulation (1/2005) that has replaced it (and other existing regulations are also noted as being inconsistently interpreted and enforced (DG SANCO, 2004c)). Additionally, as discussed above the attitudes of the various competent authorities – and the resources they have or are willing to commit to animal welfare – play a large role in how effectively compliance with animal welfare legislation is monitored and enforced (European Commission, 2000) – and this will remain a major factor even with the new regulation.

An inability to monitor or enforce maximum journey time adherence was cited as the 'most common problem' in the most recent FVO mission series overview report, and the Joint Research Council noted that (as discussed above) animal welfare during transport could, for the most part, only be checked after the fact (JRC, 2006a). To address this problem, the new regulation mandates an on-board satellite navigation system – identified as the key component of the new Regulation that would increase effective enforcement. The system would record certain data in real time, and could transmit that data to a computer where it would be stored. And though the transmission of data holds promise for animal welfare improvements (e.g. by alerting the driver when thermal conditions in the vehicle exceed acceptable limits (JRC, 2006b)), for several technical reasons it is unlikely that the requirement will address the most common problems, including enforcement of maximum journey times and observance of rest periods. As of this writing, the technical specification on the satellite navigation system is still in draft form and has yet to be finalized (JRC, 2006a) despite the implementation date of January 2007 for new vehicles. Road testing of navigation systems has yet to be completed.

Furthermore, there is as yet no central database for collection of the transmitted data (JRC, 2006b) (even when central databases are legislated, equipment to allow proper reporting is sometimes lacking (DG SANCO, 2004b)). The current proposal is for transporters to collect and retain their own data, for submission to competent authorities upon request. However, this arrangement would not allow authorities' 'verifying at any given moment during the journey that animal requirements are respected' (JRC, 2006a). Rather, the authorities will have to request the records from the individual transport companies in order to audit adherence to the regulation. Therefore, accurate real-time data collection and analysis by enforcement authorities, a significant tool in auditing compliance and enforcement activity, though mandated by the regulation will not functionally exist. Furthermore, critical information such as journey start and end times and rest periods will be entered manually by the driver, not recorded automatically by the vehicles on-board unit (JRC, 2006b), so there will be opportunity for improperly completing this information (an identified problem with route plans (DG SANCO, 2004a)).

Other problems associated with long-distance transport of animals failed to be addressed by the new regulation. Overheating of animals is noted as the primary cause of suffering in long-distance transport in the EU (European Commission, 2000). Stocking density and journey length were too contentious to be addressed in the new regulation so they will remain as they are with the only change being a temperature monitor in the animal compartment and some ventilation requirements. Furthermore, many of the projected improvements are based on an assumption that vehicles are appropriately designed, and the above reports have highlighted that this is frequently not the case (European Commission, 2000).

Because of these shortcomings, Commissioner Kyprianou was already speaking of replacing the new regulation less than a year after it was signed into law (DG SANCO, 2005).

Conclusion

Even at the present EU member size the FVO is unable to meet its legally required inspection and auditing capacity because of resource constraints (DG SANCO, 2004b). Likewise, TAIEX, the office responsible for training new and ascending member states (and veterinary officials in third countries), is unlikely to be able to adequately address animal welfare, despite the importance of training for proper implementation at BIPs (European Commission, 2000). None of their reported activities in 2005 addressed animal welfare (DG Enlargement, 2005). Furthermore, because of the burden on businesses, both European (European Commission, 2005) and national governments (e.g. the UK Hampton (2005) review) are currently focusing on refining the regulatory environment and reducing the costs associated with regulation. For example, the Commission has pledged 'to ensure that the substance of its legislative proposals is restricted to the bare essentials' (European Commission, 2002), while in the UK DEFRA has targeted to reduce staff by 2400 employees (UK Parliament, 2006).

There is, already, an insufficient level of supervision in individual member states. Together with the push for more streamlined regulations and 'risk-based' enforcement (e.g. border inspection point inspection regime above), it seems probable that enlargement of the EU is likely to exacerbate all of the issues currently affecting the enforcement of transport legislation. Current resource constraints and competing priorities are responsible for a lack of training and information for competent authority inspectors and, therefore, lead to improper inspection procedures. They also result in a low number of roadside checks of animal welfare – an important tool in detecting infringements of legislation and animal suffering during transport. And insufficient resources are responsible for a low level of inspection follow-up action, due to staff shortages and lengthy procedures. Therefore, despite the strong legislative framework concerning animal welfare during transport that exists in the EU, the FVO itself notes that 'As a result [of lax enforcement,] transporters who are tempted to disregard animal welfare probably tend to take a chance and neglect, on a regular basis, Community provisions in this field' (European Commission, 2000, p. 11). Therefore, any step that can reduce the burden on overstretched authorities and reduce the number of animals that suffer because of a disregard for their welfare should be considered. One option is to increase the amount of funding and resources dedicated to monitoring animal welfare during transport and enforcing the existing legislation. However, this seems unlikely in the face of the current regulatory landscape. Another option is to reduce the number of animals transported long distances; the FVO suggests exploring 'Measures to encourage the slaughter of animals closer to the places where they are raised' (European Commission, 2000, p. 17).

References

Appleby, M.C. (2003) The European Union ban on conventional cages for laying hens: history and prospects. *Journal of Applied Animal Welfare Science* 6, 103–121.

Council of Europe (2003) European convention on the protection of animals during international transport (revised). Available at: http://conventions.coe.int/treaty/en/Treaties/Word/193.doc

Council of Europe (undated a) Protection of animals during transport. Available at: http://www.coe.int/t/e/legal_affairs/legal_co-operation/biological_safety,_use_of_animals/transport/_Summary.asp#TopOfPage

Council of Europe (undated b) Complete list of the Council of Europe's treaties. Available at: http://conventions.coe.int/Treaty/Commun/ListeTraites.asp?CM=8&CL=ENG

Council of Europe (undated c) About conventions and agreements in the Council of Europe Treaty Series (CETS). Available at: http://conventions.coe.int/general/v3IntroConvENG.asp

Council of Europe (undated d) European convention for the protection of animals during international transport (revised) CETS No. 193. Available at: http://conventions.coe.int/Treaty/Commun/ChercheSig.asp?NT=193&CM=8&DF=10/24/2006&CL=ENG

Council of Europe (undated e) Explanatory report: European convention for the protection of animals during international transport (revised). Available at: http://conventions.coe.int/Treaty/en/Reports/Html/193.htm

Court of Justice (2006) *Court of Justice Annual Report 2005*. European Court of Justice, Brussels, Belgium.

CURIA (2005) Rules of procedure of the Court of Justice. Available at: http://curia.europa.eu/en/instit/txtdocfr/txtsenvigueur/txt5.pdf

CURIA (undated) The Court of Justice of the European Communities. Available at: http://curia.europa.eu/en/instit/presentationfr/cje.htm

DEFRA (2006a) Farmed animal welfare during transport. Available at: http://www.defra.gov.uk/animalh/welfare/farmed/transport.htm

DEFRA (2006b) *Department for Environment, Food and Rural Affairs and the Forestry Commission: Departmental Report 2006*. The Stationary Office, London.

DEFRA (2006c) *Welfare of Animals During Transport: Consultation on the Implementation of EU Regulation 1/2005*. Department for Environment, Food and Rural Affairs, Nobel House, London.

DEFRA (2006d) *The Report of the Chief Veterinary Officer: Animal Health 2005*. Department for Environment, Food and Rural Affairs, Nobel House, London.

DEFRA (2007) State Veterinary Services to Become Animal Health. Available at: http://www.defra.gov.uk/news/2007/070326b.htm

DG Enlargement (2005) TAIEX Activity Report: 2005 European Commssion, Enlargement Directorate-General, Brussels, Belgium.

DG SANCO (2001) *Summary Report on an Inspection Visit by the Food and Veterinary Office to Luxembourg from 5 to 9 November 2001 to Assess Animal Welfare on Calf, Pig and Laying Hen Holdings and During Transport*, DG SANCO/3343/2001-RS EN. European Commission Directorate General for Health and Consumer Protection, Brussels, Belgium.

DG SANCO (2002a) *Final Report of a Mission Carried Out in Austria from 14/10/2002 to 18/10/2002 Concerning Animal Welfare During Transport and at the Time of Slaughter*, DG(SANCO)8677/2002-MR Final. European Commission Directorate General for Health and Consumer Protection, Brussels, Belgium.

DG SANCO (2002b) *Final Report of a Mission Carried Out in France from 24 to 28 June in Order to Evaluate the Systems for Checks of Animal Welfare During Transport and at the Time of Slaughter*, DG(SANCO)/8554/2002-MR Final. European Commission Directorate General for Health and Consumer Protection, Brussels, Belgium.

DG SANCO (2002c) *Final Report of a Mission Carried Out in Italy from 25 February to 1 March 2002 Concerning Animal Welfare During Transport and Certain Aspects at Slaughter*, DG(SANCO)/8556/2002-MR Final. European Commission Directorate General for Health and Consumer Protection, Brussels, Belgium.

DG SANCO (2002d) *Final Report of a Mission Carried Out in Ireland from 25/11/2002 to 29/11/2002 in Order to Evaluate the System of Checks for Animal Welfare During Transport and at the Time of Slaughter*, DG(SANCO)8678/2002-MR Final. European Commission Directorate General for Health and Consumer Protection, Brussels, Belgium.

DG SANCO (2004a) *Overview of a Series of Missions Carried Out in 2003 Concerning Animal Welfare During Transport and at the Time of Slaughter*, DG(SANCO)/8506/2004-GR. European Commission Directorate General for Health and Consumer Protection, Brussels, Belgium.

DG SANCO (2004b) *General Review Report of the Missions Carried Out in Member States Concerning Veterinary Checks at Border Inspection Posts 2002–2003*, DG(SANCO)/8508/2004-GR. European Commission Directorate General for Health and Consumer Protection, Brussels, Belgium.

DG SANCO (2004c) *General Report on a Series of Missions to the Member State Regarding Control of Staging Points*, DG(SANCO)/8508/2004-GR. European Commission Directorate General for Health and Consumer Protection, Brussels, Belgium.

DG SANCO (2005) Markos Kyprianou member of the European Commission responsible for health and consumer protection: speech to the Animal Welfare Intergroup of the European parliament. Available at: http://ec.europa.eu/food/animal/welfare/speech_05_335_en.pdf

Eurogroup (2003) *Summary of Suffering II: An Investigation into the Poor Enforcement of Directive 91/628/EEC as Amended by Directive 95/29/EC on the Welfare of Animals in Transport*. Eurogroup for Animal Welfare, Brussels.

European Commission (2000) *Report from the Commission to the Council and the European Parliament on the Experience Acquired by Member States Since the Implementation of Council Directive 95/29/EC Amending Directive 91/628/EEC Concerning the Protection of Animals During Transport*, COM(2000) 809 final. European Commission, Brussels, Belgium.

European Commission (2002) *Communication from the Commission – European Governance: Better Lawmaking*, COM(2002) 275 final. European Commission Secretariat, Brussels, Belgium.

European Commission (2003) Commission proposes radical overhaul of animal transport rules. Available at: http://europa.eu/rapid/pressReleasesAction.do?reference=IP/03/1023&format=HTML&aged=0&language=EN&guiLanguage=en

European Commission (2004) Commission welcomes Council's agreement on stricter welfare rules for transport of animals. Available at: http://europa.eu/rapid/pressReleasesAction.do?reference=IP/04/1391&format=HTML&aged=0&language=en&guiLanguage=en

European Commission (2005) General report on the activities of the European Union. Section 3: better lawmaking. Available at: http://europa.eu/generalreport/en/2005/rg24.htm

European Commission (2006a) Animal welfare main community legislative references. Available at: http://ec.europa.eu/food/animal/welfare/references_en.htm

European Commission (2006b) Community legal instruments. Available at: http://europa.eu/scadplus/glossary/community_legal_instruments_en.htm

European Commission (2006c) *Commission Staff Working Document Annex to the Communication from the Commission to the European Parliament and the Council on a Community Action Plan on the Protection and*

Welfare of Animals 2006–2010 and Commission Working Document on a Community Action Plan on the Protection and Welfare of Animals 2006–2010 Strategic Basis for the Proposed Actions. Impact Assessment COM(2006)13final/COM(2006)14final, SEC(2006)65. European Commission Secretariat, Brussels, Belgium.

European Commission (2006d) Health and consumer protection chart. Available at: http://ec.europa.eu/dgs/health_consumer/chart.pdf

European Commission (2006e) Draft general budget 2007: Article 17 04 01 – Animal disease eradication and monitoring programmes and monitoring of the physical conditions of animals that could pose a public health risk linked to an external factor. Available at: http://eur-lex.europa.eu/budget/data/AP2007_VOL4/EN/nmc-titleN17E07/nmc-chapterN18100/articles/index.html#N18131

European Commission (undated a) Health and consumer protection – general. Available at: http://ec.europa.eu/dgs/health_consumer/general_info/mission_en.html

European Commission (undated b) Correct implementation of EU legislation on animal welfare. Available at: http://ec.europa.eu/food/animal/welfare/legislation/index_en.htm

European Commission (undated c) Enlargement – TAIEX. Available at: http://taiex.cec.eu.int/

European Commission (2007) Commission refers Greece to Court over animal welfare infringements. Available at: http://europa.eu/rapid/pressReleasesAction.do?reference=IP/07/379&format=HTML&aged=0&language=EN&guiLanguage=en

EUR-Lex (1991) Council Directive 91/628/EEC of 19 November 1991 on the protection of animals during transport and amending Directive 90/425/EEC and 91/496/EEC. *Official Journal of the European Union* L, 340. Available at: http://eur-lex.europa.eu/smartapi/cgi/sga_doc?smartapi!celexapi!prod!CELEXnumdoc&numdoc=31991L06288&model=guichett&lg=en

EUR-Lex (1997) Council Regulation (EC) No. 1255/97 of 25 June 1997 concerning Community criteria for staging points and amending the route plan referred to in the Annex to Directive 91/628/EEC. Available at: http://eurlex.europa.eu/smartapi/cgi/sga_doc?smartapi!celexapi!prod!CELEXnumdoc&lg=EN&numdoc=31997R1255&model=guichett

EUR-Lex (2005) Council Regulation (EC) No 1/2005 of 22 December 2004 on the protection of animals during transport and related operations and amending Directives 64/432/EEC and 93/119/EC and Regulation (EC) No 1255/97. *Official Journal of the European Union* L 003, 48. Available at: http://eur-lex.europa.eu/LexUriSeru/LexUriSeru.do?uri=OJ:L:2004:374:0001:00028:EN:PDF

EUR-Lex (2006a) Draft general budget 2007. Available at: http://eur-lex.europa.eu/budget/data/AP2007_VOL4/EN/nmc-titleN17E07/index.html

EUR-Lex (2006b) Preliminary draft budget 2007: Section IV– Court of Justice. Available at: http://eur-lex.europa.eu/budget/data/AP2007_VOL5/EN/index.html

EUR-Lex (2006c) Preliminary draft budget 2007: Chapter 22 04 – information and communication strategy. Available at: http://eurlex.europa.eu/budget/data/AP2007_VOL4/EN/nmc-titleN1A447/nmc-chapterN1AA69/index.html#N1AA69

EUR-Lex (undated) About EU law – process and players – key players in the EU legislative process. Available at: http://www.europa.eu.int/eur-lex/en/about/pap/process_and_placers3.html

FVO (2000) *Food and Veterinary Office (FVO) Annual Report 2000.* European Commission, Brussels, Belgium.

FVO (2001) *Food and Veterinary Office (FVO) Annual Report 2000.* European Commission, Brussels, Belgium.

FVO (2004a) Food & Veterinary Office Programme of Inspections 2005 January–December. Available at: http://ec.europa.eu/food/fvo/inspectprog/2005-1_jan_dec_en.pdf (accessed August 2006)

FVO (2004b) *Food and Veterinary Office – Annual Report 2003.* European Commission, Brussels, Belgium.

FVO (2005a) *Food and Veterinary Office – Annual Report 2004.* European Commission, Brussels, Belgium.

FVO (2005b) *Programme of Inspections 2006 by the Food and Veterinary Office*, DG(SANCO)660129/2005. European Commission Directorate Health and Consumer Protection, Brussels, Belgium.

Hampton, P. (2005) *Reducing Administrative Burdens: Effective Inspection and Enforcement.* HM Treasury, London.

JRC (2006a) *Technical Specifications for Navigation Systems in Long Journey Animal Transports – Draft*, G07-TRVA/JH/(2006). European Commission Directorate General Joint Research Centre, Brussels, Belgium.

JRC (2006b) *Satellite Navigation Systems in Long Journey Animal Transportation: Update on the Draft Technical Specifications.* European Commission Directorate General Joint Research Centre, Brussels, Belgium.

McRory, R. (2006) *Regulatory Justice: Sanctioning in a Post-Hampton World.* Cabinet Office, Better Regulation Executive, London.

OIE (2006) *Terrestrial Animal Health Code 2006.* World Organisation for Animal Health, Paris, France.

OIE (undated a) Office of Directorate General: welcome to the OIE. Available at: http://www.oie.int/eng/OIE/organisation/en_welcome.htm?e1d1

OIE (undated b) The OIE's initiatives in animal welfare. Available at: http://www.oie.int/eng/bien_etre/en_introduction.htm

SVS (2006) *SVS Business Plan April 2006–March 2007.* The State Veterinary Service, Worchester, UK.

UK Parliament (2006) Uncorrected transcript of oral evidence to be published as HC 1569-i. Available at: http://www.publications.parliament.uk/pa/cm/cmenvfru.htm

6 The Welfare of Livestock During Sea Transport

C.J.C. Phillips

Centre for Animal Welfare and Ethics, School of Veterinary Science, University of Queensland, Queensland, Australia

Abstract

Large numbers of livestock are reared for transport overseas, and the long duration of the journey and the changes in the animals' environments provide special challenges compared to short-distance transport. A description is provided of the most common methods of transporting live animals by sea for slaughter internationally. The biggest exporter in the world is Australia and the main markets are South-east Asia and the Middle East. The most common livestock transported are cattle and sheep, but goats, camels, buffaloes, pigs and horses may also be transported alive. It is emphasized that multiple factors impacting on animal welfare are involved before, during and after the ship voyage; these include mustering, shearing (in the case of sheep), transport to feedlots and several changes of environment that can cause fear and anxiety. Information on the welfare of exported cattle and sheep on transport ships from Australia comes mainly from a survey of expert opinion completed in 2005. This found that the major stressors on ship were believed to be clinical diseases, especially inappetence and salmonellosis in the case of sheep, heat stress, high stocking density and high ammonia levels. The reported mortality rate is considerably greater for sheep than cattle, particularly due to failure to eat in the sheep, but has tended to decline for both species over the last 5 years. Other potential stressors, about which little is known, include noise, motion sickness, changes in lighting patterns and novel environments.

Development of the Trade

The major centres of human population in the world today do not necessarily coincide with the regions of the planet with the resources to produce livestock for human food. As a result of this disparity, in the last 20 years a major trade in exporting livestock from countries with plentiful resources to the main consumption areas has become established. Many are exported live, which is cheaper and easier than the refrigerated transport necessary for the long-distance carcass trade. For ship transport, for example, vessels used previously for the transport of cars have been refitted to allow their use for live animal export. There is a considerable road transport trade between countries within the European Union, for example cattle transported between Northern Ireland and Eire, or between Canada and the USA, but this chapter is restricted to live export by ship, and principally focuses on the export trade from Australia (see Chapter 7, this volume, for road transport).

At the same time as this trade has developed, concern for animal welfare has been increasing rapidly, particularly in developed countries, with the result that the welfare impact of practices involved in the live export of animals has been regularly challenged. The welfare concerns have focused on the long duration of travel, the large range of new circumstances that the animals must learn to cope with within a relatively short period of time and the practice of maintaining a high stocking density of animals during travel, because of the high cost of the vessels/vehicles involved and their fuel requirements.

The value of world live export trade has been increasing by about 4% annually and now stands at a total of over US$10 billion. Much of this trade is between countries that are geographical neighbours, for example, in northern Europe, or between Canada and the USA. This trade does not necessarily involve any longer distances than within country transport. However, the sea trade in live animals usually does involve long distances, especially when livestock are exported from Australia and New Zealand. The Australian live export trade is the largest in the world, and Australia sends about 1 million cattle, 6 million sheep and 0.1 million goats from 18 ports to over 40 overseas destinations, mostly the Middle East and South-east Asia. Feeder cattle mainly go to Indonesia, the Philippines and Malaysia. The live sheep export industry alone was estimated to be worth AUS$268 million (or US$200 million) in 2003–2004, the largest in the world, with 95% of these animals going to the Middle East, especially Kuwait, Jordan, Bahrain and Saudi Arabia, where the prices achieved for livestock usually exceed those available in the country of origin (see Chapter 2, this volume). There is also a considerable trade in goats from Australia to Malaysia and Singapore.

Other exporters include the USA (when health restrictions allow), South America, Africa and Ireland. The UK has a very small sheep export industry, which has declined from about 100,000 animals a year in 1999–2000 to about 40,000 animals a year in 2004–2005, all travelling to continental Europe (Anon, 2006a).

New Zealand no longer exports large numbers of sheep, with the trade declining to very low levels in the 1990s, probably due to economic reasons (see Chapter 2, this volume). The last shipment was in 2003, and there was only one shipment of 43,000 in this year, with a mortality rate of 0.5% (Rickets, W., 2007, personal communication). In the previous 5 years to 2003 there were four shipments of approximately 40,000 each. In addition, the Ministry of Agriculture and Forestry placed a temporary restriction on the trade following the difficulties experienced by the Australian Government during the *Cormo Express* incident in 2003. In 1990, a *Cormo Express* shipment from New Zealand had a 12% mortality rate. Some breeding sheep are exported from New Zealand, but the trade now focuses on export of dairy heifers into South-east Asia, Latin America and more recently to China. Approximately 70,000 heifers are exported annually. The New Zealand government is currently negotiating a Memorandum of Understanding with Saudi Arabian authorities to allow the fat lamb trade to resume. Over the last 3–4 years Australia

Table 6.1. Export of cattle and sheep from the UK, 1999–2006.

Year	Total no. of animals	No. of sailings	Animals rejected at place of origin	Animals rejected at port	Destination (country)
2006	289,529 – sheep 128,028 – cattle	Unavailable	Unavailable	Unavailable	Belgium, France, Germany, Holland
2005	37,104 – sheep	5	95	407	France, Germany, Italy
2004	48,448 – sheep	28	255	3	France, Belgium, Germany, Holland
2003	68,613 – sheep	36	995	9	France, Belgium, Germany, Holland
2002	130,048 – sheep	48	1,336	608	France, Belgium, Italy, Germany, Holland
2001	109,316– sheep	19	5	3	N/A
2000	750,820 – sheep 1,230 – pigs	189	29	97	N/A
1999	269,933 – sheep 2 – goats	41	16	17	N/A

has also increased dairy cattle exports, with China as the major market, along with minor importers Mexico and the UAE.

The UK exported calves to continental Europe for fattening for veal until the European Union imposed a ban on this trade due to the disease status of cattle in that country. In May 2006, this ban was lifted and the trade resumed. It attracted many protests in the 1990s, and the death of a protestor in 1995 further increased the level of public concern. The UK also currently exports a small number of sheep and cattle to continental Europe, a number which has reduced significantly over the last 5 years (Table 6.1). Animals are either kept in their trucks during the relatively short voyage, or are offloaded into pens on the ship.

A Brief Summary of the Export Process

The export process involves much more than just the shipping of livestock, it begins with the mustering of the stock, often on remote properties, and it ends with animal slaughter in the country of destination. In between, the stock will be handled at least a further five or six times and the whole process is likely to last between 1 and 2 months. Little is known about the cumulative effects of these combined stresses on the welfare of the animals, but it is possible that multiple stressors could make the animals anxious, depressed or enter a phase of learned helplessness, in which they become unresponsive to the environment around them, and unwilling to fight for their survival. It should be emphasized, however, that there is no direct evidence of this, but only indirect evidence from human medicine (Gregory, 2004).

In contrast to sea transport, the problems faced by livestock when transported long distances by land are well understood. The journey duration is likely to be longer for sea transport, with a possibility that some animals may experience reduced stress in the latter part of the journey because they have acclimatized to the conditions. The longer duration also brings the risk of under-nutrition, which is unlikely to be a problem on land journeys. The motion patterns are also different, with a more regular rhythm at sea than experienced in trucks, where the shocks are sudden and may be less predictable.

The additional stressors before embarkation of the ship will include:

1. Mustering, usually by horsemen, helicopters, dogs, riders of motorbikes or all-terrain vehicles. In the case of sheep, this will probably be followed by shearing to reduce the amount of space used by the animals. Special care should be taken during this shearing process, since cuts to the skin could have a much greater impact on the welfare of the animal when they are one of several stressors, compared with an isolated incident. In addition, sheep being transported from the pre-assembly depots to the port are in full view of the public. Avoiding overloading of vehicles and use of correct loading procedures for sheep and cattle on to vehicles are very important.
2. Curfewing – restricting food and water from the stock overnight before they are transported, to improve the accuracy of weight measurements and to keep the trucks clean. When cattle are kept off food and water for periods longer than 24 h, the population of potentially lethal pathogens increases exponentially in the gastro-intestinal tract (Grau *et al.*, 1969; Hogan *et al.*, 2007), leading to a major risk of carcass contamination. In addition, it is likely that stress accumulates as the gastrointestinal tract progressively empties, leading to a reduction in welfare. There may also be suppression of the immune system but this has not been evaluated.
3. Road or rail transport for up to 50 h, including the stresses caused by food and water deprivation, vehicular motion, high stocking density and high temperature. This will not only affect the animals' welfare, it has food hygiene implications as well. For example, the pathogen load builds up rapidly in an animal's empty gastrointestinal tract and could contaminate the carcass if slaughter occurs before the rumen has had a chance to re-establish its normal flora and fauna. This is likely to take about 2 weeks, given suitable feed and environmental conditions. The clinical manifestations of livestock transport stress include dehydration, bruising, weight loss, salmonellosis in sheep and respiratory disease in cattle. These have been studied in relation to road transport and subsequent slaughter (see Chapter 7, this volume), but not in relation to road transport before a long ship voyage and subsequent feedlotting or slaughter.
4. Loading, unloading and transfer to a pre-export assembly depot. The depot is likely to present a new and challenging environment for the stock, and they will probably remain there for 3–10 days. It may simulate shipboard conditions, with pellet feeding and the animals constrained in buildings. This helps to prepare the animals for their travel by ship. However, some depots are outside and the animals may experience difficulties due to exposure, particularly if they have come from

warmer regions. This can predispose sheep to salmonellosis, in part because of their contact with faeces, but also probably in part because of the exposure and their failure to eat adequate quantities of feed pellets when they are in wet conditions.

5. Road transport to the port. This is likely to be brief, as most depots are situated close to the ports. It is important to have sufficient trucks and men at the port to ensure that the animals are not waiting for a long time before loading. Between unloading the vehicle and loading on to the ship, there is usually a point of inspection, which is the last chance to reject animals that are sick or unsuitable for travel. This inspection should be a rigorous process, but the time required to adequately inspect all animals is not known or controlled. In Australia, before loading animals are inspected by Australian Quarantine and Inspection Service (AQIS) veterinarians to check that they are all suitable for travel. Those that are deemed not suitable will be returned to the property of origin. Statistics are not available for the numbers of animals rejected in most countries. Animals that are rejected and returned to the property also suffer serious welfare consequences if they are not given proper medical or nutritional care that they require.

6. Loading into the ship. In Western countries this is usually well planned, with stevedores (wharf hands) available to assist in the unloading of each vehicle. However, hurrying the animals or excessive use of the electronic goad (jigger) can cause injury and may even retard the process. Animals may arrive at their pens stressed and they then react nervously to the passage of every person moving past their pen. In these circumstances, it may take several days for the animals to calm down. Sheep are unlikely to lie down for the first few days, which may indicate their fear of being trampled if they do lie down or anxiety following the loading process.

Overview of the Sea Journey

The ships are of a variety of ages and sizes, and are usually converted vessels designed for other purposes, for example, car transporters. They are of international ownership and are provided with a relatively well-paid crew from developing countries, typically the Philippines or Bangladesh in adequate numbers. The extent of training is not known. The species which are most commonly transported out of Australia are sheep and cattle, but goats, buffaloes, camels and even horses have been transported in the past. Sheep are most commonly shipped to the Middle East and cattle to South-east Asia. Pigs are rarely carried, but are transported internationally within the Americas. Typically, the livestock vessels will carry 60,000–100,000 sheep and perhaps 1000 cattle on a voyage ranging from 7 (to South-east Asia) to 23 days (to the Middle East). The new vessels used by Australian exporters are faster, smaller and better ventilated, meeting the need for the improved standards set in 2005 by the Australian Government. Vessels may collect livestock from more than one port in Australia, for example, Portland and Fremantle, and then deliver to perhaps two or three ports in the Middle East, for example Muscat and Kuwait. Sometimes the stock are transported large distances after arrival, for example up to Jordan, and this may be one of the knock-on effects of any bans on export to a particular country, e.g. Saudi Arabia, as they may then enter via another country's port.

The different species tend to be on separate decks, particularly as the cattle pens are much smaller, about 4 × 3 m, holding about five animals, than the sheep pens, which are about 8 × 4 m, holding about 100 animals. In the case of cattle, the number of animals carried cannot exactly meet the standards because the pens are small, and extra space is provided. For example, Australian standards (DAFF, 2005) require that a 500 kg cow has 1.8 m^2 of space, unless it is transported from southern Australia in winter, when the space requirement is increased to 2.1 m^2 to allow for the dramatic change in temperature as the animals enter summer conditions in the northern hemisphere. In winter, a 4 × 3 m pen could not give six animals the required space, so five animals would be carried, allowing 2.4 m^2 each. These pens are used to confine the animals for the duration of the voyage. Most vessels have full height decks of approximately 3 m for both cattle and sheep, however, some sheep decks are split so that there are two layers, doubling the number of animals that can be carried but reducing the access of stockpersons to the animals. The cattle pens are provided with some bedding, straw, sawdust or other suitable material, on long journeys, which is removed usually two or three times during the voyage after it has become soiled with excreta. More frequent cleaning could unnecessarily retard the animals' establishment in their pens, but could encourage them to lie down for longer. In the sheep pens the excreta builds up over the duration of the voyage, and providing it is not too hot or humid, a relatively soft and crumbly material develops, which is suitable for the sheep to lie on. In the event of excessive heat, additional fans are sometimes erected above specific pens. However, these are likely to be unable to cover most of the pen evenly.

The pens are usually located on about seven decks. Decks above the water level are likely to be open at the sides to assist ventilation, but those below or at water level will be fully enclosed with forced ventilation from one or two points in every pen. This must achieve one air change every 2 min. The variation in climatic conditions in the closed decks both over time and across the various parts of the deck is less than the open decks. Thus, closed decks may have a constant average to high ammonia level and open decks normally have a lower average ammonia level, but open decks are prone to very high ammonia levels when climatic conditions are conducive to such levels; for example, in the hot conditions of a destination port.

Some vessels have an additional fully open set of pens on the top deck, and exposure to the elements, in particular sea water during high seas, can be a problem if the animals become waterlogged or are standing in water.

Each pen is provided with troughs for food and water. Water provision should be checked once or twice daily, as the ballcocks may become stuck, leading to spillages of water and poor conditions for the animals. Similarly, food delivery tubes may become blocked and need checking after feeding.

Conditions and Welfare of Animals Transported by Sea for Slaughter

The welfare of livestock on the ships

A study by Pines *et al.* (2007) reported that the most serious animal welfare concerns of stakeholders in the Australian live export industry were clinical diseases (as evidenced by mortality level, clinical disease incidence and animals in hospital pens on the ship), heat stress (respiration rate and wet bulb temperature, which includes temperature and humidity), stocking density and ammonia accumulation. The key stakeholders in the trade were identified by a steering committee for the project as animal transport scientists, assembly depot operators, exporters, government officials, ship owners, stockmen, producers, animal welfare representatives and veterinarians. Other animal welfare issues that were identified in consultation with key representatives of the industry were food availability, stress, experiences in the pre-export assembly depot, noise and disturbance to the lighting patterns, but these were not chosen as the most serious animal welfare problems. The purpose of pre-export assembly is to check animals for their health and fitness prior to export, allow the aggregation of a large number of animals prior to export, accustom the animals to close confinement and food and watering conditions similar to those that they will experience on ship, and finally to provide rest following their transportation and handling over the previous few days. The failure of the key stakeholders to identify experiences in the assembly depot as of major relevant to livestock welfare does not mean that these experiences are not relevant because they may predispose the animals to further stressors, or even ameliorate some stressors.

Clinical diseases and mortality

Sheep

In the Australian live export industry, over three-quarters of all sheep deaths occur on the ship itself, and about a fifth of the deaths occur at the discharge port, with a very small number occurring in the pre-export assembly depot, mainly due to salmonellosis (Higgs *et al.*, 1993; Norris, 2005). Mortality is greatest on voyages to the Middle East, and certain ships can be identified as having higher mortality rates than others. This could be due to many factors, including variation in ventilation rate, the preparation of the livestock, the husbandry of the animals on board or the feed supply. There is also seasonal variation, with highest mortality in August, September and October and lowest mortality in March, April and May.

Sheep mortality in the trade between Australia and the rest of the world, primarily the Middle East, has declined steadily in recent years, to a level of 0.99% of transported animals, or approximately 60,000 animals, reported in 2003 (Norris, 2005). This may just be due to a reduction in the age of sheep transported or it may reflect an increased welfare provision for the animals or increased awareness of mortality as a reportable welfare measure. Proper statistical models could resolve this issue by allowing change in contributory factors, such as the type of animal carried, over time to be taken account of in the production of a weighted mean value each year. Better monitoring of sheep numbers on and off the ship would help to determine whether the mortality figure provided in the captain's reports is accurate. However, this may be practically difficult, unless it could be automated at entry to, and exit from, the ship by using video digitizing software. This could probably not be achieved without substantial research effort. Other aspects of the captain's reports may need verification or standardization – for example, the temperature recorded should be representative of all the decks on the ship, which requires careful consideration of the positioning of the thermometer. This should be specified in the government standards.

The biggest contributor to sheep mortality is persistent failure to eat (inappetence), which accounts for nearly half of all deaths, and/or salmonellosis (about a fifth of all deaths). Inappetence occurs primarily due to the animals being transferred from a pasture-based diet to a concentrate pellet, which they are not used to. Preparation and selection of suitable animals is critical, as animals that have experienced pellet feeding are more likely to adjust to the new diet faster on board. A smaller proportion of the sheep, perhaps about 10%, die through trauma such as during loading or handling procedures, and about 5% die from other diseases. Sheep that have not fed are predisposed to salmonellosis, and stress is known to be involved in lesion development (Higgs *et al.*, 1993). Open, exposed assembly depots, such as that at Portland, Victoria, may predispose the sheep to salmonellosis, particularly in wet periods. Good drainage is essential in wet conditions and good stockpeople will take preventive measures to ameliorate the effects of a less-than-perfect system, by for example running troughs down the hill, rather than across, so that any surplus water drains from them. By contrast, the feedlot at Fremantle, where sheep are kept in raised buildings with minimal faecal contact, has a much reduced risk of contributing to salmonellosis. Surprisingly, the duration of feeding sheep in the pre-export assembly depot (typically 3–13 days) does not affect their predisposition to inappetence during ship transport (Norris *et al.*, 1992).

Inappetence is partly determined by the body condition of animals at the start of the voyage. In a study by Higgs *et al.* (1991), individual sheep that were identified as fat had approximately twice the risk of death from starvation aboard ship than sheep that had not been identified as fat. There are also regions and groups of sheep that are at risk of suffering high death rates when exported by sea. In the study of Higgs *et al.* (1999), there were more high-mortality groups, and the average mortality was greater, in sheep from the zones of high rainfall and long pasture-growing season. This is probably because these sheep have not developed the ability to mobilize body fat to sustain them through a period of under-nutrition (Richards *et al.*, 1991). Sheep coming from dry pastures were nearing the end of a cycle of live weight loss and were metabolically adjusted to using adipose reserves for energy. However, when sheep had entered a period of live weight gain on green pastures, they were consequently unable to sustain lipolysis for energy supply when inappetant. The industry standards now forbid the exporting of sheep with the highest level of body condition, score five, on a scale of one to five. Lower death rate in younger wethers (castrated males) (Higgs *et al.*, 1991) can be attributed to the overriding demands of tissue growth, ensuring a stronger appetite than that seen in adult animals.

Cattle

The main causes of cattle death are heat stroke, trauma and respiratory disease (shipping fever) (Norris *et al.*, 2003). There may have been a small reduction in cattle mortality over the last 5–6 years, but this is at such a low level anyway (0.24% among 4 million cattle exported from Australia between 1995 and 2000 (Norris *et al.*, 2003) and 0.11% of transported animals in 2003), that it is hard to determine any definite trends. A greater proportion of deaths occurred on voyages to the

Middle East (0.52%) than to South-east Asia (0.13%). The risk of death on voyages to the Middle East was three times greater for cattle exported from southern ports in Australia compared to northern ports. This is because cattle coming out of southern ports in winter are adapted to cold conditions, with a longer coat and more subcutaneous fat, and find it difficult to cope with a rapid transfer to hot and humid conditions once they have passed the equator, where they will require a sleek coat and little subcutaneous fat, to aid heat loss. Cattle now only leave the southern ports in summer. Another reason for high mortality of cattle coming out of southern ports is that *Bos taurus* (European) cattle are more popular in the South Australia, where tick resistance is not required. There are more *Bos indicus* or zebu (humped) cattle coming out of the northern ports, because these cattle predominate in the Northern Territories and much of Queensland. They have a natural ability to cope with heat stress better than *B. taurus* cattle, with more sweat glands, folds of skin to increase heat loss and lower productivity.

A pernicious clinical disorder that occurs regularly on the ships and could be more controlled more effectively is infectious conjunctivitis (pinkeye). The animals are predisposed to the disease by their inability to escape from high ventilation wind movement, which often stirs up dust from the feed. Thus, feed quality, and in particular, the hardness of the pellet is of vital importance in preventing pinkeye. There is little incentive to exporters to provide a high-quality pellet, although some nutritional standards are prescribed in the Australian standards. When the pellet is made up of poor quality chaff with little binding agent, such as molasses, it frequently breaks up by the time it reaches the feeding troughs. The pellet tends to disintegrate during handling, caused by the long and many movements from the point of manufacture and the wind tunnels used to move the feed around the ship. My own observations lead me to suspect that up to 5–10% of the stock can have pinkeye in some shipments. While pinkeye is not fatal, there is no doubt that it significantly reduces the welfare of the livestock because of irritation and it is also known to reduce productivity, perhaps because of the failure of animals to get to the feed troughs or fend off other animals.

Heat stress

The high temperature load generated by the livestock and the ship's engines, combined with a high ambient temperature, makes heat stress a common problem on export vessels. It is particularly true for European type (*B. taurus*) cattle that are poorly adapted to cope with a high heat load. The upper critical temperature (at which physiological changes, such as sweating or panting, need to occur to reduce the heat load on the animal) for *B. taurus* dairy cows is only approximately 25–26°C (Berman *et al.*, 1985). As stated previously *B. indicus* or zebu breeds are able to manage high temperatures better than *B. taurus* cattle, even when their lower productivity is taken into account (which would be expected to increase heat output because of increased feed intake (Kadzere *et al.*, 2002)). Adaptation to high temperatures (through a reduction in subcutaneous fat, the development of a sleek, reflective coat) takes more than 9 weeks to develop (Kibler, 1964), therefore, a voyage of 2–3 weeks will not allow much adaptation. In many hot regions of the world cattle can cope with high temperatures during the day, because they can cool down at night. In addition, the high humidity in many shipboard conditions, because of the high density of livestock and/or a lack of adequate ventilation (which particularly occurs on open decks when the ship is close to port) severely exacerbates the heat stress risk. When the animals are at risk of heat stress the ship's master could delay entry to port until adequate crosswinds are assured, although they operate to a tight, co-coordinated schedule that would be a deterrent to such action. On other occasions, the master may have to change the direction of the ship, tacking with the winds, to ensure adequate ventilation of open decks. It should also be noted that there is very significant variation in ventilation rate within pens (Pines and Phillips, 2005), and because there is limited movements of the animals around the pen, some animals may be subjected to much greater heat stress than others. In addition, there is very significant variation around the ship, depending on the proximity to the engine, the level of natural and forced ventilation, and the type of animals enclosed within the pen.

A special risk of high humidity occurs in sheep pens if the faecal pad breaks up and becomes effectively a slurry. The floors of the ship decks are impervious and, unlike cattle pens which are cleaned out on a regular basis, the faeces and urine from sheep is allowed to accumulate for the duration of the voyage. Normally, a hard pad is formed, which makes a suitable lying surface for the sheep. However, in very humid conditions, the slurry is formed, which will probably reduce the welfare of the sheep. On other occasions, sea water may wash on to lower decks, causing the faecal pad again to turn into a slurry. This can cause mortality in the sheep, but would also reduce welfare by potentially waterlogging the animals or causing them to stand in water. Some ships are fitted with walls that can be erected to protect against the ingress of sea water on lower decks, but they reduce ventilation and are only erected in high seas. Twenty-four hour vigilance is required on the part of the master and his crew to erect these walls at the right time, and this is required by Australian government standards.

Unfortunately the ability to predict the heat tolerance of cattle on ships is currently very limited. In a fully laden cattle truck the heat output from the cattle will raise the ambient temperature by about 3°C (Winker *et al.*, 2003). A heat stress model has been developed for the Australian live export industry, which relies on historical data for mortality of a relatively small number of voyages. Although this model is used by the industry to determine the stocking density of livestock on ships, its ability to predict the variance due to ship, voyage and animal characteristics is quite limited. A more robust model is needed, which is based on scientific information on the impact of these factors on the heat stress load, not just mortality, experienced by different classes of livestock. Some of the fundamental data are available from research that has addressed the management of, in particular, cattle and sheep in hot climates. However, most assume circadian variation in temperature, which is not necessarily the case on ships. Hence there may be a need for more basic studies that address heat dissipation in conditions of continual challenge.

Heat stress has tended to cause major events on a relatively infrequent basis, but with quite high numbers of animals dying on these occasions, for example, the maiden voyage of the *Becrux* in July 2002, in which 31% of almost 2000 cattle and 2% of 63,000 sheep died. Emergency measures to control a heat stress event on ship, which normally only lasts for a day or two, have recently been devised in the form of wetting of cattle to facilitate heat loss by evapotranspiration (Gaughan *et al.*, 2005). There are concerns about increased humidity, which may eventuate if a large proportion of animals on the deck are sprayed, especially if the bedding is not removed prior to wetting. Nevertheless, it appears that body temperature can be reduced by a few degrees for perhaps 3–4 h following a substantial wetting of the cattle. The force of the water may prove a problem if the high pressure hoses are not judiciously used. These measures are probably of no value to sheep, as their fleece would become saturated.

Stocking density

Stocking density is not simply a matter of a number or weight of animals per unit area, but it includes aspects of enclosure design, the familiarity and proximity of other animals, the removal of waste products and the thermal and visual environments of the animals. One important aspect of stocking density is the contribution to social interactions between animals. Although some attempt may be made to group animals that have come from the same property into their own pen, the fact that animals are usually handled approximately six times from the time that they leave their paddock or feedlot until they arrive at their shipboard pen makes a high degree of mixing likely. The high stocking density in itself would be expected to increase the frequency of agonistic attractions, compared with the animals at pasture, provided that they have room to manoeuvre and position themselves in their head-to-head confrontations. At the highest shipboard stocking densities sheep at least may experience insufficient freedom of movement for overt aggression (charging each other), and aggression will be moderated to ritualized head swings at other animals. Land-based research indicates that until this point increasing stocking density results in increased agonistic interactions (Jarvis and Cockram, 1995). However, hauliers suggest that a high stocking density may also benefit sheep, if they give each other mutual support during sudden movements of the ship, which might otherwise result in the animals slipping (but see Chapter 7, this volume). Cattle or sheep can probably not predict all of the movements of the ship, especially near to shore where the wave motion is complex.

At times when the vessel is not fully laden, which occurs for part of the voyage if animals are taken on or discharged at multiple ports, it is unclear whether the livestock would benefit from having more space. Providing access to all the available floor space is the policy on some but not all vessels. It will increase the subsequent cleaning required and the disturbance in moving stock around may be counterproductive to their welfare, unless there are significant heat stress risks, in which case, reducing the stocking density would be expected to be wholly beneficial. However, the recognition by the staff of some vessels that it is beneficial to the animals to reduce stocking density if only for a brief period suggests that overall the benefits outweigh the disadvantages.

Because of their susceptibility to heat stress, rams are usually given more space than wethers (castrated male sheep) but this may enhance general activity, and chasing behaviour and aggression are common. The larger stock, in particular cattle, are housed in small groups, typically three to six animals, so it would be expected that each animal quickly becomes acquainted with its other penmates, and aggression rapidly diminishes once a dominance hierarchy has been established. With sheep, we have observed that individuals tend to stay within a specific area of the pen, and that movement is primarily to and from the front of the pen to get food and water (Pines *et al.*, 2007). At the front of the pen, feeding and drinking animals may run the risk of aggressive approaches to their hindquarters from other animals waiting to gain access to the troughs. However, Norris (2005) has suggested that lack of access to the troughs is not the main reason for inappetence in sheep. When inappetent sheep are found by the veterinarian, stockman or crew they are placed in a separate pen with more room and ready access to feed and water, suggesting that more space will improve their condition.

It is possible to derive space requirements from a knowledge of their requirements for lying down, defaecating, standing up, etc., but although these requirements are quite well understood for cattle at least, the data is derived from land-based measurements, without the continuous motion of the ship. Stocking density requirements can therefore theoretically be derived from the animals' need for space to perform a range of behaviours, which if not met, will result in stress and reduced productivity. A benchmark for space allowances is the space available for the cattle and sheep to lie down laterally, which is a behaviour that occurs under low stocking conditions, but when they are overcrowded only sternal recumbancy is possible. The limited evidence available for cattle and sheep suggests that productivity may decline when stocking density provides less than the amount required for lateral recumbency plus 33%.

In a number of species, but in particular cattle and sheep, the presence of horns increases the potential of the bearer to display dominance. Although the Australian standards recommend that cattle with horns longer than 12 cm or sheep and goats with horns greater than one curl should not be exported, these criteria have not been scientifically evaluated. Indeed, the stocking density standards have been essentially devised from 20 years of practical experience, but have not been empirically demonstrated to optimize welfare.

The adverse effects of high stocking densities do not just relate to agonistic behaviour. Sheep will be initially reluctant to lie down if stocking densities are too high, a phenomenon that I have observed on board ship. This may be in part because of a fear of being trampled. The extent of any adverse effect on welfare of sheep not lying down in the first few days of the voyage is yet to be determined. The type of lying may be constrained at high stocking densities, lateral recumbency being only possible at low densities. Hence animals may be forced to lie sternally recumbent, rather than their preferred posture, which is likely to adversely affect their welfare. Cattle are likely to be dirtier at high stocking densities because of a greater chance that they will directly soil each other, as well as the increased accumulation of excreta at high stocking densities, unless the excreta are removed more regularly than at low stocking densities. Even if this is the case, at the prescribed stocking densities for Australian live export (DAFF, 2005), cattle may be lying with legs overlapped, which will smear dirt on to the torso of other animals as they get up.

Feeding and drinking trough space may be inadequate at high stocking densities. Approximate space allowances of 0.20 and 0.16 m per head are required for cattle and sheep, respectively. Reduced space allowances may deter some animals, particularly the fatter ones, from performing their natural behaviours, especially feeding.

Research work on sheep transported by truck suggests that high stocking densities lead to increased bruising (Jarvis and Cockram, 1995; Menke *et al.*, 1999), but this does not necessarily relate to transport by ship, where the movement patterns are very different. Truck movements are sudden and unpredictable, ship movements are more likely to follow a regular pattern, which the livestock may be able to predict.

Virtually no experimentation has been undertaken to look at the impact of different stocking densities of livestock on ships. The most important research required is to evaluate the behavioural, physiological and production responses to a range of stocking densities, probably under simulated conditions on land initially, with the results validated on ship. Choice tests could also usefully compare the demand for space with other resources, such as food. Using a human analogy, it might be suspected that the lack of exercise associated with tight stocking might exacerbate any fear or anxiety experienced by the livestock, and this could be tested experimentally. Conversely, little is known about the effect that the close proximity of other animals has on stresses experienced by animals, such as trucking and loading into the ship. Finally it would be useful to have precise measurements of the spatial requirements for specific behaviours, such as standing and lying.

In summary, although experts believe that high stocking densities have the potential to have a major impact on the welfare of livestock on ships, there has been virtually no research completed to substantiate this belief. Standards are based on experience developed over the last 20 years, which are likely to focus on the economic optimum stocking density, rather than that which provides for optimum welfare. Research with livestock transported by road vehicles suggests that there are major disadvantages to the animals' welfare if they are kept at very high stocking densities, and this is usually for a much shorter period than ship transport. However, there has not been any experimentation conducted on shipboard stocking densities aimed at determining impact on welfare. By contrast many experiments have been done on land-based transport or housing.

Ammonia accumulation

Ammonia is a gas generated from urea in animal excreta, which has the potential to irritate the respiratory tract in animals and humans, where the excreta are not rapidly removed from the animals' accommodation. Levels of up to 50 ppm have been reported for pig and poultry units (Groot Koerkamp *et al.*, 1998; Hinz and Linke 1998) and levels exceeding 25 ppm have been reported on live export vessels (MAMIC, 2001; Costa *et al.*, 2003). Stacey (2003) found that typical levels below decks were 15 ppm, with readings commonly reaching 20–30 ppm.

Many countries use 25 ppm as a maximum for human exposure over 8 h, and indeed this is the level in force in Australia (NOHSC, 1995). A maximum of 20 ppm has been recommended by Tudor *et al.* (2003) for live export vessels, based on the observance of immune responses (increases in white blood cell count and mononucleated cell counts) at this level in cattle. A literature review by Costa *et al.* (2003) concluded that 25 ppm would be a more appropriate maximum even though they observed that 'there are initial clinical signs of inflammation at 22 ppm atmospheric ammonia'. Preference studies with pigs and poultry show that fresh air is preferred to 10 ppm ammonia, and there is evidence that welfare is reduced at levels below 25 ppm (Kristensen and Wathes, 2000; Kristensen *et al.*, 2000). The clinical symptoms of ammonia toxicosis are lacrimation (crying), coughing and sneezing and nasal discharge that can be bloody (Drummond *et al.*, 1976). The ammonia concentration in the atmosphere on ship is largely determined by the stocking density of the animals, nitrogen content of feed, ambient temperature and pH of the urine.

Other potential stressors

Motion sickness
There is anecdotal evidence of stress to cattle/sheep caused by high seas as well as evidence of motion sickness in the form of vomiting in trucked pigs and other quadrupeds (Bradshaw *et al.*, 1996). The latter, however, could be due to fear or anxiety, rather than motion sickness. Interestingly rats show *pica* not *vomiting* during motion sickness and it can therefore be speculated that motion sickness may contribute to inappetence.

Noise levels
Most of the noise generated on the ship comes from the ventilation fans. This can be attenuated, but it is expensive to do so, and currently there are no regulations requiring control of high noise levels and indeed little information on the sensitivity of cattle and sheep to noise. Some research (Agnes *et al.*, 1990) suggests that the noise associated with transport can induce as high a level of stress in calves as transport alone. In Agnes *et al.*'s work, the noise level was approximately 96 dB, and the hormonal stress response attenuated after 15–30 min, although this does not necessarily mean that the animal was not stressed after this time. Thus the continuous high levels of noise from the fans may be less stressful than sudden, infrequent noise, such as is common during loading and unloading. Most ships have a workshop, which also can generate significant sporadic noise that could reduce the welfare of nearby livestock. Interestingly, Agnes *et al.*'s work did not provide any evidence of synergistic effects of multiple stressors during transport, however, a much more complex programme of research would be needed to confirm this.

Neophobia
The responses of livestock to new situations, as opposed to familiar ones, are largely unexplored. Such neophobia is well established for novel foods (e.g. Malmkvist *et al.*, 2003), but it is unclear whether the novel environments provided by the truck, pre-assembly feedlot, ship and feedlot at destination, as well as the likelihood of a novel social context, would sufficiently stress the livestock to be of concern.

Lighting changes
All animals use the changes in photoperiod as cues to time events in the circadian rhythm of activity. The transition from the natural day length to an artificial,

controlled lighting pattern of continuous low-intensity light provided by fluorescent tubes is sudden and without precedent for most animals coming off pasture. Although in farming situations it has been determined that cattle preferences for different lighting schedules are not strong (see review by Lomas *et al.*, 1998), this abrupt change may contribute to the stress to which the animals are exposed, with adverse cumulative effects. However, maintaining continuous light may preserve welfare under these conditions of high stocking density.

The role of the veterinarian and stockpeople

Veterinary supervision during the (sea) voyage occurs only when the Australian government requires the presence of a veterinarian on a shipment, which is usually when the journey is to the Middle East from Australia. There is an argument that a veterinarian would be underutilized on shipments to Asia from Australia, and certainly mortality rate on these voyages is very low and often zero on the cattle shipments (Norris *et al.*, 2003). Given that the main causes of mortality (heat stress in cattle, inappetence and salmonellosis in sheep) are difficult to correct when detected, attention should focus on prevention (in particular proper preparation/selection of the animals and ameliorating contributory environmental conditions on board). On many voyages, it would be possible to use the veterinarian extensively in monitoring animal performance and well-being and participating in research to improve animal welfare. The veterinarian will perform at least one daily inspection of all their decks, and possibly two (personal observation). It is debatable whether they can actually inspect every animal, but they are assisted by members of the crew and the stockperson, who will point out to them any animals that appear not to be thriving. They will conduct post-mortems on any mortalities, but these are only usually likely to be of a cursory nature as the cause of death is often obvious. They will engage in discussion with the master on the health of livestock and should discuss the ship's future course and any risks to health of the stock at a daily meeting with the stockperson and master. The veterinarian or the stockperson will engage in some preventative medicine, administering antibiotics to any animals suspected of contracting an infection, spraying antiseptic on to the udders of cows to prevent mastitis and removing any stock that are not thriving to hospital pens where the stocking density is lower, and the crew and veterinarian can keep an eye on the animals. The author has not witnessed euthanasia on any ship, and animals with any chance of recovery are likely to be kept alive if the veterinarian and master have incentives to keep the mortality rate down to the lowest level possible. The vets and stockmen have both captive bolt gun and lethal injection capabilities, so humane slaughter on board is quite possible.

Disembarkation

At the port of disembarkation, it is important that the importing company provides adequate trucks and drivers, so that the offloading of the stock is not delayed. The Middle Eastern ports in particular are characterized by hot and sometimes humid conditions in summer and if the risk of heat stress is significant it would be prudent to move on to another port. If a shipment is rejected, the Australian government has a Memorandum of Understanding with several Middle East countries that shipments will be offloaded into a quarantine facility or facilities in other countries should there be any suspected problems with a particular shipment or a dispute that necessitates alternative arrangements.

In developing countries receiving livestock from Western countries, the infra-structure in the port of disembarkation is unlikely to be as well organized as in the ports of origin. Offloading ramps, inspections and truck suitability and availability are all likely to be of lower quality than in Western countries. The stevedores must be well trained in handling stock, and if the animals are passing through a feedlot before slaughter there needs to be well-trained staff managing this feedlot. There is a temptation for some of the ship's personnel to want to visit the port as soon as the ship docks, but it must be remembered that the welfare of the stock is most likely to be challenged at this point in time.

Members of animal activist groups and investigative journalists have reported on conditions in Middle Eastern ports, which supposedly show animal abuse and mistreatment. Scientific investigations on the prevalence of such alleged abuses would be worthwhile to identify the scale of the problems.

Discussion

Summary of findings

The export process is a long and involved combination of transport and handling practices, starting with the mustering of animals on properties and ending with their slaughter in the country of destination. In between, the stock will be handled at least a further five or six times and the whole process is likely to last between 1 and 2 months. There are significant challenges to the welfare of livestock all along this chain, but it can be speculated that the most important are unsuitable animals being transported, poor environmental conditions on board ship, in partic-ular, high temperatures and ammonia concentrations, high stocking densities on board ship and inadequate handling and management at the port of destination and in the abattoir. The possibility of synergistic effects of multiple stressors cannot be ignored, potentially leading to chronic stress and anxiety in the animals. The main measure of welfare currently used in Australia for the live export industry is mortality, and although this is a relatively imprecise indicator, due to lower levels in both cattle and sheep that are exported from Australia, it has been possible to show that the mortality rate for sheep at least, has declined in recent years. However, there is a need for other welfare indicators which take into account the impact on the animals' welfare of practices that will not contribute to mortality.

Current practice and its acceptability to the general public

There has been widespread public condemnation of long-distance shipboard trans-port of livestock in recent years, particularly where developed countries are involved.

Most attention is focused on sheep, perhaps because of the large numbers involved, but the welfare of less domesticated animals, such as buffalo, may be of equal or greater concern for individual animals because of the difficulties posed by the handling and penning of semi-wild animals. Over the last few years, the Australian industry has moved from self-regulation to government regulation. However, this does not go beyond what is current best industry practice in Australia, and is an attempt to weed out the rogue operators or those using unsuitable vessels and unsatisfactory practices. Nevertheless, even best practice in the industry today could still be unsatisfactory so far as some or many members of the general public are concerned. This has not yet been quantified, and should be a priority for the World Organisation for Animal Health (OIE) at least. It should be noted that this chapter has addressed welfare concerns primarily during the shipping process. Ethical concerns which deal primarily with the moral acceptability of the practices have been considered elsewhere (Phillips, 2005).

Good practice and areas of poor practice

The livestock exporters are keenly aware of the threat to their industry arising from public perceptions of some operators' inadequate attention to welfare, but it must be remembered that they are only one part of the chain of management of livestock from when they are first received until the point of departure. This contrasts with the relative lack of control over the export process when the animals are in their country of destination, which makes the practice ethically questionable (Phillips, 2005).

Some of the welfare problems on ships, in particular high stocking density and heat stress, are both poorly understood and also difficult to manipulate without having a major impact on industry profit. In contrast to this, the handling and management of animals in the country of origin that are destined for live export could potentially be at a high standard relative to other sectors in the chain. This is because they are less commercially sensitive than practices on ship, where it is inherently expensive to manipulate some of the major contributors to animal welfare. This is especially true of stocking density. Other issues, such has high ammonia concentrations, can probably be addressed by controlling feed nitrogen content and cleaning procedures, but again there would be a significant cost to the industry. Further research is undoubtedly needed on this topic, but it should be noted that the quality and quantity of feed required for different classes of stock are now specified in the government standards in Australia. A maximum crude protein concentration of feed is included in this specification, but it seems likely that further research is required before a precise model of nitrogen inputs to ammonia output can be constructed. Morbidity is being addressed by the prescription of minimum veterinary standards, and the minimum restraint and veterinary equipment that should be carried for cattle and sheep are now specified in the Australian government standards. In other countries, the standards are generally not as well developed, but the evolution of OIE standards (OIE, 2005) may enable a worldwide improvement in the welfare of livestock transported by sea.

Situations in which welfare is conducive to financially viable production methods

The precise relationship between welfare and production output is still not properly understood. For example, we know that if we have high stocking densities the weight gain of livestock during a voyage is likely to be reduced, but it is unclear whether this is compensated for by a greater number of animals transported. The standards which have evolved over more than 20 years of experience are those that lead to maximum financial gain for the exporter and not optimum welfare of the stock. Exporters are driven by financial objectives, so this is not surprising, but it emphasizes the role of government control and the pressure lobby provided by NGOs and the general public as a counterbalance to the commercial pressures on the industry.

Positive examples of best practice in sea transport

To achieve good welfare animals have to be properly prepared well in advance, and there needs to be a much better understanding of practices that will improve animal welfare on the ships. To date, only a very limited amount of research by a small and restricted group of researchers, mainly in Australia, has been completed. This contrasts markedly with the situation regarding road transport, where there has been extensive research conducted in the last 15 years, mainly in Europe, as a result of public concerns about the inhumane treatment of animals transported by road.

Acknowledgements

I am grateful to the live export industry of Australia for the opportunity to witness the trade first-hand.

References

Agnes, F., Sartorelli, P., Abdi, B.H. and Locatelli, A. (1990) Effect of transport loading or noise on blood biochemical variables in calves. *American Journal of Veterinary Research* 51, 1679–1681.

Anon. (2006a) Live animal exports statistics. Available at: http://www.defra.gov.uk/animalh/welfare/farmed/transport/dover/dover.htm

Berman, M., Folman, M., Kaim, M., Mamen, Z., Herz, D., Wolfenson, A. and Graber, Y. (1985) Upper critical temperatures and forced ventilation effects for high-yielding dairy cows in a tropical climate. *Journal of Dairy Science* 68, 488–495.

Bradshaw, R.H., Parrott, R.F., Forsling, M.L., Goode, J.A., Lloyd, D.M., Rodway, R.G. and Broom, D.M. (1996) Stress and travel sickness in pigs: effects of road transport on plasma concentrations of cortisol, beta-endorphin and lysine vasopressin. *Animal Science* 63, 507–516.

Costa, N., Accloly, J. and Cake, M. (2003) *Determining Critical Atmospheric Ammonia Levels for Cattle, Sheep and Goats – A Literature Review*. Meat & Livestock Australia, Sydney, Australia.

DAFF (Deptartment of Agriculture, Fisheries and Forestry) (2005) *Standards for the Export of Livestock*. Draft Version 1.02. DAFF, Canberra, Australia.

Drummond, J.G., Curtis, S.E., Lewis, J.M., Hinds, F.C. and Simon, J. (1976) Exposure of lambs to atmospheric ammonia. *Journal of Animal Science* 1343.

Gaughan, J., Lott, S. and Binns, P. (2005) Wetting cattle to alleviate heat stress on ships. *Proceedings of the Live Export R and D Forum, 2005*. Meat & Livestock Australia and Livecorp, Sydney, Australia, pp. 38–43.

Grau, F.H., Brownlie, L.E. and Smith, M.G. (1969) Effects of food intake on numbers of salmonellae and *Escherichia coli* in rumen and faeces of sheep. *Journal of Applied Bacteriology* 32, 112–117.

Gregory, N.G. (2004) *Physiology and Behaviour of Animal Suffering*. Blackwell, Oxford, UK.

Groot Koerkamp, P.W.G., Metz, J.H.N., Uenk, G.H., Phillips, V.R., Holden, M.R., Sneath, R.W., Short, J.L., White, R.P., Hartung, J., Seedorf, J., Schroder, M., Linkert, K.H., Pedersen, S., Takai, H., Johnsen, J.O. and Wathes C.M. (1998) Concentrations and emissions of ammonia in livestock buildings in northern Europe. *Journal of Agricultural Engineering Research* 70, 79–95.

Higgs, A., Norris, R.T., Love, R.A. and Norman, G.J. (1999) Mortality of sheep exported by sea: evidence of similarity by farm group and of regional differences. *Australian Veterinary Journal* 77, 729–733.

Higgs, A.R.B., Norris, R.T. and Richards, R.B. (1991) Season, age and adiposity influence death rates in sheep exported by sea. *Australian Journal of Agricultural Research* 42, 205–214.

Higgs, A.R.B., Norris, R.T. and Richards, R.B. (1993) Epidemiology of salmonellosis in the live export industry. *Australian Veterinary Journal* 70, 330–335.

Hinz, T. and Linke, S. (1998) A comprehensive experimental study of aerial pollutants in and emissions from livestock buildings. Part 2: results. *Journal of Agricultural Engineering Research* 70, 119–129.

Jarvis, A.M. and Cockram, M.S. (1995) Some factors affecting resting behaviour of sheep in slaughterhouse lairages after transport from farms. *Animal Welfare* 4, 53–60.

Kadzere, C.T., Murphy, M.R., Silanikove, N. and Maltz, E. (2002) Heat stress in lactating dairy cows: a review. *Livestock Production Science* 77, 59–91.

Kibler, H.H. (1964) Thermal effects of various temperature–humidity combinations on Holstein cattle as measured by eight physiological responses. *University of Missouri Agricultural Experiment Station Research Bulletin* No. 862.

Kristensen, H.H. and Wathes, C.M. (2000) Ammonia and poultry welfare: a review. *World's Poultry Science Journal* 56, 235–245.

Kristensen, H.H., Burgess, L.R., Demmers, T.G.H. and Wathes, C.M. (2000) The preferences of laying hens for different concentrations of atmospheric ammonia. *Applied Animal Behaviour Science* 68, 307–318.

Lomas, C.A., Piggins, D. and Phillips, C.J.C. (1998) Visual awareness. *Applied Animal Behaviour Science* 57, 247–257.

Malmkvist, J., Herskin, M.S. and Christensen, J.W. (2003) Behavioural responses of farm mink towards familiar and novel food. *Behavioural Processes* 61, 123–130.

Menke, C., Waiblinger, S., Fölsch, D.W. and Wiepkema, P.R. (1999) Social behaviour and injuries of horned cows in loose housing systems. *Animal Welfare* 8, 243–258.

MAMIC (2001) *Investigation of the Ventilation Efficacy on Livestock Vessels – Final Report*. Meat & Livestock Australia, Sydney, Australia.

NOHSC (1995) Exposure standards for atmospheric contaminants in the occupational environment. In: *National Occupational Health and Safety Commission 1003*. Australian Government, Canberra, Australia, pp. 75.

Norris, R.T. (2005) Transport of animals by sea. *Revue Scientifique et Technique Office International des Epizooties* 24, 673–681.

Norris, R.T., Richards, R.B. and Norman, G.J. (1992) The duration of lot-feeding of sheep before sea transport. *Australian Veterinary Journal* 69, 8–10.

Norris, R.T., Richards, R.B., Creeper, J.H., Jubb, T.F., Madin, B. and Kerr, J.W. (2003) Cattle deaths during sea transport from Australia. *Australian Veterinary Journal* 81, 156–161.

OIE (2005) Guidelines for the transport of animals by sea. Available at: http://www.oie.int/eng/normes/mcode/en_chapitre_3.7.2.htm

Phillips, C.J.C. (2005) Ethical perspectives of the Australian live export trade. *Australian Veterinary Journal* 83, 558–562.

Pines, M. and Phillips, C. (2005) *Developing Alternative Methods of Measuring Animal Welfare on Ships and in Pre-export Assembly Depots. Stage 1a – Evaluation of the Concentrations and Effects of Potentially Noxious Gases on Two Ship Voyages. Report on Project Live 222, Stage 1A.* Meat & Livestock Australia, Sydney, Australia.

Pines, M., Petherick, C., Gaughan, J. and Phillips, C. (2007) The opinion of stakeholders in the Australian live export industry concerning welfare indicators for sheep and cattle exported by sea (in press).

Richards, R.B., Hyder, M.W., Fry, J., Costa, N.D., Norris, R.T. and Higgs, A.R. (1991) Seasonal metabolic factors may be responsible for deaths in sheep exported by sea. *Australian Journal of Agricultural Research* 42, 215–226.

Stacey, C. (2003) *Development of a Heat Stress Risk Management Model.* Meat & Livestock Australia, North Sydney, Australia.

Tudor, G., Accioly, J., Pethick, D., Costa, N., Taylor, E. and White, C. (2003) *Decreasing Shipboard Ammonia Levels by Optimising the Nutritional Performance of Cattle and the Environment on Ship During Live Export.* Meat & Livestock Australia, North Sydney, Australia.

Winker, I., Gegresenbet, G. and Nilsson, C. (2003) Assessment of air quality in a commercial cattle transport vehicle in Swedish summer and winter conditions. *Deutsche Tierarztliche Wochenschrift* 110, 100–104.

7

The Welfare of Livestock During Road Transport

D.M. BROOM

Cambridge University Animal Welfare Information Centre, University of Cambridge, Madingley Road, Cambridge, UK

Abstract

The welfare of animals during transport should be assessed using a range of behavioural, physiological and carcass quality measures. In addition, health is an important part of welfare so the extent of any disease, injury or mortality resulting from, or exacerbated by, transport should be measured. Many of the indicators are measures of stress in that they involve long-term adverse effects on the individual. Some of the key factors affecting the welfare of animals during handling and transport which are discussed are: attitudes to animals and the need for training of staff; methods of payment of staff; laws and retailers' codes; genetics especially selection for high productivity; rearing conditions and experience; the mixing of animals from different social groups; handling procedures; driving methods; stocking density; journey length; increased susceptibility to disease; and increased spread of disease. In order that welfare can be good during transport, it is important that all of those involved are properly informed about the animals and how to assess their welfare. There should be careful planning of journeys and suitable vehicles should be selected. Space allowances should be sufficient for the animals to lie in most species, to stand in the case of all horses, and cattle and sheep on short journeys, and to move to get food and water if the journey is long enough for this to be necessary. Vehicle design and space allowance should allow for adequate inspection of each animal on the vehicle or, if this is not possible, journey time should be kept short. Long journeys, the term long having different meanings for different species as explained below, should be avoided wherever possible and much better conditions are needed if journeys are long. Vehicles should be driven more carefully than vehicles with human passengers and sudden turns and braking should be avoided, especially on roads with sharp bends or at right angle turns into other roads. Ventilation management and other efforts to avoid harmful physical conditions are important. Transport should be managed so that disease susceptibility is not high and disease spread is minimized.

Introduction and Definitions

The handling, loading, transporting and unloading of animals can have very substantial effects on their welfare and meat quality. The welfare of an individual is its state as regards its attempts to cope with its environment (Broom, 1986). Hence, welfare is a measurable characteristic of an animal at a particular time or during a period. Animal protection is a human activity that should lead to better welfare. Coping means having control of mental and bodily stability and welfare includes both the extent of failure to cope and the ease or difficulty in coping. Welfare varies on a scale from very good to very poor. Health is an important part of welfare while feelings, such as pain, fear and various forms of pleasure, are components of the mechanisms for attempting to cope and so should be evaluated where possible in welfare assessment (Broom, 1998, 2001b, 2006a; and see Chapter 1, this volume). Where an individual is failing to cope with a problem, it is said to be stressed. Stress is an environmental effect on an individual, which overtaxes its control systems and reduces its fitness or appears likely to do so (Broom and Johnson, 2000). There are many different systems in the body whose function is to maintain the individual in conditions that it can tolerate. When one or more of these systems cannot correct for the environmental impact, they are said to be overtaxed, the individual starts to become stressed. Reduced fitness means that the individual is more likely to die or be unable to spread its genes in the population. Some of the effects of the environment may simply be stimulation or useful experience or a physiological response involving the adrenal gland, which has no adverse consequences, and the individual is not stressed. All stress involves poor welfare. However, there can be poor welfare without stress because the poor welfare has no long-term consequences, for example temporary pain or distress. Pain is an unpleasant feeling that is generally strongly avoided, so any pain, or indeed any other negative feeling, means that welfare is poor. These issues are discussed further in several papers in Broom (2001a).

In this chapter, after presentation of a summary of information about land transport, the factors, which can result in stress during transport, are introduced. The methodology for assessing the welfare of the animals during handling and transport is then explained. Finally, some of the various factors, which affect the likelihood of stress, are discussed with examples from work on cattle and sheep.

Information About the Amount of Land Transport

Of all species of animals transported for slaughter, by far the largest numbers are of chickens reared for meat production. With more than 48 billion chickens produced each year in the world and transport occurring at least twice, the total number of journeys in the world in 2005 was about 96 billion (FAOSTAT, 2006). Comparable figures for other species are shown in Table 7.1.

A typical journey

The Federation of Veterinarians of Europe has stated (FVE Position Paper, 2001) that it has always been of the opinion that the fattening of animals should take place within or near the place of birth. Animals should be slaughtered as near the point of production as possible. The journey time for slaughter animals should never exceed the physiological needs of the animal for food, water or rest; the long-distance transport

Table 7.1. Numbers of farmed animals produced in the world in 2005, most of which would be transported to slaughter. (From FAOSTAT, 2006.)

Animal	Millions of animals
Chickens reared for meat production	48,000
Hens for egg production	5,600
Pigs	1,310
Rabbits	882
Turkeys	689
Sheep	540
Goats	339
Cattle for meat production	296
Cattle for milk production	239
Fish of many species are farmed but live transport often does not occur	

of animals for slaughter should be replaced, as much as possible, by a carcass-only trade. Animals going from farm to slaughter may pass through a market. Based on data from the Meat and Livestock Commission, Murray *et al.* (2000) reported that, in the UK in 1996, 56% of cattle, 65% of sheep and 5% of pigs passed through a live-stock auction market. In a study of 16,000 sheep travelling to slaughter in south-west England, the median journey duration was 1.1 h, with very few journeys of more than 5 h, and when the sheep went directly from farm to abattoir 7.8 h, with over a third of journeys 10–17 h, when they went via a livestock market (Murray *et al.*, 2000). Fewer animals were going to auctions in 2006 but the exact number is not known.

The first stage of a journey is the selection of animals that will travel and there should be inspection of these animals to check that they are fit to travel. Animals are prepared for the journey and this preparation will depend on the species and the length of the journey envisaged. For shorter journeys preparation can include fasting before collection and possibly movement away from the main herd to pro-tect its health status. For longer journeys, where watering and feeding will be nec-essary on the vehicle, it can be an advantage to collect the animals involved 2–3 days before the transport, so that they can be prepared for the journey and become accustomed to the feed that will be offered en route.

The animals then have to be loaded on to the transport vehicle. This may be a road or rail vehicle or a boat. Loading is discussed later in relation to welfare. During transport, animals gradually relax to some extent as they start to become accustomed to the new environment so there is some recovery from the stress of loading. Animal comfort during transport is highly dependent on vehicle design, driving technique and the roads being traversed. Requirements for vehicles will therefore depend on the length of the journey. All things being equal, the demands for transport vehicles will become more stringent as transport distances increase and weather conditions become more extreme, whether this is very cold or very hot weather. Drivers should always be conversant with the needs of animals during transport and should receive formal training, unless previous experience can be proven for the type of transport envisaged.

During transport, offloading and on-loading of animals for rest periods, feeding and watering is often recommended. However, it may increase stress levels and the risk of injury. Moreover, there will be a risk of spreading disease if the animals come into contact with other animals. Except in the circumstance where there can be a substantial period lasting a day or more, when the animals can rest off the vehicle, a vehicle of such a standard that animals can remain on it during the resting, watering or feeding period may be advantageous.

At the end of the journey to slaughter, animals are unloaded at the slaughterhouse. Good facilities are required for this. A period of lairage may occur before slaughter. For most species, welfare is better if any lairage period is very short.

The vehicles used

Horses and cattle are normally carried in vehicles with one tier but pigs and sheep may be carried in two- or three-tier vehicles. Poultry and rabbits are usually carried in crates. These may be stackable crates, of such a size that a person can lift one of them, or may be modular units that have to be lifted with a forklift vehicle. On the vehicle, there should be air spaces between the rows of crates or modules.

Some vehicles are air-conditioned, some have well-designed openable areas on the sides for ventilation, some have poorly designed possibilities for ventilation, and some of the simplest vehicles have open bars on the sides. The latter can provide good ventilation but little protection from injury in crash situations or from poor weather. In developed countries and in some developing countries, all but the worst vehicles provide some shade and shelter from rain or other inclement weather.

The suspension system in animal transport vehicles varies from very good to negligible. The provision for loading animals also varies greatly. Some vehicles are adapted for use with well-designed ramps while others have hydraulic lifts on tailgates or floors. Many have very steep loading ramps that cause poor welfare in all animals loaded.

Factors Which Can Result in Poor Welfare During Animal Handling and Transport

In the section below, major factors that result in poor welfare are underlined. People are sometimes cruel to one another but generally believe that other people are aware and sentient and so are likely to feel some guilt if they have been cruel. Non-human animals are regarded as aware and sentient by some people but as objects valued only according to their use by others (Broom, 2003, 2006b). Hence, there is a wide range of attitudes to animals and these have major consequences for animal welfare. During handling and transport, these attitudes may result in one person causing high levels of stress in the animals while another person doing the same job may cause little or no stress. People may hit animals and cause substantial pain and injury because they are trying to do the work very quickly, or because they do not consider that the animals are subject to pain and stress, or because of lack of knowledge about animals and their welfare. Training of staff can substantially alter attitudes to, and treatment of, animals.

Laws can have a significant effect on the ways in which people manage animals. Within the European Union, the Council Regulation (EC) No 1/2005 'On the protection of animals during transport and related operations' takes up some of the recommendations of the EU Scientific Committee on Animal Health and Animal Welfare Report 'The welfare of animals during transport (details for horses, pigs, sheep and cattle)' (March, 2002) and of the European Food Safety Authority 'Report on the welfare of animals during transport' (2004) which deals with the other species. Laws have effects on animal welfare provided that they are enforced and the mechanisms for enforcement within EU member states are the subject of discussion in 2006. Adequate enforcement requires training and there is variation among countries in both training and willingness to enforce laws about animal welfare during transport (see Chapter 5, this volume). The existence of the OIE Guidelines on animal transport should lead to greater uniformity of laws and of their enforcement.

Codes of practice can also have significant effects on animal welfare during transport. The most effective of these, sometimes just as effective as laws, are retailer codes of practice since retail companies need to protect their reputation by enforcing adherence to their codes (Broom, 2002).

Some animals are much better able to withstand the range of environmental impacts associated with handling and transport than are other animals. This can be because of genetic differences, associated with the breed of the animal or with selection for production characteristics. Differences among individuals in coping ability also depend on housing conditions and the extent and nature of contact with humans and conspecifics during rearing. If pigs are handled gently on 2–3 days during development, they are easier to handle and their welfare is better during transport.

Since physical conditions within vehicles during transport can affect the extent of stress in animals, the selection of an appropriate vehicle for transport is important in relation to animal welfare. Similarly, the design of loading and unloading facilities is of great importance. Although seemingly far removed, the person who designs the vehicle and facilities has a substantial influence, as does the person who decides which vehicle or equipment to use.

Before a journey starts, there must be decisions about the stocking density of animals on the vehicle and the grouping and distribution of animals on the vehicle. If there is withdrawal of food from animals to be transported, this can lead to poor welfare. For all species, tying of animals on a moving vehicle can lead to major problems and for cattle and pigs any mixing of animals can cause very poor welfare.

The behaviour of drivers towards animals while loading and unloading and the way in which people drive vehicles are affected by the method of payment. If people are paid more if they load or drive fast, welfare could be worse, so such methods of payment should not be permitted (Broom, 2000). Payment of handling and transport staff at a higher rate if the incidences of injury and poor meat quality are low improves welfare. Insurance against bad practice resulting in injury or poor meat quality should not be permitted.

All the factors mentioned so far should be taken into account in the procedure of planning for transport. Planning should also take account of temperature, humidity, weather conditions and the risks of disease transmission. Disease is a major cause of poor welfare in transported animals and has significant consequences for trade in animals and animal products (see Chapter 3, this volume). Planning of routes should take account of the needs of the animals for rest, food and water. Drivers or other persons responsible should have plans for emergencies including a series of emergency telephone numbers to use to obtain veterinary assistance in the event of injury, disease or other welfare problems during a journey.

The methods used during handling, loading and unloading can have a great effect on animal welfare. The quality of driving can result in very few problems for the animals or in poor welfare because of difficulty in maintaining balance, motion sickness, injury, etc. The actual physical conditions, such as temperature and humidity may change during a journey and require action on the part of the person responsible for the animals. A journey of long duration will have a much greater risk of poor welfare and some durations inevitably lead to problems. Hence, good monitoring of the animals with inspections of adequate frequency, and in conditions which allow thorough inspection, is important.

Assessing Welfare

A variety of welfare indicators, which can be used to assess the welfare of animals that are being handled or transported, are listed in Table 7.2. Some of these measures are of short-term effects while others are more relevant to prolonged problems. Where animals are transported to slaughter it is mainly the measures of short-term effects such as behavioural aversion or increased heart rate, which are used but some animals are kept for a long period after transport and measures such as increased disease incidence or suppression of normal development give information about the effects of the journey on welfare. When animals are transported, there may be various stressors that affect them and the response of the animals

Table 7.2. Generally used measures of welfare. Details of these and other measures may be found in Broom (1998), Broom and Johnson (2000) and Broom and Fraser (2007). (From Broom, 2000.)

Physiological indicators of pleasure
Behavioural indicators of pleasure
Extent to which strongly preferred behaviours can be shown
Variety of normal behaviours shown or suppressed
Extent to which normal physiological processes and anatomical development are possible
Extent of behavioural aversion shown
Physiological attempts to cope
Immunosuppression
Disease prevalence
Behavioural attempts to cope
Behaviour pathology
Brain changes, e.g. those indicating self-narcotization
Body damage prevalence
Reduced ability to grow or breed
Reduced life expectancy

depends on the duration and intensity of these stressors (Fazio and Ferlazzo, 2003). The factors may interact in an additive or multiplicative way to determine the magnitude of the total effect on the animal (Broom, 2001a).

Behavioural measures

Changes in behaviour are obvious indicators that an animal is having difficulty coping with handling or transport. Some of these help to show which aspect of the situation is aversive. The animal may stop moving forward, freeze, back off, run away or vocalize. The occurrence of each of these can be quantified in comparisons of responses to different races, loading ramps, etc. Examples of behavioural responses such as cattle stopping when they encounter dark areas or sharp shadows in a race and pigs freezing when hit or subjected to other disturbing situations may be found in Grandin (1980, 1982, 1989, 2000).

Certain behavioural responses are often shown in painful or otherwise unpleasant situations. Their nature and extent vary from one species to another according to the selection pressures, which have acted during the evolution of the mechanisms controlling behaviour. Human approach and contact may elicit antipredator behaviour in farm animals. However, with experience of handling, these responses can be greatly reduced in cattle (Le Neindre et al., 1996). Social species, which can collaborate in defence against predators, such as pigs or man, vocalize a lot when caught or hurt. Species which are unlikely to be able to defend themselves, such as sheep, vocalize far less when caught by a predator, probably because such an extreme response merely gives information to the predator that the animal attacked is severely injured and hence unlikely to be able to escape. Cattle can also be relatively undemonstrative when hurt or severely disturbed. Human observers sometimes wrongly assume that an animal that is not squealing is not hurt or disturbed by what is being done to it. In some cases, the animal is showing a freezing response and in most cases, physiological measures must be used to find out the overall response of the animal.

Within species, individual animals may vary in their responses to potential stressors. The *coping strategy* adopted by the animal can have an effect on responses to the transport and lairage situation. For example, Geverink et al. (1998) showed that those pigs which were most aggressive in their home pen were also more likely to fight during pre-transport or pre-slaughter handling but pigs which were driven for some distance prior to transport were less likely to fight and hence cause skin damage during and after transport. This fact can be used to design an on-farm test, which reveals whether or not the animals are likely to be severely affected by the transport situation (Lambooij et al., 1995).

The procedures of loading and unloading animals into and out of transport vehicles can have very severe effects on the animals and these effects are revealed in part by behavioural responses. Species vary considerably in their *responses to loading procedures*. Any animal which is injured or frightened by people during the procedure can show extreme responses. However, using efficient and careful loading procedures, sheep and cattle can be loaded without severe effects, although pigs and poultry always show more disturbance. Broom et al. (1996) and Parrott et al. (1998b) found that sheep show largely physiological responses and these are associated with the unfamiliar situation encountered in the vehicle rather than the loading procedure.

Once journeys start, some species of farm animals explore the compartment in which they are placed and try to find a suitable place to sit or lie down. Sheep and cattle try to lie down if the situation is not disturbing but stand if it is. After a period of acclimatization of sheep and cattle to the vehicle environment, during which time sheep may stand for 2–4 h looking around at intervals and cattle may stand for rather longer, most of the animals will lie down if the opportunity arises. Pigs lie down much more rapidly. Unfortunately for the animals, many journeys involve so many lateral movements or sudden brakings or accelerations that the animals cannot lie down.

An important behavioural measure of welfare in experimental studies when animals are transported is the amount of fighting. For example, when male adult cattle are mixed during transport or in lairage, they may fight and this behaviour can be recorded directly (Kenny and Tarrant, 1987). Calves of 6 months of age may also fight (Trunkfield and Broom, 1991). The recording of such behaviour should include the occurrence of threats as well as the contact behaviours that might cause injury.

A further valuable behavioural method for welfare assessment of farm animals during handling and transport involves using the fact that the animals remember aversive situations in experimentally repeated exposures to such situations. Any stock-keeper will be familiar with the animal, which refuses to go into a crush after having received painful treatment there in the past or hesitates about passing a place where a frightening event such as a dog threat occurred once before. These observations give us information about the welfare of the animal in the past as well as at the present time. If the animal tries not to return to a place where it had an experience then that experience was clearly aversive. The greater the reluctance of the animal to return, the greater the previous aversion must have been. This principle has been used by Rushen (1986a,b) in studies with sheep. Sheep, which were driven down a race to a point where gentle handling occurred, traversed the race as rapidly or more rapidly on a subsequent day. Sheep, which were subjected to shearing at the end of the race on the first day, were harder to drive down the race subsequently and those subjected to electro-immobilization at the end of the race were very difficult to drive down the race on later occasions. Hence, the degree of difficulty in driving and the delay before the sheep could be driven down the race are measures of the current fearfulness of the sheep and this in turn reflects the aversiveness of the treatment when it was first experienced.

Some behavioural measures are clear indicators that there will be a long-term effect on the animal, which will harm it, so they indicate stress. As explained later, other behavioural measures provide evidence of good or poor welfare but not necessarily of stress.

Physiological measures

The physiological responses of animals to adverse conditions, such as those which they may encounter during handling and transport, will be affected by the anatomical and physiological constitution of the animal as mentioned later. Some physiological measures are detailed in Table 7.3 and described in detail below.

Whenever physiological measurement is to be interpreted it is important to ascertain the *basal level* for that measure and how it fluctuates over time (Broom, 2000). For example, plasma cortisol levels in most species vary during the day and tend to be higher during the morning than during the afternoon. A decision must be taken for each measure concerning whether the information required is the difference from baseline or the absolute value. For small effects, e.g. a 10% increase in heart rate, the difference from baseline is the key value to use. The large effects where the response reaches the maximal possible level, for example, cortisol in plasma in very frightening circumstances, the absolute value should be used. In order to explain this, consider an animal severely frightened during the morning and showing an increase from a rather high baseline of 160 nmol/l but in the afternoon showing the same maximal response which is 200 nmol/l above the lower afternoon baseline. It is the actual value which is important here rather than a difference whose variation depends on baseline fluctuations. In many studies, the value obtained after the treatment studied can usefully be compared with the maximum possible response for that measure. A very frightened animal may show the highest response of which it is capable.

Heart rate can decrease when animals are frightened but in most farm animal studies, tachycardia, increase in heart rate, has been found to be associated with disturbing situations. Heart rate increase is not just a consequence of increased activity; heart rate can be increased in preparation for an expected future flight response. Baldock and Sibly (1990) obtained basal levels for heart rate during a variety of activities by sheep and then took account of these when calculating responses to various treatments. Social isolation caused a substantial response but the greatest heart rate increase occurred when the sheep were approached by a man with a dog. The responses to handling and transport are clearly much lower if the sheep have previously been accustomed to human handling. Heart rate is a useful measure of welfare but only for short-term problems such as those encountered

Table 7.3. Commonly used physiological indicators of poor welfare during transport. (Modified after Knowles and Warriss, 2000.)

Stressor	Physiological variable
	Measured in blood or other body fluids
Food deprivation	↑ FFA, ↑ ß-OHB, ↓ glucose, ↑ urea
Dehydration	↑ Osmolality, ↑ total protein, ↑ albumin, ↑ PCV
Physical exertion	↑ CK, ↑ lactate
Fear, lack of control	↑ Cortisol, ↑ PCV
Motion sickness	↑ Vasopressin
	Other measures
Fear, physical effects	↑ Heart rate, heart rate variability ↑, ↑ respiration rate
Hypothermia/hyperthermia	Body temperature, skin temperature

FFA, free fatty acids; ß-OHB, ß-hydroxybutyrate; PCV, packed cell volume; CK, creatine kinase.

by animals during handling, loading on to vehicles and certain acute effects during the transport itself. However, some adverse conditions may lead to elevated heart rate for quite long periods. Parrott *et al.* (1998a) showed that heart rate increased from about 100 beats per min to about 160 beats per min when sheep were loaded on to a vehicle and the period of elevation of heart rate was at least 15 min. During transport of sheep, heart rate remained elevated for at least 9 h (Parrott *et al.*, 1998b). Heart rate variability has also been found to be a useful welfare indicator in cattle and other species (van Ravenswaaij *et al.*, 1993).

Observation of animals can provide information about physiological processes in animals without any attachment of recording instruments or sampling of body fluids. *Breathing rate* can be observed directly or from good-quality video recordings. The metabolic rate and level of muscular activity are major determinants of breathing rate but an individual animal, which is disturbed by events in its environment, may suddenly start to breathe fast. *Muscle tremor* can be directly observed and is sometimes associated with fear. *Foaming at the mouth* can have a variety of causes, so care is needed in interpreting the observations, but its occurrence may provide some information about welfare.

Changes in the *adrenal medullary hormones* adrenaline (epinephrine) and noradrenaline (norepinephrine) occur very rapidly and measurements of these hormones have not been used much in assessing welfare during transport. However, Parrott *et al.* (1998a) found that both hormones increased more during loading of sheep by means of a ramp than by loading with a lift.

Adrenal cortex changes occur in most of the situations, which lead to aversion behaviour or heart rate increase but the effects take a few minutes to be evident and they last for 15 min to 2 h or a little longer. An example comes from work on calves (Kent and Ewbank, 1986; review by Trunkfield and Broom, 1990; Trunkfield *et al.*, 1991). Plasma or saliva glucocorticoid levels gave information about treatments lasting up to 2 h but were less useful for journeys lasting longer than this.

Saliva cortisol measurement is useful in cattle. In the plasma, most cortisol is bound to protein but it is the free cortisol which acts in the body. Hormones such as testosterone and cortisol can enter the saliva by diffusion in salivary gland cells. The rate of diffusion is high enough to maintain an equilibrium between the free cortisol in plasma and in saliva. The level is ten or more times lower in saliva but stimuli which cause plasma cortisol increases also cause comparable salivary cortisol increases in humans (Riad-Fahmy *et al.*, 1982), sheep (Fell *et al.*, 1985), pigs (Parrott *et al.*, 1989) and some other species. The injection of pilocarpine and sucking of citric acid crystals, which stimulate salivation, has no effect on the salivary cortisol concentration. However, any rise in salivary cortisol levels following some stimulus is delayed a few minutes as compared with the comparable rise in plasma cortisol concentration.

Animals, which have substantial adrenal cortex responses during handling and transport, show increased body temperature (Trunkfield *et al.*, 1991). The increase is usually of the order of 1°C but the actual value at the end of a journey will depend upon the extent to which any adaptation of the initial response has occurred. The body temperature can be recorded during a journey with implanted or superficially attached temperature monitors linked directly or telemetrically to a data storage system. Parrott *et al.* (1999) described deep body temperature in eight sheep.

When the animals were loaded into a vehicle and transported for 2.5 h, their body temperatures increased by about 1°C and in males were elevated by 0.5°C for several hours. Exercise for 30 min resulted in a 2°C increase in core body temperature which returned rapidly to baseline when the exercise finished. It would seem that prolonged increases in body temperature are an indicator of poor welfare.

The measurement of oxytocin, although a useful indicator in the longer term, has not been of particular value in animal transport studies (e.g. Hall *et al.*, 1998a). However, plasma beta-endorphin levels have been shown to increase during loading (Bradshaw *et al.*, 1996b). The release of corticotrophin releasing hormone (CRH) in the hypothalamus is followed by release of pro-opiomelanocortin (POMC) in the anterior pituitary, which quickly breaks down into components, including adrenocorticotrophic hormone (ACTH) which travels in the blood to the adrenal cortex and beta-endorphin. A rise in plasma beta-endorphin often accompanies ACTH increases in plasma but it is not yet clear what its function is. Although beta-endorphin can have analgesic effects via mu-receptors in the brain, this peptide hormone is also involved in the regulation of various reproductive hormones. Measurement of beta-endorphin levels in blood is useful as a back-up for ACTH or cortisol measurement.

Creatine kinase is released into the blood when there is muscle damage, e.g. bruising, and when there is vigorous exercise. It is clear that some kinds of damage that effect welfare result in creatine kinase release, so it can be used in conjunction with other indicators as a welfare measure. Lactate dehydrogenase (LDH) also increases in the blood after muscle tissue damage but increases can occur in animals whose muscles are not damaged. Deer, which are very frightened by capture, show large LDH increases (Jones and Price, 1992). The isoenzyme of LDH, which occurs in striated muscle (LDH5), leaks into the blood when animals are very disturbed so the ratio of LDH5 to total LDH is of particular interest.

On long journeys animals are generally unable to drink for many times longer than the normal interval between drinking bouts. This lack of control over interactions with the environment may be disturbing to the animals and there are also likely to be physiological consequences. The most obvious and straightforward way to assess this is to measure the osmolality of the blood (Broom *et al.*, 1996). When food reserves are used up there are various changes evident in the metabolites present in the blood. Several of these, for example beta-hydroxy butyrate, can be measured and indicate the extent to which the food reserve depletion is serious for the animal. If chickens reared for meat production were deprived of food for 10 h prior to 3 h of transport, when compared with undeprived birds, their plasma had higher thyroxine and lower tri-iodothyronine, triglyceride, glucose and lactate concentrations, indicating negative energy balance and poor welfare. Another measure, which gives information about the significance for the animal of food deprivation, is the delay since the last meal. Most farm animals are accustomed to feeding at regular times and if feeding is prevented, especially when high rates of metabolism occur during journeys, the animals will be disturbed by this. Behavioural responses when allowed to eat or drink (e.g. Hall *et al.*, 1997) also give important information about problems of deprivation.

The haematocrit, a count of red blood cells, is altered when animals are transported. If animals encounter a problem, such as those which may occur when they are handled or transported, there can be a release of blood cells from the spleen and a higher cell count (Parrott *et al.*, 1998b). More prolonged problems, however, are likely to result in reduced cell counts (Broom *et al.*, 1996).

Increased adrenal cortex activity can lead to immunosuppression. One or two studies in which animal transport affected T-cell function are reviewed by Kelley (1985), but such measurements are likely to be of most use in the assessment of more long-term welfare problems. The ability of the animal to react effectively to antigen challenge will depend upon the numbers of lymphocytes and the activity and efficiency of these lymphocytes. Measures of the ratios of white blood cells, for example the heterophil to lymphocyte ratio, are affected by a variety of factors but some kinds of restraint seem to affect the ratio consistently so they can give some information about welfare. Studies of T-cell activity, e.g. *in vitro* mitogen stimulated cell proliferation, give information about the extent of immunosuppression resulting from the particular treatment. If the immune system is working less well because of a treatment, the animal is coping less well with its environment and the welfare is poorer than in an animal which is not immunosuppressed. Examples of the immunosuppressive effect of transport are the reduction in four different lymphocyte subpopulations after 24 h of transport in horses (Stull *et al.*, 2004) and the reduction in phytohaemagglutinin stimulated lymphocyte proliferation in *Bos indicus* steers during the 6 days after they had been transported for 72 h (Stanger *et al.*, 2005).

As with behavioural measures, some physiological measures are good predictors of an earlier death or of reduced ability to breed while others are not measures of stress because the effect will be brief or slight.

Carcass and mortality measures

Measures of body damage, major disease condition or of increased mortality are indicators of long-term adverse effects on animal welfare. A bruise or cut will result in a slight or a substantial degree of poor welfare depending on its magnitude. Death during handling and transport is usually preceded by a period of poor welfare. Mortality records during journeys are often the only records, which give information about welfare during the journey and the severity of the problems for the animals are often only too clear from such records.

Carcass measures are discussed in detail in Chapter 4, this volume.

Experimental procedures for the scientific study of welfare during transport

As Hall and Bradshaw (1998) explain, information on the stress effects of transport is available from five kinds of study:

1. Studies where transport, not necessarily in conditions representative of commercial practice, was used explicitly as a stressor to evoke a physiological response of particular interest (Smart *et al.*, 1994; Horton *et al.*, 1996).
2. Uncontrolled studies with physiological and behavioural measurements being made before and after long or short commercial or experimental journeys (Becker *et al.*, 1985; Dalin *et al.*, 1988, 1993; Becker *et al.*, 1989; Knowles *et al.*, 1994).

3. Uncontrolled studies during long or short commercial or experimental journeys (Lambooij, 1988; Hall, 1995).

4. Studies comparing animals that were transported with animals that were left behind to act as controls (Nyberg *et al.*, 1988; Knowles *et al.*, 1995).

5. Studies where the different stressors that impinge on an animal during transport were separated out either by experimental design (Bradshaw *et al.*, 1996c; Broom *et al.*, 1996; Cockram *et al.*, 1996) or by statistical analysis (Hall *et al.*, 1998b).

Each of these methods is of value because some are carefully controlled but less representative of commercial conditions while others show what happens during commercial journeys but are less well controlled.

Discussion of Some Key Factors

Animal genetics and transport

Cattle and sheep have been selected for particular breed characteristics for hundreds of years. As a consequence, there may be differences between breeds in how they react to particular management conditions. For example, Hall *et al.* (1998c) found that introduction of an individual sheep to three others in a pen resulted in a higher heart rate and salivary cortisol concentration if it was of the Orkney breed than if it was of the Clun Forest breed. The breed of animal should be taken into account when planning transport.

Farm animal selection for breeding has been directed especially towards maximizing productivity. In some farm species there are consequences for welfare of such selection (Broom, 1994, 1999). Fast-growing broiler chickens may have a high prevalence of leg disorders and Belgian Blue cattle may be unable to calve unaided or without the necessity for Caesarean section. Some of these effects may affect welfare during handling and transport. Some beef cattle, which have grown fast, have joint disorders, which result in pain during transport and some strains of high-yielding dairy cows are much more likely to have foot disorders. Modern strains of dairy cows, in particular, need much better conditions during transport and much shorter journeys if their welfare is not to be poorer than the more traditional breeds of dairy cow used 30 years ago.

Rearing conditions, experience and transport

If animals are kept in such a way that they are very vulnerable to injury when handled and transported, this must be taken into account when transporting them or the rearing conditions must be changed. A notable example of such an effect is the osteopenia and vulnerability to broken bones, which is twice as high in hens in battery cages than in hens which are able to flap their wings and walk around (Knowles and Broom, 1990). Calves are much more disturbed by handling and transport if they are reared in individual crates than if they are reared in groups, presumably because of lack of exercise and absence of social stimulation in the rearing conditions (Trunkfield *et al.*, 1991).

Human contact prior to handling and transport is also important. If young cattle have been handled for a short period just after weaning they are much less disturbed by the procedures associated with handling and transport (Le Neindre *et al.*, 1996). All animals can be prepared for transport by appropriate previous treatment.

Mixing social groups and transport

If pigs or adult cattle are taken from different social groups, whether from the same farm or not, and are mixed with strangers just before transport, during transport or in lairage there is a significant risk of threatening or fighting behaviour (McVeigh and Tarrant, 1983; Guise and Penny, 1989; Tarrant and Grandin, 2000). The glycogen depletion associated with threat, fighting or mounting often results in dark firm dry meat, injuries such as bruising and associated poor welfare. The problem is sometimes very severe, in welfare and economic terms, but is solved by keeping animals in groups with familiar individuals rather than mixing strangers. Cattle might be tethered during loading but should never be tethered when vehicles are moving because long tethers cause a high risk of entanglement and short tethers cause a high risk of cattle being hung by the neck. Mixing of pigs on vehicles causes a substantial increase in aggression (Shenton and Shackleton, 1990) and cortisol levels in transported pigs were higher if there was mixing of pigs from different origins (Bradshaw *et al.*, 1996a,b).

Handling, loading, unloading and welfare

Many studies have shown that loading and unloading are the most stressful part of transport (Hall and Bradshaw, 1998). The physiological changes indicative of stress occur at loading and last for the first few hours of transport. Then, the stress response can gradually decline, depending on driving quality and other factors, as the animals become accustomed to transport (Knowles *et al.*, 1995; Broom *et al.*, 1996). The large effect that loading may have on the welfare of the animals results from a combination of several stressors that impinge upon the animals in a very short period of time. One of these stressors is forced physical exercise as the animals are moved into the vehicle. Physical exertion is particularly important when animals have to climb steep ramps. Second, psychological stress is caused by the novelty of being moved into unknown surroundings. Also, loading requires close proximity to humans and this can cause fear in animals that are not habituated to human contact. Finally, pain may result from mishandling of animals at loading. For example, beating or poking animals with a stick, especially in sensitive areas like the eyes, mouth, anogenital regions or belly and catching sheep by the fleece will cause pain. The use of electric goads will be painful as well.

The slope of ramp is an important aspect when loading or unloading animals. This can be measured in degrees (e.g. 20°) or as percentage gradient (e.g. 20%). The percentage gradient indicates the increase in height in metres over 100 horizontal metres distance. For example, a gradient of 20% means a slope of 20 in 100 (i.e. 1 in 5) and is equivalent to 11°.

There are important differences between species in their response to handling and loading and these should be taken into account when choosing appropriate loading procedures. For example, pigs have more difficulties than sheep or cattle in negotiating steep ramps.

Despite all these differences between and within species, several general recommendations can be made. For example, even illumination and gently curved races without sharp corners facilitate the movement of the animals. Non-slip flooring and good drainage to prevent pooling of water are also important. As animals prefer to walk slightly uphill rather than downhill, floors should be flat or slope upwards. On the other hand, however, ramps should not be too steep (Grandin, 2007), i.e. not more than 20°. If the floor of the loading ramp is not slippery, there still remain differences between species in the steepness of slope which they can climb or descend safely.

Well-trained and experienced stock-people know that cattle can be readily moved from place to place by human movements that take advantage of the animal's flight zone, i.e. the point during an approach when the animal will flee (Kilgour and Dalton, 1984; Grandin, 2000). Cattle will move forward when a person enters the flight zone at the point of balance and can be calmly driven up a race by a person entering the flight zone and moving in the opposite direction to that in which the animals are desired to go. Handling animals without the use of sticks or electric goads results in better welfare and less risk of poor carcass quality.

The handling and loading of poultry and rabbits is very different from that for the larger mammals. Chickens reared for meat production are often collected by human catching teams and sometimes by broiler collector machines. The welfare is almost always worse when human catching is involved (Duncan *et al.*, 1986). Laying hens are usually also collected and put into crates or modules by people and show substantial adrenal responses when caught. Bone breakage is common in hens during catching, especially if the birds have had insufficient exercise because they have been kept in small cages, or have leg disorders, as in fast-growing broiler meat chicken strains.

Thorough knowledge of animal behaviour and the presence of appropriate facilities are important for good welfare during handling and loading. This point is stressed in the OIE Guidelines.

Temperature and other physical conditions during transport

Extremes of temperature can cause very poor welfare in transported animals. Exposure to temperatures below freezing has severe effects on small animals including domestic fowl. However, temperatures that are too high are a more common cause of poor welfare with poultry, rabbits and pigs being especially vulnerable. For example, De la Fuente *et al.* (2004) found that plasma cortisol, lactate, glucose, creatine kinase, lactate dehydrogenase and osmolarity were all higher in warmer summer conditions than in cooler winter conditions in transported rabbits. In each of these species, and particularly in chickens reared for meat production, stocking density must be reduced in temperatures of 20°C or higher or there is a substantial risk of high mortality and poor welfare.

A period of rest during a journey can be important to animals, especially those that are using up more than the usual amount of energy during the journey because of the position that they have to adopt or because they have to show prolonged or intermittent adrenal responses, which mobilize energy reserves. One way of judging how tired animals become during a journey is to observe how strongly they prefer to rest after the journey. Another way is to assess any emergency responses or adverse effects on their ability to cope with pathogen challenge. For example, Oikawa *et al.* (2005) found that horses on a 1500 km journey showed less adrenal response and less sign of harmful inflammatory responses if they had longer rests and had their pen on the vehicle cleaned during the journey.

Vehicle driving methods, stocking density and welfare

When humans are driven in a vehicle, they can usually sit on a seat or hold on to some fixture. Cattle standing on four legs are much less well able to deal with accelerations such as those caused by swinging around corners or sudden braking. Cattle always endeavour to stand in a vehicle in such a way that they brace themselves to minimize the chance of being thrown around and avoid making contact with other individuals. They do not lean on other individuals and are substantially disturbed by too much movement or too high a stocking density. In a study of sheep during driving on winding or straight roads, Hall *et al.* (1998b) found that plasma cortisol concentrations were substantially higher on winding than on straight roads. Tarrant *et al.* (1992) studied cattle at a rather high, an average and a low commercial stocking density and found that falls, bruising, cortisol and creatine kinase levels all increased with stocking density. Careful driving and a stocking density which is not too high are crucial for good welfare.

The amount of space allowed for an animal during transport is one of the most important factors affecting animal welfare. In general, smaller space allowances lead to lower unit costs of transport since more animals can be carried in a vehicle of any particular size. Space allowances have two components. The first component is the floor area available to the animal to stand or lie in. This equates to what is usually referred to as stocking density. The second component is the height of the compartment in which the animal is carried. With multi-decked road vehicles, this may be especially important because there are practical constraints on the overall maximum height of the vehicles, for example to enable them to pass under bridges. There is thus a commercial pressure to reduce the vertical distance between decks (deck height), and therefore, the volume of space above the animals' heads. This reduction may adversely affect adequate ventilation of the inside of the compartment in which the animals are held.

Absolute minimum space allowances are determined by the physical dimensions of animals, but this will not be sufficient to allow for good welfare. Acceptable minimum allowances will be dependent on other factors as well. These include the ability of the animals to thermoregulate effectively, ambient conditions, particularly environmental temperature and whether the animals should be allowed enough space to lie down if they so wish. Whether animals want to lie down may depend on journey length, transport conditions, especially whether it is comfortable to do so and the care exercised in driving the vehicle and its suspension characteristics in relation to the quality of the road surface. A very important consideration in establishing practical minimum space requirements is if the animals are to be rested, watered and fed on the vehicle. Resting, watering and feeding on the vehicle will require lower stocking densities to enable the animals to access feed and water. Space allowances may need to be greater if vehicles are stationary for prolonged periods to promote adequate ventilation, unless this is facilitated and controlled artificially.

When four-legged animals are standing on a surface subject to movement, such as a road vehicle, they position the feet outside the normal area under the body in order to help them to balance. They also need to take steps out of this normal area if subjected to accelerations in a particular direction. Hence, they need more space than if standing still. When adopting this position and making these movements on a moving vehicle, cattle, sheep, pigs and horses make considerable efforts not to be in contact with other animals or the sides of the vehicle. Provided that vehicles are driven well, up to a space allowance larger than that used in animal transport, the greater the space allowance, the better the welfare of the animals. However, if vehicles are driven badly and animals are subjected to the substantial lateral movement that results from driving too fast around corners, or to violent braking, close packing of animals may result in less injury to them. The best practice is to drive well and stock in a way which gives space for the animals to adopt the standing or lying position which is least stressful to them.

A separate problem, which is linked to space allowance, is aggression or potentially harmful mounting behaviour. Pigs and adult male cattle may threaten, fight and injure one another. This results in poor welfare and increased percentage of DFD meat. Rams and some horses may also fight. Such fighting is minimized or avoided by keeping animals in the social groups in which they lived on the farm or by separating animals that might fight. Groups of male animals may mount one another, sometimes causing injuries in doing so. At very high stocking densities, fighting and mounting is more difficult and injuries due to such behaviour may be reduced. However, such problems can be solved by good management of animals and keeping animals at an artificially high stocking density in an attempt to immobilize them will result in poor welfare.

Floor space allowances need to be defined in unambiguous terms. In particular, stocking densities must be defined as m^2 floor area per animal of a specified live weight, e.g. $m^2/100\,kg$, or kg live weight per m^2 floor area (kg/m^2). Stocking rates, such as m^2 per animal, are not an acceptable way of defining floor space requirements since they take no account of variation in animal weight. Definitions of acceptable space allowances must consider the whole range of animal sizes (live weights) to be encountered. A problem is that information applicable to very small or very large animals is sometimes not available. Moreover, the relationship between minimum acceptable space allowance and animal weight is often not linear. Determining appropriate minimum acceptable space allowances for transported animals relies on several types of evidence. These include evidence based on first principles using measurements of the dimensions of animals, evidence based on behavioural observations of animals during real or simulated transport conditions and evidence based on the measurement of indices of adverse effects of transport. An example of the latter kind of evidence would be the amount

Table 7.4. Recommended minimum floor space allowances – examples.

Species	Body weight (kg)	Travel duration	Floor space allowance
Pigs	100 kg	Up to 8 h	0.42 m²
		More than 8 h	0.60 m²
Sheep – shorn	40 kg	Up to 4 h	0.24 m²
		4–12 h	0.31 m²
		More than 12 h	0.38 m²
Sheep – unshorn	40 kg	Up to 4 h	0.29 m²
		4–12 h	0.37 m²
		More than 12 h	0.44 m²
Cattle	500 kg	Up to 12 h	1.35 m²
		More than 12 h	2.03 m²

of bruising on the carcass or the activity of enzymes such as creatine kinase (CK) in the blood that indicate bruising or severe disturbance.

For an animal of the same shape, and where body weight is W, linear measurements will be proportional to the cube root of W ($\sqrt[3]{W}$). The area of the surface of the animal will be proportional to the square of this linear measure ($\left(\sqrt[3]{W}\right)^2$). Algebraically this is equivalent to the cube root of the weight squared ($\sqrt[3]{W^2}$), or weight to the power of two-thirds ($W^{0.67}$). The minimum acceptable area for all types of animal is:

$$A = 0.021 \ W^{0.67}$$

where A is the minimum floor area required by the animal in m² and W is the weight of the animal in kg. The number in the equation (0.021) is a constant for a given shape of the animal, in particular the ratio of its body length to its body width.

As a result of a review of the literature on the effects of space allowance on welfare, the EU SCAHAW (2002) recommended equations for calculating space allowance for pigs, sheep and cattle and examples of the results of such calculations are as shown in Table 7.4. Similarly, the EFSA Scientific panel AHAW (2004) recommended equations to be used for the calculation of space allowances for other farm animal species. In many countries, much less space per animal is provided on vehicles.

Feeding and watering during transport

Drinking is stimulated when the lack of water causes the blood concentration to increase. Animals vary according to species in how often they drink in a 24 h period and horses may only drink once or twice a day. It is difficult to provide water continuously and many animals will not drink during vehicle movement, so frequent stops of sufficient duration for drinking may be necessary if adequate drinking is to occur when water is provided on the vehicle.

Research aiming to characterize progressively dehydration, stress responses and water consumption patterns of horses transported long distances in hot weather and to estimate recovery time after 30 h of commercial transport concluded that transporting healthy horses for more than 24 h during hot weather and without water will cause severe dehydration; transport for more than 28 h, even with periodic access to water, will likely be harmful due to increasing fatigue (Friend, 2000).

Brown *et al.* (1999) compared constant transport for 8, 16 and 24 h without resting periods or watering/feeding and observed the need for pigs to drink and feed during a 6 h lairage period. The results showed that even though the environmental temperatures were relatively mild (14–20°C), all pigs drank and ate during the lairage period and that in particular pigs transported for 8 h, ate and drank immediately after arrival before they rested. It is clear that they had already become dehydrated and hungry. Sheep often do not eat during vehicle movement. Nevertheless, after 12 h of deprivation sheep become very eager to eat (Knowles, 1998).

As for water deprivation sheep seem to be well adapted to drought, as they are able to produce dry faeces and concentrated urine. In addition, their rumen can act as a buffer against dehydration. The effects of water deprivation seem to be largely dependent on ambient temperatures. For example, Knowles *et al.* (1993) found no evidence of dehydration during journeys of up to 24 h when ambient temperatures were not above 20°C. However, when ambient temperatures did rise above 20°C for a large part of the journey, there were clear indications that animals became dehydrated (Knowles *et al.*, 1994). Very many daytime journeys in the world are at higher temperatures than this.

If resting periods within the journey are considered as a means to prevent the effects of food and water deprivation several points have to be taken into account. First, short resting periods – of 1 h, for example – are insufficient and may even have detrimental effects on welfare. Hall *et al.* (1997) studied the feeding behaviour of sheep after 14 h of deprivation and concluded that few sheep obtained adequate food and water within the first hour. Knowles *et al.* (1993) found that recovery after long journeys took place over three phases and that after 24 h of lairage sheep seemed to have recovered from short-term stress and dehydration. It has been suggested that at least 8 h of lairage are needed to gain any real benefit (Knowles, 1998). A further problem is that sheep will not readily drink from unfamiliar water sources, even after prolonged periods of water deprivation (Knowles *et al.*, 1993). Therefore, it is likely that during short resting periods, sheep will not drink and the food they eat may lead to an increased water deficit, particularly if given concentrates (Hall *et al.*, 1997).

A second problem is that feeding during resting periods may cause competition between animals, and the stronger individuals may exclude the weaker ones (Hall *et al.*, 1997). It is therefore important that feeding and drinking space is enough for all animals to have access to food and water simultaneously. Recommended trough space for sheep is $0.112 W^{0.33}$ m (Baxter, 1992). This means 30 cm for sheep of 20 kg body weight and about 34 cm for sheep of 30 kg body weight.

Finally, sheep can be reluctant to eat during lairage, particularly adult animals that are unfamiliar with the feed. Hay has been found to be the most widely accepted form of feed (Knowles, 1998), although Hall *et al.* (1997) found that only small amounts of hay were eaten by sheep after 14 h of food deprivation.

Journey duration and welfare

For all animals except those very accustomed to travelling, being loaded on to a vehicle is a particularly stressful part of the transport procedure. Furthermore, as journeys continue, the duration of the journey becomes more and more important in its effects on welfare. Animals travelling to slaughter are not given the space and comfort that a racehorse or showjumper are given. They are much more active, using much more energy, than an animal that is not transported. As a result they become more fatigued, more in need of water, more in need of food, more affected by any adverse conditions, more immunosuppressed, more susceptible to disease and sometimes more exposed to pathogens on a long journey than on a short journey.

In a survey of records of the transport of 19.3 million broilers killed in four processing plants in the UK, Warriss *et al.* (1990) found an average time, from loading to unloading, of 3.6h, with a maximum of 12.8h. Comparable average times for 1.3 million turkeys killed at two plants were 2.2 and 4.5h, with maxima of 4.7 and 10.2h (Warriss and Brown, 1996). Although there seem to be no published data, spent hens are thought to travel very long distances to slaughter in the UK because of the very small number of plants willing to process them. This long transport must be a cause of some considerable concern. Because poultry held in crates or drawers cannot be effectively fed and watered during transport, journeys must be considerably shorter than for red meat species. Mortality is increased progressively with longer transport times (Warriss *et al.*, 1992). These authors recorded the number of broilers dead on arrival in a sample of 3.2 million birds transported in 1113 journeys to a poultry processing plant. Journey times ranged up to 9h with an overall average time of 3.3h. Total time, from the start of loading birds on to the vehicle to the completion of unloading at the processing plant, ranged up to 10h with an average of 4.2h. The overall mortality for all journeys was 0.194%. However, as journey time increased, so did mortality rate. In journeys lasting less than 4h the prevalence of dead birds was 0.16% while for longer journeys the incidence was 0.28%. In all journeys longer than 4h mortality was therefore on average 80% higher than in all journeys shorter than this.

Birds that have previously suffered painful traumatic injuries such as broken bones and dislocations, which are not uncommon, will suffer progressively more in longer journeys. Animals may also become progressively more fatigued. Liver glycogen, which provides a ready source of metabolic fuel in the form of glucose, is very rapidly depleted after food withdrawal. Warriss *et al.* (1988) found depletion to negligible levels within 6h. Broilers transported 6h had only 43% the amount of glycogen in their livers compared with untransported birds (Warriss *et al.*, 1993).

Rest periods are impracticable and counterproductive for poultry since, as mentioned above, birds can neither realistically be offered food and water, nor can they be effectively inspected by veterinary authorities because of their close confinement in the transport receptacles. Moreover, with current systems of passively ventilated transport vehicles, the reduction in airflow, likely if vehicles stop without unloading the birds, is likely to lead to an increase in temperature within certain parts of the load and possibly cause the development of heat stress (hyperthermia) in the birds.

Horses stand during transport and have to make balance correction movements throughout any vehicle movement. In a study, comparing the effects of road transport ranging from <50 to 300km, a range of physiological measures showed the extent to which the animals had to adapt. The levels of certain lymphocytes were increased in horses after journeys of 150–300km. Plasma concentration of myocardial depressant factor (MDF) peptide fraction was significantly lowered by road transport in journeys exceeding 100 km. It has been reported that road transportation of Sanfratellani horses over distances of 130–200 km resulted in significant elevations in serum creatinine and creatinine kinase (CK). Similar changes were recorded after journeys of 130–350 km in 16 untrained horses of various breeds in aspartate amino transferase (AST), lactate dehydrogenase (LDH), alanine aminotransferase (AAT) and serum alkaline phosphatase (SAP) (Ferlazzo, 1995).

Disease incidence can increase on longer journeys and this has important implications for animal welfare and trade. For example, it is well known that an increased incidence of equine respiratory disease follows prolonged transport. Predisposition to respiratory disease after transport may be due to a marked increase in the numbers and, in viral-infected horses, the activity of pulmonary alveolar macrophages. Therefore, it is evident that transits of 8–12 h or more tend to be more measurably stressful and consideration should be given to monitoring welfare and pathology indicators.

A number of experiments have investigated the effects of journey length on cattle welfare. The majority of authors state that, with increasing duration of the transport, the negative effect on the animals increases as well, as represented through various physiological parameters such as body weight CK, NEFA, BHB, total protein, etc. A period of food and water deprivation of 14 h results in vigorous attempts to obtain food and water when the opportunity arises, although deprivation of 24 h is required before blood physiology changes in calcium, phosphorus, potassium, sodium, osmolarity and urea are apparent (Chupin *et al.*, 2000). However, food and water deprivation during a journey are likely to have much greater and more rapid effects. The extent of energy deficit when cattle were transported for two successive journeys of 29 h with a 24 h rest between them was quantified by Marahrens *et al.* (2003). After 14 h of transport, a break of 1 h for feeding and watering of the animals does not give ruminants enough time for sufficient food and water intake but just prolongs the total duration of the journey. Cattle become more fatigued as journeys continue and there are more frequent losses of balance.

References

Baldock, N.M. and Sibly, R.M. (1990) Effects of handling and transportation on heart rate and behaviour in sheep. *Applied Animal Behaviour Science* 28, 15–39.

Baxter, M.R. (1992) The space requirements of housed livestock. In: Phillips, C. and Piggins, D. (eds) *Farm Animals and the Environment*. CAB International, Wallingford, UK, pp. 67–81.

Becker, B.A., Neinaber, J.A., Deshazer, J.A. and Hahn, G.L. (1985) Effect of transportation on cortisol concentrations and on the circadian rhythm of cortisol in gilts. *American Journal of Veterinary Research* 46, 1457–1459.

Becker, B.A., Mayes, H.F., Hahn, G.L., Neinaber, J.A., Jesse, G.W., Anderson, M.E., Heymann, H. and Hedrick, H.B. (1989) Effect of fasting and transportation on various physiological parameters and meat quality of slaughter hogs. *Journal of Animal Science* 67, 334.

Bradshaw, R.H., Hall, S.J.G. and Broom, D.M. (1996a) Behavioural and cortisol responses of pigs and sheep during transport. *Veterinary Record* 138, 233–234.

Bradshaw, R.H., Parrott, R.F., Forsling, M.L., Goode, J.A., Lloyd, D.M., Rodway, R.G. and Broom, D.M. (1996b) Stress and travel sickness in pigs: effects of road transport on plasma concentrations of cortisol, beta-endorphin and lysine vasopressin. *Animal Science* 63, 507–516.

Bradshaw, R.H., Parrott, R.F., Goode, J.A., Lloyd, D.M., Rodway, R.G. and Broom, D.M. (1996c) Behavioural and hormonal responses of pigs during transport: effect of mixing and duration of journey. *Animal Science* 62, 547–554.

Broom, D.M. (1986) Indicators of poor welfare. *British Veterinary Journal* 142, 524–526.

Broom, D.M. (1994) The effects of production efficiency on animal welfare. In: Huisman, E.A., Osse, J.W.M., van der Heide, D., Tamminga, S., Tolkamp, B.L., Schouten, W.G.P., Hollingsorth, C.E. and van Winkel, G.L. (eds) *Proceedings of the 4th Zodiac Symposium: Biological Basis of Sustainable Animal Production.* Wageningen Publications, Wageningen, The Netherlands, pp. 201–210.

Broom, D.M. (1998) Welfare, stress and the evolution of feelings. *Advances in the Study Behavior* 27, 371–403.

Broom, D.M. (1999) The welfare of dairy cattle. In: Agaard, K. (ed.) *Proceedings of the 25th International Dairy Congress.* Danish National Committee of I.D.F., Aarhus, Denmark, pp. 32–39.

Broom, D.M. (2000) Welfare assessment and problem areas during handling and transport. In: Grandin, T. (ed.) *Livestock Handling and Transport,* 2nd edn. CAB International, Wallingford, UK, pp. 43–61.

Broom, D.M. (ed.) (2001a) *Coping with Challenge: Welfare in Animals Including Humans.* Dahlem University Press, Berlin, Germany.

Broom, D.M. (2001b) Coping, stress and welfare. In: Broom, D.M. (ed.) *Coping with Challenge: Welfare in Animals Including Humans.* Dahlem University Press, Berlin, Germany, pp. 1–9.

Broom, D.M. (2002) Does present legislation help animal welfare? *Landbauforschung Völkenrode* 227, 63–69.

Broom, D.M. (2003) *The Evolution of Morality and Religion.* Cambridge University Press, Cambridge.

Broom, D.M. (2006a) Behaviour and welfare in relation to pathology. *Applied Animal Behaviour Science* 97, 71–83.

Broom, D.M. (2006b) The evolution of morality. *Applied Animal Behaviour Science* 100, 20–28.

Broom, D.M. and Fraser, A.F. (2007) *Domestic Animal Behaviour and Welfare.* CAB International, Wallingford, UK.

Broom, D.M. and Johnson, K.G. (2000) *Stress and Animal Welfare.* Kluwer, Dordrecht, The Netherlands.

Broom, D.M., Goode, J.A., Hall, S.J.G., Lloyd, D.M. and Parrott, R.F. (1996) Hormonal and physiological effects of a 15 hour journey in sheep: comparison with the responses to loading, handling and penning in the absence of transport. *British Veterinary Journal* 152, 593–604.

Brown, S.N., Knowles, T.G., Edwards, J.E. and Warriss, P.D. (1999) Behavioural and physiological responses of pigs being transported for up to 24 hours followed by six hours recovery in lairage. *Veterinary Record* 145, 421–426.

Chupin, J.M., Savignac, C., Aupiais, A. and Lucbert, J. (2000) Influence d'un jeûne hydrique et alimentaire prolonge sur le comportement, la denutrition, la dehydratation et le confort des bovins. *Rencontres Recherches Ruminants* 7, 79.

Cockram, M.S., Kent, J.E., Goddard, P.J., Waran, N.K., McGilp, I.M., Jackson, R.E., Muwanga, G.M. and Prytherch, S. (1996) Effect of space allowance during transport on the behavioural and physiological responses of lambs during and after transport. *Animal Science* 62, 461–477.

Dalin, A.M., Nyberg, L. and Eliasson, L. (1988) The effect of transportation/relocation on cortisol. CBG and induction of puberty in gilts with delayed puberty. *Acta Veterinaria Scandinavica* 29, 207–218.

Dalin, A.M., Magnusson, U., Haggendal, J. and Nyberg, L. (1993) The effect of transport stress on plasma levels of catecholamines, cortisol, corticosteroid-binding globulin, blood cell count and lymphocyte proliferation in pigs. *Acta Veterinaria Scandinavica* 34, 59–68.

De La Fuente, J., Salazar, M.I., Ibáñez, M. and Gonzalez de Chavarri, E. (2004) Effects of season and stocking density during transport on live-weight and biochemical measurements of stress, dehydration and injury of rabbits at time of slaughter. *Animal Science* 78, 285–292.

Duncan, I.J.H., Slee, G.S., Kettlewell, P.J., Berry, P.S. and Carlisle, A.J. (1986) Comparison of stressfulness of harvesting broiler chickens by machine and by hand. *British Poultry Science* 27, 109–114.

EFSA Scientific panel AHAW (2004) *The Welfare of Animals During Transport*. European Food Safety Authority Scientific Panel on Animal Health and Welfare. Available at: http://www.efsa.eu.int/science/ahaw/ahaw_opinions/424/opinion_ahaw_01_atrans_ej44_en1.pdf

FAOSTAT (2006) see FAO. Available at: https:// faostat.fao.org

Fazio, E. and Ferlazzo, A. (2003) Evaluation of stress during transport. *Veterinary Research Communications* 27, 519–524.

Fell, L.R., Shutt, D.A. and Bentley, C.J. (1985) Development of salivary cortisol method for detecting changes in plasma 'free' cortisol arising from acute stress in sheep. *Australian Veterinary Journal* 62, 403–406.

Ferlazzo, A. (1995) Animali da reddito: indicatori di benessere e "linee guida" per la loro applicazione nel trasporto. In: *Conferenza gli indicatori scientifici del benessere animale*. Brescia, Italy.

Fraser, A.F. and Broom, D.M. (1997) *Farm Animal Behaviour and Welfare*. CAB International, Wallingford, UK.

Friend T.H. (2000) Dehydration, stress, and water consumption of horses during long-distance commercial transport. *Journal of Animal Science* 78, 2568–2580.

FVE (2001) *Transport of Live Animals, Position Paper 01/043*. Federation of Veterinarians of Europe, Brussels, Belgium.

Geverink, N.A., Bradshaw, R.H., Lambooij, E., Wiegant, V.M. and Broom, D.M. (1998) Effects of simulated lairage conditions on the physiology and behaviour of pigs. *Veterinary Record* 143, 241–244.

Grandin, T. (1980) Observations of cattle behaviour applied to the design of cattle handling facilities. *Applied Animal Ethology* 6, 19–31.

Grandin, T. (1982) Pig behaviour studies applied to slaughter plant design. *Applied Animal Ethology* 9, 141–151.

Grandin, T. (1989) Behavioural principles of livestock handling. *Professional Animal Scientist* 5, 1–11.

Grandin, T. (2000) Behavioural principles of handling cattle and other grazing animals under extensive conditions. In: Grandin, T. (ed.) *Livestock Handling and Transport*, 2nd edn. CAB International, Wallingford, UK.

Grandin, T. (2007) Handling facilities and restraint of range cattle. In: Grandin, T. (ed.) *Livestock Handling and Transport*, 3rd edn. CAB International, Wallingford, pp. 90–108.

Guise, J. and Penny, R.H.C. (1989) Factors affecting the welfare, carcass and meat quality of pigs. *Animal Production* 49, 517–521.

Hall, S.J.G. (1995) Transport of sheep. *Proceedings of the Sheep Veterinary Society* 18, 117–119.

Hall, S.J.G. and Bradshaw, R.H. (1998) Welfare aspects of transport by road of sheep and pigs. *Journal of Applied Animal Welfare Science* 1, 235–254.

Hall, S.J.G., Schmidt, B. and Broom, D.M. (1997) Feeding behaviour and the intake of food and water by sheep after a period of deprivation lasting 14 h. *Animal Science* 64, 105–110.

Hall, S.J.G., Forsling, M.L. and Broom, D.M. (1998a) Stress responses of sheep to routine procedures: changes in plasma concentrations of vasopressin, oxytocin and cortisol. *Veterinary Record* 142, 91–93.

Hall, S.J.G., Kirkpatrick, S.M., Lloyd, D.M. and Broom, D.M. (1998b) Noise and vehicular motion as potential stressors during the transport of sheep. *Animal Science* 67, 467–473.

Hall, S.J.G., Kirkpatrick, S.M. and Broom, D.M. (1998c) Behavioural and physiological responses of sheep of different breeds to supplementary feeding, social mixing and taming, in the context of transport. *Animal Science* 67, 475–483.

Horton, G.M.J., Baldwin, J.A., Emanuele, S.M., Wohlt, J.E. and McDowell, L.R. (1996) Performance and blood chemistry in lambs following fasting and transport. *Animal Science* 62, 49–56.

Jones, A.R. and Price, S.E. (1992) Measuring the response of fallow deer to disturbance. In: Brown, R.D. (ed.) *The Biology of Deer*. Springer, Berlin, Germany, pp. 211–216.

Kelley, K.W. (1985) Immunological consequences of changing environmental stimuli. In: Moberg, G.P. (ed.) *Animal Stress*. American Physiological Association, Bethesda, Maryland, pp. 193–223.

Kenny, F.J. and Tarrant, P.V. (1987) The reaction of young bulls to short-haul road transport. *Applied Animal Behaviour Science* 17, 209–227.

Kent, J.F. and Ewbank, R. (1986) The effect of road transportation on the blood constituents and behaviour of calves. III. Three months old. *British Veterinary Journal* 142, 326–335.

Kilgour, R. and Dalton, C. (1984) *Livestock Behaviour: A Practical Guide.* Granada, St Albans, UK.

Knowles, T.G. (1998) A review of road transport of slaughter sheep. *Veterinary Record* 143: 212–219.

Knowles, T.G. and Broom, D.M. (1990) Limb bone strength and movement in laying hens from different housing systems. *Veterinary Record* 126, 354–356.

Knowles, T.G. and Warriss, P.D. (2000). Stress physiology of animals during transport. In: Grandin, T. (ed.) *Livestock Handling and Transport*, 2nd edn. CAB International, Wallingford, UK, pp. 385–407.

Knowles, T.G., Warriss, P.D., Brown, S.N., Kestin, S.C., Rhind, S.M., Edwards, J.E., Anil, M.H. and Dolan, S.K. (1993) Long distance transport of lambs and the time needed for subsequent recovery. *Veterinary Record* 133, 286–293.

Knowles, T.G., Warriss, P.D., Brown, S.N. and Kestin, S.C. (1994) Long distance transport of export lambs. *Veterinary Record* 134, 107–110.

Knowles, T.G., Brown, S.N., Warriss, P.D., Phillips, A.J., Doland, S.K., Hunt, P., Ford, J.E., Edwards, J.E. and Watkins, P.E. (1995) Effects on sheep of transport by road for up to 24 hours. *Veterinary Record* 136, 431–438.

Lambooij, E. (1988) Road transport of pigs over a long distance: some aspects of behaviour, temperature and humidity during transport and some effects of the last two factors. *Animal Production* 46, 257–263.

Lambooij, E., Geverink, N., Broom, D.M. and Bradshaw, R.H. (1995) Quantification of pigs' welfare by behavioural parameters. *Meat Focus International* 4, 453–456.

Le Neindre, P., Boivin, X. and Boissy, A. (1996). Handling of extensively kept animals. *Applied Animal Behaviour Science* 49, 73–81.

McVeigh, J.M. and Tarrant, V. (1983) Effect of propanolol on muscle glycogen metabolism during social regrouping of young bulls. *Journal of Animal Science* 56, 71–80.

Marahrens, M., Von Riehthofen, I., Schmeiduch, S. and Hartung, J. (2003) Special problems of long-distance road transports of cattle. *Deutsche Tierärztliche Wochenschrift* 110, 120–125.

Murray, K.C., Davies, D.H., Cullenane, S.L., Edisson, J.C. and Kirk, J.A. (2000) Taking lambs to the slaughter: marketing chanels, journey structures and possible consequences for welfare. *Animal Welfare* 9, 111–112.

Nyberg, L., Lundstrom, K., Edfors-Lilja, I. and Rundgren, M. (1988) Effects of transport stress on concentrations of cortisol, corticosteroid-binding globulin and glucocorticoid receptors in pigs with different halothane genotypes. *Journal of Animal Science* 66, 1201–1211.

Oikawa, M., Hobo, S., Oyomada, T. and Yoshikawa, H. (2005) Effects of orientation, intermittent rest and vehicle cleaning during transport on development of transport-related respiratory disease in horses. *Journal of Comparative Pathology* 132, 153–168.

Parrott, R.F., Misson, B.H. and Baldwin, B.A. (1989) Salivary cortisol in pigs following adrenocorticotrophic hormone stimulation: comparison with plasma levels. *British Veterinary Journal* 145, 362–366.

Parrott, R.F., Hall, S.J.G. and Lloyd, D.M. (1998a) Heart rate and stress hormone responses of sheep to road transport following two different loading responses. *Animal Welfare* 7, 257–267.

Parrott, R.F., Hall, S.J.G., Lloyd, D.M., Goode, J.A. and Broom, D.M. (1998b) Effects of a maximum permissible journey time (31 h) on physiological responses of fleeced and shorn sheep to transport, with observations on behaviour during a short (1 h) rest-stop. *Animal Science* 66, 197–207.

Parrott, R.F., Lloyd, D.M. and Brown, D. (1999) Transport stress and exercise hyperthermia recorded in sheep by radio-telemetry. *Animal Welfare* 8, 27–34.

Ravenswaaij, C.M.A. van, van Kollée, L.A.A., Hopman, J.C.W., Stoelinga, G.B.A. and van Geijn, H. (1993) Heart rate variability. *Annals of Internal Medicine* 118, 427–435.

Riad-Fahmy, D., Read, G.F., Walker, R.F. and Griffiths, K. (1982) Steroids in saliva for assessing endocrine function. *Endocrinology Review* 3, 367–395.

Rushen, J. (1986a) The validity of behavioural measure of aversion: a review. *Applied Animal Behaviour Science* 16, 309–323.

Rushen, J. (1986b) Aversion of sheep for handling treatments: paired choice experiments. *Applied Animal Behaviour Science* 16, 363–370.

SCAHAW (2002) *The Welfare of Animals During Transport (Details for Horses, Pigs, Sheep and Cattle).* Scientific Committee on Animal Health and Animal Welfare, European Commission Health and Consumer Protection Directorate General, Brussels, Belgium.

Shenton, S.L.T. and Shackleton, D.M. (1990) Effects of mixing unfamiliar individuals and of azaperone on the social behaviour of finishing pigs. *Applied Animal Behaviour Science* 26, 157–168.

Smart, D., Forhead, A.J., Smith, R.F. and Dobson, H. (1994) Transport stress delays the oestradiolinduced LH surge by a non-opioidergic mechanism in the early postpartum ewe. *Journal of Endocrinology* 142, 447–451.

Stanger, K.J., Ketheesan, N., Parker, A.J., Coleman, C.J., Lazzaroni, S.M. and Fitzpatrick, L.A. (2005) The effect of transportation on the immune status of *Bos indicus* steers. *Journal of Animal Science*, 83, 2632–2636.

Stull, C.L., Spier, S.J., Aldridge, B.M., Blanchard, M. and Stott, J.L. (2004) Immunological response to long-term transport stress in mature horses and effects of adaptogenic dietary supplementation as an immunomodulator. *Equine Veterinary Journal* 36, 583–589.

Tarrant, V. and Grandin, T. (2000) Cattle transport. In: Grandin, T. (ed.) *Livestock Handling and Transport*, 2nd edn. CAB International, Wallingford, UK, pp. 151–173.

Tarrant, P.V., Kenny, F.J., Harrington, D. and Murphy, M (1992) Long distance transportation of steers to slaughter, effect of stocking density on physiology, behaviour and carcass quality. *Livestock Production Science* 30, 223–238.

Trunkfield, H.R. and Broom, D.M. (1990) The welfare of calves during handling and transport. *Applied Animal Behaviour Science* 28, 135–152.

Trunkfield, H.R. and Broom, D.M. (1991) The effects of the social environment on calf responses to handling and transport. *Applied Animal Behaviour Science* 30, 177.

Trunkfield, H.R., Broom, D.M., Maatje, K., Wierenga, H.K., Lambooij, E. and Kooijman, J. (1991) Effects of housing on responses of veal calves to handling and transport. In: Metz, J.H.M. and Groenestein, C.M. (eds) *New Trends in Veal Calf Production*. Wageningen Publications, Wageningen, The Netherlands, pp. 40–43.

Warriss, P.D., Bevis, E.A. and Brown, S.N. (1990) Time spent by broiler chickens in transit to processing plants. *Veterinary Record* 127, 617–619.

Warriss, P.D. and Brown, S.N. (1996) Time spent by turkeys in transit to processing plants. *Veterinary Record* 139, 72–73.

Warriss, P.D., Kestin, S.C., Brown, S.N. and Bevis, E.A. (1988) Depletion of glycogen reserves in fasting broiler chickens. *British Poultry Science* 29, 149–154.

Warriss, P.D., Bevis, E.A., Brown, S.N. and Edwards, J.E. (1992) Longer journeys to processing plants are associated with higher mortality in broiler chickens. *British Poultry Science* 33, 201–206.

Warriss, P.D., Kestin, S.C., Brown, S.N., Knowles, T.G., Wilkins, L.J., Edwards, J.E., Austin, S.D. and Nicol, C.J. (1993) The depletion of glycogen stores and indices of dehydration in transported broilers. *British Veterinary Journal* 149, 391–398.

8 Africa

K. MENCZER

Natural Resources/Biodiversity Consultant, USA

Abstract

This account of long-distance transport in Africa covers North Africa, East Africa, Southern Africa and West Africa. The chapter examines the long-distance transport of cattle, goats and sheep throughout Africa and, to a lesser extent, the transport of horses, donkeys and camels. Pigs and chickens are often produced in relatively close proximity to the place of slaughter, and therefore, are less likely to be subject to long-distance transport.

The main long-distance transport routes in Southern Africa are from Namibia, via Botswana to South Africa (by road 2–5 days, with distances covered varying from 1000 to 2000 km) and the export of animals by sea from ports in South Africa and Mozambique to Mauritius (by sea 7–10 days). In West Africa, long-distance transport can take several days, with routes from Niger and Mali to Togo, Benin, Ghana and Nigeria the longest, taking, on average, 3–6 days, and covering up to 2000 km. However, the number of days in transit depends on the number of stops at markets, and can take much longer than 6 days. In East Africa, one of the longest routes is in Southern Sudan where the journey can take 3 days from Rumbek to the Uganda border; however, cattle often travel through Uganda from there, another 2–3 days. In North Africa land transport can be 7–9 h. The exception is for sheep imported from Australia. Sheep are shipped directly on vessels from Australia to the Suez port in Egypt and can take about 3 weeks. Egypt imported annually around 50,000–100,000 head of sheep. Trekking is a common means of transporting livestock in Africa, and in East Africa some of the longest treks are recorded – about 75 days. These Southern, West and East African long-distance routes are described further in this chapter.

The chapter discusses the cultural, religious and economic factors influencing the livestock trade in Africa; describes transport by trekking, trucking, rail and ship; and the welfare issues associated with each type of transport. Trucking and trekking are the most common means of long-distance transport in Africa. Animal welfare issues common to long-distance transport include poorly developed and degraded infrastructure; lack of enforcement of national legislation, where legislation governing livestock transport exists; and inhumane handling of livestock throughout the production chain.

Specific cases of good practices, poor practices and opportunities to impact long-distance transport are presented. A good practice is illustrated by South Africa, where NGOs are having an impact on livestock transport and slaughter – providing oversight of the livestock industry. Poor practices include cruel treatment of animals during loading, unloading, transport and slaughter. Cruel treatment includes gouging out eyes before slaughter, using fire, twisting tails and beating exhausted animals to load and offload animals on to trucks and slaughtering animals with cuts across the throat that are incomplete and slaughtering in front of other animals.

Opportunities include the potential for international trade in meat and other livestock products, which, if the importing market demands it, could encourage improved production practices, including more humane transport and slaughter; and the increasing presence and strength of animal welfare NGOs, which can have a positive impact on livestock handling and transport.

Introduction

This chapter covers: background information on the livestock industry in Africa, including an overview of the stakeholders; a summary of key national legislation; economic, cultural and religious factors that influence the livestock trade; detailed information on the handling and condition of animals transported; the seasonal and infrastructure issues related to transport; an overview of some long-distance routes and specific animal welfare issues and concerns; information on livestock slaughter in the region; and a discussion that includes research findings including good practices, poor practices and opportunities for interventions based on how the livestock sector is expected to change over the next 5–10 years.

For the purposes of researching and compiling the information in this chapter, the Africa Region was divided into: the East Africa subregion, which covers Burundi, Djibouti, Eritrea, Ethiopia, Kenya, Rwanda, Somalia, Sudan, Tanzania and Uganda (combined human population about 230 million); the Southern African subregion, covering South Africa, Zimbabwe, Zambia, Botswana, Lesotho, Swaziland, Angola, Mozambique and Mauritius – the last included here because of the long-distance sea transport of livestock from Southern African ports; Francophone West Africa covering Benin, Burkina Faso, Mali, Niger, Togo and Ivory Coast (human population of about 230 million); Anglophone West Africa, covering Ghana and Nigeria (human population of about 90 million); and Northern Africa, with a focus on Morocco (about 35 million people as of July 2006) and Egypt (human population of about 70 million as of 2003).

Types of transport in the region

In general, livestock transport in Africa occurs by foot (trekking), by vehicles designed to carry livestock, by makeshift vehicles, including lorries that may also be used to carry crops and other goods that are not designed specifically for livestock transport, and to a lesser extent by rail and boat. Trekking and motorized overland transport are most common; motorized transport is most often by vehicles not designed to carry livestock.

Trekking is steadily decreasing due to increased urbanization, land-use conflicts and initiatives to encourage pastoralists to become sedentary. Transport by rail is decreasing as well due to deteriorating railway infrastructure. Transport by road is increasing. Transport by ship occurs mainly for export and is seen most prominently in the Southern Africa subregion and in East Africa (mainly in Southern Sudan for export to the Gulf).

Livestock transport follows well-established routes, including road networks and historically used trekking routes. Some of these routes are described in detail below.

Types of animals transported

Throughout Africa, cattle, goats and sheep are transported long distances. Long-distance transport is considered as a minimum of 1 day; the maximum identified in Africa is the sea routes from South Africa to ports in Mauritius and Mozambique. In West Africa, East Africa and Southern Africa, overland transport routes may be up to 6 days, however, with stops at markets these journeys can be several days longer.

Pigs and chickens are more often slaughtered in closer proximity to where they are raised, often by 'backyard breeders'. Transport of live chickens and pigs usually is from one location in a village to another, and is usually no longer than a few hours. In Francophone West Africa, for example, pigs are not subject to long-distance transport and their consumption is mainly restricted to some areas of Ivory Coast, Benin, Togo and a limited part of Burkina Faso. Similar situations regarding pigs are found in many parts of Africa, and are related to the religious community most prevalent in the area – most abattoirs are operated by Muslims, and pigs are not allowed to be slaughtered there. In addition, pigs are usually not raised in Muslim majority communities. Therefore, it is more likely that pigs will be slaughtered within or near to the community in which they are raised. There are exceptions however, notably Namibia, Zambia and South Africa, where pigs are transported long distances.

Poultry also are often slaughtered closer to the area in which they are raised, and most often are slaughtered informally, not in abattoirs, but by families who raise the birds or by those who buy a live chicken. Although poultry may be transported long distances, the threat of avian flu has had a significant impact on transport of poultry in Africa, and imports of live poultry are now banned in some countries.

Donkeys, camels and horses are transported long distances in specific areas of Africa. Mainly in the north and west of the continent, donkeys and camels are transported long distances; transport may be for slaughter or for use as draught animals. Horses are transported long distances in Southern Africa and North Africa. Ostriches may be transported long distances, but this is mainly restricted to South Africa and Namibia.

While goat, sheep and other meat products are regularly consumed throughout Africa, for the most part, cattle drive the livestock trade. This is largely because the profit margin for cattle is greater than for other livestock, and because the demand is mainly in urban areas, but most cattle are raised in rural areas (whereas goats, sheep and chickens are raised in semi-urban and urban environments as well as in rural). This chapter focuses on long-distance transport of cattle, goat, sheep, and to a lesser extent donkeys, camels, horses and chickens. However, the main animal welfare issues related to transport for all draught and food animals (including ostriches, horses and pigs) are similar.

Types of animal welfare issues that arise

In general, the animal welfare issues attributed to livestock transport are common throughout Africa.

Infrastructure: because of limited and deteriorating roads and the need to move animals as quickly as possible (because of the perception that 'time=money'), livestock are not allowed to rest, drink water and feed during long-distance transport. For the most part, there are no facilities along main stock routes that allow for rest and offloading. A lack of appropriate infrastructure at markets (e.g. ramps, shelter) results in injury and distress during loading and offloading. Poor road infrastructure results in prolonged journeys and accidents involving livestock vehicles.

Animal handling: because of the often sub-standard nature of the trucks, live-stock may be overloaded; species may be mixed together rather than separated into compartments and smaller animals may be trampled. Livestock are not protected from extreme heat or chill. The floors of trucks do not prevent slippage, and animals can get gored by horns or become injured when they fall; large gaps in trucks allow horns, heads and legs to stick out, and animals often break horns or legs in transit.

In many countries in the region, bureaucracy and corruption prolong already long trips, and increase exposure to heat and sun and the amount of time spent without food, water and rest. Overland trips can take 7 days, and can be even longer if the truck stops at more markets along the way, has a breakdown or is stopped at borders for incorrect permits or for harassment. The multiplicity of control points, taxes and fees, complicate trade routes, and truck drivers may choose to use less convenient routes (more degraded and/or more circuitous) to avoid additional fees. Illicit fees are a serious problem in Ghana and Nigeria where they significantly increase the cost of transport.

In Egypt, the main animal welfare problems that occur during transport are exposure to direct sun, slippery truck floors resulting in injury, overcrowding and mixing of species and loading and unloading using cruel techniques, such as those recorded at Cairo's Bassatin abattoir during an investigation by Animals Australia in January 2006. This investigation demonstrated violent treatment to animals during transport and pre-slaughter, which did not comply with halal slaughter. Breaches included the cutting of Achilles tendons, gouging of eyes, slaughtering an animal in front of others and killing animals with several incomplete cuts of the animals' throats.

In East Africa, as in other subregions, animal welfare concerns arise from the use of inappropriate means and modes of transport and handling. Many animals may be abused while being transported, during loading, offloading and during trekking. Injuries include bruises, fractures, broken horns, suffocation, cases where animals abort their fetus, and even death. Also common in East Africa is transporting different species in the same vehicle compartment resulting in trampling, overloading, speeding and motor vehicle accidents.

Animal welfare concerns vary with the season – different concerns arise during the wet/rainy seasons and dry seasons. In the dry season/drought periods a lack of pasture and water, especially in the arid and semi-arid areas, makes these areas even more difficult to traverse. Animals trekked or trucked over long distances during this period suffer dehydration, heat exhaustion, hunger and thirst. This is made worse by overcrowding in poorly ventilated and inappropriate transport vehicles. Animals trucked during the rainy/wet seasons suffer wind chill and injuries from slipping and falling in vehicles or suffer from vehicle accidents. Many roads over which livestock vehicles pass are murram (laterite); during rains, trucks get stuck in the mud and animals remain on board for long periods, exposing them to stress and exhaustion. They are usually not offloaded due to lack of facilities, fed or given water.

Rough terrain is also a problem. In East Africa, this is the case especially in Eritrea, Sudan, Djibouti, Somalia and Ethiopia. In some of these countries, roads are so rutted that it is impossible for animals to remain upright in the vehicle.

Livestock placed in dhows, long, flat sailing vessels that are lateen-rigged and found in the Indian Ocean along the east coast of Africa, and in other boats, as occurs in Southern Sudan for livestock exported to the Gulf, are often overcrowded and suffocation may occur. On the journey, there is also a lack of water, feed and basic care.

In Uganda, the long-horned Ankole cattle are tied by their horns to overhead truck bars, and often during transport the rope will fall around the neck of the cow and the animal will hang during the long overland trip, resulting in slow death, as has been observed many times by the author.

Similar situations occur in the Southern Africa subregion. They include overcrowding during sea voyages; transport of livestock in unsuitable vehicles causing injury and death; delays at border posts; and speculators moving livestock long distances in an uncontrolled fashion to meet needs of informal markets (set up ad hoc) and auction sales. For Southern Africa, additional animal welfare concerns were noted: lack of codes of practice on transport of equines for slaughter and difficulties with traceability of animals once they are established in the new country – after import they 'change nationality'.

While animal welfare concerns vary with seasons, they also vary with the type of transport. In this respect, West Africa is typical of most subregions in Africa.

Trekking

Trekking speed can be very high and sometimes, exhausted or slow animals are beaten to make them run. The average trekking distance varies from 50 km over about 95 days to 241 km during 210 days (Jost, 2002). The cattle drivers walk alongside their cattle. Food and water is sometimes scarce on the journey, and animals arrive at camp exhausted and unhealthy. Animals often have sores and wounds on their skin caused by crossing thorny pastures.

For example, in Mali, Niger and Burkina Faso (as well as in other parts of Africa), vegetation is thorny, water points and pasture are scarce, and herders often run with their animals when they have to cross these difficult areas. Owners attempt to cut the time of stay in hostile areas; however, carcasses of animals too weak to go on, suffering from lack of water, are frequently found on the savannah.

In coastal areas, animals get stuck in wet soils, and insects and diseases specific to humid areas, such as Tryponosomiasis and Dermatophilosis, affect the welfare and health of livestock.

By vehicle

In Francophone West Africa, road journeys can be up to 500 km and last 2 days or more. However, due to poor and in many places, deteriorating road conditions and because of the checkpoints, a 500 km journey may take up to 12 h or more. Animal welfare issues related to vehicle transport are discussed above, and they are common to all Africa's subregions. The longer the distance and the more complicated the route (which are affected by the number of markets the driver stops at), the more the animals suffer – often being beaten during loading and unloading, rarely getting sufficient food and water to gather strength before the next leg of the journey.

By rail

Animals transported by rail are usually in a healthier condition upon arrival because they can rest for some days in holding pens, have some water and are fed before being slaughtered. This method also has the advantage that it avoids the many illegal 'checkpoints', where animals are often forced to withstand hours, even days, of hot sun with no food or water or area in which to rest.

Animal welfare issues also vary according to the party responsible for transport. As found in Francophone West Africa, when animals are conveyed by dealers and traders (landlords, middlemen customs officers, transporters), the conditions of transport are much worse, as management is solely for profit with no regard for an animal's well-being.

By ship

Animals are often overloaded, and during the journey are unable to feed, have water or rest.

Background information

Long-distance transport in Africa is mainly by road (mainly substandard trucks, as well as trucks designed for livestock) and trekking; and to a lesser extent rail and boat. Reasons for live animal transport include sale in markets, slaughter, search for pasture and water, rearing/replacement of breeding stock, fattening in ranches and feedlots and export.

Some examples of the condition of infrastructure in the region include the following:

Southern Africa

In Southern Africa, transport is mainly by road, in modern carriers or substandard make-do lorries and trailers. There is some seasonal trekking of cattle and goats within Botswana, Angola and Namibia. There is also cross-boundary movement of livestock between Malawi and Mozambique, as well as between Namibia and Angola, but this is limited by the threat of transboundary diseases in the region. In Southern Africa, there is very little, if any movement of livestock by train.

Northern Africa

In Egypt, transport of livestock for slaughter is exclusively overland by trucks. In urban areas, transport from farm to market and from market to abattoir is the responsibility of wholesalers or retailers, usually using substandard trucks, and there is typically no governmental oversight. In rural areas, transport of livestock from market to abattoir is done by the butchers, usually on small, substandard trucks, not equipped for carrying livestock. One or two head of cattle are often loaded with insufficient space and improper restraining.

East Africa

In East Africa, the main methods of live animal transport are trekking, trucking and shipping (dhows) for those destined for export markets. Animals are also carried by hand, tied to bicycles, in trunks of cars and on roof tops, especially chickens and other small stock. Trekking is the most common mode of transport, particularly for livestock from pastoral areas. Trucking is preferred from secondary to terminal markets, is used most during periods of heightened demand, and is used for goats and sheep and finished cattle to big consumption areas and export centres. Rail is common in Tanzania and to a small extent in Kenya. Shipping is used for animals destined for export markets, mainly from countries in the Horn of Africa (Somalia, Djibouti and Eritrea) to markets in Gulf countries; from Kenya to Mauritius; and from Tanzania to the Comoros, among other destinations. Because of high transport costs (especially in Kenya, Somalia, Sudan and Ethiopia), livestock producers and traders in the region prefer trekking livestock if security permits.

West Africa

In Ghana and Nigeria, chickens and goats are often seen tied to the rooves of buses travelling long distances across country. Special carrying baskets may be used and tied to the roof or the animals may be strapped down directly on the roof. Cattle, goats and sheep are most often conveyed long distances, and trucks specifically designed for livestock and less appropriate vehicles are the most common means of transport. In Francophone West Africa, as in other subregions, transport is by trekking, by truck and where railway networks exist, transport may be by train. Rail transport may be up to 48 h, and distances can be as high as 1000 km. Animals are transported by rail for export or for slaughtering in urban areas. Transportation by train is frequent in areas where agro-pastoral systems exist, particularly in coastal areas.

Francophone West Africa is typical of regions where trekking still occurs. This practice is diminishing in most countries, with increasing urbanization, conflicts between herders and other land users, disease concerns and increasingly limited natural resources available for livestock. Many countries are also providing incentives for herders to settle in designated areas, and give up their seasonal movement.

In general throughout Africa, the same animal may be transported by foot, train and truck, and may change owners numerous times, being sold and resold at different markets, until the best price is obtained and they are slaughtered.

Across Africa, live animal transport is a daily activity that typically increases on market days, public holidays and during religious festivities. Countries across the region collect different information on livestock movement, and collect and collate the data with varying degrees of accuracy. In addition, individual subregional reports provide data differently; and therefore, it is impossible to identify the number of animals by species transported for slaughter across the continent.

Overview of the livestock industry

Key stakeholders

In general, key stakeholders in the livestock industry include producers; livestock traders in the domestic and export markets; animal handlers at all levels of the marketing chain; processors; transporters; veterinary services and animal production personnel (often Ministry of Agriculture and district-level staff); regulatory and law enforcement agencies including the courts, police and customs personnel; local government authorities; and animal welfare organizations.

Southern Africa

More specifically, in Southern Africa's formal commercial livestock production system, chief stakeholders are: private, state or parastatal meat board or controlling company; a state-sponsored veterinary control system for quality control in line with OIE, EU and WTO legislation; private, state or parastatal abattoirs; private shipping or transport companies; feedlots; auctioneers; retail butchers; and private, state-subsidized or parastatal livestock producers.

However, in Southern African countries, the informal livestock sector is 20–80% of production, marketing and slaughtering. Stakeholders in the informal sector are: speculators; auctioneers, shipping and transport agents (who do not always comply with traceability standards); commercial farmers who supply to speculators; informal livestock traders; communal and small-scale livestock farmers; stock thieves; caterers who buy directly off-farm for cultural occasions such as weddings and funerals; informal livestock transporters – often one animal is transported at a time in a small truck or trailer or tied on top of a bus or taxi; and illegal informal butchers and meat traders.

West Africa

Francophone West Africa has a similar set of stakeholders, with some specific to the West Africa situation in which tradition, culture and ethnicity figure strongly:

- Breeders or owners of livestock are generally *Peulh* transhumant or farmers or from other ethnic groups close to the *Peulh* (e.g. *Gando* in Benin).
- Livestock traders include two groups of traders: retailers and wholesale traders. Retailers purchase animals in primary markets and sell them to wholesale dealers in secondary markets. Animals are then escorted on foot or by truck. Wholesale traders buy animals in secondary markets. Animals (at least 20–30 per dealer) are sold at terminal markets for slaughtering in the country or for export. They are escorted by road or by train from the terminal market.
- Butchers are traditionally Muslim (*Hausa* or *Yoruba*), and less often, from other ethnic groups. Middlemen are very numerous and represent a traditional institution in all West African countries. They are present during all purchase and sale transactions, and they collect a traditional tax from the breeder (seller) and the trader (customer), as well as other informal taxes.

National Legislation Pertaining to Animals During Transport

In Africa, some countries have well-developed legislation that covers animal welfare and livestock transport (South Africa, Kenya); while other countries have fairly well-developed legislation (Uganda, Ghana); and others have a weak legislative framework. However, whatever the status of legislation, the commonality is that enforcement is very weak; compliance is very low; and knowledge of the legislation is lacking even among key stakeholders. In general, the political will to enforce legislation is minimal.

Examples of national legislation and its status include:

1. The South African Development Community's (SADC) livestock sector regulates trade in livestock commodities in the Southern Africa region. Most of the relevant country-specific legislation in the region is based on South Africa's, particularly in Namibia and the Commonwealth countries, and covers animal welfare, animal disease control, marketing, meat safety and traceability. Due to colonization, Mauritius, Mozambique and Angola have legislation based more on European models. Enforcement of the legislation varies from country to country, depending on infrastructure, political will and available funding.

In South Africa, guidelines on transport of animals by land and by sea have been developed, and the National Society for the Prevention of Cruelty to Animals plays a significant role, in partnership with government, in enforcing animal welfare standards for long-distance transport in South Africa, including at border crossings. Fortunately, in the Southern Africa subregion, movement permits are mandatory and failure to comply with animal health regulations is a severe offence that can be used as motivation for improved transport.

2. Although since 1902, Egypt has had regulations on animal welfare that prohibit cruel handling of animals, punishment of violators is not included in the Law of Agriculture No. mm 53/1966. The welfare of livestock during transport is dependent on the ethics, religious aspects, awareness and understanding of the individuals responsible for livestock transportation. Egypt's legislation cannot be relied on for the well-being of livestock during transport. North African countries are not livestock exporting countries, and therefore are not subject to the pressure that an exporting country might experience from its importers.

3. In Francophone West Africa, rules and regulations regarding animal transport have not yet been implemented in the countries concerned in this study. Although livestock transport is sometimes briefly mentioned in some legislation, there was no specific legislation identified covering livestock transport for Francophone West Africa countries. There are reports that, spurred on by economic and trade interests in the subregion, regional regulations are being created (Renard *et al.*, 2004). Since the colonial period, regional organizations have been involved in livestock transport, and have focused on economic, sanitary and environmental aspects, as well as activities directly related to transport. The OCLALAV (*Organisation Commune de Lutte anti Acridienne et de Lutte anti Aviaire*), ITC (*Inter States Trypanosomiases Center*), PARC (*Panafrican Rinderpest Campaign*) and PACE (*Programme Africain de lutte contre les Epizooties*) focus on sanitary activities; CIRDES (*Centre international de Recherche-Développement de l'Elevage en zone subhumide*) is concerned with research; and CILSS (*Comite Permanent Inter-Etats de Lutte Contre Secheresse*) focuses on environmental issues.

4. In Nigeria, the Criminal Code of Federal Nigeria (amended 1991) includes regulations on animal protection. Criminal Code Part 7, Chapter 50, Cruelty to Animals, states that '. . . *any person who conveys or carries, or being the owner, permits to be conveyed or carried any animal in such manner or position as to cause such animal unnecessary suffering. . . . For such purposes of this section, an owner shall be deemed to have committed cruelty within the meaning of this Chapter if he failed to exercise reasonable care.*'

Owners who are guilty of cruelty to animals are first given the option of a fine, and are not liable to imprisonment unless they are unable or unwilling to pay the fine. Reports from local sources indicate that these regulations are rarely – if ever – enforced.

5. In Ethiopia, livestock transport is governed by the Livestock Markets and Stock Routes Regulations No. 50/1968 and also general requirements for the transport of live animals on hoof which require among other things that: a movement permit be obtained (this is not very effective), animals must be transported within the specified time period in the permit, sick animals while in transit must be inspected and treated and only continue to be transported upon the permission of a veterinarian and animals must be rested at appropriate intervals and provided with feed and water. The regulations stipulate that any conveyance must be designed and constructed such that it can withstand the weight of the animals and ensure the safety and comfort of the animals during transport.

Other conditions require that: during the journey, if conditions do not permit the unloading of animals, they should be rested and provided with feed and water on board the vehicle, the vehicle should be cleaned and disinfected before loading and after unloading, the height of the receptacle must be in accordance with the type and size of the animals, the receptacle must be well ventilated and its floor non-slippery and the safety and welfare of animals should be ensured at all times. Enforcement and compliance with any existing regulations are very low.

6. Kenya's main legislation that provides for protection of animals during transport and in markets is The Prevention of Cruelty to Animals Act, Chapter 360 of 1960 (revised in 1983), of The Laws of Kenya (GoK, 1983). The legislation sets out what constitutes offences and the penalties. In relation to transport of animals, subsidiary legislation, under Legal Notice No. 119 of 1984 (GoK, 1984), provides guidelines on appropriate transport of animals on land, sea and by air. General provisions of Chapter 360 make it an offence to: ill-treat, override, overdrive . . . an animal; convey, carry, confine or impound an animal in a manner or position that causes that animal unnecessary suffering and; without sufficient cause, starve, underfeed or deny water to an animal. Proposed revisions (in 2000) are designed to make the legislation more effective. These have yet to be enacted.

Currently, the Kenya Society for the Protection and Care of Animals (KSPCA) works in partnership with government to promote compliance with the legislation. However, compliance remains very low. Constraints to enforcement include: lack of awareness of existing legislation by stakeholders; lack of financial and human capacity (DVS and KSPCA); inadequate legal framework; and poor infrastructure.

7. Eritrea has two statutes that are designed to protect animals during transport, in markets and at slaughter facilities: the Veterinary Act which defines the roles and functions of actors in the veterinary profession including responsibilities in ensuring adherence to animal welfare provisions; and the Proclamation of Animal Diseases Control, which provides for the welfare of animals in controlling outbreaks of diseases. Both Acts also provide varying levels of jail terms and/or monetary fines for contravening the Acts.

Compliance is fairly low throughout the production chain beginning from sale at farm gate through primary, secondary and tertiary markets before animals are offloaded at slaughterhouses or ports of Massawa and Assab. The main constraints to enforcement include inadequate financial and human resources; a poor economy; and low level of formal education of livestock sector stakeholders.

Cultural/Religious Factors Important for Animal Management or the Perception of Animals

The most significant religious factor influencing the perception and management of animals in the region is the existence of Christian and Muslim populations. Most African countries have a mix of Christian and Muslim populations (as well as traditional religions, which are often practised along with Christianity and Islam) and a small minority of Hindu. The Muslim religion bans ownership and handling of pigs; in Muslim dominated areas of Africa, pigs are largely absent. In countries with even minority Muslim communities, pigs are not placed with other livestock for transport, and they are slaughtered at separate slaughterhouses and often slaughtered informally, i.e. on-farm or at the place of sale.

Many Muslims believe it is forbidden to sell donkeys for meat. In Nigeria, the Council of Northern Emirs forbids trade in donkey meat, and several northern states have passed ordinances forbidding the large-scale shipping of donkeys. According to Blench *et al.* (undated) large-scale trade in donkeys still occurs. However, in many markets, trading and slaughter of donkeys is not allowed.

In the East Africa subregion, where there are both Christian and Muslim livestock producers, their strong cultural beliefs and practices influence the management of livestock. These communities tend their animals with a lot of care and affection, and as mentioned above, this balance is disturbed when livestock traders, who treat animals as commodities, enter the equation.

In the Southern Africa subregion, Christianity is the main religion, but Muslim, Hindu, Buddhist and Jewish populations, as well as myriad other religions and cultural beliefs also exist. Practising Jews and Muslims consume only kosher and halal foods, respectively.

South Africa, with the largest and most diverse population in the region, has facilities for religious slaughter at certain abattoirs. This does not involve long-distance transport as the animals are sourced locally and the meat is transported all over Southern Africa.

By far the greatest cultural influence on the perception and management of livestock is the extent of poverty within the country. It is widely held that countries that have high incidences of poverty are less developed, are unable to provide adequate care for their human populations and afford the luxury of providing good care for animals. However, contradicting this is the attitude of small farmers, who are often the poorest among the population, to their livestock. The attitude goes beyond simply caring for the animal because of the price it will bring at market; there is true animal husbandry involved, which arguably takes time and money that could be spent elsewhere.

Economic Factors Influencing Live Transport

The major factors influencing live animal transport are economics (from economies of the individual countries down to economic factors influencing individual stakeholders' decisions) and health aspects, which are playing an increasingly large role in influencing live animal transport. Some examples follow.

Producers/consumers

In West Africa, the northern countries (Mali, Niger and Burkina Faso) are cattle producing countries, and Ghana and Nigeria, with their large urban centres, limited land
for cattle production and relatively prosperous populations, are livestock importing
countries. South Africa, with its large and relatively prosperous population, imports
from Namibia, a country sparsely populated and with good-quality rangeland.

The availability of inexpensive meat limits the cash that traders are willing to
invest. For example, in South Africa, old equines have little value and can be bought
for far less than their slaughter value. The markets for their meat are a considerable
distance from their point of origin, and the price differential is not sufficient (horse
meat is cheap) to motivate the supplier to transport the meat in a refrigerated van.
Compounding this is the small number of horse abattoirs available in the region
(only two in South Africa) and the lack of transport guidelines for horses.

Cross-border trade

West Africa is a good example of the influence of economic factors on live animal
transport. The Economic Community of West African States (ECOWAS) signed a
protocol in 1982 to facilitate the free movement of goods and people within the
West Africa region. The protocol is designed to ease transport of goods from countries that are partners in the protocol because customs documents, vehicle licensing
and method of packing are to be harmonized (http://skyscrapercity.com, updated
by West Africa Trade Hub staff, 2006, personal communication). However, livestock trade policies still differ widely among countries in West Africa (ILRI Briefs,
Oct 2005), and livestock importing countries (Ghana) promote policies that protect
their own livestock producers. Because of bureaucratic bottlenecks and licit and
illicit fees, ILRI (2003) found that cross-border transportation can cost four times
the equivalent movement of beef from Europe to West Africa's coast.

To date, some of the intentions of ECOWAS to relieve the bureaucratic and
financial burden of inter-country trade have failed to materialize. ILRI (2003)
found that the expansion of the livestock trade will require that trade policies are
streamlined and harmonized to cut down on administrative requirements.

Licit and illicit fees

Throughout Africa, illicit taxes (known as 'dash' in West Africa) are a major
impediment to decreasing the time in transit for livestock trucks; trucks wait in line
until they pay the fees and are often harassed or threatened with harassment unless
they pay. While the trucks wait and negotiate the dash, animals are left in the
truck. In addition, these fees work against improving animal welfare since extra
cash goes towards paying the fees, alternative and often more arduous routes are
taken to avoid checkpoints and the fees imposed, and transporters, in an effort to
make up for time lost at checkpoints, drive more recklessly and animals in the truck
become injured, and accidents are more likely to occur.

The most often mentioned economic factors influencing live transport are the high official duties and taxes (4.1% of the final price of animals along the Burkino Faso–Ghana corridor (ILRI, 2003)); and the unofficial extortion fees – transporters are subject to bribes. The ILRI (2003) report found that cross-border traders' margins were more than double the price of traders operating in the domestic segment of the marketing chain. Domestic sales are more profitable because they do not have to pay the taxes and other fees at the border. Transportation and handling was the single highest cost item, accounting for 48% of the total cost from Bittou market in Burkina Faso to Ghana (ILRI, 2003).

Farm ownership

In the Southern Africa subregion, as for the continent in general, complying with livestock transport regulations is a challenge, especially for small farmers, because cash to purchase adequate transport is non-existent. As most often happens, small farmers sell to middlemen who have access to vehicles or they rent a vehicle, which is less often the case. Throughout Africa, cattle producers have close ties to their livestock; however, once an animal is sold to middlemen, the bond is broken between the animal and the primary producer. Typically, the middlemen, other livestock brokers and livestock transporters view the animal as a commodity, and there is no longer an animal husbandry aspect in the human–animal relationship.

Absence of electricity and refrigeration

Limited power availability plays a part in the movement of animals from where they are raised to where they are slaughtered. It is more efficient to transport the live animal rather than to transport meat when the potential exists for spoilage without the cold storage facilities available in more developed regions. In addition, illegal fees and taxes extorted from transporters and poor road infrastructure means that the length of a trip is often unpredictable, and, without refrigeration, live animal transport is deemed to be more cost-effective.

Disease

In East Africa, the impact of disease outbreaks on economies of the subregion is well illustrated by the 1997/1998/2000 import bans imposed on livestock and livestock products from the region by Saudi Arabia and the Gulf states. This had significant economic impacts on Sudan, Somalia and other Horn of Africa countries.

Extent of Trade in Meat Products in Comparison with Live Animals and Factors Which Influence This

Southern Africa

South Africa, with its large urban populations and many abattoirs, provides a good market for excess animals, whereas Namibia has the roads, land and climate to produce more than it consumes, thus creating an export–import trade between these countries. Furthermore, when there is sudden drought, and/or where the number of available animals exceeds the slaughter capacity, live animals will be exported. If it becomes cheaper to export meat than animals, the pendulum will swing away from the export of live animals. There are import tariffs in place for meat in Namibia and export tariffs for live animals so as to motivate producers to use the abattoirs (MeatCo) within the country. Only ungulates from foot-and-mouth disease free zones in Namibia can be slaughtered at export abattoirs. This, coupled with excess capacity in some South African abattoirs and the ready market in the informal sector (often at higher prices than offered in the commercial sector), provides a serious constraint to the idea that meat, rather than live animals, should be moved.

East Africa

In East Africa, regional export markets are limited, although Sudan and Djibouti and other Horn of Africa countries have started to realize increasing export volumes, mainly into the Gulf countries. However, exporting countries have failed to attain certified quality and health standards required for live animal exports to the Gulf countries, and this led to the ban on live animals and meat by Saudi Arabia, Yemen and several Gulf countries following the outbreak of Rift Valley Fever in 1997.

Sudan is more advanced than Kenya and Ethiopia in terms of its organizational set up and its volume of live animal exports and this is attributable to its long tradition of exporting live animals to the Gulf states, the efforts made by the country's Ministry of Animal Resources in setting up a relatively efficient quarantine system, and its successful negotiations with existing and potential livestock importing countries as witnessed by getting the ban on live animal exports lifted by Saudi Arabia.

West Africa

Ghana imports mainly frozen low-quality front quarter cuts with a high fat content. This is preferred by lower-income groups, who are unable to afford more expensive, higher-quality West African beef. Ghana's current policy of trade liberalization facilitates the import of livestock from Burkina Faso, which has been displacing imports from overseas.

Officially, in Nigeria, the import of frozen meat is banned. According to official data, all trade is live animal trade because it is Government policy to encourage national production of livestock in Nigeria (Iro, 1994). This type of 'protectionist' legislation is becoming less common, and free market systems are driving the trade in meat versus live animals. Health factors are playing a significant role in determining whether live animal trade across borders is acceptable, and the acceptance by different cultures of frozen meat also plays a role.

Overview of the Long-distance Transport Chain for Slaughter Animals

Livestock marketing systems in East Africa typify long-distance transport chains throughout the African region. The livestock marketing chain is based on a series of primary, secondary and terminal markets. Between the livestock producer and terminal markets (or abattoir), there are a number of intermediaries, described in detail below under 'stakeholders'.

In general, the chain begins with primary producers (livestock owners, breeders or herders), who may sell to traders and middlemen (also known as brokers). Some producers take their animals directly to market, but this is rare, and is mostly the case for larger producers. For the most part, transactions at all stages are carried out through visual estimates ('eyeballing') of the live weight of animals, and the price of the animal increases as they move up in the marketing chain. The livestock marketing chain includes government and local authority taxes and fees, which are levied (sometimes several times over and sometimes illegally) at every step of the marketing chain.

While the transport chain in West Africa (especially in Ghana, Nigeria, Burkina Faso, Mali and Niger) conforms to the general features of commodity chains found throughout Africa, its features are rooted in tradition in which specific ethnic groups have played specific roles, and these roles are often out of bounds to other groups. While this is changing in some areas, the tradition is so ingrained it is unlikely that significant changes will be realized over the foreseeable future. The transport chain in Nigeria, one of the more complex examples, is as follows (Mamman, 2005).

The first step in the livestock trading chain involves the connection between producers and small traders in the cattle producing areas in the north of the country, who identify cattle owners, usually with five to ten animals. Small traders purchase these animals and trek with them or load them on to a vehicle, depending on the distance, and travel to a local market where they are sold locally or for transport farther south (to additional consuming centres). At the market, *dillali* (dealers) and *yan kwamisho* (transport commission agents) collect a sufficient number of cattle for transport to the south.

At the market, the *dillali* make commissioned sales to a consumer, a wholesaler, a butcher or a long-distance trader. The value of an animal is determined by visual and tactile examination: scales are rare, and sellers and buyers appraise and determine price of animals by physical appearance. The *dillali* report their transactions to the market landlord. Drovers (*yan'kora*) transport purchased livestock by foot to the abattoir, the buyer's residence or the long-distance trade section of the market for loading to other markets.

Cattle dealers' associations, the National Union of Road Transport Workers (NURTW) and the Association of Cart Pushers perform regulatory roles, such as overseeing behaviour of their members and settling quarrels. Many of these associations rely on tradition, having been in operation for many years.

Once cattle are purchased for long-distance transport, the *falke* (long-distance trader) informs his transport commission agent of his transport requirements, and the commission agent organizes transportation. The agent sets the price, date and time the cattle are to be transported. Members of NURTW may also assist the *yan kwommission* in sourcing vehicles. Most livestock vehicles are owned by transporters. If the vehicle is not fully loaded, the agent will approach or is approached by other traders who also have cattle to transport, and the animals are combined, with each trader marking his cattle. The price is paid, the vehicle is loaded by loaders, and the journey is scheduled.

A full load, for standard vehicles, is 28–30 cattle (for smaller cows, the trucks may hold up to 35). To reach this number, it could take an agent up to 3 days (Mamman, 2005) to gather cows at the market for transport south. Usually departure to southern Nigeria begins in the evenings when temperatures are cooler. Animals are guarded by *maigadi* until the truck is full and ready to travel.

Regional infrastructure

The quality of road infrastructure varies from country to country, and within each country. In general, primary roads are in fairly good condition in some Southern African countries such as South Africa, Botswana and Namibia. In West and East Africa, main roads may be in fairly good condition but secondary and tertiary roads are often highly degraded. Southern Sudan, Somalia, Angola and Nigeria have notoriously poor road infrastructure. Southern Sudan and Angola even have landmines along roads, which can be hazardous to livestock and humans. Infrastructure, such as inspection and veterinary checkpoints and market facilities, also varies from country to country, as well as within individual countries.

Some examples of the condition of infrastructure in the region include the following.

In Kenya, stock transport routes are in a state of disrepair, with inadequate grazing, holding and watering facilities (AGREF, 2005). However, in most areas/markets, ramps for loading/offloading animals exist (even where no permanent structures exist, makeshift ramps are made from heaps of soil, sand or they use roadside embankments), trucks are loaded with sand to make the truck floor nonslippery (adequacy of sand and effectiveness of this is another matter) and most trucks have canvas/tarpaulin covers to shield the animals from adverse weather conditions (however, they are often not used).

In Eritrea, inadequate infrastructure at market places in the lowland areas is a major problem, especially for producers who travel many days to reach them, thus exposing the animals to distressing conditions. The poor road network in Eritrea, especially in production areas, exposes animals to injuries. The quality of trucks used to transport livestock, including the truck floor (not a non-slip surface), are substandard.

Horn of Africa countries have inadequate infrastructure at quarantine and port facilities, such as the holding facilities at the Port of Djibouti and the quarantine facility at Port Sudan.

In Ethiopia, three of the quarantine facilities are operational (Dire Dawa, Ditcheotto, Nazareth) and six (Humera, Mekele, Metma, Bahir, Dar Kombolcha and Jijiga) are proposed to be established. Construction of four control/checkpoints (Aysha, Teferi Ber/Awbere, Addis Ababa (Bole) and Moyale) has been completed, while two more (Kuraz – South Omo and Almehal) are proposed. Functioning quarantine facilities provide veterinary services to animals destined for the export market (including deworming, tick control, vaccinations). However, these services are inadequate.

In Tanzania, there are about 64 livestock markets scattered around the country, with the highest concentration being in the Arusha region. The markets are primary, secondary and tertiary (terminal) markets. Tanzania's primary markets have animal collection pens, weigh bridges, auction rings, buyers' stands and buyers' holding pens. However, the current conditions of these facilities are unknown. In addition to these facilities, there may be a cattle dip, water trough, vaccination crush and loading ramps. Most of the primary markets do not have perimeter fencing. Secondary markets have the same facilities as primary, but are much bigger in size and have perimeter fencing. Tanzania's market infrastructure is more highly developed than most of the other countries on the continent.

In Nigeria, the rail system is in a dilapidated state, and due to poor maintenance, many of the roads have also deteriorated. Nigeria's slaughterhouse infrastructure is totally inadequate, cold van/refrigeration is almost non-existent (as is electricity to run them), accommodation for livestock at markets is non-existent or inadequate. Basic hygiene is lacking at slaughterhouses.

Conditions and welfare of animals transported for slaughter

Livestock handling: at farm, market and abattoir

Animals raised on-farm in Africa are often handled well, considered symbols of wealth, and may even be treated as part of the family. While 'factory farms' are found in some African countries, this type of livestock production is still uncommon, and most livestock is raised on-farm rather than under industrial-scale conditions.

In some cultures, livestock are subject to long-distance trekking, which can be inhumane due to the lack of rest, food and water along the way. In Ethiopia, as in most other countries where trekking is still practised, hired trekkers drive animals at high speed while flogging them repeatedly and causing a lot of bruising. The hired hands' major concern is not animal welfare, but delivering the animals and getting paid.

Once animals are sold to traders, handling usually gets rougher. In most countries in Africa, men are hired to sit on top of livestock trucks to make sure animals do not get nervous and bolt or slip during the trip.

While farmers and herders often are aware of proper handling, once an animal is sold, those handling the animal during the remaining chain of custody are unaware or uncaring of proper animal handling techniques. Animal handlers are often illiterate, and there are no educational or skill requirements.

The East Africa subregion is typical of Africa in general in regard to livestock markets. There are many primary, secondary and terminal markets along the livestock marketing routes. However, most markets lack loading and offloading facilities, watering facilities, feeding areas, shelter for attendants, scales and veterinary inspection facilities. Market facilities, where they exist for example in Ethiopia, Kenya and Sudan, are run down. Local authorities and councils who operate market facilities are not keen to plough back part of the revenue they collect from livestock traders into livestock markets to maintain and improve infrastructure.

While many countries throughout Africa have well-defined livestock routes, facilities along these routes including resting and holding grounds, quarantine facilities, watering pans, animal health facilities, crushes, loading and offloading ramps either do not exist at all, or where they do, they are in a state of disrepair. In Kenya, Uganda, Ethiopia and Sudan, there are ongoing efforts to rehabilitate these facilities, and even establish new ones (Ethiopia).

Loading/unloading at farm, market and abattoir is a particular point of stress for livestock. As described, many markets throughout Africa have degraded or non-existent loading and unloading structures. Cruel measures such as tail twisting, beating and fire are used to force livestock on to and off trucks. After long journeys without food and water, this treatment is especially inhumane. In addition, many animals arrive at markets and abattoirs injured, unable to get out of the vehicle, however, these cruel measures to force the animal to move are used regardless of the condition of the animal.

The case in Kenya illustrates livestock handling in much of Africa. Livestock are subject to different forms of suffering during trekking and trucking.

During trekking, the following were observed or reported: starvation, dehydration, stress and exhaustion from long distances travelled without feeding, watering or resting; bruises, broken tails, lacerations and other types of wounds from beatings, especially by hired trekkers, broken tails as they are twisted to make animals move faster; spontaneous abortions by pregnant animals who abort along the way due to stress and exhaustion; stress due to disease and lack of veterinary care.

During motorized transport the following were observed or reported: fractures, broken horns, broken tails and various other types of wounds suffered during loading/offloading when animals are beaten or their tails twisted to force them on and off vehicles; fractures resulting from poorly constructed ramps or lack of ramps, slipping and falling when truck floors are slippery (because of lack of non-slip material on truck floors or due to rain or animal urine and faeces); falling when transport vehicles come to abrupt stops or in motor vehicle accidents and animals trampling on each other; broken horns from getting caught in the side bars of trucks; suffocation of goats and sheep due to overcrowding and trampling as vehicles manoeuvre over rough roads or stop overnight without offloading in poorly ventilated transport vehicles; poultry suffocate due to overcrowding while being transported in poorly ventilated vehicles/car boots; wind chill especially when animals are transported overnight (despite a ban on this in Kenya) in open trucks in the rainy season; distress, starvation, dehydration especially when animals remain aboard trucks overnight at stopovers or when transport vehicles breakdown or in the holding pens in the secondary and terminal markets as they await sale or slaughter (which sometimes takes days, without adequate food or water).

Opportunities for rest, feeding, watering and holding facilities en route

Common throughout Africa during long-distance livestock transport is the lack of opportunities en route for rest and feeding and to drink water. Narrowly focused economic reasons force transporters to push themselves and the animals as fast as possible, and rest stops cut into the earnings of livestock workers. In addition, for the most part, infrastructure does not exist that would allow animals to offload and rest at points along the journey. Attitudes about livestock probably contribute to this – livestock handlers view the animals as commodities rather than beings that suffer if they do not receive adequate food and water.

In Ghana, quarantine stations no longer serve the quarantine function, which used to allow for a period of rest (up to 48 h), and now only serve as entry points into Ghana. With Ghana rinderpest-free, drivers are not forced to unload, rest, feed and water their animals. During the trip from across the border to southern Ghana markets, animals receive no food or water and are not let out of the truck to graze and rest except at markets. Traders claim that there is no place along this route to allow cattle to disembark and graze (Ashaiman Market, April 2006, personal communication).

In Eritrea, the traditional stock routes lack feed and water facilities and veterinary facilities as well as security. This is similar to Ethiopia, where currently the existing stock routes do not have animal resting facilities and inspection posts. Stock routes are poorly developed and transport infrastructure along these routes is very poor.

Due to the presence of livestock markets in two major territories in Egypt (Delta and Upper Egypt), the transport of livestock between farm and markets usually takes from 2 to 6 h, and never more than 8 h. There is no rest or feeding during transport.

Average morbidity and mortality on the major long-distance routes

Statistics on injuries and death during transport are unavailable (because these statistics are not collected in most countries). However, injuries are common along the main transport routes. In Tanzania, injuries from horns gouging out eyes or piercing the abdomen are fairly common when animals with long pointed horns are transported. This is also the case in Uganda, Kenya and other countries where long-horn cattle are transported.

Other common injuries result from stumbling and falling in trucks while in transit. In Tanzania, it is noted that often the animal attendants travelling with the animals assist to minimize losses and major injuries from falls. However, in Uganda and Ghana, animal attendants were not observed assisting animals to regain their standing positions, once they had fallen.

Heat stress and dehydration are often the causes of livestock deaths during transport, especially for pigs and poultry. Animals also suffer exhaustion while being transported or trekked as evidenced by a recumbent cow being assisted to stand at the Pugu Market, observed in April 2006.

In Tanzania, mortalities are common when animals are transported unattended in trucks and, in particular, in rail wagons. A case in point is when a train wagon arrived with 32 animals dead at Pugu railway station in 2003. Litigation involving the transporters (the railway company) and the owner of the animals followed and the Ministry was tasked to investigate and arbitrate the case.

In Nigeria, most livestock mortality during long-distance transport is a result of road accidents.

In Ghana, no information is available on the number, extent and type of injuries – police do not keep records of livestock injuries and mortality, and police are the ones who respond to accidents. However, from conversations with traders and veterinary officers at markets, it was determined that most common injuries to cattle occur from: (i) road accidents; (ii) improper loading (overloading or under-loading) causing animals to fall during transport; (iii) exhaustion from the long jour-ney without water, food or rest, as well as enduring high temperatures; and (iv) unsafe driving resulting in animals becoming injured. Once down, the animal is often left on the floor of the truck and may be trampled by other cows. Grass is placed on truck floors to make them less slippery; however, by the number of downed cows witnessed, grass was not having a significant impact.

In February 2002, the NSPCA took a video of a truck containing 49 horses from Namibia being unloaded at Chamdor abattoir in Gauteng. The truck had allegedly been under way for 5 days without the animals being watered or fed. Only 23 horses were standing, the rest were lying underneath them: 35 horses were even-tually unloaded, 10 were destroyed on the truck and 4 were dead on arrival.

Sea travel to Mauritius has resulted in dehydration of animals due to lack of water and injuries due to loading and poor facilities on the ships. Contamination of bedding, food and water with sea water also occurred. Ships sometimes dock at Durban port in South Africa with dead livestock on board. For example in 2003, a ship loaded with sheep needed urgent repairs, and made headlines because of the foul stench from the ship. Neither the office of the Durban Harbour Master nor the shipping companies could give details of what had occurred on board (Veary, 2006). This uncontrolled and unrecorded criss-crossing of the ocean by livestock, where the effluent and dead animals land in the ocean, has implications for animal welfare, and also presents a risk for the transmission of animal diseases.

Climatic and seasonal aspects that affect animal welfare during transport

RAINY SEASONS. Some locations in Africa have one rainy season, others have two. Within the same country, the wet season can occur at different times at different locations in the country. The rainy season brings common problems for animal welfare during transport. Since many of the road networks over which livestock are transported are murram (also known as laterite, a soil often found in humid tropi-cal climates with a clayey consistency, generally brick red or yellowish brown in colour – because of its prevalence it is often used for road construction) or what-ever fill material may be available at a site, the rains make passage difficult and trucks can get mired in mud and stuck for days. Livestock are rarely offloaded when a truck is stuck, and the animals will remain on board without food or water until the truck is freed.

During the rainy season, when animals are being trekked, their hooves may get caught in mud, and they will be unable to move, remaining there until they are able to pull themselves out or until they die of dehydration.

HOT, DRY SEASONS. During the hot, dry seasons, there is little protection for livestock against heat stress while they are transported in trucks. This can be especially problematic at police checkpoints and at border crossings where trucks may be stopped for several hours to deal with customs and fees. Most livestock vehicles have no protection from the elements, and with the sun beating down directly on the animals, temperatures in the truck can become unbearable. Animals can get heat exhaustion, sun stroke and can die from the heat.

Primary Routes

Transport routes on the continent are complex and numerous, and include traditional and regulated routes, as well as informal and illegal routes (usually used when trying to avoid checkpoints, especially at borders). Support facilities, such as border crossings, checkpoints, markets and slaughterhouses, are found in every country, and are too numerous to delineate separately. Below is a summary of some key livestock routes; these illustrate the typical livestock transport routes and problems found throughout the continent.

In several of the larger countries in the Southern Africa subregion, long-distance transport takes place within the country. For example, long-distance movement of livestock within South Africa from west to east and north to south is chiefly along the main highways. Long-distance road transport also occurs across Botswana and Namibia, mainly east to west; and in Mozambique from the Tete Province to Beira and Maputo.

International trade in meat and meat products, except for the overland Namibia–South Africa route and the sea export to Mauritius, far outweighs the trade in live animals for slaughter in the region. Due most probably to the danger of transboundary diseases, animals are not transported from one country to another for fattening, as they are in Europe.

Main routes over which long-distance transport occurs and reasons for transport are: South–north transport within South Africa and northwards into the rest of Africa – animals are transported for sales, breeding and recreation (racing, riding, etc.). East coast: movement north–south of slaughter animals in Mozambique by road and transhumance into Malawi. West–east transport: animals move from Namibia and Botswana into South Africa by road, mainly for slaughter or auction sales. This involves mainly cattle and goats but may also involve horses. The distances covered vary from 1000 to 2000 km and may take 3–4 days. Vehicles used for transport of livestock are usually of commercial standard, but those used for horses are not suitable for horse transport. Sea routes – from South Africa and Mozambique to Mauritius (7–10 days).

The last two routes are the most important for long-distance transport and are discussed in detail below.

Namibia/South Africa routes

Up until 3 years ago, a large number of cattle, sheep and horses were exported to South Africa for slaughter. However, in the last 3 years, the Namibian government revised legislation to motivate producers to slaughter locally and not export live animals.

This has resulted in, according to informal discussions with state veterinarians in South Africa, a decrease of approximately one-third in the number of cattle previously imported from Namibia and has had a direct influence, according to local suppliers in the Western Cape, on the closure of the abattoir at Maitland. Sheep previously slaughtered at Maitland are now slaughtered at low-throughput abattoirs in the Western Cape that have inadequate capacity for slaughtering the Namibian imports.

However, data supplied by the Namibian Meat Board (Meat Chronicle, January, 2006) directly contradict this: '*On the hoof marketing of cattle increased by 45.91% from 144573 exported in 2004 to 210945 in 2005*'. Independent corroboration of these figures was obtained from the meat board in Namibia and it appears that the confusion may have resulted because cattle were mainly routed to Kwa-Zulu Natal, Gauteng and the Eastern Cape rather than the Western Cape. Figures supplied by the Namibian Meat Board and SAMIC, as well as interviews indicate that goats are being exported to South Africa.

Previous to the incident at Chamdor abattoir described above, approximately 200 horses from Namibia were being slaughtered per week. Since that time, it appears that horses from Namibia are not being slaughtered at the two South African abattoirs that still slaughter horses. There are unsubstantiated reports that they are being slaughtered in Namibia but no figures on importation of horse meat or live horses to South Africa are currently available. However, from hearsay, it appears that the horses are entering South Africa via Botswana, through the border post in North-west Province.

Southern Africa livestock transported by sea: the Mauritius route

The NSPCA has been, for many years aware of and monitoring sea transport of animals exported to Mauritius for slaughter, as well as sea transport vessels en route to other destinations that put in at South African ports. As a direct result of discussions with the National Department of Agriculture, guidelines based on those of the OIE on sea transport of animals were drafted in August 2004. However, shippers do not always comply with these guidelines.

The normal length of the voyage is a minimum of 7 and up to 10 days in adverse weather or rough seas. Voyages take place as often as every 2 months, usually from the port at Durban. The shipments out of Durban ceased when foot-and-mouth disease broke out in Kwa-Zulu Natal (circa, 2000). Vessels were then loaded instead at East London, Maputo and Beira.

Livestock are sourced from an area less than 15 h away from the port by road as prolonged journeys prior to the sea voyage affect the welfare of the animals. The number of animals transported varies from 100 to 500 for sheep and 100 to 800 for cattle, depending on the size of the vessel. On arrival at Port Louis in Mauritius, animals are offloaded and sent to feedlots, from where they are returned to holding pens at the abattoir for slaughter. The live animal demand is mainly due to the requirements of the Muslim community on Mauritius for halal slaughter.

West Africa

In West Africa, the main trade direction is from the dry Sahelian interior areas to the humid sub-equatorial coastal countries (mainly Ivory Coast, Ghana and Nigeria), but livestock trade also occurs between the dry countries.

The main routes by road are as follows:

Route 1: Mali–Burkina Faso–Ivory Coast: Bamako–Sikasso (Mali)–Bobo-Dioulasso (Burkina Faso)–Ferkessedougou;

Route 2: Mali–Burkina Faso: Mopti–Ouahigouya (1); Segou–Bobo-Dioulasso (2); Sikasso–Bobo-Dioulasso (3);

Route 3: Mali–Burkina-Faso: San (Mali)–Nouna, Ouagadougou (Burkina Faso);

Route 4: Mali–Burkina Faso: Koury (Mali)–Bobo-Dioulasso (Burkina Faso);

Route 5: Mali–Burkina Faso: Sikasso (Mali)–Niangoloko, Bittou (Burkina Faso);

Route 6: Mali–Burkina Faso: Gorom-Gorom (Burkina Faso)–Gao (Mali);

Route 7: Mali–Burkina Faso:Kaya, Ouhigouya (Burkina Faso)–Djenne (Mali);

Route 8: Mali–Ivory Coast: Sikasso (Mali)–Ferkessedougou (Ivory Coast);

Route 9: Mali–Ivory Coast: Niena, Sikasso–Port Bouet;

Route 10: Burkina Faso–Ghana: Bittou–Accra;

Route 11: Burkina Faso–Benin: Fada N'Gourma(Burkina Faso);KoloKonde, Natitingou (Benin);

Route 12: Burkina Faso–Niger: Dori (Burkina Faso), Tera, Tillaberi–Niamey, Dosso, Gaya (Niger);

Route 13: Burkina Faso–Niger: Fada N'Gourma (Burkina Faso); Niamey (Niger);

Route 14: Niger–Benin: Dosso, Gaya (Niger)–Mallanville, Kandi, Parakou (Benin);

Route 15: Burkina Faso, Togo: Ouagadougou, Tengkodogo (Burkina Faso)–Dapaong, Kara, Sokode, Atakpame, Tsevie, Lome (Togo).

Control points, markets, means of transport, borders

In Francophone West Africa, official control points are determined by national and regional regulation (CILSS, CEDEAO – *Communaute Economique Des Etats de L'Afrique de L'OUEST* /ECOWAS, UEMOA – *Union Monetaire Ouest Africaine/ Western African Monetary Union*), at the borders and inside each country. These control points can be veterinary control points, commercial control points, police and/or customs. Despite the existence of official control points, breeders keep on using traditional routes for cultural reasons, or informal variable routes for economic reasons to avoid the taxes. The reliability of data from control points is therefore very low. However, in Western Africa, the *International Certificate of Transhumance* of CEDEAO/ECOWAS is officially in force.

The number of markets varies in each country. In Mali, there are 334 primary markets and 25 regional markets. In Burkina Faso, there are 132 primary markets and 10 regional markets (ILRI, Abidjan Proceedings, 2001). The animals are led from the local markets to the regional or the international markets outside the country. These markets often are the best business places for important dealers in big towns or at the borders of a country. Examples are: Mali: Mopti, Sikasso, Bamako; Ivory Coast: Ferkessedougou, Bouake, Yamoussoukro; Burkina Faso:

Bobo-Dioulasso, Gorom-Gorom, Fada N'Gourma, Ouahigouya, Ouagadougou; Niger: Gaya, Maradi, Niamey, Gazaoua, Tessaoua, Lazaret, Guidan Ider, Tounfafi, Tchitabarad, Zinder, Bakirbirgi; Togo: Dapaong, Kara, Sokode, Atakpame; Benin: Parakou, Bohicon.

Consumption markets or terminal markets are large markets located mainly in urban areas. Trade movement ends at terminal markets in coastal towns such as Abidjan (Ivory Coast), Accra (Ghana), Lome (Togo) and Cotonou (Benin).

Official border posts, although often not used in an effort to escape extortion fees and legal customs fees, are located between: Mali–Ivory Coast; Mali–Burkina Faso; Burkina Faso–Benin; Burkina Faso–Niger; Benin–Niger; Ivory Coast–Ghana; Ivory Coast–Burkina Faso; Ghana–Togo; and Togo–Benin.

Livestock entering Ghana from Burkina Faso may legally cross at five border crossings, Hamale, Tumu, Paga, Mogonori or Pusiga. From there the following routes may be taken.

1. The main transportation route originates in northern Ghana (from Burkina Faso, and these animals may have come as far as Mali or Niger); the route continues on to Techiman Market; and from Techiman to Kumasi Market (where most animals are bought for slaughter); or from Techiman to Ashaiman Market (where most animals are bought for slaughter). Some animals may be transported from northern Ghana straight to Kumasi Market.

Animals bought at Ashaiman are destined for slaughter (not for fattening) at the Tema slaughterhouse, Accra abattoir, other numerous slaughterhouses and slabs in the Accra/Tema area or for informal slaughter at household level.

The trip from Mali to Burkina and on to Ashaiman (in the south-east) may take 1 week (this does not include trekking or trucking to central markets in Mali or Burkina before entering Ghana). From Ghana's border with Burkina to Ashaiman Market takes up to 2 days. From Ghana's border to Techiman Market will take about 1 day; from Techiman to Kumasi, a few hours; and from Ghana's border to Kumasi 1 day.

Recently, Ghana has improved the roads, but before 2000, the trip from Mali to Burkina to southern Ghana used to take up to 2 weeks for what is now a 1-week trip.

2. From Burkina Faso animals are transported to Techiman Market, and then are transported to Cape Coast. Animals not sold in Cape Coast market may be transported farther down the coast road to Sekondi-Takoradi market.

From Ghana's border with Burkina, this route can take 2 days to Cape Coast, and about 2 h more to Sekondi-Takoradi. Animals that are sold in Cape Coast or at the Sekondi-Takoradi market are mostly destined for slaughter, which could take place at any of the numerous slabs, slaughterhouses or informal settings in the region.

Another route, which has been more commonly used since the unrest in Ivory Coast, crosses into Ghana from Burkina Faso and at Elubo, crosses into Ivory Coast: in Ghana, the journey length is 2 days. Even with the improved security conditions in Ivory Coast, this route to markets in Ivory Coast continues to pass through Ghana (http://www.fews.net/centers/innerSections.aspx?f=bf&m=10008 61&pageID=monthliesDocMigratory movements and trends in livestock prices).

There are two main cattle trade routes (where goats and sheep are also traded) from the northern production areas in Nigeria, or from Niger, Chad and Cameroon to southern markets; 75% of the animals follow these fixed routes (Iro, 1994):

1. From Sokoto in the north-west, animals are transported to Ibadan and Lagos in the south-west, through Jebba. This route is about 1400 km. The Sokoto region is the largest supplier of livestock produced in Nigeria. The market in Ibadan (Bodija) is one of the major southern cattle markets and centres of meat consumption. This journey will take about 2 days if travelling straight to the southern markets; with stops at interim markets, the journey can take several more days, depending on the number of markets visited.
2. From Maiduguri in the north-east, animals are transported to Calabar in the south-east going through Kano and Zaria. This route is about 2000 km. This journey will take 2–3 days if travelling straight to the southern markets. If stopping along the way at markets, it could take several more days.

There are numerous markets all along the routes; at some, only live cattle are sold, and at others, live and slaughtered animals are sold. For example, at Akinyele only live cattle are sold; at the Bodija market, a terminal market in Ibadan, cows are sold live or slaughtered (Mamman, 2005). At all markets, live or slaughtered goats and sheep may be sold.

There are numerous police and veterinary checkpoints along all routes. In the short distance on the approach into Benin City, Edo State, coming from Aduwawa cattle market, Ekahuan cattle market and the Benin Technical College cattle market into Benin City, there are three checkpoints: Okpella Veterinary Post is the largest. There are quarantine posts, however, these are significantly understaffed, and most are barely operating. The team was told that daily, more than 20 trailers from each of the two main northern markets (Sokoto and Maidugiri) pass to the south loaded with animals. Each trailer holds up to 33 cattle.

The majority of Nigeria's donkeys are imported from Niger, although a few come from northern Cameroon and Chad (Blench *et al.*, undated). As donkeys become more popular as draught animals, more are being raised in northern Nigeria. Donkeys are usually brought to markets along the Niger/Nigeria border by traders who specialize in the donkey trade. Nigerian traders then buy donkeys from the specialized traders or directly from breeders, and then they take them to the border markets. The principal markets where donkeys are sold are:

Kano State markets: Maigatari, Mai-Adua, Babura and Garki;
Katsina State: Garken-Daura and Zango;
Sokoto State: Shinkafe, Kauran Namoda and Sabon Birni;
Borno State: Gashua and Nguru.

From these markets, donkeys will be transported to southern markets (the same used for other animals, travelling the same routes) and sold for slaughter there.

According to Blench *et al.* (undated) donkeys used to be considered a free resource – people would take a wild donkey when needed, and return the animal when they were not in use. However, due to theft and the growing taste for donkey meat, the donkey trade has grown more formal and common.

In the East Africa subregion, major ports are used for shipping animals to export destinations. These include Mombasa (Kenya), Dar es Salaam (Tanzania),

Berbera and Bosasso (Somalia), Port Sudan (Sudan), Djibouti (Djibouti) and Massawa and Assab (Eritrea).

Cross-border livestock trade, especially cattle trade from Ethiopia and Somalia into Kenya, plays an important role within the East Africa subregion. Other major cross-border trade patterns are between southern Sudan and Uganda/Kenya, Tanzania and Kenya and Somalia/Djibouti/Ethiopia. Another notable feature of livestock trade in this region is that unofficial cross-border trade in live animals exceeds by far the value of official exports of live animals and meat, especially from Somalia and Djibouti to Saudi Arabia, Yemen and the Gulf states.

Export markets for countries in the East Africa subregion are limited, although Sudan and Djibouti have started to increase exports. Gulf countries are the major export destinations for animals from the Horn of Africa. In the subregion, net exporters of live animals include Sudan (mainly sheep), Ethiopia and Somalia.

Below is a description of transport routes in Tanzania and Uganda, two East African countries with long-distance transport routes that are of interest since while there is currently little or no monitoring for welfare, they may be most amenable to monitoring of all the East African Countries – with the exception of Kenya, whose KSPCA already implements a monitoring programme.

In Tanzania, from the production areas, livestock transport by road follows main trunk roads: Dodoma to Dar es Salaam through Morogoro; Singida to Arusha; Shinyanga to Arusha and Mwanza; Moshi (Weruweru secondary market) to Tanga and Dar es Salaam via Korogwe. Livestock transport routes by train on the Central Railway Line of Tanzania. Railway Corporation (TRC) shipments are from Mpanda to Dar es Salaam via Tabora, joining wagons from Mwanza and Shinyanga at Tabora, those from Singida at Manyoni and, those from Dodoma at Dodoma. For the Tanzania Zambia Railway Authority Line, the route is from Mbeya Region Stations to Dar es Salaam. Transport by ship is from Dar es Salaam to Zanzibar, the Comoros, Lindi and Mtwara. Some shipment of animals also takes place on Lake Victoria; however, details of this are unavailable.

Overland livestock entry and exit points are numerous. Livestock entry and exit locations at airports include Kilimanjaro International Airport, Mwalimu Nyerere International and the airport at Mwanza.

Harbour/sea port entry and exit points include: Bukoba Port, Kemondo Bay Port, Mwanza, Musoma, Kigoma, Kirando, Kabwe, Kasanga, Itungi Port, Mbamba Bay, Mtwara, Kilwa, Lindi, Dar es Salaam, Bagamoyo, Pangani and Tanga.

In Uganda, the major livestock routes and distances are: Mbarara–Masaka–Kampala (270 km) – end points: Kampala abattoirs, four in number; Masindi–Nakasongola–Kampala (213 km); Kiboga–Wakiso–Kampala (113 km); Kampala–Jinja (80 km); Kampala–Tororo (217 km); Moroto–Soroti–Mbale (272 km); Moroto–Mbale (226 km); and Pallisa–Mbale–Kumi–Soroti (135 km).

Discussion

Cattle, goats and sheep are the main animals that are transported long distances throughout Africa. Horses, donkeys and camels are also transported long distances between specific countries. Pigs and chickens are usually raised in close proximity to where they are slaughtered, and are usually not subject to long-distance transport (with exceptions for pigs, such as in Namibia, Zambia and South Africa). Ostriches are transported long distances in some Southern African countries (South Africa and Namibia).

Trekking and trucking are the main forms of long-distance transport. Rail and ships are also used to transport livestock long distances.

In general, across the continent, infrastructure that the livestock sector is dependent upon is deteriorating or already fairly degraded. Roads, especially secondary and tertiary, are in poor condition. Many markets lack basic infrastructure including shelter, watering facilities and ramps; and abattoirs are basic, lacking ramps, pens and other devices that can ease the slaughtering process for animals and humans. Tanzania, South Africa and Botswana have better market and slaughterhouse infrastructure than that found in most other parts of Africa.

Legislation varies across the continent; in some countries, it is detailed and complete (South Africa, Kenya), and in other countries (Ghana, Nigeria) it is weak. Compliance and enforcement are challenges – most countries do not have the resources necessary to implement the legislation.

Issues related to animal welfare during long-distance transport are common throughout the continent; these are: lack of food, water and rest opportunities during long-distance trucking, trekking, rail and ship transport; inhumane handling during transport, loading, unloading and slaughter; limited infrastructure, resulting in journeys made more arduous than necessary (degraded roads forcing erratic driving, resulting in accidents; no protection on trucks or at markets from heat and sun); and lack of enforcement of existing legislation that could improve the livestock transport process.

Situations in which welfare is conducive to financially viable production methods

Attention to animal welfare can complement economically viable livestock production. For example, in the Southern African subregion, if there is quality control and good meat inspection at registered abattoirs, bruised, injured and diseased animals will be condemned. This will be an incentive, coupled with action by traffic authorities and animal welfare inspectors, to use recommended (more humane) transport methods.

As has been well documented, bruised animals produce lower-quality meat, hides and skins (see Chapter 4, this volume). Markets in Africa will need to demand good-quality beef, not only for export but also for consumption by local populations.

In some ways South Africa is moving backwards: deregulation of the meat industry has resulted in the construction and use of feedlots, and speculators, abattoir owners and meat inspectors all employed by the same company. Bruised and diseased carcasses can slip through the system and there is no incentive, except death prior to unloading, to adhere to regulations.

Export to Western countries, who are increasingly demanding humanely raised and slaughtered meat, is an incentive to improve animal handling. Export of livestock products (hides, skins, as well as meat) can increase the profit margin for livestock producers and traders.

Examples of best practice in the region

The meat trade in South Africa, with a high level of input by animal welfare organizations, is much more profitable and better organized than in neighbouring countries where animal welfare standards are not well respected. A livestock trade that has quality control at all points, and that is concerned with the welfare of animals has greater potential for export to the EU and other regions, including the Gulf and other parts of Africa and eventually to Western countries, especially for skin and hides.

The agreement reached between the NSPCA and Spoornet, the state-subsidized rail network, that livestock will not be transported by rail in South Africa is a good example of best practice that can be a model for other countries.

Woolworths, a major food and clothing retailer in South Africa (also found in Botswana, and expanding to other countries, such as Ghana), is responding to consumer concern about farm animal welfare and their labelling reflects this.

Another example of best practice from Southern Africa is the effect that public pressure (CAFW) has had on the structure of the vessels used for livestock transport to Mauritius. There may also be some economic imperative, as the exporters are apparently not paid for animals that do not reach the Port Louis abattoir in a condition fit for slaughter. The survival and condition of animals reaching the island has improved significantly.

Overland transport within Egypt is conducted by the private sector in well-equipped trucks, which nearly meet international standards. The trucks have two rows; each row has a suitable number of openings for ventilation. This method of transport is a good model, and has been supported by the private sector.

In Morocco, the livestock cooperative associations provide refrigerated trucks for transport of carcasses from abattoirs to retailers or butcher shops; this promotes a carcass trade rather than a live animal trade.

In East Africa, there are a few examples of successful meat trade establishments in the region, mainly Farmer's Choice and Kenchic in Kenya. Kenchic, a leading poultry producer, transports live chickens in special crates and in trucks designed for that purpose. Each crate is designed for ten chickens. To avoid long-distance transport, the firm contracts farmers within 100 km of its slaughter facilities.

Currently, in Kenya, the technology for refrigerated meat transport can be used only by the larger, commercial producers. Therefore, it is unlikely to replace live animal transport. Commercial producers currently using refrigerated transport are Farmers' Choice and Kenchic. Large cold stores are found at Kenya Meat Commission's (KMC) Athi River plant and its Nairobi depot (currently not in use but set to be operational following the reopening of KMC).

Opportunities for improving welfare

In East Africa, as in other parts of Africa, a number of developments in animal transport are expected in the next 5–10 years. These include improved policy and legislation environment (including development of animal transport guidelines in line with OIE requirements); improved transport infrastructure, market infrastructure, improved railway infrastructure and its increased use in live animal transport; increased interest in practices that improve animal welfare; and implementation of practices that help control transboundary animal disease transmission, which will help limit long-distance live animal transport.

African countries also expect to have increased access to international markets for livestock and livestock products. With consumers increasingly concerned with the quality of their food and the way it is produced, international trade could help promote the improved treatment of livestock from farm to abattoir.

The Southern Africa region provides a view into what we might expect from the rest of the continent over the next 5–10 years. As has been happening in South Africa, advocates (NGOs, and other private sector advocacy initiatives) promoting quality control in the transport, handling and slaughter of animals would result in a 'win–win' solution for livestock welfare, livestock markets, food safety, food security and disease control in livestock. This approach is far more likely to succeed in the long run than the animal-centred 'policeman' approach of the 1990s. Most injuries occur in transport, therefore if transport is to continue special attention should be paid to the vehicles' flooring, space between floor and side panels, sharp points and corners, partitions, ventilation, shelter, exhaust fumes and floor space. Supervision and driver knowledge is also important.

These issues will become critical as African beef is in greater demand (with the loss of rangeland in more developed countries and with breaking down of trade barriers), and as Africans themselves demand improved quality of meat and improved practices in animal care.

As indicated in this chapter, some countries are more developed regarding livestock sector infrastructure, livestock management, livestock community-based organizations, animal welfare NGOs and/or oversight of livestock transport and slaughter. The degree of private sector versus public sector involvement in livestock sectors across the region varies. A competitive private sector involved in livestock production, transport and marketing of livestock products could provide a springboard for improved transport and slaughter practices. A strong NGO community and press that can freely advocate for improved conditions are key to making changes across the continent.

Continent-wide opportunities for improving animal welfare during transport include: strengthening enforcement of legislation, and in countries with weak legislation, improving the legal framework. Specific areas of concentration could include.

1. Improving the issuance of export certificates for live animals so that they are better recorded and controlled; improving transboundary traceability of livestock; improving the control, documentation and observation of vessels carrying live animals; including equines (horses, donkeys and mules) as meat animals so that the slaughter, meat inspection, traceability and quality control of their meat can be used to improve transport and slaughter methods; improving control of auction sales by speculators who move animals from one location to another – the same animal may be sold at different sales in different areas over a matter of days or weeks.

2. Assisting countries to implement animal welfare legislation by raising awareness of decision makers and other stakeholders, educating the public and providing the means (vehicles, funds, improved inspection services and other resources) needed to implement regulations.

3. Raising awareness of OIE guidelines and providing the means to put guidance into practice (and promulgate national legislation based on OIE guidance).

4. Training of veterinarians, para-veterinarians, community animal workers; animal handlers at abattoirs, markets and on trucks; and of police, judges and customs officials in proper animal handling, regulations, and as appropriate to their roles, other aspects of livestock welfare.

5. Identifying good practices being implemented in Africa, and raising awareness among stakeholders of the economic incentives involved in improved handling of livestock during transport and slaughter.

6. Promoting the establishment of satellite slaughter facilities to reduce the trekking and trucking time for animals to get from farm to abattoir. These slaughter facilities should be of high quality and operated according to acceptable standards. Kenya could serve as an example, where several projects by private sector investors are ongoing to establish satellite slaughter facilities. These should be attractive to livestock sector stakeholders since they will reduce transaction costs and also, will hopefully improve returns, as animals will be reaching slaughter facilities in better body condition.

7. Funding the construction and promoting the use of cold storage facilities to minimize long-distance transport of live animals and encourage carcass trade.

8. Combatting corruption in the form of illegal fees extorted from transporters at borders and within individual countries could have an indirect positive effect on livestock transport. Increasing the transparency and collection of fees could result in less waiting time at borders and police checkpoints, thereby resulting in less time spent in transport for livestock. This could be a model for other regions, and could also provide an entry point to working on improved handling and management activities with livestock organizations.

Acknowledgements

This chapter was compiled from regional reports commissioned by the Global Long-Distance Transport Coalition.

References

AAALAC (2003) Overcoming the Challenges of Animal Transportation. Available at: www.aaalac. org/publications/Connection/Spring_2003_lowres.pdf

Abdullahi, A.M. (1993) Somali's livestock economy: export and domestic markets and prices. In: Baumann, M.P.O., Janzen, J. and Schwartz, H.J. (eds) *Pastoral Production in Central Somalia*. GTZ (German Agency for Technical Cooperation), Eschborn, Germany, pp. 265–288.

Abe, Labbee, De (2005) *Evaluation environnementale strategique du secteur élevage au Benin. Etat des lieux et analyse diagnostic. Rapport provisoire*. République du Benin.

African Development Bank (2002) *ADB Programme Performance Evaluation Report: Agricultural Sector Rehabilitation Programme.* African Development Bank, Accra, Ghana.

AGREF (2005) *Draft Report on: The Study of Livestock Marketing ond Support to the Establishment of Livestock Disease Free Zones/Livestock Disease Free Export Zones in Kenya.* Arid Lands Resource Management Project (ALRMP) II. January 2005. Agricultural Research Foundation, Nairobi, Kenya.

Agriconsortium (2003) *Livestock and Livestock Products Production and Marketing System in Kenya, Final Report.* Kenya/European Commission, Brussels, Belgium.

Agricultural Geo-Referenced Information System (2006) Dynamic Maps > Agric Infrastructure Atlas. Available at: http://www.agis.agric.za/agisweb/agis.html

Aklilu, Y. (2004) Pastoral livestock marketing groups in Southern Ethiopia: some preliminary findings. Paper presented at a CORDAID workshop entitled *'Access to Markets'*, held 2–3 November 2004 at the Mekonnen Hotel, Nazareth (Adama), Ethiopia.

Aklilu, Y., Irungu, P. and Reda, A. (2002) An audit of livestock marketing status. In: *Kenya, Ethiopia and Sudan – Volume I and II.* Community-based animal health and participatory epidemiology unit, Pan African Programme for the Control of epizootics (PACE), African Union/InterAfrican Bureau for Animal Resources, Nairobi, Kenya, p. 85.

Al-Awwal (2006) Muslim Judicial Council S.A. Available at: www.mjc.org.za/hguidelines.htm

Al-Hassan, R. M. (2005) *Review of Agricultural Policies and Strategies in Ghana, 1984–2004.* Department of Agricultural Economics and Agribusiness, University of Ghana, Legon, Ghana.

All Africa (2006) Mmegi/The Reporter Newspaper (Gaborone) March 22, 2006. Available at: http://allafrica.com/stories/200603220613.html

All Africa (2006) Stories. Available at: http://allafrica.com/stories/200601120281.html

Asante, A. (2004) *Assessment of Food Import and Food Aid Against Support for Agricultural Development: The Case of Ghana.* FAO Regional Office, Accra, Ghana.

Asuming-Brempong, S. and Asafu-Adjei, K. (2001) *Estimates of Food Production and Food Availability in Ghana: The Case of Year 2000.* MoFA and University of Legon, Ghana.

AU/IBAR (2004) *Pan African Animal Health Yearbook, 2004.* African Union/Interafrican Bureau for Animal Resources, Nairobi, Kenya.

Bada-Alambedji, R., Kaboret, Y. and Akapko, A.J. (2006) *Les maladies emergentes et le role du veterinaire en Afrique, Discours de la Neuvieme Rentree solennelle de l'Ecole Inter – Etats des Sciences et Medecine Veterinaire EISMV.*Dakar.

Banque de France (2003) *Rapport annuel.*

Bardi, G. (2005) Adoption of cattle management practices among cattle merchants in Benin metropolis. Thesis for Bachelor's Degree, University of Benin, Benin City, Nigeria.

Bendsen, H. and Meyer, T. (2002) The Dynamics of the Land Use Systems in Ngamiland, Botswana: Changing Livelihood Options and Strategies. Available at: http://66.249.93.104/search?q=cache: roeVNa5AanoJ:www.ees.ufl.edu/homepp/brown/hoorc/docs/5%2520Papers%2520%26% 2520Paper%2520Abstracts/Bendsen%2520and%2520Meyer%25205.2%2520.doc+livestock+disease+ control+cordon+in+Namibia+%22cordon+fence%22+ap+namibia&hl=en&gl=za&ct=clnk&cd=21

Boutonnet, J.P., Griffons, M. and Viallet, D. (2000) *Competitivite des Productions Animales en Afrique Sub-Saharienne et a Madagarcar. Synthese Generale.* Ministere des Affaires Etrangeres, France.

Bowles, D., Paskin, R., Gutierrez, M. and Kasterine, A. (2005) Animal welfare and developing countries opportunities for trade in high-welfare products from developing countries. *Revue Scientifique et Technique* 24, 783–790.

Campher, J.P. (1995) Livestock Improvement in South Africa: Performance Driven. Proceedings of the International Symposium on Livestock Production Through Animal Breeding and Genetics. Available at: http://www.ilri.cgiar.org/InfoServ/Webpub/Fulldocs/AnGenResCD/docs/ ProceedAnimalBreedAndGenetics/LIVESTOCK_IMPROVEMENT.htm

CBS (2000) Kenya Economic Survey, 2000. National Bureau of Statistics, Ministry of Planning and National Development, Kenya.

Chambers, P. and Grandin, G. (2001) *Guidelines for Humane Handling, Transport and Slaughter of Livestock.* FAO Regional Office for Asia and the Pacific, Rome.

Commission of the European Communities (2001) *Final Report of a Mission Carried Out in Zimbabwe from 16 to 24 January 2001 to Evaluate the Operation of Controls over Animal Health* (Foot-And-Mouth Disease) and Fresh Meat (Council Directive 72/462/EEC). Available at: http://europa.eu.int/comm/food/fs/inspections/vi/reports/zimbabwe/vi_rep_zimb_3178-2001_en.pdf

Compassion in World Farming (2003) *The Science on Animal Transport – a Short Note on Journey Times.* Compassion in World Farming, Surrey, UK.

Conley, A.H. (1996) *A Synoptic View of Water Resources in Southern Africa. Monograph No 6: Sink or Swim?* Institute for Security Studies, Cape Town, South Africa.

Cook, A. (2003) *Development of a Regional Market Information System for Agricultural and Livestock Commodities.* Abt Associates, Bethesda, Maryland.

Delgado, C., Rosegrant, M., Steinfeld, H., Ehui, S. and Courbois, C. (1999) *Livestock to 2020. The Next Food Revolution.* International Food Policy Research Institute, Washington, DC.

Department of Finance (1997) *Southern Africa: A Growth Opportunity.* 1997 Southern Africa Economic Summit, Harare, Zimbabwe 21–23 May. SADC Finance and Investment Sector Coordinating Unit, Department of Finance, South Africa.

Department of Land Affairs (2005) *Strategic Plan 2005–2010 – Sustainable Land and Agrarian Reform: A Contribution to Vision 2014.* Department of Land Affairs, South Africa.

Department of Trade and Industry (2002) Zambia Overview. Available at: http://www.dti.gov.za/econdb/raportt/zambiaOverview.html

Derah, N. and Mokopasetso, M. (2005) The control of foot and mouth disease in Botswana and Zimbabwe. *Tropicultura*, Special Issue 2005: Epidemiosurveillance for Animal Diseases in Southern Africa.

Directorate of Agricultural Statistics (2005) Economic review of the South African agriculture. In: *The Republic of South Africa*, The Department of Agriculture, South Africa.

Ewbank, R., Kim-Madslien, F. and Hart, C.B. (eds) (1988) *Management and Welfare of Farm Animals: The UFAW Farm Handbook*, 3rd edn. Universities Federation of Animal Welfare, Wheathampstead, UK.

FAO (1998) *FAO Production Yearbook, 1998.* FAO, Rome.

FAO (2003) *FAO Production Yearbook*, Volume 57. FAO, Rome.

FAO (2004) *FAO Statstical Yearbook*, Volume 1. FAO, Rome.

FAO (undated) *Trypanotolerant Livestock in West and Central Africa. Volume 2 Country Studies.* International Livestock Centre for Africa, Addis Ababa, Ghana.

FAO/AGAL (2005a) *Livestock Sector Brief – Djibouti.* Livestock Information, Sector Analysis and Policy Branch (AGAL), Food and Agriculture Organization of the United Nations, Rome.

FAO/AGAL (2005b) *Livestock Sector Brief, United Republic of Tanzania.* Livestock Information Sector Analysis and Policy Branch (AGAL), Food and Agriculture Organization (FAO), Rome.

FAO/AGAL (2005c) *Livestock Sector Brief, Ghana.* Livestock Information Sector Analysis and Policy Branch (AGAL), Food and Agriculture Organization (FAO), Rome.

FAO/AGAL (2005d) *Livestock Sector Brief, Zimbabwe.* Livestock Information Sector Analysis and Policy Branch (AGAL), Food and Agriculture Organization (FAO), Rome.

FAO Corporate Document Repository (2003) Zimbabwe. In: *WTO Agreement on Agriculture: The Implementation Experience – Developing Country Case Studies.* FAO, Rome.

FAO/World Bank/EU (2004) *Somalia: Towards a Livestock Sector Strategy*, Report No. 04.001 IC-SOM. The World Bank, Washington, DC.

Frempong, A., Frempong, S. and Frempong, I. (2003) *New Directions for Livestock Policy in Ghana.* The IDL Group, Ghana.

GoK (1983) *The Prevention of Cruelty to Animals Act*, Chapter 360 of The Laws of Kenya. The Government of Kenya, Government Printers, Nairobi, Kenya.

GoK (1984) *Legal Notice No. 119: The Prevention of Cruelty to Animals (Transport of Animals) Regulations, 1984.* The Government of Kenya, Government Printers, Nairobi, Kenya.

GoK (1989) *The Animal Diseases Act*, Chapter 364 of The Laws of Kenya. The Government of Keneya, Government Printers, Nairobi, Kenya.

GOSE (1998) *1997–1998 Livestock Census Results and Supporting Estimates.* Government of State of Eritrea, Ministry of Agriculture, Department of Animal Resources Development, Eritrea.

Government Communication and Information System (2003) *Agriculture, Forestry and Land. A Pocket Guide to South Africa.* GCIS, South Africa.

Government Communication and Information System (2005) Agriculture and land affairs. *South Africa Yearbook 2004/05.* GCIS, South Africa.

Hamade, K. (2002) *Burkina Faso.* Country Pasture profile/Forage Resources Profiles, FAO, Rome.

Hancock, P. (2005) Conservation and development in Botswana. Two sides of the same coin. *Regional Perspectives* Chapter 12.

Harris, T. (2005) Animal transport and welfare: a global challenge. *Revue Scientifique et Technique* 24(2), 647–653.

Holleman, C.F. (2002) *The Socio-Economic Implications of the Livestock Ban in Somaliland.* FEWS NET/ Somalia, Nairobi, Kenya.

IAC (1996) *Management of Waste from Animal Product Processing.* Livestock, Environment and Development (LEAD), FAO, Rome.

Ibeabuchi, F.J. (1998) Marketing system of cattle in Oredo and Ikpoba-Okha local government areas of Edo state. Thesis for Bachelor's Degree, University of Benin, Benin City, Nigeria.

ILRI (2001) Quelles politiques pour améliorer la compétitivité des petits éleveurs dans le corridor central de l'afrique de l'ouest : Implications pour le commerce et l'intégration régionale. *Proceedings of the workshop held in Abidjan.* pp. 83. ILRI, CIRES, Ministère des Productions Animales et des Productions, Halieutiques, Cote d'Ivoire.

International Marketing Council of South Africa (2006) About South Africa, Geography & Climate. Available at: www.southafrica.info/ess_info/sa_glance/geography/geography.htm

International Marketing Council of South Africa (2006) Doing Business, Economy, Key Sectors. Available at: www.southafrica.info/doing_business/economy/key_sectors/agricultural-sector. htm

International Research Institute for Climate and Society (2006) Data Library UEA CRU New CRU05 Climatology c6190 Mean Temperature. Available at: http://iridl.ldeo.columbia.edu/SOURCES/ .UEA/.CRU/.New/.CRU05/.climatology/.c6190/.mean/.temp/

Janzen, J. (1993) Mobile livestock keeping in Somalia. In: Baumann, M.P.O., Janzen, J. and Schwartz, H.J. (eds) *Pastoral Production in Central Somalia.* GTZ (German Agency for Technical Cooperation), Eschborn, Federal Republic of Germany, pp. 17–32.

Jost, C. (2002) Facilitating the survival of African pastoralist in the face of climate change: looking back to move forward. MA thesis, Fletcher School of Law and Diplomacy, Tufts University, Medford, Maryland, 78 pp., pp. 10–55.

Kamau, C. (2004) *Food Processing Sector in Kenya.* Office of Agricultural Affairs, American Embassy, Nairobi, Kenya.

Katorogo, G. (1999) *A Study of Cruelty (Abuses and Trauma) Inflicted on Cattle Delivered at Abattoirs in Kampala City.* Makerere University, Kampala, Republic of Uganda.

Kayouli, C., Tesfai, T. and Tewolde, A. (2002) *Eritrea.* Country Pasture/Forage Resource Profiles, Food and Agriculture Organization of the United Nations, Rome.

Kiambi, D. (1999) *Assessment of the Status of Agrobiodiversity in Djibouti.* International Plant Genetic Resources Institute (IPGRI), Sub-Saharan Africa Office, Nairobi, Kenya.

King, A. and Mukasa-Mugerwa, E. (2002) *Livestock Marketing in Southern Sudan, with Particular Reference to the Cattle Trade Between Southern Sudan and Uganda.* Community-Based Animal Health and Participatory Epidemiology Unit, Pan African Programme for the Control of Epizootics (PACE), African Union/Interafrican Bureau for Animal Resources, Nairobi, Kenya.

Knips, V. (2004) *Review of the Livestock Sector in the Horn of Africa (IGAD Countries).* Livestock Information, Sector Analysis and Policy Branch (AGAL), Food and Agriculture Organization of the United Nations (FAO), Rome.

L'Afrique de l'Ouest et le Club du Sahel et de l'Afrique de l'Ouest (2005) (CSAO_OCDE/OECD)

Leeflang, P. (1993) Some observations on ethnoveterinary medicine in Northern Nigeria. *Indigenous Knowledge and Development Monitor* 1(1), 17–19.

Legall, F. (2003) *The Role of Animal Disease Control in Poverty Alleviation, Food Safety, Market Access, and Food Security Objectives in Africa.* 5th Conference of the OIE Regional Commission for Africa, Maputo, Mozambique.

Les Grandes Orientations de la Politique Agricole de l'UEMOA (2002) Rapport definitif. vol 1: Rapport principal, 63pp. La Commission UEMOA, Ouagadougou.

Les Grandes Orientations de la Politique Agricole de l'UEMOA (2002) Rapport definitif. vol 2: Annexes. La Commission UEMOA, Ouagadougou.

Livestock Marketing Authority (2001) *A Study on Different Payments on Live Animals, Meat and Hides and Skins Export Trade.* Market Research and Promotion Department, Addis Ababa, Ethiopia.

MAAIF (2002) *Guidelines for Transportation of Livestock and Livestock products,* Ministry of Agriculture, Animal Industry and Fisheries (Directorate of Animal Resources), Republic of Uganda.

MARA/ARMA (2004) Mapping Malaria Risk in Africa: Slide Show. Available at: www.mara.org.za/SlideShow/slide11.html

Masiga, W.N. and Munyua, S.J.M. (2005) Global perspectives on animal welfare: Africa. *Revue Scientifique et Technique* 24(2), 579–586.

Mathes-Scharf, A. (2006) Kashrut. Available at: www.kashrut.com

Mathias, E. and Munday, P. (2005) *Herds Movements: The Exchanges of Livestock Breeds and Genes between North and South.* League for pastoral peoples and Endogenous livestock development, Ober-Ramstadt, Germany.

McCrindle, C.M.E (1998) The community development approach to animal welfare: an African perspective. *Applied Animal Behaviour Science* 59, 227–233.

Meat Board of Namibia (2006) Market Statistics. Available at: www.nammic.com.na/stats.php

Meat Board of Namibia (2006) Meat Chronicle, January 2006. Available at: www.nammic.com.na/news.php

MoARD (2000) *Report of the Strategic Workshop of Stakeholders on the Review of Policy and Laws Relating to Delivery of Veterinary Services and the Management of Animal Health, Products and Marketing.* Ministry of Agriculture and Rural Development, Kenya.

MoFA (2000) *Ghana Report on Food Production and Consumption.* Ministry of Food and Agriculture, Accra, Ghana.

Mogoa, E.G.M., Wabacha, J.K., Mbithi, P.M.F. and Kiama, S.G. (2005) An overview of animal welfare issues in Kenya. *The Kenya Veterinarian* 29, 48–52.

Mohammed, I. and Hoffmann, I. (2006) *Management of Draught Camels in Crop-Livestock Production Systems in Northwest Nigeria.* CIPAV, Cali, Colombia.

MoPND (2004) *Statistical Abstracts.* Ministry of Planning and National Development, Central Bureau of Statistics, Government Printers, Nairobi, Kenya.

National Department of Agriculture (2000) Foot-and-Mouth Disease. Available at: www.nda.agric.za/docs/fmd/fmd.htm

National Department of Agriculture (2000) Red meat marketing. *Paper no. 7 on Livestock Marketing,* Department of Agriculture, South Africa.

National Department of Agriculture (2003) Animal production. *Trends 2003.* Department of Agriculture, South Africa.

National Department of Agriculture (2003) The value chain for red meat. *Report of the Food Pricing Monitoring Committee.* Department of Agriculture, South Africa.

National Department of Agriculture (2006) Animal Welfare Aspects at Abbattoirs. Available at: www.nda.agric.za/vetweb/Food%20Safety/FS_RM_Manual/12%20Animal%20welfare.pdf

National Geographic (2006) Map Machine, Country Profile, South Africa. Available at: http://plasma.nationalgeographic.com/mapmachine/profiles/sf.html

National Treasury (1999) *Namibia, General Information. Southern Africa Economic Summit, Durban, South Africa, 4–6 July, 1999.* SADC Finance and Investment Sector, National Treasury, South Africa.

Nyenzi, B.S., Garanganga, B. and Raboqha, S.P. (2000) SADC drought monitoring centre review of regional climate outlook forums: meteorological aspects for Southern Africa. In: *Coping with the Climate: A Way Forward.* The International Research Institute for Climate and Society.

Nyiransabimana, B. (2005) *Etude du circuit traditionnel de commercialisation du lait et des produits laitiers au nord du Benin: cas de l'Atacora et Donga.* Memoire de Master en acteur du Developpement Rural du CNEARC.ProGCRN/GTZ.

Oku, S. (2001) *Present State of Food and Agricultural Statistics in Ghana.* MoFA, Accra, Ghana.

Onigbon, P. (2004) *Capitalisation et evaluation des marches a betail auto-geres au Nord du Benin. Articulation avec le Developpement local.* Centre Technique de Cooperation Agricole et Rurale/Ministere des Affaires Etrangeres, France/Inter –Reseaux/ Developpement Rural

Observatoire des opportunités d'affaires du Benin (2004) *Expansion du commerce intra et interrégional entre les pays de la CEMAC et de l'UEMOA. Etude de l'offre et de la demande sur les produits alimentaires.* OBOPAF, Cotonou, Benin.

OIE (2005) *Terrestrial Animal Health Code.* World Organization for Animal Health, Paris, France.

OIE (2005) Animal welfare: global issues, trends and challenges. *Revue Scientifique et Technique* 24(2).

OIE (2006) OIE Members and Official Delegates. Available at: http://www.oie.int/eng/ OIE/PM/en_PM.htm

Olumese, I.M. (2004) The cooperative movement and livestock development in Edo state, Nigeria. Thesis for Bachelor's Degree, Univeristy of Benin, Benin City, Nigeria.

Otieno, J. (2006) Sale of chickens hurt by news of bird flu. *Daily Nation,* Tuesday 14th February, 2006, News page 5.

PELUM (2006) Small holder agriculture: ignored gold mine? *PELUM Zambia Policy Brief – Issue No. 1.* Participatory Ecological Land-Use Management, South Africa.

Pemberton-Pigott, C. (2005) Development in Swaziland. *Regional Perspectives* Chapter 13.

Population Reference Bureau (2005) *World Population Data Sheet.* Population Reference Bureau, Washington, DC.

Renard, J.-F., Cheick, Ly. and Knips, V. (2004) *L'élevage et l'intégration régionale en Afrique de l'Ouest.* Ministère des Affaires Etrangères, République Française (MAE)/ FAO (éd CIRAD), 39 pp.

Reusse, E. (1982) Somalia's nomadic livestock economy: its response to profitable export opportunity. *World Animal Review* 43, 2–11.

SAMIC (2005) *Newsletter, 16 September 2005.* South African Meat Industry Company, South Africa.

SARCOF (2004) Statement from the Eighth Southern Africa Regional Climate Outlook Forum Held in Harare, Zimbabwe from 1–2 September 2004. Available at: www.cpc.ncep.noaa.gov/products/ african_desk/rain_guidance/SARCOF_2004_STATEMENT.pdf

SA-Venues (2006) South Africa Climate & Weather. Available at: www.sa-venues.com/no/weather. htm

Schwartz, H.J. (1993) Pastoral production systems in the dry lowlands of Eastern Africa. In: Baumann, M.P.O., Janzen, J. and Schwartz, H.J. (eds) *Pastoral Production in Central Somalia.* GTZ (German Agency for Technical Cooperation), Eschborn, Federal Republic of Germany, pp. 1–16.

Shimshony, A. and Chaudry, M.M. (2005) Slaughter of animals for human consumption. *Revue Scientifique et Technique* 24(2), 693–710.

Seng, P.M. and Laporte, R. (2005) Animal welfare: the role and perspectives of the meat and livestock sector. *Revue Scientifique et Technique* 24(2), 613–623.

South African Consulate General (2005) Agriculture and Land Affairs. Available at: www.southafrica-newyork.net/consulate/landaffairs.htm

South African Meat Industry Company (2006) Imports/Exports. Available at: www.samic.co.za/

Southern African Development Community (2002) *Promotion of Regional Integration in Livestock Sector Community Secretariat – Technical Dossier.* Secretariat of the Southern African Development Community, South Africa.

Southern African Marketing Company (2005) Transport. Available at: http://www.sadcreview.com/ country_profiles/mozambique/moz_transport.htm

Steffan, P., Shirwa, A.H. and Addou, S.I. (1998) *The Livestock Embargo by Saudi Arabia: A Report on the Economic, Financial and Social Impact of Somaliland and Somalia.* FEWS/Somalia, Nairobi, Kenya.

Stockbridge, M. (2004) *The Marketing System for Livestock Exports: Monitoring and Analysis.* Consultancy report on livestock marketing survey in Somalia. Terra Nuova, Nairobi, Kenya.

Tatcher and Leteneur. (2000) *Quelles Productions Animales en Afrique Sub-Saharienne dans une Generation?* CIRAD/EMVT, 39 pp.

Tatcher *et al.* (2001) *Le secteur des productions animales en Afrique Subsaharienne des Independences a 2020* chap 2. Approche des echanges par zones sous-regionales.

Tatcher *et al.* (2001) *Le secteur des productions animales en Afrique Subsaharienne. chap1.* La place de l'Afrique subsaharienne dans les echanges mondiaux et evolution du secteur de l'elevage.

Terra Nuova (2004) *Expanding the Network for the Provision of Supportive Veterinary Services for the Somali Livestock Industry. Itinerant Training Programme for Somali Veterinary Professionals. Phase III. Seventh progress report,* April–June 2004. Terra Nuova, Nairobi, Kenya.

Terra Nuova (2005) *Expanding the Network for the Provision of Supportive Veterinary Services for the Somali Livestock Industry. Itinerant Training Programme for Somali Veterinary Professionals. Phase III. Final report.* Terra Nuova, Nairobi, Kenya.

The East African (2002) Ban on Beef is Unfair, Kenya Tells Tanzania. *The East African,* Business, Monday, 21 January 2002.

The Veterinary Record (2004) International disease monitoring, January to March 2004. *The Veterinary Record* 15 May 2004.

Thiermann, A.B. and Babcock, S. (2005) Animal welfare and international trade. *Revue Scientifique et Technique* 24(2), 747–755.

Toe, J.Y. (2003) *Droit Francophone: Doctrine: Afrique de L'Ouest: Quel Ordre juridique dans les Etats de l'Afrique de l'Ouest? In Seminaire sous-regional de sensibilisation sur le droit francophone.* UEMOA, 18 pp.

URT (2002) United Republic of Tanzania: *Presidential Circular No. 1 of 2002: Prevention of the Spread of Animal Diseases in the Country,* Kiswahili Version. Government Printers, Dar es Salaam, Tanzania.

URT (2003a) *The Animals (Protection) Ordinance,* CAP.153, 2003. United Republic of Tanzania, Government Printer, Dar es Salaam, Tanzania.

URT (2003b) *The Animal Diseases Act,* Number 17, 2003. United Republic of Tanzania, Government Printer, Dar es Salaam, Tanzania.

URT (2005) *Ministry of Water and Livestock Development Budget Speech 2005/2006.* United Republic of Tanzania, Government Printers (KIUTA), Dar es Salaam, Tanzania.

URT (2006) *The Meat Industry Act, 2006.* United Republic of Tanzania, Government Printer, Dar es Salaam, Tanzania.

Vallat, B. (2005) Animal welfare: global issues, trends and challenges, Preface. *Revue Scientifique et Technique* 24(2), 467–468.

Vordzorgbe, S.D. and Caiqo, B. (2001) *Report on Status Review of National Strategies for Sustainable Development in Ghana.* IIED, London.

Webster, J. (2005) The assessment and implementation of animal welfare: theory into practice. *Revue Scientifique et Technique* 24(2), 723–734.

Williams, T.O., Spycher, B. and Okike, I. (2003) *Improvement of Livestock Marketing and Regional Trade in West Africa.* International Livestock Research Institute, Nairobi, Kenya.

World Bank (2000) *Project Appraisal Document on the Agricultural Services Sub-sector Investment Project.* The World Bank, Washington, DC.

World Trade Organisation (2002) *Zambia Consolidates Liberalisation, Says WTO.* Trade Law Centre for Southern Africa, Stellenbosch, South Africa.

Zimplats (2006) Zimbabwe Profile. Available at: www.zimplats.com/about/profile.htm http://skyscrapercity.com/archive/index.php/t-330394.html, (http://www.fews.net/centers/innerSections.aspx?f=bf&m=1000861&pageID=monthliesDocMigratory movements and trends in livestock prices).

9 North America

M. Engebretson

Born Free USA united with Animal Protection Institute, 1122 S Street, Sacramento, California, USA

Abstract

Transportation is one of the most stressful events in a farmed animal's life. Because nearly all of the billions of farmed animals raised in North America are subjected to transportation at some point during their lives, transportation is also one of the most important welfare issues affecting farmed animals.

This chapter reviews the available information on farmed animal transport in North America with particular reference to road transport of animals destined for slaughter. Because animal welfare generally declines with increasing journey length the chapter will emphasize problems associated with long-distance transport. While 'long-distance transport' has not been concretely defined or agreed upon in the scientific literature or by advocacy groups, for the purpose of this chapter journeys exceeding 8–12 h of non-stop travel are generally regarded as 'long-distance'. This characterization is consistent with the recommendations for travel and rest periods for horses, pigs, sheep and cattle, set forth by the European Scientific Committee on Animal Health and Welfare.

Deprivation of food and water, overcrowding, lack of opportunities for rest and prolonged exposure to extreme heat or cold are commonly cited welfare concerns associated with long-distance transport. Poor and abusive handling of animals during loading and unloading and at auctions also increases animal stress and suffering during the transport process.

Despite this, animals may be transported across multiple states, regions or provinces and across national borders for fattening and slaughter. Some may even be moved across national borders for slaughter only for the meat then to be shipped back to their countries of origin for consumption.

The exact number of animals affected by long-distance transport and the distances commonly travelled are difficult to determine because data on livestock movements in the USA, Canada and Mexico are highly fragmented and limited in scope. The primary reason for this is lack of accurate record keeping at the state, province and country level.

The transport of livestock in North America is influenced by many factors. The economic costs of transporting animals (which tend to be lower than transporting feed) and geographical differences in feed and forage availability and prices, as well as the development and location of feedlots and slaughterhouses largely determine where animals will be transported and at what stage of production. Live animal movements across the national boundaries between the USA and Canada and the USA and Mexico is normal practice as animals move to pastures, feedlots or to more available slaughter facilities.

The setting of travel time limits, rest periods and provisioning of food and water for livestock during road transport seems to be fundamental in addressing animal welfare concerns.

While Canada, the USA and Mexico each have varying laws, codes and regulations governing the transport of farmed animals, there are significant shortcomings in scope and enforcement which present significant challenges for ensuring animal welfare in each of these countries.

Because animal protection legislation stems largely from civil society pressure, public education and advocacy will be fundamental in effecting lasting change for transported farmed animals in North America.

Introduction

The vast majority of animals raised for food in North America are transported at some time during their lives with their final ride ending at a slaughter facility. On-the-ground investigations and scientific literature suggest that compared to transport for further fattening, transport to slaughter is accompanied by the most welfare problems (Van Houwelingen and Vingerling, 1989; Perry, 1990; Knowles, 1998; Animals' Angels, 2003; Animal Protection Institute/Compassion in World Farming, 2005; Ecostorm Investigations, 2006, personal communication). However, welfare problems associated with the transport of cattle and calves to feedlots have also been shown to be significant in the USA and abroad (Tarrant, 1990; Knowles, 1999; Lonergan et al., 2001; Ishmael, 2005). Welfare issues associated with transport may also be compounded if the journey involves transit through auction (Animal Protection Institute/Compassion in World Farming, 2005).

This chapter summarizes animal welfare issues associated with long-distance transport in North America and examines some of the reasons farmed animals are transported for food production. To supply some background information, the chapter begins with an overview of the modern livestock and meat industries of Canada, the USA and Mexico as well as national legislation and enforcement, or lack thereof, within each of these countries.

Trends in the North American meat and livestock industries are characterized by a move towards consolidation, vertical integration and separation between different stages of production which often includes transport. The relative low cost of transporting animals compared to the cost of transporting feed and geographical differences in feed prices or grazing pasture availability, as well as the location of feedlots, largely determine where animals will be transported for fattening (Shields and Mathews, 2003). Market demand and prices and the location, capacity and ownership of slaughterhouses also play a large role in determining where animals will be transported for slaughter. Regulatory differences between states, regions or countries such as waste disposal and labour costs also affect animal movement for slaughter and feeding (Shields and Mathews, 2003; Animal Protection Institute/ Compassion in World Farming, 2005).

The second section of the chapter examines trade between North American countries, national and international routes and the conditions and welfare of transported animals including driver and handler requirements.

Measurements of injuries, bruises, mortality, morbidity and carcass quality can be used as indicators of welfare during handling and transport (Broom, 2000; and see Chapter 4, this volume). Mortality records give information about welfare during the journey while bruises, scratches, blemishes, broken bones and incidences of PSE/DFD pork and dark cutting beef or lamb provide information about the welfare of the animals during handling, transport and in holding areas (Broom, 2000).

Data or access to data on the deaths attributed to transport stress in North America are limited but what little information exists sheds some light on the issue. For example, industry data suggest that 17 pigs in every 10,000 marketed in Ontario, Canada die on the truck (Dewey *et al.*, 2004), a recent study estimated 12 cattle die for every 1000 transported to feedlots (Lonergan *et al.*, 2001) and a report prepared by the European Commission found that 5% of horses transported to Mexican slaughterhouses were dead on arrival (European Commission, 2000). While death of animals may be a significant economic consideration, it is important to point out that many animals can and do experience extremely poor welfare during transport even if a majority of animals survive the journey.

The chapter ends with a brief discussion of how the issues raised in the chapter might best be approached and remedied in each of the three countries that make up North America taking into consideration the political and cultural characteristics of these countries.

Background

The USA

The USA is a leading beef and pork producer (Shagam, 1998). The USA (along with Hungary) also leads the world in global meat consumption per capita (Shagam, 1998). In addition, people in the USA spend the lowest percentage of their per capita income on food than any country in the world. Meat including poultry accounts for less than 2% of the disposable income (Boyle, 2004). The primary reason for this is that while prices of many goods and services have risen sharply over the last 50 years due to inflation, the cost of meat by comparison has lagged far behind.

Vertical integration within the industry and transition from small family farms to concentrated animal feeding operations (CAFOs) or intensive farms enabled by government subsidies and the subsidized large-scale production of corn have played a major role in reducing the cost of meat and in giving large-scale livestock operators an advantage over smaller-scale farms (Nierenberg, 2006; Pollan, 2006). The US federal treasury spends up to 5 billion a year to subsidize corn, most of which is used for animal feed (Pollan, 2006).

The US meat packing (slaughter and processing) industry has also become increasingly consolidated. Three major companies control the American meat packing industry: Excel – a subsidiary of the agribusiness giant Cargill; ConAgra – also a huge agribusiness company; and Tyson IBP, which is the biggest meat packing company in the world (Boyle, 2004). Most of the companies involved in the US meat industry are represented by one or more meat trade and lobbying organizations: the American Meat Institute, the National Meat Association and the National

Cattlemen's Beef Association. The American Meat Institute is the largest national trade association for the meat industry – representing nearly 300 companies. Collectively, these companies account for about 95% of the beef, pork, lamb and veal produced in the USA (Boyle, 2004).

Transport is a fundamental component of most livestock production in the USA. According to a report produced by the United States Department of Agriculture (USDA) Economic Research Service, 'The economic reality in the livestock industry is the geographic separation between livestock in one stage of production and the feed/forage resources or facilities needed for successive stages of production' (Shields and Mathews, 2003). For example, on average, 58% of the calves born in the USA each year are shipped to another destination for feeding or breeding. Approximately 85% of those cattle move through at least one auction (Juday, 2005).

In addition, pigs are frequently shipped from farrowing operations in North Carolina to nursery facilities or grower/finisher facilities in Iowa where they are fed to market weights, then moved again to California for slaughter, according to the report. This trend appears to be escalating; the number of pigs crossing state lines increased from 5 million in the early 1990s to 26.9 million in 2001 (Shields and Mathews, 2003).

The greatest movement of cattle is into and within the Northern and Southern Plains which include North Dakota, South Dakota, Nebraska, Kansas, Oklahoma and Texas. For pigs the greatest movement is into the Corn Belt which includes Iowa, Illinois, Indiana and Ohio. Colorado is a major feeding and slaughtering destination for sheep many of which are likely transported from Utah, South Dakota, Montana and Wyoming – states that produce far more sheep than they slaughter (Shields and Mathews, 2003).

The top four cattle feeding states are Texas, Kansas, Nebraska and Colorado, accounting for more than 65% of the US feeder cattle (cattle to be fed for beef production) supply (Shields and Mathews, 2003). More than two-thirds of all cattle are also slaughtered in these states (Shields and Mathews, 2003). North Carolina, Iowa, Minnesota and Illinois account for half of US pig production, 60% of pig fattening or 'market hog inventory' and 56% of the nation's pig slaughter (Shields and Mathews, 2003). Given that the highest beef and pork producing states also rank high in processing, it seems that long-distance transport of animals for slaughter within the USA is largely unnecessary from a logistics standpoint.

Canada

Meat and meat products represent the largest sector of the Canadian food production industry (Canada Food Inspection Agency, 2004). Like the USA, Canada is also among the top meat consuming nations in the world. In 2002, Canada's per capita meat consumption was 238.32 lb (108.10 kg) (see Table 9.1 below) (Food and Agriculture Organization of the United Nations, 2004).

Due to the large size of the country and its high level of integration with the USA's beef and pork industries, long-distance transport of livestock for fattening and slaughter is a significant component of Canada's livestock industry. The export

Table 9.1. Summary of North American slaughter and meat consumption.

	Canada	USA	Mexico
Livestock slaughtered[a]	28,616,000	140,092,000	30,907,414
Poultry slaughtered[a]	668,380,000	9,572,092,000	1,292,530,000
Slaughterhouses[b]	178 (121 federal)	2,893 (806 federal)	2,249 (80 federal)
Meat consumption[c]	238.32 lb/108.10 kg	275.14 lb/124.80 kg	129.19 lb/58.60 kg

[a]2005, excluding poultry (from FAO Stat Database) (http://faostat.fao.org/site/569/DesktopDefault. aspx?PageID = 569).
[b]2004, 2006, 2005 data, respectively. For Canada total number includes 57 provincial plants inspected by federal agency. Numbers do not include poultry.
[c]Per capita 2002 data (includes poultry).

of Canadian livestock to the USA for further fattening and slaughter has histori-cally been economically advantageous since the USA has greater feedlot and slaughter capacity than Canada.

However, in May 2003, the export of Canadian cattle into the USA was dra-matically reduced following the discovery of a bovine spongiform encephalopathy (BSE) infected cow on an Alberta farm prompting the USA to ban imports of Canadian cattle, costing Canada more than 7 billion dollars in lost trade (Gregerson, 2005, 2006). Prior to BSE, Canada typically exported approximately 1.1 million head of cattle to the USA each year (Standing Senate Committee on Agriculture and Forestry, 2005) (see also Chapter 2 on Economics, this volume). Canada's hog industry is also highly dependent on trade with the USA for slaughter because its pig production capacity exceeds its slaughter capacity (Economic Research Service, 2004). In addition, demand for pigs in the USA, particularly feeder pigs, often exceeds domestic supply thereby creating a market for Canadian pigs.

Canada's meat industry is represented by several trade organizations that advocate on behalf of industry's interests. In 1993, Alberta's major livestock groups formed the Alberta Farm Animal Care (AFAC) Association whose stated purposes include to, 'promote responsible humane animal care within the livestock industry', and to 'monitor and participate in issues and legislation that affect animal care'. Other notable industry groups include the Canadian Pork Council, the Canadian Cattlemen's Association and the Canada Beef Export Federation.

Mexico

While livestock represents only about 1.1% of Mexico's gross domestic product (Commission for Environmental Cooperation, 2003), livestock production (mostly for cattle grazing) occupies more than half of its 758,000 mi^2 (1,940,000 km^2) of land area and involves more than 3 million producers (Commission for Environmental Cooperation, 2003).

Mexico appears to be moving in the same direction as its northern neighbours with local producers of animal products being replaced by large companies that engage smaller farmers and contract growers (Nierenberg, 2005). Half of Mexico's milk production, half of its pork production and 90% of its egg production now come from intensive, mechanized operations. Although cattle are still extensively grazed, beef production from feedlots increased by 40% between 1980 and 2000 (Commission for Environmental Cooperation, 2003).

Trade in live animals between Mexico and the USA consists largely of Mexican feeder cattle destined for US feedlots as the USA has a larger feed grain production and more and larger grain-feeding facilities than Mexico which relies primarily on grazing to bring cattle to slaughter weights. About 1 million head of cattle per year enter the USA through ten ports of entry in Arizona, New Mexico and Texas (Skaggs *et al.*, 2004).

Hogs, cull ewes and horses (see Table 9.1.2A) are also exported from the USA for slaughter in Mexico. The trade in animals destined for slaughter often involves substantial travel distances and is accompanied by many animal welfare concerns.

To meet a growing demand for pork, pork production and imports of live hogs from the USA are expected to increase in the future. Imports of live hogs are forecast to reach 225,000 head in 2007 (USDA Foreign Agricultural Service, 2006). In addition, ongoing negotiations between the USA and Mexico are expected to allow imports of US dairy cull cattle in the near future as Mexico continues to loosen bovine import restrictions imposed in 2003 following the detection of BSE in a US cow (USDA Foreign Agricultural Service, 2006). This is particularly concerning from an animal welfare perspective as in the USA dairy cattle make up 75% of all 'downer' cattle (McNaughton, 1993). One cow in four in the average US dairy herd is culled each year due to decreased milk production, lameness, illness or reproductive problems (Ensminger, 1993). If export of live cattle to Mexico resumes, the trade in spent dairy cows will likely resume too, with serious implications for animal welfare.

Like the USA and Canada, Mexico also has industry organizations that advocate for, and politically represent, livestock and meat packing interests. The Consejo Mexicano De La Carne (Mexican Council of Meat) comprises about 70 members, representing importers, exporters, distributors, meat packers and special suppliers to the meat industry. Mexican cattle producers are also represented by the regional cattle growers' association, Union Ganadera and the Unión Ganadera Regional de Chihuahua (Chihuahua regional cattle grower's association) which owns and operates both sides of the Santa Teresa, New Mexico port through which 25% of all Mexican cattle imports enter the USA and through which the majority of US slaughter horses are exported to Mexico.

National legislation

Canada, the USA and Mexico have varying laws, codes and regulations governing the transport of farmed animals. However, in each of these countries the legislation is limited in both scope and enforcement presenting significant challenges for ensuring animal welfare.

Regulations that could have a beneficial effect on the welfare of animals transported by road include requirements for driver education and licensing, handler education and oversight, vehicle and ramp design, stocking density, temperature controls, sanitation, bedding, access to food and water at appropriate intervals, veterinary inspection, prohibition of transporting injured or otherwise unfit animals, animal and vehicle tracking mechanisms and journey limits.

While some of the above referenced parameters are addressed in industry or government recommendations, voluntary guidelines or official norms, without the force of law and or mechanisms for enforcement such standards are of questionable value and their affect on the welfare of transported animals is uncertain.

Similarly, lack of enforcement can also make legal requirements meaningless. Animal advocacy organizations have demonstrated that laws governing the welfare of farmed animals are often under-enforced or not enforced at all. However, the existence of such laws is none the less important as they provide an avenue for concerned citizens, advocacy organizations or industry representatives to hold government agencies accountable for enforcing the legally mandated regulations and prohibited acts.

Of the North American countries, Canada has the most comprehensive national laws governing the transport of farmed animals, with the USA and Mexico falling far behind. Canada has set forth pre-transport, handling and onboard vehicle requirements while the USA (with the exception of horses) and Mexico have no mandatory handling requirements or minimal standards for onboard vehicle conditions such as space, ventilation and temperature control. However, none of the three countries have set forth overall transport limits. As a result, animals may be legally subjected to endless transport intervals.

In addition, the transport intervals allowed before provision of food, water and rest are required to allow between 38 and 52 h of transport in Canada and 28–36 h of transport in the USA. Animals may be denied food and water much longer if pre-transport provisions are not required. For example, under the current Canadian regulations cattle, sheep and goats in transit may be denied water for up to 57 h, 5 h before loading and 52 h in transit. It is also possible for animals to be denied feed for up to 81 h, 5 h before loading, 52 h on the truck and 24 h after arrival at a federally registered slaughter plant.

In 2006, the Canada's Food Inspection Agency (CFIA) began considering regulatory changes to the animal transportation requirements (Canadian Food Inspection Agency, 2006). Changes are being considered by the CFIA to reduce these intervals and to better reflect animal needs related to water, food and rest.

In North America the longest trips endured by livestock are those that cross national boundaries but national regulation and transport limits currently do not apply across national borders. For example, Canadian drivers transporting pigs from Canada to California where they will be shipped to Hawaii are not governed by Canadian transport regulations after crossing into the USA. Similarly, the USA's 28 h time limit fails to apply to livestock such as pigs and cull ewes transported to Mexican slaughterhouses after they cross the Mexican boarder. Table 9.1.1A lists national/federal legislation governing transport in North America.

Canada

In Canada, on-farm animal welfare regulations fall under the jurisdiction of the individual provinces while the responsibility of ensuring humane transportation falls under the CFIA – a federal regulatory body.

In general, change in addressing farm animal welfare concerns in Canada has been slow compared to other developed countries (with the exception of the USA, which also lags behind internationally), however, the CFIA has begun the process of revising its regulations with regard to transport (Canadian Food Inspection Agency, 2006).

HEALTH OF ANIMALS ACT.   One of the statutes under the Department of Justice is the Health of Animals Act – 'an Act respecting diseases and toxic substances that may affect animals or that may be transmitted by animals to persons, and respecting the protection of animals'. The Act is implemented by the Health of Animals Regulations. Part XII of these regulations applies to, 'the transportation of animals entering or leaving Canada or within Canada' and was introduced in 1975. The CFIA, created in 1997, is now the enforcing body. Prior to 1997, the enforcing agency was Agriculture Canada.

The Part XII regulations come into effect when the animal is loaded for transport and continue throughout its time in transit, including refuelling periods and market auction stays, until the animal is unloaded at the final destination. If the animal is unloaded at a federally inspected slaughter plant, the CFIA's oversight includes its humane handling and slaughter regulations under the Meat Inspection Act (see 'Meat Inspection Act').

COMPROMISED ANIMALS POLICY.   In June 2005, an amendment to the Health of Animals Regulations was implemented that clarified the requirements related to the shipping of non-ambulatory livestock that are unfit for transport (Canadian Food Inspection Agency, 2006). The result is the Compromised Animals Policy, which clarifies the restrictions and options for dealing with non-ambulatory or 'downed' animals.

If the CFIA is able to determine that a non-ambulatory animal was loaded and/or unloaded, they may charge anyone who was part of the chain that contributed to the offence, including the driver, the trucking company and/or the producer. A fine is determined through the Administrative Monetary Penalties (AMP) process. Fines can be as high as $4000 per contravention. Severe cases may be sent to court instead of being handled with an AMP. The CFIA Enforcement and Investigation Services determine how each case will be handled.

CODES OF PRACTICE.   Recommended Codes of Practice for the Care and Handling of Farm Animals are a series of nationally developed species-specific voluntary guidelines intended to encourage welfare-oriented farm animal management and handling practices. The Canadian Agri-Food Research Council is responsible for managing the codes.

The 'Transportation' code of practice was developed in 2001, through the combined effort of two government and 20 industry groups. The codes set forth voluntary housing and management practices for farmed animals, as well as transportation and slaughter. The transportation code of practice is a 75-page document covering various recommendations such as that animals should not be rushed during loading and unloading, electric prods should be discouraged, vehicles should be cleaned and sanitary and sufficient floor space should be provided to allow for adequate ventilation and that transporters engaged in long-haul transport (defined as longer than 6 h) should consider weather conditions, emergency procedures and expected delays.

With regard to transport times the codes of practice are somewhat contradictory. On page 2, it states that the longer animals are transported the greater the risk
of injury and death and defines long-haul transportation as 'generally longer than
6 h' and notes that journeys exceeding this length may expose animals to significant
environmental changes and/or increase the length of time they are exposed to risk
factors. On page 3, the document states that 'current knowledge of animal nutrition
requirements prior to and during transportation and during the antemortem period,
in general is substantially incomplete. The current recommendations for feed water
and withdraw allowance are not substantiated by research data'. Despite the earlier
admissions of the risks of transport exceeding 6 h and the admission that research
data do not support their recommendations, the codes set forth liberalize maximum
transport times that are strikingly similar to the standards of the Health of Animals
Regulations – allowing for 36 and 48 h of transport before providing feed, water and
rest for pigs and chickens, and cattle and sheep, respectively.

MEAT INSPECTION ACT.   Federally inspected establishments are subject to operational
policies and regulations established under the CFIA's Meat Inspection Act, which
prescribes the humane handling and slaughter of food animals. Provisions of the
Meat Inspection Regulations cover the unloading, holding and movement of animals in slaughter facilities.

The federal act sets standards for the humane handling and slaughter of food
animals and plants whose products may be exported or sold interprovincially or
internationally. Provinces and territories have similar legislation covering some or
all plants whose products will be sold only in the home province.

Under authority of the Meat Inspection Act, the Meat Inspection Regulations
prohibit the handling of a food animal in a manner that subjects the animal to
avoidable distress or avoidable pain. Subsection 67(6) of the Meat Inspection
Regulations requires that if an operator or inspector who is not an official veterinarian suspects in the course of the ante-mortem examination of a food animal that
it shows a deviation from normal behaviour or appearance, the food animal shall
be held and referred to an official veterinarian for a detailed inspection and instructions regarding its disposition.

Some 94% of domestically produced beef is inspected by the CFIA, 95% of
pork is also inspected by the CFIA as is 84% of chicken and 93% of turkey. The
remaining meat is inspected by provincial meat inspection systems or by the local
municipality (Canada Food Inspection Agency, 2004).

CRIMINAL CODE OF CANADA.   The cruelty to animals provisions of Canada's Criminal
Code prohibits the willful and malicious hurting or killing of animals as well as
willful neglect, which causes unnecessary suffering, although most common and
customary farming practices are not unlawful.

CANADA ANIMAL TRACKING AND IDENTIFICATION REQUIREMENTS.   The Canadian Cattle
Identification Program is an industry-initiated and industry-established trace-back
system designed for the containment and eradication of animal disease. This programme is not currently used to monitor journey times or compliance with transport regulations.

Under the programme, all cattle in Canada must be ear tagged with a Canadian Cattle Identification Agency (CCIA)-approved ear tag by the time they leave their herd of origin. The unique number of each individual animal is maintained to the point of export or carcass inspection where the animal is either approved for consumption or condemned. The programme began on 1 January 2001 with monetary penalties beginning on 1 July 2002. The CFIA enforces the Canadian Cattle Identification Program with penalties for non-compliance ranging from $500 to $4000.

The agency is led by a Board of Directors made up of cattle industry representatives including The Canadian Cattlemen's Association, Livestock Marketing Association of Canada, Canadian Meat Council, Canadian Veterinary Medical Association, Dairy Industry, The Canadian Bison Association, Alberta Beef Producers, Alberta Cattle feeders Association, Livestock Order Buyers of Canada, Manitoba Cattle Producers Association and Ontario Cattlemen's Association. CFIA and Agri-Food Canada are ex officio members.

While this tracking system could be used to provide valuable information on livestock movements across Canada, the information covered in the programme is not readily available. The reason for this is that the CCIA controls access to the information stored in the database. Information is provided to the CFIA only in response to a request from this agency in the event of a health or safety concern. All other requests for access, including government departments other than CFIA, are made through a legal process and the petitioner is required to justify the need for access. CCIA is not listed in the schedule (Schedule 1) to the Access to Information Act and, as such, the Act does not apply to the CCIA. However, if CFIA requests and receives information from the CCIA, the received information would fall under the provisions of the Act (Canadian Food Inspection Agency, USA, 2006, personal communication).

CANADA ENFORCEMENT. As previously mentioned, compliance with the Health of Animals Regulations is mandatory. CFIA officials monitor compliance with the regulations through routine inspection at strategic locations (e.g. ports of entry, registered establishments and auction markets) and by following up on reports of non-compliance. Actual rate of recorded violation, issued citation or other correctional action taken to ensure compliance with the law is unknown.

USA

At the state level, anti-cruelty laws or transportation laws of most states and the District of Columbia require that transport of animals be conducted in a humane manner. However, fewer than a third of the states have adopted laws that restrict the amount of time animals may be confined during transportation and many states exempt customary agricultural practices from laws governing humane treatment of animals.

Where laws do exist at the state level, time limits are liberal and requirements for provision of food and water non-existent. The shortest maximum time period an animal can be transported without food, water and rest is 18h for trucks in Vermont. States often allow 36h, if requested, for both railroad and trucks, and Washington allows a total of 2 days for the transport of animals without food,

water or rest on the railroad. Fines for violating such transport laws are low, averaging approximately $500 (Wolfson, 1999).

As most states with transport-time restrictions can be crossed in less than the
maximum amount of time allowed, such liberal time limits at the state level are rather
inconsequential. Transport limits are best addressed at the federal level to regulate
transport across state lines. The federal 28-Hour Law (described below) does address
transport across state lines but it is questionable whether a 28 h limit is sufficient
to ensure good welfare and the enforcement of this law is effectively non-existent.

Generally regarded as companion or service animals, horses are not consumed
as human food in the USA. Commercial Transportation of Equine for Slaughter
Act, an amendment to the Humane Slaughter Act, regulates the transport of horses
to slaughter. However, as with livestock, horses may still be transported for 28 h
without food, water or rest under the Act.

HUMANE METHODS OF SLAUGHTER ACT.   The Humane Methods of Slaughter Act requires
that animals be rendered unconscious or insensible to pain prior to slaughter, with
exemptions for ritual or kosher slaughter. The USDA has not applied the law to
birds and rabbits and exotic 'game' animals, such as bison and antelope are not
covered unless the producer opts for voluntary, fee-for-service meat inspection by
the USDA. The law does not cover transportation; however, the humane handling
regulations under the Act apply to activities taking place on the premises of slaughter plants, such as unloading from trucks and the handling of downers – animals
too sick or injured to move on their own.

28-HOUR LAW OF **1906**.   Amended in 1994, the 28-Hour Law of 1906 covers the
interstate transport of animals for sale or slaughter. Generally, it requires that livestock being transported across state lines be humanely unloaded into pens for food,
water and at least 5 h of rest every 28 h. The statute does not apply to animals
transported in a vehicle or vessel in which the animals have food, water, space and
an opportunity to rest. Sheep may be confined for an additional 8 consecutive
hours when the 28 h period of confinement ends at night, and animals may be
confined for 36 consecutive hours upon the request of the owner or the person
having custody of the animals. Time spent in loading and unloading animals is not
included as part of a period of confinement.

Until recently, the USDA had operated on the assumption that the law does
not apply to road vehicles. According to a USDA web site offering guidance to
livestock transporters, 'Federal law requires that livestock in interstate commerce
be in transit for no more than 28 h without food, water, and rest. However, this
law applies only to rail shipments' (USDA, Transportation Services Branch, 1999).
In October 2005, several animal advocacy groups (Animals Angels, Compassion
Over Killing, Farm Sanctuary and the Humane Society of the United States) filed
a federal rulemaking petition seeking to limit truck transport of animals to no more
than 28 h without unloading for food, water and rest – as mandated by the federal
28-Hour Law (Brandt, 2005). The USDA changed its position (but took no enforcement action) in 2003, but did not make the decision public until September 2006,
when it responded to the legal petition (Cody, 2006). However, the enforcement of
the law is still problematic.

The Animal and Plant Health Inspection Service (APHIS) is allowed to exercise functions of the Secretary of Agriculture under the authority of the 28-Hour Law. The law is also listed as falling under the auspices of the Veterinary Services programme of APHIS. However, there are neither provisions in the code or the regulation that speak to how this agency should monitor the transportation of animals, nor is there mention of the 28-Hour Law or any implementing programmes on the Veterinary Services, APHIS or USDA web sites. APHIS has stated that complaints for violations of the 28-Hour Law are to be referred to the Justice Department (Brasher, 2006).

COMMERCIAL TRANSPORTATION OF EQUINES FOR SLAUGHTER ACT. In March 1996, Congress passed the Commercial Transportation of Equines for Slaughter Act. While the USDA was directed to write regulations to enforce the Act, the regulations were not released until January 2002. Unfortunately, the regulations allowed the use of double-deck trailers, known to cause welfare problems, until December 2006. The regulations require that cargo space of a vehicle transporting equines must:

- Be designed, constructed and maintained 'in a manner that at all times protects the health and well-being of the equines (e.g. provides adequate ventilation, contains no sharp protrusions, etc.)';
- Separate stallions and aggressive equines;
- Have sufficient height;
- Be equipped with doors and ramps safe for unloading/loading; and
- Be single tiered.

The Act further requires that:

- Immediately prior to transportation, the equine must have 'appropriate food, water, and rest' for 6h and a USDA backtag (a form of identification applied to the back of the animal) must be applied.
- Equines must be loaded so that each has floor space so as not to cause injury or discomfort.
- The driver of the vehicle must drive in a manner to avoid causing injury to equines.
- Any equine that has been on conveyances for 28 consecutive hours must be unloaded for feed, water and 6h of rest.
- The use of electric prods is prohibited.

Penalty for violating the above regulations is a fine of up to $5000 per equine.

Because the provisions of the Commercial Transportation of Equines for Slaughter Act only recently came into effect (in 2002 and 2006) compliance and enforcement is unknown. However, an August 2006 incident suggests that state anti-cruelty codes may provide better protection for transported horses than the current federal law. In the incident, a man hauling 19 horses through Arkansas en route to a Texas slaughterhouse was charged with five counts of animal cruelty under Arkansas state law when employees at an auto repair shop called local police after noticing several horses with lacerations to their face and body (U.S. Newswire, 2006).

GUIDELINES. In 1997, the US Department of Agriculture Transportation Services Branch released the Cattle and Swine Trucking Guide for Exporters, in an effort to 'help maintain the market value of the stock and reduce claims due to injury and mortality' (USDA Agricultural Marketing Service, 1997).

Issues covered include preparation for transport, feeding and watering, veterinary care, stress reduction, rest stops, trailer design, loading and unloading and handling downed animals. However, these guidelines are not overseen by any regulatory body to ensure or measure compliance; therefore, it is not possible at this stage to assess the effect of the guidelines on transporters' practices and animal welfare. Moreover, like Canada's voluntary guidelines, the USDA recommendations seem contradictory when it comes to transport times and rest periods. It is recommended that transporters, 'keep transit time to a minimum', and notes that 'studies have shown that animals continue to lose weight as long as they are in a truck', however, the recommendations go on to suggest a maximum transport time of 30–40 h for cattle and 36 h for pigs (the document claims that the 28-Hour Law does not apply to trucks). The section concludes with a recommendation that drivers use 'good judgment' when deciding how long animals should be on the road and that rest period decisions must be made on a 'case by case basis'. Such vague recommendations make compliance and animal welfare even more difficult to measure.

The animal agriculture industry in the USA has developed voluntary animal care quality assurance programmes in response to pressure from the retail food industry and to avoid government regulation and third-party audits (Farm Sanctuary, 2005). However, the area of transportation is among the least well-addressed welfare issues of all aspects of farmed animal agriculture, according to Farm Sanctuary, which completed a comprehensive assessment of industry quality assurance programmes in 2005.

The National Pork Board launched a Trucker Quality Assurance programme in 2002 and has certified 10,000 truckers to date (Vansickle, 2005). The training for this programme takes place at slaughter plants and consists of a 2 h educational session and written test (National Pork Board, 2001). Truckers must complete re-certification classes every 3 years (Vansickle, 2005).

In addition, individual producers may market themselves as 'USDA Process Verified' if they meet particular animal care standards and submit to auditing of their practices by the USDA. As of 2006, four US pork producers, including the three largest, were participating in the Process Verified programme. The content of these standards programmes is not available to the public (Farm Sanctuary, 2005).

Currently, there are four independent, third-party food certification programmes in the USA that include standards for the care and handling of animals – Certified Organic (administered by the USDA), Certified Humane (administered by Humane Farm Animal Care), Free Farmed (administered by the American Humane Association) and Animal Welfare Approved (administered by the Animal Welfare Institute). With the exception of the guidelines being developed by Whole Foods, these independent certification programmes do not limit transit times.

Whole Food's 'Animal Compassionate Standards' – which will be required for producers who wish to sell meat products with the 'Animal Compassionate' label in Whole Foods stores – have set forth specific transport requirements for loading and unloading, and transport conditions including transport limits for most animals. For cattle and sheep, if transport exceeds 8 h the animals must be rested off the truck for 24 h before continuing their journey, unless the destination can be reached in a total of 12h (Whole Foods, 2005a,b). For pigs, there is an overall transport limit of 12h (Whole Foods, 2005c). Ducks must travel no longer than 2h from departure to destination (Whole Foods, 2005d). There are no transport limits for broiler chickens (Whole Foods, 2006). Standards for laying chickens and dairy animals have not yet been developed.

The Whole Food's Animal Compassionate Standards are the most progressive care standards with regard to transport and seem to balance more evenly the welfare of animals and interests of the industry than other industry and independent standards and far exceed current legislative requirements in both the USA and Canada. The Animal Compassionate standards are also more consistent with and exceed the recommendations for travel and rest periods for pigs, sheep and cattle, set forth by the European Scientific Committee on Animal Health and Welfare (SCAHAW, 2002), compared to other standards and national legislation in North America.

US ANIMAL TRACKING AND IDENTIFICATION REQUIREMENTS.   While some US states require certificates of veterinary inspection for interstate commerce for animals destined for feeding or breeding purposes, they do not require certificates for each individual nor do they require health certificates for animals destined for immediate slaughter. Further complicating matters, there is no consistency between the states in how certificates are filed or what data, if any, are collected and recorded in state databases. Some states do not require health certificates at all (Economic Research Service, 2003).

The inconsistent or non-existent tracking method has raised public health and safety concerns. A 2003 report by the non-profit Trust for America's Health stated, 'Without sufficient funding to track animal health, the U.S. is missing the chance to detect a zoonotic disease early, and control if not prevent its spread. This is troubling given that many bioterror agents are zoonotic' (Trust for America's Health, 2003).

The inability of the US government to successfully track diseased livestock was demonstrated in December 2003 following the first confirmed case of BSE in an American cow. The diseased cow was part of a herd of 80 cattle, which could have been infected. However, despite assurances from the USDA that the location of the other suspected cattle would be traced within a few days, after 7 weeks only one-third of the animals were found (Hightower, 2004).

In recognition of the need to quickly and accurately identify and trace animals, Agriculture Secretary, Ann Veneman announced in December 2003 that the USDA would expedite the implementation of a national animal identification programme. The USDA's Animal Plant Health Inspection Service received more than $18 million to begin a programme capable of detecting any contact a diseased animal has had in its life within 48h of an outbreak. However, in January 2007 the USDA gave up its plans for a mandatory animal identification programme. The programme, if ever fully developed, will be strictly voluntary (Hahn, 2007).

ENFORCEMENT.   As noted above, there is effectively no federal law governing the transport of farmed animals in the USA, and, as such, no enforcement (see '28-Hour Law').

Advocacy groups, requesting that the USDA apply the 28-Hour Law to transport by truck (referenced above), noted in their rulemaking petition that the agency did not respond to requests for records pertaining to violations of the law and USDA enforcement of the law. The petitioners also noted that there were no reported USDA administrative decisions involving USDA enforcement of the law between 1977 and the filing of the petition (Brandt, 2005), suggesting that the law is not enforced.

Also as noted above, because the provisions of the Commercial Transportation of Equines for Slaughter Act only recently came into effect (in 2002 and 2006) compliance and enforcement is unknown (see 'Commercial Transportation of Equines for Slaughter Act').

Mexico

Environmental and animal protection laws in Mexico are relatively new. While such laws are rooted in federal legislation, enforcement is increasingly a function of state and local governments (Commission for Environmental Cooperation, 2003). According to an article in *Review of Policy Research*, enforcement of these laws is very poor (Norman and Hernandez, 2005). In general, authorities are not given adequate financial or human resources to enforce environmental and animal protection laws. It has been suggested that the Mexican government has 'spread itself too thin' and may not be able to afford effective implementation of existing regulations for a long time (Norman and Hernandez, 2005) suggesting that live transport to, from and within Mexico is likely to involve significant animal welfare problems for the foreseeable future.

FEDERAL ANIMAL HEALTH LAW.   The 63 articles of the Federal Animal Health Law (Ley Federal de Sanidad Animal – LFSA) are for the most part limited to diagnosis, prevention, control and eradication of animal diseases and pests (Commission for Environmental Cooperation, 2003).

Of particular relevance to the trade and transport of livestock, Article 24 of the LFSA requires that animal health certificates must contain at least the following information:

1. Name and address of the owner, holder or importer.
2. Specific data concerning the place of origin and destination of the animals, animal products and animal by-products or biological, chemical, pharmaceutical and feed products for use in, or consumption by, animals, being moved or offered for importation, as well as a description of these.
3. Mention made of the appropriate norm being complied with.
4. Date of issuance of the animal health certificate.
5. Expiration date of the certificate.

For public health matters the federal veterinary services in Mexico belong to the Secretaria de Agricultura, Ganderia, y Desarollo Rural (SAGAR) – (Secretariat of Agriculture, Livestock and Rural Development). SAGAR is represented in each state by Delegations. The State Delegations serve as arms of SAGAR and are responsible for coordination and implementation of national programmes (European Commission, 2000).

The General Directorate of Animal Health (Direccion General de Salud Animal) is subdivided into three directorates, two of which deal with national health programmes such as eradication schemes, while the third is the Directorate of Import/Export, Services and Certifications (Direccion de Importacion, Exportacion, Servicios y Certificacion). Within this directorate, the Department of Establishments TIF (Departamento Establicimientos TIF y Rastros) is in charge of approval and coordination of approval, control and supervision of federally inspected TIF establishments (TIF = Tipo de Inspeccion Federal). In 2000, there were about 180 federally approved establishments in Mexico. Without TIF approval, establishments are not eligible for export (European Commission, 2000).

FEDERAL LAW FOR ANIMAL HYGIENE. In essence, this law states that animals should receive 'humane treatment, in order to avoid unnecessary pain during capture, transportation, exhibition, quarantine, commercialization, consumption, use for entertainment and sacrifice' (Chapter 2, Article 2) (Norman and Hernandez, 2005). What constitutes humane treatment in each case is detailed. Chapter 3, Article 17 stipulates: '[f]arm or companion animals should be vaccinated against transmissible diseases, provided with food, hygiene, transportation and shelter with proper ventilation to assure their health.' Slaughterhouses must have at least one veterinarian, in certain cases on call from the health department (Chapter 4, Article 20) (Norman and Hernandez, 2005).

STATE LEGISLATION. State legislation is intended to provide flexibility for the diversity of the nation while providing coherence of the general principles provided by federal law (Norman and Hernandez, 2005). At least one state addresses humane slaughter. Puebla State mandates that animals for slaughter should have an instant death, not be previously tortured, beaten or suffer any other action that can torture them (Chapter 4, Articles 35–43).

OFFICIAL MEXICAN NORMS. The Mexican legal system also contains comprehensive technical standards known as Official Mexican Norms (Normas Oficiales Mexicanas) intended to standardize good practice in the treatment of animals throughout Mexico (Norman and Hernandez, 2005).

Norm 33 describes proper desensitization and slaughter methods for a variety of domestic and wild animals intended to avoid distress and guarantee a fast and painless death (Norman and Hernandez, 2005). Norm 51 defines humane treatment as a 'group of measures to diminish the trauma and pain to animals during capture, mobilization, exhibition, quarantine, commercialization, consumption, entertainment and sacrifice' (Norman and Hernandez, 2005).

Of relevance to transport, Norm 51 indicates that transportation cars should not be overloaded and ventilation should be provided according to the weather. Animals must never be tied by the feet or by any method that could strangle them. Maximum hours of travel for each domestic and wild animal are also specified. For example, equines should be transported '18 hours maximum and should have eight hours rest. If the transport is of more than 18 hours, they should rest 12 hours, allowing stops every six hours, letting animals rest for one hour' (Norman and Hernandez, 2005). Norm 45 provides comprehensive guidelines for all animals at events such as fairs and auctions. Beating or any other action that is contrary to the humane treatment of animals is explicitly prohibited (Norman and Hernandez, 2005).

The creation of these comprehensive laws and norms demonstrates a commitment to improvements in sanitation, disease control and animal treatment and provides a good legal framework on paper. However, the implementation and enforcement of these laws and norms is a significant problem (European Commission, 2000; Norman and Hernandez, 2005). For example, the Mexican Health Department estimates that of 2249 slaughterhouses in Mexico, only 80 have been accredited with the federal inspection certification awarded to establishments whose practices are within all the federal norms (Norman and Hernandez, 2005).

ANIMAL TRACKING AND IDENTIFICATION REQUIREMENTS.   The registration of farms *(granjas/ranchos)* in Mexico is mandatory under Ley federal de Sanidad Animal, Article 18 (SAGARPA, 1993). Farms are also required to participate in disease eradication and control programmes (European Commission, 2000).

Contrary to other humane requirements, horses must be identified with registered hot iron farm brands to prevent theft. Because horses usually originate from a variety of different farms, and are traded via dealers, traceability can be problem, especially because the hot brands are usually not identified on the certificado zoosanitario (the official movement permit) but only (and not always) on the Guia de Transito issued by the local livestock association (Asociacion Ganadera Local). Further the certificado zoosanitarios do not always contain all the required information (identification marks, purpose, etc.) (European Commission, 2000). A similar lack of accurate record keeping for cattle exported to the USA has also been reported (Skaggs *et al.*, 2004).

ENFORCEMENT.   As mentioned above, Norm 51 sets forth basic humane handling and transportation procedures for animals including a recommendation for transport limits and rest periods for horses. However, these standards are not mandatory and as such, not enforceable.

Conditions and Welfare of Animals Transported for Slaughter

There are five types of vehicles commonly used for livestock transport in North America, the 'possum belly', 'goose neck', 'straight trailer', 'pup trailer' and 'straight truck'. The most commonly used for long-haul trips is the 'possum belly'. 'Possum belly' trailers are 46–53 ft (14.0–16.2 m) long and about 8 ft (2.4 m) wide and are configured with two to four internal decks depending on the species and size of animals to be transported. The trucks are equipped with interior ramps for animals to walk to, and from, decks. 'Straight trailers' are also commonly used and are about the same length and width as possum belly trucks and contain one to three decks and interior ramps. Laws and regulations governing weight and size of trucks, use of trailers hooked in tandem, and how many hours a driver may be on the road vary between states and countries.

In Canada, the Commercial Vehicle Drivers Hours of Service Regulations set out the driving and rest requirements for drivers of commercial vehicles (Transport Canada, 1987). The regulations essentially require that 13 h of driving time are followed by at least 8 consecutive hours of off-duty time, or that after accumulating 15 h of 'on-duty' time (i.e. on-duty but not driving) the driver rest for at least 8 consecutive hours. In addition, drivers are not allowed to drive after accumulating 60h of on-duty time during a period of 7 consecutive days or after accumulating 70h of on-duty time during a period of 8 consecutive days (Transport Canada, 1987). Drivers are required to keep written logs of work and rest periods and distance travelled.

The US Department of Transportation's Federal Motor Carrier Safety Administration (FMCSA) issues 'Hours-of-Service' (HOS) regulations that spell out the length of time commercial drivers can operate trucks before they are required to take a break. Drivers must keep a record of duty status, in duplicate, for each 24h period or use an automatic onboard recording device to record duty status including rest periods, driving time and miles travelled (United States Department of Transportation, 2005) The HOS regulations apply directly only to interstate commerce. However, most states have adopted intrastate regulations which are identical or very similar to the federal HOS regulations.

Some livestock industry officials assert that it is common practice (particularly in the pork industry) to assign two drivers to each load for long hauls in order to comply with required breaks while avoiding stops thereby reducing the amount of time animals spend in transit (Brasher, 2006). However, in the investigation by the Animal Protection Institute and Compassion in World Farming, investigators documented single drivers exceeding the 11 hour maximum, and interviews with truck drivers indicated that these laws are regularly broken (Animal Protection Institute/ Compassion in Word Farming, 2005).

A 2002 report on livestock transportation in Alberta, notes that about 10–15 commercial livestock loads per year are involved in accidents resulting in trucks rolling over (Alberta Farm Animal Care, 2002). In 2004, a survey of media archives of US transport incidents conducted by Farm Sanctuary found the most common type of animal transport accident to be single-vehicle rollover. The number of live-stock transportation accidents in Mexico is unknown; however, it has been observed that livestock drivers in Mexico frequently travel at high speeds (Ecostorm, 2005, 2006, USA, personal communication). High speeds, failure to negotiate curves coupled with shifting of animals' weight were among the most common causes for accidents cited in the Farm Sanctuary survey (Farm Sanctuary, 2004).

In 2005, industry group AFAC, livestock handling specialist Jennifer Woods, Hartford Insurance and Canadian Farm Insurance established the Livestock Emergency Response (LER) hotline. The province of Alberta was split into 12 regions, with each assigned a team leader (Canadian Cattlemen, 2006).

The stated aim of the hotline is to enable Woods to assist with accidents and locate resources anywhere throughout Canada and the USA. The AFAC database tracking farmed animal incidents including transport accidents, however, is confidential and cannot be accessed by the public.

In Canada, the Transportation section of the voluntary 'Recommended Code of Practice for the Care and Handling of Farm Animals' includes guidance on responding to highway accidents involving farmed animals. In addition AFAC has produced an Alberta Livestock Incident Response Plan and created trained teams throughout the province to assist with situations where animals are trapped, injured or in distress due to truck rollover (Farm Sanctuary, 2004).

In the USA, there is no formal tracking or monitoring of livestock transport accidents by the government or industry (Farm Sanctuary, 2004). Moreover, no government or industry policies and procedures exist for dealing with animal transport accidents, with the exception of the National Pork Board's Trucker Quality Assurance training programme, which was revised in 2005 to include material on emergency response plans (Farm Sanctuary, 2004).

As truck driving requires additional skills and may require special licensing, pay for truck driving is considerably more than pay for animal handlers who do not operate large vehicles. In the USA and Canada local drivers are typically paid by the hour, while long-distance drivers or 'over-the-road' (OTR) drivers are usually paid by the mile; the pay schedule is likely similar in Mexico. Payment by the mile versus payment by the hour may provide little incentive for drivers to drive more slowly and cautiously or to take needed or required breaks either to rest or to check on the welfare of the animals and to take the time to provide needed care such as water for cooling or drinking, food and bedding.

Slaughterhouse Location, Capacity and Workforce

The final ride most farmed animals take ends at a slaughter facility. While in general, animals travel shorter distances for slaughter than for feeding, consolidation of the slaughter industry is leading to longer trips. This is illustrated by the movement of animals out of areas that no longer have federally inspected plants and the movement of animals over national borders for slaughter.

Canadian slaughterhouses

There are approximately 121 federal slaughter facilities registered with the CFIA. The number of registered slaughter facilities has declined over the last 7 years by 21% (from 153 to 121) as a result of consolidation or specialization for niche markets (Canada Food Inspection Agency, 2004).

Consolidation in pork and beef processing and packing plants in Canada has resulted in a handful of large-scale plants processing the majority of Canadian cattle and hogs with small- to medium-sized facilities finding it increasingly difficult to compete. However, according to the Canadian Meat Council, this consolidation has not been entirely negative, as it has reportedly enabled the Canadian industry to compete internationally by allowing processors to increase efficiency and ultimately profitability (Standing Senate Committee on Agriculture and Forestry, 2005).

Four Canadian facilities are responsible for processing close to 80% of the Canadian production of fed (slaughter-ready, non-cull) cattle, and two facilities process 90% of cull animals (Standing Senate Committee on Agriculture and Forestry, 2005). Similarly about 74% of the pigs are slaughtered in four of the largest Canadian swine abattoirs (Canada Food Inspection Agency, 2004).

Slaughterhouse wages tend to be a bit higher in Canada than the USA due to lower recruitment rates as fewer Canadians are willing to fill the positions for low pay. High turnover rate is also a problem for Canadian processors with only one out of ten new workers remaining for a second year. Official and detailed employment information is difficult to access but according to a report by the Canadian Press, some Canadian processors bring in workers from Mexico to fill the low-wage positions (Edmonds, 2002). Line workers starting at $8 per hour generally hit a wage ceiling of just over $11 per hour by the second year of employment (Edmonds, 2002) although at least one plant offered wages up to $14 per hour after 6 months of employment in 2002 (Edmonds, 2002).

US slaughterhouses

Widespread consolidation within the USA meat packing industry has placed slaughter and processing operations, particularly pork and beef, under the control of just a handful of larger companies. For example, in 2001, four giant competitors – IBP (now owned by Tyson Foods), ConAgra, Excel (owned by Cargill) and Farmland National Beef (now known as National Beef Packing) – collectively controlled over 85% of the US beef market (Olson, 2002). Large US slaughterhouses can process 300–400 cattle an hour (Schlosser, 2005).

Consolidation of the US meat industry has taken place over the last four decades. Data furnished by the USDA National Agricultural Statistics Service, show a trend of plants switching from state inspection to federal inspection in the early 1970s, and then a continual and consistent decline in the number of plants under both federal and state inspection from the mid-1970s through to the present. In 2002, of the 918 federally inspected slaughterhouses, just 49 (mostly located in the Midwest), accounted for approximately 80% of total meat production in the country (General Accounting Office, 2004).

Animals may be transported long distances if they have been fed in a state with only a few or no federally inspected slaughter plants for that particular species. For example, the following states have no, or only one, federally inspected plant for the slaughter of cattle: Louisiana, Mississippi, Nevada, New Mexico, South Dakota and Wyoming (National Agricultural Statistics Service, 2006). In addition, some large-scale slaughter facilities do not buy from independent producers, so animals may be shipped for several hours despite availability of nearby slaughterhouses (Krause, 2006). In Missouri for example, nearly all of the hogs must be shipped out of state for this reason (Missouri Swine Audit, 2005).

Information on the origin of animals slaughtered at these plants is not collected at the state or federal level, therefore, it is impossible to determine without exhaustive on-the-ground research, the average distance travelled by animals to federally inspected slaughterhouses.

We were unable to find reliable information on slaughterhouse demographics in Mexico.

Infrastructure, Primary Routes and Crossing Points

Canada internal transport

The size of Canada coupled with the consolidation of meat packers suggests that in many cases, livestock may be transported long distances to reach slaughterhouses.

However, specific data on the number and destination of live animals transported across provinces either for further fattening or slaughter are not collected (Agriculture and Agri-Food Canada, 2006, USA, personal communication). A 2004 report on livestock transportation in Alberta, the fifth largest cattle feeding region in North America (Alberta Farm Animal Care, 2002), revealed that about 480 commercial livestock trucks are on Alberta roads each day hauling cattle, horses, hogs, poultry, sheep and other animals, however, the report did not provide information on specific routes taken. The report did note that in any given week, loaded commercial livestock trucks travelling Alberta's roadways included nearly 500,000 mature cattle destined for slaughter in the USA. This report was prepared prior to the US ban on cattle due to BSE, and the number of live cattle exported to the USA for slaughter since 2003 is zero (Alberta Agriculture Food and Rural Development, 2004). However, imports and exports of live cattle to other Canadian provinces continue (Alberta Agriculture Food and Rural Development, 2004). The 2002 report, produced by AFAC, reported that more than 100,000 mature cattle were transported, destined for slaughter in British Columbia, Saskatchewan, Manitoba or Ontario, and that 550,000 hogs were destined for US slaughter and 50,000,000 broiler chickens, 1,615,000 turkeys and 50,000 horses were destined for slaughter in undetermined destinations (Alberta Farm Animal Care, 2002).

Canada routes

While little information exists on specific routes taken by long-distance livestock haulers driving across provinces, it seems logical that livestock trucks travelling long distances between provinces would travel in great part on the Trans-Canada Highway – a federal provincial highway system that joins all ten provinces of Canada. Together, Falcon and West Hawk Manitoba (situated approximately 75 mi (120 km) east of Winnipeg along the four-lane Trans-Canada Highway) form the gateway to eastern and western Canada, as all traffic must pass through the area when travelling east or west. It appears that there is a weigh station and veterinary inspection at the Manitoba/Ontario border for livestock haulers.

Portions of the route inevitably include the major highways leading to cities containing sizeable processing plants such as major beef packing plants in Brooks and High River, Alberta and large pork packing plants in Red Deer, Alberta, Saskatoon, Saskatchewan, Brandon, Manitoba and Burlington, Ontario.

US interstate transport

While some US states require certificates of veterinary inspection for interstate commerce for animals destined for feeding or breeding purposes, they do not require certificates for each individual nor do they require health certificates for animals destined for immediate slaughter. Further complicating matters, there is no consistency between the states in how certificates are filed or what data, if any, are collected and recorded in state databases. Some states do not require health certificates at all, and as such, no data on the number of animals entering the state. (Forde *et al.*, 1998; Economic Research Service, 2003).

A 2003 report, produced by the US Economic Research Service, compiled available state data on state imports and exports of sheep, pigs and cattle not destined for slaughter. The report demonstrates that large numbers of animals are moved across state lines for further fattening. The report did not, however, indicate the trip journey times or whether animals were rested, fed or watered during transit.

The Economic Research Service Data Sets for Interstate Livestock Movements (2003) show the following animal outflows from these states for 2000/01[a]:

1,941–115,000	cattle from Louisiana to four states
26,316–355,000	cattle from Mississippi to six states
10,987–49,444	cattle from Nevada to four states
3,480–841,000	cattle from New Mexico to four states
12,788–576,522	cattle from South Dakota to nine states
6,718–325,761	cattle from Wyoming to eight states

In addition, both South Dakota and Wyoming shipped out significant numbers of sheep to California, Iowa, New Mexico and Texas in 2000/01 (Economic Research Service, 2003). With the continuing consolidation of the slaughter industry, in the future it is likely even more animals will be travelling long distances to slaughter.

Recent investigations indicate that trucks transporting livestock within the USA frequently comply with the US 28-Hour Law – if the law were to be applied and enforced (Animal Protection Institute, 2005; Compassion Over Killing, 2005). However, some US livestock drivers claim that they regularly transport animals in excess of 28h without offering food or water or unloading for rest (Compassion Over Killing, 2005), but actual documentation of such transport is lacking.

Movement of cattle occurs in most regions of the country with the largest volumes moved into and within the Northern and Southern Plains (North and South Dakota, Nebraska, Kansas, Oklahoma and Texas) (Shields and Mathews, 2003). US beef cattle can move up to 10 times through various stages of production and trading (Juday, 2005).

Major cattle feeding states contain the largest cattle slaughter plants (Shields and Mathews, 2003). Fed cattle from large feedlots are shipped an average of 100mi (160km) to slaughter, while fed cattle from smaller feedlots are shipped an average of 144mi (230km) (United States Department of Agriculture, 2000) although some travel closer to 250mi (400km) (Krause, 2006). Dairy cattle are moved longer distances to slaughter than beef cattle, since dairies are generally located further from slaughter plants and some cattle slaughter establishments do not handle cull cows.

Distances travelled for hog slaughter appear to be similar to those for beef cattle slaughter. It has been estimated that over two-thirds of hogs marketed in the USA are transported less than 150mi (240km) to slaughter (Missouri Swine Audit, 2005).

However, some pigs are transported much further. According to a report by the Economic Research Service, pigs are frequently shipped from farrowing operations in North Carolina to nursery facilities or grower/finisher facilities in Iowa.

[a] There is a lack of specificity for each importing state because the source presented the data graphically on a map and not numerically in a table. Note that multiple states are involved in receiving shipments, so there is a range of values. If there are shipments from one state to nine different states, for example, some receiving states will be at low-end of range while others are at high-end.

According to the National Pork Producers Council transporting pigs from North Carolina to the Midwest for feeding typically takes 20–24 h (Cody, 2006).

In addition, according to the report, the pigs are frequently moved again, from these grower/finisher facilities in Iowa to California for slaughter (Shields and Mathews, 2003) – a distance of approximately 1700 mi (2700 km). Depending on frequency of refuelling, breaks and traffic conditions this journey could take between 24 and 30 h, if travelling without taking extended rest breaks or stopping overnight.

US routes

As previously indicated, data on interstate livestock movements in the USA are highly fragmented and limited in geographic and historic scope (Forde *et al.*, 1998). The primary reason for this is lack of accurate record keeping by individual states and the lack of a national animal tracking system. A 1998 study on interstate cattle movement found that only 18 of the 50 states recorded updated import and export information using certificates of veterinary inspection and occasionally entry permits for verification (Forde *et al.*, 1998). Moreover, not all of these states classified their import records based on the reason for import (e.g. feeding, grazing, breeding or slaughter) (Forde *et al.*, 1998; Shields and Mathews, 2003).

In spite of the lack of official data on numbers of animals, distances and journey times, some further information has been obtained from investigations by animal protection organizations and interviews with livestock haulers and buyers. Interviews with livestock drivers conducted at livestock auctions and rest stops have shed some light on the state of origin and state of final destination of some transported livestock. The longest routes for cattle appear to be those destined for feedlots. For example one driver claimed that he regularly transports feeder cattle from a farm in Virginia to a feedlot in Kansas and that the journey lasts 24–30 h (Compassion Over Killing, 2005). The likely route for this trip would be via Interstate 70/64. Another driver indicated that he transports hogs from a location near Kansas City, Missouri to Modesto, California for slaughter and that the journey lasts at least 35 h (Compassion Over Killing, 2005). This journey would likely involve travel on Interstate 80 (or state highway 30 which parallels and intersects Interstate 80) through Nebraska, Wyoming, Utah, Nevada and on into California transitioning to Interstate 5 into Modesto.

Reliable information is not available on livestock transport routes in Mexico.

International transport

Trade liberalization under the Canada–US Free Trade Agreement (FTA, implemented in 1989) and the North American Free Trade Agreement (NAFTA, implemented in 1994) has been a leading factor in the integration of North American livestock production. The agricultural economies of Canada, Mexico and the USA have been characterized as behaving as one market. For example, many North American pastures and feedlots contain animals that have lived in more than one NAFTA country (Zahniser, 2005). While the NAFTA countries have reached mutual agreements on sanitary regulations over the last decade, no such agreements have been reached with regard to animal welfare during rearing, transport or slaughter.

Trade in live animals between Mexico and the USA consists largely of feeder cattle from Mexico destined for US feedlots as well as export of hogs, cull ewes and horses exported from the USA for slaughter in Mexico. This often involves considerable travel distances and is accompanied by considerable animal welfare concerns.

Live animal movement across the national boundaries between the USA and Canada and the USA and Mexico is normal practice as animals move to pastures, feedlots or to more available slaughter facilities. For example, feeder cattle from Mexico and Canada move into US feedlots at a rate of about 95,000 and 30,000 head per month, respectively (United States Department of Agriculture, 2006).

A large majority of farmed animals moved between North American countries are being transported for reasons other than immediate slaughter, as illustrated by the following examples: (i) less than 1 million of the nearly 18 million birds (chickens, turkeys, ducks, geese and guinea fowl) exported from Canada to the USA in 2005 were of slaughter weight; (ii) although more than 1 million beef cattle were exported from Mexico to the USA in 2005, less than 5000 were designated as animals for slaughter; and (iii) of 8.1 million pigs exported from Canada to the USA in 2005, only one-third (2.7 million) were destined for immediate slaughter (US Department of Agriculture (USDA) Foreign Agricultural Service Trade Statistics, 2005).

Any animals transported across international borders must stop at export and inspection points. Sometimes health and safety precautions can compromise animal welfare as demonstrated in 2002 when exports of ruminants from Canada to the USA were halted in 2003 following the finding of BSE in Canadian cattle. In 2005, when the USDA reopened the border to certain ruminants, the agency required that the animals be moved to the designated feedlot or slaughterhouse 'in a sealed means of conveyance' (USDA APHIS, 2005a). In November 2005, the USDA amended this regulation to broaden who is authorized to break the official seal on the means of conveyance. In making the change, the USDA noted that it had come to the agency's attention that restrictions on who may break the official seal 'can create a situation that is not conducive to the humane treatment of livestock' (USDA APHIS, 2005b). Prior to the emergency rule change, animals arriving at feedlots or slaughter plants at night or on weekends had to be held on the conveyance until an authorized individual became available to release them.

In late June 2006, nearly 150 pigs died when the trailers they were transported in were abandoned at a Texas Department of Agriculture (TDA) export facility for 3–4 days. The animals allegedly died while in full view of TDA employees, who have claimed that the pigs could not have been unloaded because of inadequate paper work. The pigs were among 2644 pigs ordered from Ohio by the pig subsidiary of the British biotechnology distributor PIC International, destined for Queretaro, Mexico. The surviving neglected pigs were shipped on to the intended destination in Queretaro (Animal Legal Defense Fund, 2006).

US and Canada international routes

Major ports of exit from Canada to the USA are in southern Alberta (cattle and hogs), Saskatchewan (cattle), Manitoba (cattle and hogs) and Ontario (cattle and hogs) (Agriculture and Agri-Foods Canada, 2004).

Approximately 95% of Canadian feeder pigs (pigs transported for further fattening) are shipped to two US regions: Region 7 (Iowa, Kansas, Missouri and Nebraska) and Region 5 (Illinois, Indiana, Michigan, Minnesota, Ohio and Wisconsin) (Economic Research Service, 2004). Canadian hogs destined for slaughter in the USA are shipped more widely to slaughterhouses in regions that produce fewer pigs. In 2004, about 32% of slaughter hogs transported to slaughterhouses in Region 8 (Colorado, Montana, North Dakota, South Dakota, Utah and Wyoming) and about 25% to Region 5 (Illinois, Indiana, Michigan, Minnesota, Ohio and Wisconsin), around 10% went as far as Region 9 (Arizona, California, Hawaii and Nevada) and approximately 10% went to Region 4 (Alabama, Florida, Georgia, Kentucky, Mississippi, North Carolina, South Carolina and Tennessee) (Economic Research Service, 2004).

More than half of all Canadian hogs destined for slaughter enter the USA through North Dakota, and probably head towards a large slaughterhouse in South Dakota (Economic Research Service, 2004). Another 27% enter through Montana and Idaho suggesting destinations west of the Rocky Mountains such as California and 18% enter through Michigan likely headed for slaughter in Indiana, Kentucky and Pennsylvania (Economic Research Service, 2004).

Routes between Mexico and the USA

Thousands of live horses, pigs, goats and sheep are exported to Mexico each week primarily for slaughter (see Table 9.1.2A). Slaughter hogs make up the majority of animals exported to Mexico, followed by slaughter ewes and horses. Some cull ewes may be kept for a few more breeding cycles prior to slaughter and wool-type sheep may be held in order to receive a final shearing prior to slaughter (Shields and Mathews, 2003). The Mexican market provides an outlet for cull ewes due to lack of demand for mutton in the US market (Tescher, 1992).

Texas and New Mexico are the primary states through which livestock move between the USA and Mexico. The port at Eagle Pass, Texas is the primary livestock export port and is the port where all US hogs other than breeding stock transit to Mexico (National Pork Producers Council, 2004). Slaughter ewes destined for Mexico pass through the port at Del Rio, Texas (USDA-APHIS-VS, USA, 10 July 2007, personal communication), although some also pass through at Eagle Pass (USDA Agricultural Marketing Service, 2006).

The port at Santa Theresa, New Mexico is the primary port of entry for imported Mexican cattle, and is likely the primary port of export for US slaughter horses, although some may exit through nearby Columbus, New Mexico. Interstate 10 and 25 and US highway 70 and 180 are likely routes to, and from, these New Mexico ports

While some records are maintained for Mexican cattle, they are maintained by APHIS in confidential form, none are intended for aggregate analysis of US–Mexico cattle trade (Skaggs *et al.*, 2004). Moreover, data kept by state livestock inspections are typically handwritten and in some cases illegible. Inspection records have also been found to contain obviously erroneous information such as declaring cattle as being shipped to metropolitan areas or ghost towns (Skaggs *et al.*, 2004).

According to the USDA Agricultural Marketing Service's weekly livestock trade database, the USA exports up to 138 horses per week for slaughter in

Mexico. The majority of Mexican-bound US horses cross the border in New Mexico at the Santa Teresa border crossing. The horses transported to this crossing may come from all over the USA. Border crossing officials have indicated that horses may come from as far away as Washington state (stockyard worker at Santa Teresa, New Mexico, 2006, personal communication), which is approximately 1630 mi (2600 km) from the Santa Teresa, New Mexico border. This journey would take approximately 25 h if driven continuously without rest stops or stops to load additional animals. The frequency of stops and provision of food and water and compliance with the Commercial Transportation of Equine for Slaughter Act is unknown.

Once horses arrive at the border crossing they are unloaded into holding pens for inspection. Following inspection, the horses are walked across the border where they are loaded into Mexican transport trucks and driven to the slaughterhouse or markets (USDA Market News Service, Las Cruces, New Mexico, 2006, personal communication).

It is possible that some of these horses will be slaughtered in European Union (EU)-approved slaughterhouses in Mexico, processed into meat for human consumption and reimported into the USA for export to the EU (European Commission, 2000), but this trade route has not yet been confirmed.

There are two horse slaughterhouses approved for export to the EU (European Commission, 2000). These establishments are located in the cities of Jerez and Fresnillo (European Commission Food and Veterinary Office, 2006), Mexico, both of which are located in the state of Zacatecas. Zacatecas city is the capital of the state of Zacatecas located 613 km from Mexico city, by the federal highways 57 and 45.

The report noted that animals may be derived from 'the whole of the Mexican territory' (European Commission, 2000), but did not mention whether any of the horses may have originated in the USA. Also according to the report, fresh horse meat intended for export to the EU is sometimes shipped from Mexican ports but more commonly it is shipped via the USA to Europe due to greater vessel availability (European Commission, 2000).

Welfare of Transported Animals

Time spent in transit is stressful both physically and mentally for farmed animals. Transported animals are subjected to social challenges from unfamiliar animals, handling by people who are sometimes untrained and unskilled, loaded on to and off vehicles, exposed to vibration, noise, unfamiliar surroundings, long journeys and deprivation of food and water. Other problems during transport include overcrowding, lack of bedding, lack of opportunities for rest and exposure to extreme heat or cold (Warriss, 1998; Knowles, 1998; see Chapters 6 and 7, this volume). Poor and abusive handling of animals during loading and unloading and at auctions also increases animal stress and suffering during the transport process (Allen, 2005). With the exception of cattle, whose welfare appears to be poorer during transport to feedlots (Tarrant, 1990; Knowles, 1999; Lonergan *et al.*, 2001; Ishmael, 2005), welfare appears to be poorest in animals destined for slaughter, especially those transported for slaughter across national borders.

It is also important to keep in mind that research studies typically look at animals being transported in good conditions and in good physical health at the beginning of the journey which is contrary to the way many animals are transported. One of the most variable welfare factors affecting transported livestock is climate, including ambient temperature and humidity. In North America these variables can vary widely between regions and seasons. Animals transported very long distances may endure prolonged exposure to extreme heat or cold or may endure radical climate changes that can exacerbate transport stress. According to the 'Livestock Weather Safety Index' in the Cattle and Swine Trucking Guide produced by the US Department of Agriculture Transportation Services Branch, temperatures above 100°F (38°C) are considered 'dangerous', and transporting animals in relative humidity above 25% is considered an 'emergency' situation. The guidelines recommend hauling 20% fewer hogs when temperature and humidity levels are in the 'danger zone' and recommend postponing hog shipments in the 'emergency zone' until conditions moderate (USDA Transportation Services Branch, 1999).

Canada

In 2001, 7969 pigs out of approximately 4.8 million hogs marketed in Ontario, Canada died during transport, this amounts to about 17 pigs in every 10,000 marketed (Dewey et al., 2004). Assuming an average market value of $100 per pig, this amounts to nearly an $800,000 loss to the Ontario pork industry.

In Alberta, it is estimated that approximately 13% of pigs produce loin muscles that are PSE (pale, soft, exudative), and that this condition reduces the value of the pig carcass by about $5 (Murray, 2000). PSE is linked to genetics and stress before slaughter (Tarrant, 1989). Poor welfare can also lead to DFD (dark, firm, dry) meat. Unlike PSE, DFD is not influenced by genetic factors, it occurs when pigs are fatigued and when their glycogen or energy store is exhausted at the time of slaughter resulting in no acidification. In short, anything which increases a pig's body temperature and/or dramatically decreases its glycogen levels decreases meat quality and therefore, profitability (Lambooij and van Putten, 1993). Figures for losses attributed to DFD were not found.

In August 2003, German-based Animals' Angels documented a randomly selected long-distance truck journey of cull cows from the beef and dairy industries from Medicine Hat, Alberta to the Colbex slaughter plant east of Montreal, near St-Cyrille-de-Wendover, Québec. The investigators documented animals in poor physical condition often appearing visibly thin. In addition, a badly limping cow was included in the load.

The cattle travelled 67 h, for a total of 2220 mi (3575 km) with one stop of 8 h for feed and water in Thunder Bay, Ontario. Canada's animal transportation regulations under the Health of Animals Act allow ruminants to be transported 52 h without water, food or rest within Canada. Time without water and food, however, is often longer since animals are not typically watered and fed immediately prior to transport.

Animals' Angels has documented the journey of Canadian pigs transported through the USA to be slaughtered in California and Hawaii (island of Oahu).

According to Animals' Angels, every week 400 Canadian pigs, collected from all over the province of Alberta, are transported from a collecting station run by Perlich Brothers in Lethbridge, Alberta to a slaughterhouse in Oahu, Hawaii (Animals' Angels, 2003).

One documented trip of 331 pigs travelling from Alberta to California took 35 h. During this time the pigs appeared to be stressed and desperate for water (Animals' Angels, 2003). While water was sprayed on the pigs to cool them, none was provided for drinking. Temperatures outside the truck reached 100°F (38°C); temperatures inside the truck were likely much higher. When the pigs were unloaded at the slaughterhouse in California many appeared sunburned (Animals' Angels, 2003). In addition there was a single driver who drove without stopping for 28 h – a violation of both Canada and US driver regulations (Animals' Angels, 2003).

Upon arriving in Vacaville, the pigs were loaded on to and held in rusted iron containers for 1.5 days. Then they were brought to the port of Oakland and shipped to Oahu, Hawaii. This sea journey lasted another 4 days. After being unloaded from the ship, the animals were then trucked to a slaughterhouse. The total journey was observed to take 8.5 days (Animals' Angels, 2003).

The USA

Despite the considerable stress that transport causes, farmed animals in the USA are typically moved several times during their lives, often over large distances. As a result many animals become sick or injured or even die on the way to their final destination.

Pigs are especially sensitive to transport stress and many pigs arrive fatigued, injured or dead upon arrival at US slaughterhouses. According to data from a survey of packer members of the National Institute of Animal Agriculture (Bowling Green, Kentucky) approximately 80,000 hogs die annually during the transit process in the USA (Grandin, 1992). Seventy per cent of these losses occur on the truck during transportation, and 30% occur shortly after arrival and are directly attributable to the transportation process (Grandin, 1992). Estimates from the Agriculture Department's Food Safety and Inspection Service (FSIS) report a higher death rate – 0.26%, which translates to 260,000 pigs per year.

It has also been estimated that 0.08% of pigs taken to market in the USA arrive as 'fatigued' – out of breath and unable to get off the truck on their own. This would translate to approximately 82,400 'fatigued' pigs in 2005 (Kelley, 2005). Using the aforementioned estimated mortality of 80,000 hogs each year, if an average market value of $100 per animal is assumed, such losses equate to an $8 million annual loss to the US pork industry (Speer *et al.*, 2001). Using the statistics from the FSIS this would equate to a loss of $26 million. Even if the animals survive, poor welfare and handling during the journey can still impact profitability. Bruising for example, costs the pork industry an estimated $6 to $7 million annually (Speer *et al.*, 2001). In addition, stress before slaughter leads to an increased breakdown of glycogen and rapid acidification which can lead to PSE meat depending on the genotype of the pig (Tarrant, 1989). It is estimated that PSE costs the US pork industry nearly $30 million annually (Speer *et al.*, 2001).

Transport is considered one of the most stressful events that cattle will undergo during their lives (Swanson and Morrow-Tesch, 2001). Calves tend to be more susceptible to the effects of transport than healthy adult cattle. Shipping fever or bovine respiratory disease is generally considered to be caused by stress-induced changes in the immune system during transport (Tarrant and Grandin, 1992). This disease has been estimated to cost the US beef industry $500 million annually (National Agricultural Statistics Service, 1996).

Further, it is estimated that 1% of cattle destined for feedlots die as a consequence of transport stress and its aftermath (Irwin and Gentleman, 1978). This amounts to approximately 120,000 cattle per year based on a 2006 estimate of 12 million cattle and calves on US feedlots (National Agriculture Statistics Service, 2006b). It has also been estimated that for every 1000 cattle entering feedlots 12 die (Lonergan *et al.*, 2001). Dan Thomson of Kansas State University's College of Veterinary Medicine has characterized this mortality as an 'economic and animal welfare tragedy' (Ishmael, 2005).

While transport mortality in mature beef cattle (i.e. those destined for slaughter) is reported as low (Knowles, 1999; Speer *et al.*, 2001), actual data on the number of downed and dead cattle on arrival at US processing plants are lacking (Swanson and Morrow-Tesch, 2001). Moreover, other welfare problems appear to be increasing. According to beef-quality audits by the National Cattlemen's Beef Association the incidence of stifled, arthritic or structurally unsound cattle significantly increased from 1994 to 1999, with beef cows experiencing a fourfold increase from 2.9% to 11.9% and dairy cows experiencing a threefold increase from 4.7% to 14.5% (Speer *et al.*, 2001).

In the summer and autumn of 2005, the Animal Protection Institute (API) and Compassion in World Farming (CIWF) carried out an investigation into the transport of live farmed animals throughout the USA and export of pigs to Mexico for slaughter. Investigators documented animals in the USA arriving at, and proceeding through, auction with broken legs, infected eyes, foaming mouths and bleeding cuts and sores. Dead and downed animals were also seen at the auctions. In addition, investigators filmed the unloading of cull sows (breeding females from intensive production) destined for slaughter. Many of these pigs had difficulty walking, on account of having spent nearly their entire lives in confinement according to the driver delivering the pigs (Animal Protection Institute/Compassion in World Farming, 2005).

API and CIWF investigators also followed a truck leaving the Joplin, Regional Stockyards in Carthage, Missouri, with a truck load of cattle destined for slaughter in San Angelo, Texas. A bull loaded on to the truck was noticeably lame and had blood running down his leg. The total journey time was at least 12 h. The driver made three confirmed stops during the journey. The first two stops at gas stations were for approximately 15 min. The last stop was at a roadside turn-off near the town of Munday, Texas for approximately 1 h. Cattle were not unloaded and no food or water for the animals was seen to be offered during these stops. Dr Ned Buyukmihci, veterinarian and professor emeritus at the University of California, concluded after viewing the footage of this bull that the animal was in severe pain and that transport would increase the animal's level of pain and discomfort (Animal Protection Institute/Compassion in World Farming, 2005).

Mexico

The welfare of live animals exported to Mexican slaughterhouses raises serious concerns as animals may be transported for long distances and far beyond the reach of the USA's Humane Slaughter Act and 28-Hour Law. The welfare of animals at Mexican slaughterhouses has been documented to be poor (European Commission, 2000; Animal Protection Institute/Compassion in World Farming, 2005). According to a 2000 report by the European Commission Health and Consumer Protection Directorate-General on EU-approved slaughter and packing establishments in Mexico, at one Mexican horse slaughterhouse up to 5% of the animals were dead on arrival due to long transport distances (European Commission, 2000).

Mexico banned imports of US beef and live cattle in December 2003 following the detection of BSE. In March 2004, Mexico announced that it would accept US boneless beef from cattle less than 30 months of age, but the ban on live cattle remains (United States Trade Representative, 2005). Before the ban, cull dairy cows were regularly shipped to Mexico for slaughter (Shields and Mathews, 2003). This is particularly concerning from an animal welfare perspective as in the USA, dairy cattle make up 75% of all 'downer' cattle (McNaughton, 1993). During the 2005 investigation by API and CIWF, investigators documented pigs being transported over the USA–Mexican border on crowded trucks through many hours of baking desert heat without food, water or rest, to be killed at the end of the journey in Mexican slaughterhouses (Animal Protection Institute/Compassion in World Farming, 2005). Much of the meat is then exported back to the USA for consumption (Animal Protection Institute/Compassion in World Farming, 2005; Reynolds, 2007).

Since 1994 after the passage of the North America Free Trade Agreement (NAFTA) US companies have been using 'approved plants' – plants that meet US slaughter and processing standards – in Mexico for slaughtering pigs and processing pork products (Reynolds, 2007). At one Mexican slaughterhouse, CIWF investigators were informed that more than 400 pigs per day were slaughtered; most of them from the USA. On average about 94,000 US hogs per year are slaughtered in Mexican slaughter plants that are 'approved' to export pork products to the USA for consumption (Reynolds, 2007).

Conversations between investigators and pig dealers in the USA revealed that low labour costs and the lack of government oversight in Mexico are considered the primary (if not the only) benefits of sending live pigs to Mexico for slaughter. Indeed labour costs in pork processing plant are nine times lower in Mexico than the USA (Reynolds, 2007). On average production workers in Mexico make $2.24 per hour compared with $18.61 per hour in the USA (Reynolds, 2007).

Discussion

Due to the large size of the three countries that make up North America's land mass (Canada, the USA and Mexico), transport routes are widely dispersed and largely unregulated. While regulations, legislation or codes addressing the transport of animals exist to some degree in each of these countries, they are generally limited in scope and poorly or never enforced. Moreover, current regulations in each of these countries, even if properly enforced and applied to all types of animals and modes of transport, would be insufficient to protect animal welfare during transport because they fail to set reasonable journey limits, stocking densities, rest and thermal requirements.

The issue of farmed animal transport has been of growing interest to the livestock industry as evidenced by the amount of scientific literature published on the subject including those cited at the end of this chapter and in other chapters in this volume. This interest appears to be based largely on financial concerns associated with livestock loss and decline in meat quality as the result of travel conditions and has not yet resulted in significant or enforceable changes in how farmed animals are transported in North America.

The economic risks associated with long-distance transport are well documented and include reduction in meat quality (bruising, PSE and DFD), increased disease susceptibility and exposure, and increased transit mortality (see Chapter 4, this volume). Moreover, longer transport distances, inter- and intra-auction movements and movement across national boundaries bring an increased potential for wider spread of disease (Shields and Mathews, 2003). Despite this, animals are frequently transported long distances and may even be moved across national borders for slaughter, with the meat shipped back to their countries of origin for consumption.

Evidence suggests that restrictions on the transport of live animals will be compensated by increased trade in fresh or frozen meat. For example, in response to the 2004 US ban on the import of live cattle from Canada, Canada augmented its cattle slaughter capacity (Standing Senate Committee on Agriculture and Forestry, 2005).

Concomitantly, Canada increased its boxed beef exports (Minter, 2005). The Canadian Beef Export Federation (2006) has indicated that it supports a 'cut-in-Canada' strategy, under which Canada would reduce its exports of live cattle to the USA to zero by 2015, further indicating that Canada's livestock industry can be successful without long-distance transport of live animals to the USA. In addition, with Canada's largest export market closed for more than 2 years, many drivers who used to transport the animals south moved on to other jobs and some companies sold some of their trucks. This coupled with high fuel prices and a strong Canadian dollar is expected to make the cost of transporting cattle more expensive than it was prior to the closure (CBC News, 2005).

Transport restrictions could also result in an increase in smaller and more dispersed processing facilities or increase the use of mobile abattoirs that process animals on the farm and could provide an incentive for retailers and restaurants to purchase from local producers, many of which may be smaller, independent or organic farms.

OIE guidelines do recommend specific documentation requirements. Such documentation, if applied by OIE member countries, could aid in enforcing journey limits and resting intervals and in monitoring other animal welfare objectives during transit, as well as aiding in tracing disease and preventing outbreaks.

Another possible enforcement mechanism is the animal tracking and identification requirements already in place in Canada and recently attempted in the USA. Such tracking systems could be used to monitor journey times and enforce transportation limits and rest periods in addition to providing an expedient means of tracking animals exposed to or infected with dangerous disease. To be most effective, identification systems should be mandatory.

Ultimately, the most effective way of addressing farmed animal transport in Canada and the USA is through federal legislation. While legislative advocacy on the federal level demands a significant commitment, the benefit of a federal law far outweighs the costs when compared to legislative advocacy on a state-by-state or province-by-province basis. In addition, because many states and provinces can be crossed in a relatively short amount of time, state and provincial transport laws may be ineffective unless time travelled outside state/province borders is included. In the USA, it is possible that state laws restricting the transport of animals could be struck down for burdening interstate commerce. In Mexico, however, federal legislation addressing farmed animal transport, even if passed, is unlikely to be enforced in a meaningful manner due to current economic, social and political factors.

As animal protection legislation stems largely from civil society pressure, public education and public opinion will be key in effecting lasting change for transported farmed animals. In the USA and Canada, consumer opinions and expectations regarding animal welfare have proven to be a major driving force in improving the welfare of farmed animals – as evidenced by food chains, grocers and restaurants that have developed their own animal welfare standards that their suppliers must meet. It is foreseeable that such standards could be extended to include maximum transport limits.

The situation in Mexico, however, is quite different. As Norman and Hernandez (2005) explain: 'Animals in general do not have a high priority on the public agenda. Given this, it is unsurprising that, for all the government rhetoric, it [animal protection] suffers a similarly low position on the political agenda.'

In contrast, public opinion surveys conducted in the USA have demonstrated that Americans are quite concerned about animal pain and suffering. In a 1999 poll conducted by Decision Research for the Animal Protection Institute, 76% of respondents agreed that 'an animal's right to live free of suffering should be just as important as a person's right to live free of suffering'. This concern is not limited to companion animals and wildlife, but also applies to animals raised for food as demonstrated in a national survey conducted by Opinion Research Corporation (1995) for Animal Rights International. In this survey, 93% of respondents agreed that 'animal pain and suffering should be reduced as much as possible, even though the animals are going to be slaughtered'. A Canadian poll conducted by Decima Research for the World Society for the Protection of Animals and the Canadian Coalition for Farm Animals in 2005, similarly found that 94% of Canadians felt it important that farm animals be treated humanely.

Trade and transport in live animals for slaughter could readily be replaced by trade in fresh or frozen meat throughout North America. In the past, fresh meat could not be transported over long distances because of the lack of refrigeration and proper distribution channels. Today, however, new technologies allow for chilled meat to be shipped to many locations and remain fresh. Advertising and other marketing efforts can improve the acceptance of frozen and packaged meat to overcome traditional purchasing preferences for live animals.

Consumer research shows that customers place a high value on locally produced products and that they associate the term 'local' with reduced transport distance from farm to store. Moreover, customers are typically willing to pay 5–15% more for locally raised meat items than for similar items that are not locally produced.

Passing federal legislation to limit transport times and substituting the long-distance transport of live animals with a meat-only trade in North America will require a high degree of cooperation among members of the concerned public, industry, retailers, legislators and advocacy organizations. Given the large number of animals affected, the degree of suffering endured and the clear alternatives available, ending the long-distance transport of animals for slaughter is a goal well-worth pursuing.

Acknowledgements

Detailed information on long-distance transport in North America is not well recorded in official of scientific sources. For this reason many essential details can only be obtained through investigations by animal protection NGOs or similar organizations.

I would like to thank Dena Jones, World Society for the Protection of Animals, Liam Slattery, Compassion in World Farming and Jim Wickens, Ecostorm Investigations for their invaluable contributions that made this chapter possible.

References

Agriculture and Agri-Foods Canada (2004) Government announces strategy to reposition Canada's livestock industry. News Release, September 2004. Available at: www.agr.gc.ca/cb/index_e.php?S1=n&S2=2004&page=n40910a

Alberta Agriculture Food and Rural Development (2004) Update on Alberta trade in beef and live cattle, 2002 and 2003. Available at: www1.agric.gov.ab.ca/$department/deptdocs.nsf/all/sdd8642

Alberta Farm Animal Care (2002) *Livestock Transportation in Alberta*. Alberta Farm Animal Care, Calgary, Canada.

Allen, D.M. (2005) Handle with care. Animal handling and welfare are issues whose time has come. Question is, are processors ready? *Meat Marketing Technology* May, 85–86.

Animal Legal Defense Fund (2006) Demand Texas Department of Agriculture take action. Available at: http://www.aldf.org/article.asp?cid=574

Animal Protection Institute/Compassion in World Farming (2005) Driving pain: the state of farmed-animal transport in the U.S. and across our borders. Available at: www.api4animals.org/a6a_transport.php

Animals' Angels (2003) North America Summer Investigation Schedule and video, Lesley Moffat, 47 pages, Medicine Hat, Alberta to St-Cyrille-de-Wendover, Québec, 19–22 August 2003. Available from Animals' Angels, Frankfurt, Germany.

Boyle, B. (2004) Frontline Interview with Patrick Boyle. *Modern Meat*. Available at: www.pbs.org/wgbh/pages/frontline/shows/meat/interviews/boyle.html

Brandt, P. (2005) Loop Hole on Wheels: Trucks and the 28-Hour Law. Available at: www.hsus.org/farm_animals/farm_animals_news/trucks_and_the_28-hour_law.html

Brasher, P. (2006) USDA says rule on livestock stops applies to trucks. *Des Moines Register*, 29 September 2006.

Broom, D.M. (2000) Welfare assessment and welfare problem areas during handling and transport. In: Grandin, T. (ed.) *Livestock Handling and Transport.* CAB International, Wallingford, UK, pp. 43–61.

Canadian Beef Export Federation (2006) About the Canadian Beef Export Federation. Available at: www.cbef.com/about_cbef.htm

Canadian Cattlemen (2006) When animals are the victims. Available at: www.agcanada.com/custompages/stories_story.aspx?mid=31&id=802

Canadian Food Inspection Agency (2004) Performance Report Meat Hygiene Program. Available at: www.inspection.gc.ca/english/corpaffr/ar/ar04/meavia_anne.shtml

Canadian Food Inspection Agency (2006) Advanced notice of possible changes to animal transportation regulations in Canada. Available at: www.inspection.gc.ca/english/anima/heasan/transport/notavie.shtml

CBC News (2005) Potential trucker shortage looms as U.S. border reopens to cattle. Available at: www.cbc.ca/story/business/national/2005/07/18/cattletruck-050718.html

Cody, C. (2006) USDA livestock rule changed, kept quiet. *Arkansas Democrat-Gazette,* 30 September 2006.

Commission for Environmental Cooperation (2003) Comparative standards for intensive livestock operations in Canada, Mexico and the United States. Available at: www.cec.org/files/pdf/LAWPOLICY/Speir-etal_en.pdf

Compassion Over Killing (2005) COK investigation exposes farmed animal suffering during interstate transport. Available at: www.cok.net/feat/usti_notes.php

Dewey, C., Haley, C., Widowski, T. and Friendship, R. (2004) Factors associated with in-transit losses. In: Murphy, J.M., Kane, T.M. and de Lange, C.F.M. (eds) *Proceedings of the London Swine Conference: Building Blocks for the Future.* London Swine Conference, London, Ontario, pp. 51–54.

Economic Research Service (2003) Data interstate livestock movements. Available at: www.ers.usda.gov/Data/InterstateLivestockMovements/dataqanda.htm

Economic Research Service (2004) Market integration in the North American hog industries. Outlookreport No.(LDPM12501). Available at: http://www.ers.usda.gov/publications/ldp/nov04/ldpm12501/

Edmonds, S. (2002) Canadian meat packers find low pay leads to recruitment problems. Available at: www.beyondfactoryfarming.org/documents/Meat_Packers_2002.pdf

Ensminger, M.E. (1993) *Dairy Cattle Science,* 3rd edn. Prentice-Hall, New Jersey.

European Commission (2000) *Final Report of a Mission Carried Out in Mexico from 3 to 14 April 2000, in Order to Evaluate the Implementation and Enforcement of Council Directives 72/462/EEC and 77/99/EEc and Council Decision 95/408/EC and to Review the Animal Health Situation,* 1168/2000-MR Final. European Commission Health and Consumer Protection Directorate General, Brussels, Belgium.

European Commission Food and Veterinary Office (2006) Third country establishments' lists. Available at: http://forum.europa.eu.int/irc/sanco/vets/info/data/listes/list_all.html

Farm Sanctuary (2004) *U.S. Highway Accidents Involving Farm Animals.* Farm Sanctuary, Watkins Glen, New York.

Farm Sanctuary (2005) *Farm Animal Welfare: An Assessment of Product Labeling Claims, Industry Quality Assurance Guidelines and Third Party Certification Standards.* Farm Sanctuary, Watkins Glen, New York.

Food and Agriculture Organization of the United Nations (2004) FAOSTAT On-line Statistical Service. Available at: http://apps.fao.org

Forde, K., Hillberg-Seitzinger, A., Dargatz, D. and Wineland, N. (1998) The availability of state-level data on interstate cattle movements in the United States. *Preventive Veterinary Medicine* 37, 209–217.

General Accounting Office (2004) *Humane Methods of Slaughter Act: USDA Has Addressed Some Problems but Still Faces Enforcement Challenges.* GAO-04-247. United States General Accounting Office, Washington, DC.

Grandin, T. (1992) Livestock management practices that reduce injuries to livestock during transport. In: *Livestock Trucking Guide.* Livestock Conservation Institute, Bowling Green, Ohio.

Gregerson, J. (2005) Canada's lesson. *The Meatingplace* October, p. 6.

Gregerson, J. (2006) Canada's comeback. *The Meatingplace* February, p. 12–17.

Hahn, A. (2007) *USDA Halts Mandatory Animal ID Program*. Brenham Banner-Press, Brehnam, Texas.

Hightower, J. (2004) BushCo's mad, mad, mad, mad, Mad Cow policy. *The Hightower Lowdown* 6(4), 1–4.

Irwin, M.R. and Gentleman, W.R. (1978) Transportation of cattle in rail car containing feed and water. *South Western Veterinarian* 31, 205–208.

Ishmael, W. (2005) More than money. *Beef Magazine*. Available at: www.beef-mag.com/mag/beef_money/index.html

Juday, D. (2005) A looming economic threat to the Kansas beef industry: federal country of origin labeling regulations. *Kansas Policy Review* 27(1), Spring 2005.

Kelley, T. (2005) Don't let stress, heat be a downer for pigs. *Pork* May 2005, 16–18.

Knowles, T.G. (1998) A review of the road transport of slaughter sheep. *Veterinary Record* 143, 212–219.

Knowles, T.G. (1999) A review of the road transport of cattle. *Veterinary Record* 144, 197–201.

Krause, J. (2006) I'll pay you to kill my steer. It's not so easy for small farmers to get their animals slaughtered. *Chow*. Available at: http://www.chow.com/stories/10190

Lambooij, E. and van Putten, G. (1993) Transport of pigs. In: Grandin, T. (ed.) *Livestock Handling and Transport*, 2nd edn. CAB International, Wallingford, UK, pp. 213–231.

Lonergan, G.H., Dargatz, D., Morley, P.S. and Smith, M.A. (2001) Trends in mortality ratios among cattle in US feedlots. *Journal of the American Veterinary Medical Association* 219, 1122–1127.

McNaughton, M.T. (1993) Not for sale, mobile slaughterers: the meat industry's grey trade. *Meat and Poultry* September, 28–44.

Minter, J. (2005) Cattle economic situation and outlook. Department of Agricultural Economics – Kansas State University. Available at: www.beef.org/uDocs/cattleeconomicsituation.pdf

Missouri Swine Audit (2005) An analysis of Missouri's competitive position in the swine industry. July 2005. Available at: http://agebb.missouri.edu/commag/swine/audit/htmlindex.htm

Murray, A.C. (2000) Reducing losses from farm gate to packer. *Advances in Pork Production* 11, 175.

National Agricultural Statistics Service (1996) Agriculture Statistics Board, United States Department of Agriculture, Washington, DC.

National Agricultural Statistics Service (2006) Livestock Slaughter: 2005 Summary, Released March 2006. Available at: http://usda.mannlib.cornell.edu/MannUsda/viewDocumentInfo.do?document ID=1096

National Pork Board (2001) Trucker Quality Assurance Brochure. Available at: http://www.porkboard.org/TQA/intro.asp

National Pork Producers Council (2004) Mexico is unfairly restricting U.S. hog exports. Available at: www.nppc.org/public_policy/Mexico8_2_04LiveHogBorderTlkpts.pdf

Nierenberg, D. (2006) *Happier Meals: Rethinking the Global Meat Industry*. World Watch Institute, Oxon Hill, Maryland.

Norman, E. and Hernandez, N.C. (2005) Like butter scraped over too much bread: animal protection policy in Mexico. *Review of Policy Research* 22(1), 59–76.

Olson, K. (2002) The shame of meatpacking. *The Nation*, 16 September 2002, 11. Available at: www.thenation.com/docprint.mhtml?i=20020916&s=olsson

Opinion Research Corporation (1995) Nationwide Telephone Survey. Available from Animal Rights International, Connecticut.

Perry, G. (1990) Introduction. *Applied Animal Behavior Science* 28, 1–2.

Pollan, M. (2006) *The Omnivore's Dilemma. A Natural History of Four Meals*. The Penguin Press, New York.

Reynolds, G. (2007) Regulator proposes re-opening border to Mexican pork. Available at: www.truthabouttrade.org/print.asp?id=6859

SAGARPA (1993) Mexican Federal Animal Health Law. Available at: http://www.sagarpa.gob.mx/legislacion/docs/leyes/09_LEY%20Fed%20de%20San%20Animal.pdf

SCAHAW (2002) *The Welfare of Animals during Transport (Details for Horses, Pigs, Sheep and Cattle)*. Scientific Committee on Animal Health and Animal Welfare, European Commission Health and Consumer Protection Directorate General, Brussels, Belgium.

Schlosser, E. (2005) Frontline Interview. *Modern Meat.* Available at: www.pbs.org/wgbh/pages/frontline/shows/meat/interviews/schlosser.html

Shagam, S. (1998) World meat consumption and trade patterns. *National Food Review* January–March.

Shields, D.A. and Mathews, K.H. (2003) Interstate livestock movements. Electronic outlook report from the Economic Research Service United States Department of Agriculture. Available at: www.ers.usda.gov/publications/ldp/jun03/ldpm10801/ldpm10801.pdf

Skaggs, R., Acuna, R., Torell, L. and Southard, L. (2004) Live cattle exports from Mexico into the United States: where do the cattle come from and where do they go? Available at: www.choices-magazine.org/scripts/printVersion.php?ID=2004-1-05

Speer, N.C., Slack, G. and Troyer, E. (2001) Economic factors associated with livestock transportation. *Journal of Animal Science* 79, E166–E170.

Standing Senate Committee on Agriculture and Forestry (2005) Cattle slaughter capacity in Canada: Interim report. Available at: www.parl.gc.ca/38/1/parlbus/commbus/senate/Com-e/agri-e/rep-e/repintmay05-e.pdf

Swanson, J.C. and Morrow-Tesch, J. (2001) Cattle transport: historical, research and future perspectives. *Journal of Animal Science* 79, E102–E109.

Tarrant, P.V. (1989) The effects of handling, transport, slaughter and chilling on meat quality and yield in pigs – a review. *Irish Journal of Food Science Technology* 13, 79–107.

Tarrant, P.V. (1990) Transportation of cattle by road. *Applied Animal Behavior Science* 28, 153–170.

Tarrant, P.V. and Grandin, T. (1992) Cattle transport. In: Grandin, T. (ed.) *Livestock Handling and Transport*, 2nd edn. CAB International, Wallingford, UK.

Tescher, E. (1992) Mexico's call for American lamb: a boon or a threat? *National Wool Grower*, April 1992.

Transport Canada (1987) Motor Vehicle Transportation Act 1987. Available at: www.tc.gc.ca/acts-regulations/GENERAL/M/mvta/regulations/mvta001/mvta1.html

Trust for America's Health (2003) Animal borne epidemics out of control. Threatening the nation's health. Available at: http://healthyamericans.org/reports/files/Animalreport.pdf

United States Department of Agriculture (2000) *National Animal Health Monitoring System. Part II: Baseline Reference of Feedlot Health and Health Management 1999.* USDA Animal Plant Health Inspection Service, Fort Collins, Colorado, p. 46.

United States Department of Agriculture (2006) Livestock, Dairy, and Poultry Outlook, Report No LDP-M-140. USDA Economic Research Service. Available at: www.ers.usda.gov/publications/so/view.asp?f=livestock/ldp-mbb/2006/

United States Department of Transportation (2005) Federal motor carrier safety administration: hours of service of drivers. Available at: www.fmcsa.dot.gov/rulesregulations/administration/fmcsr/fmcsrguidedetails.asp?rule_toc=764&section_toc=764

United States Trade Representative (2005) Report on Mexico trade. Available at: www.ustr.gov/assets/Document_Library/Reports_Publications/2005/2005_NTE_Report/asset_upload_file467_7483.pdf

USDA Agricultural Marketing Service (1997) USDA trucking guide assists cattle and swine exporters. Available at: www.ams.usda.gov/NEWS/166.htm

USDA Agricultural Marketing Service (2006) Livestock trade database. Available at: www.ams.usda.gov/lsmnpubs/mexico.htm

USDA Foreign Agricultural Service (2005) Trade Statistics. Available at: www.ams.usda.gov

USDA APHIS (2005a) Bovine spongiform encephalopathy; minimal-risk regions and importation of commodities; final rule. *Federal Register* 70, 459–553.

USDA APHIS (2005b) Bovine spongiform encephalopathy; minimal-risk regions and importation of commodities; unsealing of means of conveyance and transloading of products; interim rule and request for comments. *Federal Register* 70, 71213–71218.

USDA Foreign Agricultural Service (2006) Mexico Livestock and Products Annual 2006. Available at: www.thepigsite.com/articles/7/markets-and-economics/1754/mexico-livestock-and-products-annual-2006

USDA Transportation Services Branch (1999) Cattle and swine trucking guide for exporters. Available at: www.ams.usda.gov/tmd/livestock/Truck%20Guide.htm

U.S. Newswire (2006) Man hauling horses to slaughter charged with animal cruelty. Society for Animal Protective Legislation. News release, 10 August 2006.

Van Houwelingen, P. and Vingerling, P. (1989) *Field Study into the International Transport of Animals and Field Study Concerning the Stunning of Slaughter Animals. Studies Carried Out on Behalf of the Commission of the European Communities.* Office for Official Publications of the European Communities, Luxembourg.

Vansickle, J. (2005) Trucker program revisited. *National Hog Farmer* 15 September 2005.

Warriss, P.D. (1998) Choosing appropriate space allowances for slaughter pigs transported by road: a review. *Veterinary Record* 142, 449–454.

Whole Foods (2005a) Whole Foods market natural meat program and animal compassionate standards for beef cattle. Available at: www.wholefoodsmarket.com/issues/animalwelfare/cattle.pdf

Whole Foods (2005b) Whole Foods market natural meat program and animal compassionate standards for sheep. Available at: www.wholefoodsmarket.com/issues/animalwelfare/sheep.pdf

Whole Foods (2005c) Whole Foods market natural meat program and animal compassionate standards for pigs. Available at: www.wholefoodsmarket.com/issues/animalwelfare/pigs.pdf

Whole Foods (2005d) Whole Foods market natural meat program and animal compassionate standards for ducks. Available at: www.wholefoodsmarket.com/issues/animalwelfare/ducks.pdf

Whole Foods (2006) Whole Foods market natural meat program and animal compassionate standards for broiler chickens. Available at: www.wholefoodsmarket.com/products/meat-poultry/qualitystandards.html

Wolfson, D.J. (1999) *Beyond the Law. Agribusiness and the Systematic Abuse of Animals Raised for Food or Food Production.* Farm Sanctuary, Watkins Glen, New York.

Zahniser, S. (2005) North America moves toward one market. *Amber Waves,* June 2005.

Appendix 9.1

Table 9.1.1A. National/Federal[a] transport regulations.

	Canada	USA	Mexico
Overall transport limits required	N	N	N
Food, water and rest periods required	Y	Y	N[a]
	36 h	28 h	
	48 h	(36 h)	
	(52 h)		
	Horses and pigs – 36 h limit before 5 h rest, food and water, unless animals are fed, watered and rested on suitable vehicle. Cattle, sheep and goats – 48 h limit before 5 h rest unless fed, watered and rested on suitable vehicle or if they can reach their final destination within 52 h.	Cattle, sheep, goats, pigs – 28 h limit before 5 h rest, food and water unless animals are fed, watered and rested on the vehicle. Animals may be confined for 36 consecutive hours upon the request of the owner or the person having custody of the animals. Sheep may be confined for an additional 8 consecutive hours when the 28 h period of confinement ends at night. Horses – 28 h limit before 6 h rest, food and water.	[a]Official Mexican Norms set forth recommendations but are not enforced as mandatory.
Pre-transport requirements	Y	Y[a]	N
	Animals may not be loaded for a trip of more than 24 h without first providing food and water within 5 h before loading.	[a]For horses only (see 'Commercial Transportation of Equines for Slaughter Act' for details).	
Handling requirements	Y	N[a]	Y[a]
	'No person shall beat an animal being loaded or unloaded in a way likely to cause injury or undue suffering to it.'	Humane handling requirements under the Humane Methods of Slaughter Act only apply to activities taking place on the premises of slaughter plants.	Beating or any action contrary to humane treatments is explicitly prohibited

Continued

Table 9.1.1A. *Continued.*

	Canada	USA	Mexico
Stocking densities required	Y	Non-specific. Animals may not be crowded to such an extent as to cause 'injury or undue suffering'. Animals must be allowed to stand in 'natural position'.	N	Nᵃ Norm 51 indicates that transportation cars should not be overloaded and ventilations should be provided according to the weather. But it is not enforced as mandatory. Animals must never be tied by the feet or by any method that could strangle them.
Transport of downed animals prohibited	Y	It is illegal to load, transport and unload non-ambulatory animals. 'Non-ambulatory animals, animals with a body-condition score indicating emaciation or weakness or animals with severe lameness, [sic] would endure additional suffering during the transportation process must not be transported except for veterinary treatment or diagnosis. This is true of any condition associated with pain that will be aggravated by transport.'	Nᵃ ᵃFederal law to prohibit transport of downed animals has been introduced	N

Continued

Transport of sick and injured animals prohibited	Y	Animals may not be transported if they are sick or injured where 'undue suffering will result'. In addition, animals liable to give birth may not be transported.	N		N
Regulations for the transport of sick, injured or downed animals	Y	Non-ambulatory animals must be euthanized on the farm. Animals that become non-ambulatory during transit may be: (i) euthanized on the truck; (ii) stunned and bled out on the truck; or (iii) stunned on the truck and unloaded and transferred to the bleeding area.	Y[a]	[a]Downed animals unloaded at slaughter plants are covered under the Humane Methods of Slaughter Act but are not covered during loading at the farm or auction or during transit. Regulations for horses (see 'Commercial Transportation of Equines for Slaughter Act' for details).	N
On-board vehicle requirements	Y	Animals must be segregated if they are of different species or substantially different weights and ages or if they are incompatible by nature. Transport vessels must be: (i) strewn with sand or fitted with secure footholds; and (ii) be littered with straw, wood shavings or other bedding material. [a]If transported for not more than 12h the transport vessel only needs to comply with a or b.	Y[a]	[a]For horses only (see 'Commercial Transportation of Equines for Slaughter Act' for detail).	N

Table 9.1.1A. *Continued.*

	Canada	USA	Mexico
Accident/ emergency procedures	Y Drainage or absorption of urine required.	N Voluntary codes include guidance for dealing with highway accidents involving farmed animals. Response teams trained, and accidents tracked.	N
Driver driving and rest requirements	Y 13h	Y 11h 13h of driving must be followed by at least 8 consecutive hours of off-duty time or after 15h on-duty whether driving or not. Drivers must keep written logs of work, rest and distance travelled.	U[a] Drivers may not drive more that 11 consecutive hours or work longer than 14h in a shift and must rest for at least 10h between shifts. Drivers must keep a record of duty status, in duplicate, for each 24h period or use an automatic on-board recording device to record duty status including rest periods, driving time and miles travelled. [a]Unknown
Animal tracking required	Y	N Ear tags for cattle only. Industry-initiated and industry-established programme. Not used to track or enforce transport limits.	Y Horses must be identified with registered hot iron farm brands to prevent theft

| Enforcement mechanisms | Y | CFIA officials are charged with monitoring compliance of regulations. Monitored through routine inspections at ports of entry, registered establishments and auction markets and following reports of non-compliance. | N[a] | No agency officially charged with enforcement of 28-Hour Law. Complaints of non-compliance referred to the Justice Department. [a]USDA is charged with enforcing horse transport requirements. | N |

[a]State and provincial laws may vary and are not included in this summary.

Table 9.1.2A. The US import and export statistics with Canada and Mexico for 2005, as reported by the US Department of Agriculture (USDA) Foreign Agricultural Service Trade Statistics.[a]

US live animal import from Canada	US live animal exports to Canada	US live animal imports from Mexico	US live animal exports to Mexico
Pigs greater than 50 kg (110 lb) 'for immediate slaughter' 2,695,128	Pigs greater than 50 kg (110 lb) 1,655		Pigs greater than 50 kg (110 lb) 122,104
Steers for slaughter 207,961		Steers for slaughter 3,878	
Bulls for slaughter 476		Bulls for slaughter 665	
Cows for slaughter 2,799			Cattle 0
Heifers for slaughter 107,883			
Chickens 2,000 g 979,974	Chickens greater than 2,000 g 76,227		Chickens 185–2000 g 18,093
Sheep 798	Sheep 382	Sheep 1,105	Sheep 71,824
Horses 'for immediate slaughter' 7,533	Horses not pure bred[b] 20,730		Horses not pure bred[b] 11,010

[a]The USDA Foreign Agricultural Service import and export statistics database does not consistently distinguish between animals for immediate slaughter and those being transported for feeding or breeding purposes. Therefore, the statistics given above represent estimates of the number of animals imported and exported for slaughter based on animal weight at the time of transport.
[b]The export statistics database provided by the USDA Foreign Agricultural Services does not contain a category for 'horses for immediate slaughter' as it does for import. It is suspected that at least some if not all of the horses designated as 'not pure' on the export records may be destined for slaughter. The USDA Agricultural Marketing Service's weekly livestock trade database does, however, distinguish slaughter horses and based on information gathered from this database, the USA exports up to 138 horses per week for slaughter in Mexico.

10 South America

C.B. GALLO[1] AND T.A. TADICH[2]

[1]*Instituto de Ciencia Animal y Tecnología de Carnes, Facultad de Ciencias Veterinarias, Universidad Austral de Chile, Chile;*
[2]*Programa Doctorado en Ciencias Veterinarias, Escuela de Graduados, Facultad de Ciencias Veterinarias, Universidad Austral de Chile*

Abstract

This chapter gives an overview of the long-distance transport of animals for slaughter in South America, with special regard to inherent sociocultural, livestock production, climatic and geographic conditions, as well as trade, stakeholders, legal framework, research and training in the field. Information for Argentina, Brazil, Bolivia, Ecuador, Chile, Colombia, Paraguay, Perú, Uruguay and Venezuela is presented.

Within South America cattle and sheep production are characterized mainly by grazing, usually in extensive systems; pig and poultry production are significant intensive activities in some countries. In this region there are some of the world's most important beef production and exportation countries (Brazil, Argentina); there are also some countries where, even with a small cattle population, meat exports are an important part of the economy (Uruguay) or have access to high-meat-price markets because of good animal health conditions (Chile). Meat exportation provides a good opportunity to make improvements in quality assurance schemes and good livestock practices that consider animal welfare as a component in the production chain on farm, during transport and at slaughter. Livestock producers as well as veterinary services in each country are aware that international commercial agreements not only require them to fulfil sanitary and animal health regulations but also other requirements of consumers, among them ethical considerations for product positioning.

The evidence presented here shows that exports of live animals occur mainly for breeding and only very occasionally for slaughter. The transport of live animals for slaughter within each country occurs mostly for relatively short distances (300–500 km), but it also occurs over long distances (1000–1500 km) or for long durations. There is a great variation in the conditions of the transport of farm animals within South American countries: if commonly transport duration is between 1 and 12 h, it can occasionally reach up to 60 h. This is due mainly to a combination of bad roads, bad weather conditions and the existence of several intermediate dealers. In some cases the animals transported are facing very bad conditions, causing strong risk of welfare deterioration and in others, the risk is minor and the welfare of transported animals is not under strong pressure. Bad practices during loading, transport and unloading of animals are common, as well as overstocking the trucks. There is a large difference between countries from the southern part of the region (Brazil, Uruguay, Paraguay, Argentina, Chile), and those from the central and northern part of the region, the latter being less developed, with more sociocultural problems which tend to take priority over animal welfare issues.

Most South American countries are members of the World Organisation for Animal Health (OIE), and delegates exist in each country, who have already been given the responsibility for animal welfare issues and to bring national regulations into line with OIE recommendations. Increasing regional research and training of human resources at all levels of the meat chain is seen as an important tool to achieve this goal.

Introduction

In order to gain knowledge about the situation of the transportation of farm animals in countries of South America, information was collected from the following geographical areas: Andean Region – Bolivia, Ecuador, Perú and Venezuela – (Gallego, 2006), Brazil (Paranhos da Costa *et al.*, 2006), Colombia (Delgado, 2006), Chile (Gallo, 2006a) and Mercosur – Argentina, Uruguay and Paraguay (Ponce, 2006) (Fig. 10.1).

South America is a large, geographically and climatically diverse region, which is surrounded by the Pacific and Atlantic Oceans and the Caribbean Sea. Some of its main topographical landmarks are the Amazonian forest, the Andes mountain range, the Pampas, Patagonia and the Atacama Desert (Fig. 10.1).

All the above mentioned characteristics give South America a great variety of climates and temperatures, between, as well as within, countries, ranging from the Amazonian tropical rainforest to Patagonia. The geographic conditions of South America have strongly influenced the cattle production systems, which are extensive and allow generally better welfare conditions than those provided by intensive systems; however, in many cases seasons are extreme (either very dry or very wet) with low forage production, and consequently drastic reductions in the body condition scores of the animals can be seen during these periods.

The socio-economic and cultural situation in many South American countries is characterized by low incomes, high poverty indexes and in many cases still high numbers of illiterate individuals (The World Bank, 2006). The human population is more concerned about food safety and a fair price for the meat than about maintaining animal welfare standards. Considering this background, the interest in animal welfare when it comes to production animals is based more on economic reasons (the loss of quality and quantity of the meat due to mistreatment) rather than on ethical reasons.

When consumer's incomes increase or there are possibilities of getting better prices for beef at farm and industry levels, animal welfare standards become more visible in a society. This is the case in the meat exporting countries, where there is increasing awareness about animal welfare among consumers, as well as among people working throughout the meat chain. In the case of these countries, the increasing importance given to animal welfare has been due mainly to the relationship between good handling practices and meat quality, as well as because of the demands established directly from the importing countries.

On the other hand, the cultural background of the population in South America has also been a limitation on the growth of the long-distance transport of

Fig. 10.1. Political division and important geographic landmarks of South America. (From World Atlas, 2007.)

meat rather than the transport of live animals, because meat from newly slaughtered animals is traditionally preferred to matured and stored meat. So, most slaughter still occurs close to the main cities, which are often far from the main production areas, at least in the case of cattle.

In the region, the higher the income of a family, the higher is the proportion spent on meat and dairy products. So there is a relationship between meat consumption and a better living standard. In South America beef production is important, and will keep its traditional first place among farm species. Meat consumption is in general growing in most South American countries, it is highest in Argentina, Uruguay and Brazil and lowest in Perú. Although beef is the most traditional meat consumed, pork and especially poultry consumption have increased rapidly in most countries as part of a more varied and usually cheaper offer for the consumer.

Livestock Production Characteristics and Meat Trade

Cattle are the main species produced in many South American countries (Table 10.1). In Brazil, Colombia and Venezuela, production is based mainly on *Bos indicus* cattle, whereas in Perú, Bolivia and Ecuador, it is *B. indicus* and *Bos taurus*, and in Argentina, Uruguay and Chile mainly *B. taurus*, with criollo, dual-purpose and British breeds such as Hereford and Aberdeen Angus. In most cases, cattle are produced in large farms (haciendas), but a considerable number are kept on smallholdings, which produce only for subsistence. Because beef is produced in extensive pasture conditions, animals have little contact with humans and can be difficult to handle and easily frightened and fearful when gathering and loading for transport. In countries where mainly *B. indicus* cattle are reared, these have been observed to be more easily stressed by handling than *B. taurus*, hence making the handling during loading and transport more difficult (ABCZ, 2005). On the other hand, dual-purpose and dairy breeds have closer contact with people, because they are either supplemented with silage and grains during winter or, in some cases (less than 15%), finished in feedlots. Newer techniques in animal reproduction and production are leading to a reduction in the slaughter age of cattle in most of the region.

Farming activities are of major importance in the Brazilian and Argentinean economy (Table 10.1). Meat production has a prominent role in this scenario; beef, poultry and pork are very important export items in Brazil (ANUALPEC, 2005). Argentina is the largest horse meat producer in South America; most of the horse meat is exported, mainly to the Netherlands, Russia, France, Italy, Japan, Switzerland and Belgium (Ponce, 2006). A considerable number of horses are also slaughtered for meat consumption in Brazil and Chile; most of them are horses that did not show a good enough performance in sports, or are injured, old or not tamed.

Uruguay raises both cattle and sheep; livestock production takes up the majority of the surface area of the country, supplying national consumption; it provides the raw material for the meat industry. At least 50% of the meat produced, mainly beef, is exported and it generates a great deal of employment among the rural population. There is little pig production overall, although there is a high density of pigs in certain areas, for example in Canelones, San José and Colonia (Ponce, 2006).

Chile has the smallest cattle population within South America (Table 10.1), but has very good sanitary conditions favoured by the natural limits of its territory (Andes Mountains on the east and Pacific Ocean on the west). Chile does not export live animals for slaughter to other countries, only animals for breeding or sports purposes. There are no imports of live animals for slaughter. At the moment Chile is exporting meat from various species (pork, chicken, lamb, cattle) to many countries (México, Japan, UK, USA, etc.) in increasing amounts.

Sheep are produced mainly in the southern part of South America, in the Patagonian area. The main producing countries are Brazil, Perú, Uruguay and Argentina and conditions are extensive, on grazing land. Brazil also has the highest goat population; goats are produced in most South American countries just for consumption in the home. Hence, the formal market, with goats being transported to slaughterhouses and the meat sold in butcher shops, is small. A similar situation happens with camelids, which are produced mainly in Perú and Bolivia.

Table 10.1. Summary of the main geographic, demographic and livestock population ('000 head) indexes in South America. (From FAO, 2005.)

Country	Land area ('000 km²)	Population	Cattle	Sheep	Pigs	Goats	Poultry[a]	Equids[b]
Argentina	3,761	38,747	50,768	12,450	1,490	4,200	100,395	3,938
Bolivia	1,084	9,182	6,822	8,550	2,984	1,501	450	1,040
Brazil	8,515	186,405	207,000	15,200	33,200	10,700	1,119,750	8,080
Chile	756	16,295	4,200	3,400	3,450	735	121,500	711
Colombia	2,070	45,600	25,699	3,333	1,724	4,105	150,000	3,124
Ecuador	277	13,228	4,971	1,053	1,281	144	104,217	706
Paraguay	407	6,158	9,838	500	1,600	155	17,930	410
Peru	1,280	27,968	5,241	14,822	3,005	1,957	99,255	1,640
Uruguay	175	3,463	11,956	9,712	257	16	14,365	437
Venezuela	882	26,749	16,615	525	3,264	1,342	110,000	1,012

[a]Including chickens, ducks, geese, guinea fowls and turkeys.
[b]Including asses, mules, hinnies and horses.

In the case of pigs and poultry the meat chain is short in most South American countries; producers are usually also owners of the slaughterhouses (integrated production). Production is very intensive with modern technology at all levels of the meat chain. On the other hand, in Colombia egg production is distributed among a large number of producers with scales ranging from 500 birds to over 500,000 layers, mainly being placed in barn systems (74%) or in battery cages (26%), with large variations regarding technology, equipment and handling systems. In Uruguay and Paraguay poultry and pig production are not significant.

In South America there are some of the main meat producing and exporting countries in the world, such as Brazil and Argentina (FAO, 2003). Colombia is ranked as South America's third livestock producer, but it is focused mainly for internal consumption. Chile, Ecuador and Venezuela are importers of meat for national consumption, but Chile has also lately become a meat exporter (beef, pork, poultry) to specific high-price markets.

The importance of the EU as a market for beef and lamb meat (better prices) and commercial agreements between South American countries and the EU are the main reasons for animal welfare concern. Some countries (Uruguay, Chile, Brazil and Argentina) have approached the animal welfare issue as a tool for improving meat quality.

Stakeholders

In South America there is a complex system of beef trade, where the producer seldom sells directly to the final consumer. The main components of the meat chain are producers, livestock markets, livestock dealers, livestock transporters, slaughterhouses, supermarkets and butchers.

Producers

Cattle producers can be divided into those who breed and fatten beef (complete cycle), those who only produce weaned calves and those who buy calves and other cattle for fattening and selling. Cattle producers in most South American countries sell their stock through cattle dealers (middlemen, also called 'gestores', 'corredores de ganado', 'consignatarios', depending on the country) to the slaughterhouses and to livestock markets. However, in large beef exporting countries, the situation differs and cattle producers sell mainly directly to the abattoirs, as they have to comply with varying requirements according to the demands of each country of destination (such as traceability, no use of growth promoters, carcass weight and quality).

Producers for export usually get paid on the basis of carcass weight and quality; because problems such as bruising, dark colour and high-pH meat often arise due to bad handling during production, collection and transport and this would reduce the price paid for the animal, producers are more concerned about animal welfare. But the most common situation in many countries still is that farmers sell through dealers for the internal market; dealers are usually unaware of the relationship between handling and quality of the final product, they base the sale price on live weight (sometimes merely on the appearance and type of the animal) and are not concerned about the final product.

In the case of small ruminants, producers tend to control the complete rearing and production cycle and sell the lambs directly at weaning. Again, lambs, goats and camelids can go directly to slaughterhouses, to livestock markets and to intermediate dealers. In the case of goats and camelids, the market is small and scarcely developed, meaning that this meat is rarely found for sale in butcher shops or supermarkets, and there are no further processing systems available such as jointing, packaging or freezing.

The pig and poultry producers not only have the complete production cycle, but in most cases production is very intensive and highly technological; there is an integrated chain, and producers also own the slaughterhouses and market the final products themselves. Poultry farm owners are in many cases also pig farm owners; they have their producer associations and employ the latest technology in terms of production and slaughter.

Some horses, sheep and even home-grown pigs and poultry also go through livestock markets in many South American countries.

Transporters

The livestock transporters are usually not organized in any of the countries studied. But in some countries (Chile, Bolivia) a national register for meat and livestock trucks is being introduced, in order to keep better control of animal transportation. In most cases the livestock transport is paid by live weight; interestingly in Chile, due to the existent regulations, the transporters are made responsible for the animals during the journey and therefore, if there are bruises in the cattle transported for slaughter, transporters have to pay for any losses. Due to the existing carcass grading system, bruised carcasses are downgraded and economic losses occur, which have to be paid by the transporters. This has been influencing decisions about stocking densities and about how far animals are transported.

Slaughterhouses

All slaughterhouses in South America are required to carry out a health inspection of the animals before slaughter and of the meat produced (ante- and post-mortem), this is compulsory by law in all countries and is done by official veterinarians, accredited for that task. However, it has been observed that in some countries inspections do not always take place in small premises, due to the fact that there are not enough trained veterinarians to carry out the inspection (Gallego, 2006).

In meat exporting countries (Argentina, Brazil, Uruguay, Paraguay and Chile) there are big slaughterhouses accredited for export. These main slaughterhouses (slaughtering hundreds to thousands of head per day) are usually associated with, and in many cases belong to, large enterprises that own several premises and slaughter the majority of the commercial livestock in the countries concerned. These industries usually organize their cattle suppliers in groups (30–40) and provide them with talks given by specialized professionals on subjects such as what to produce, how to produce, how to improve pastures, productivity and product quality; some aspects regarding the transport and handling of cattle and the consequences on meat quality and welfare are being introduced as well. In most South American countries municipal or communal slaughterhouses are common, belonging either to the state or privately owned. These are small premises, for local consumption (only serving small communities) and usually have deficiencies in terms of infrastructure, equipment, personnel and handling of animals and products. One of the reasons that the registered and export slaughterhouses are sometimes discouraged from continuing to implement improvements in technology, infrastructure, personnel, etc. is the existence of illegal and sometimes clandestine slaughtering places, not registered by the health authorities; they compete directly and with lower costs within the national market.

Another problem related to animal welfare in South American slaughterhouses is the long periods for which animals are held prior to slaughter (lairage times). Although there has been a trend in recent years to reduce lairage time at slaughterhouses, still all South American countries have a compulsory minimum lairage time; this has been established by legislation in order to give time for resting the animals after transport, for ante-mortem inspection and for emptying the guts. Depending on each country, the minimum lairage time can vary from 6 to 12 h; but usually this period is extended much longer than the minimum because of operational reasons and/or inadequate planning. Lairage time usually includes access to water but not to food for the waiting animals. The conditions in lairage pens, especially in small slaughterhouses, are also often inadequate. Another problem observed is the lack of facilities for unloading and humanely culling those animals that arrive injured or sick; these exist only in some export slaughterhouses. There is still much improvement to be done in public (municipal) and small village slaughterhouses. They need training of personnel, improvement of infrastructure and to solve environmental problems related to waste disposal.

Supermarkets and butcher shops

There are already some butcher shops and supermarket chains which are aware of the need to adopt quality standards and they have set up their own quality assurance schemes from 'farm to shop'. The big supermarkets buy meat directly from the abattoirs; some of them even buy the cattle directly from selected farms that meet their requirements. But the traditional butcher shop in many South American countries (Ecuador, Bolivia, Perú) is still characterized by a low or non-existent capacity for cold storage.

Consumers

Consumers, and more specifically urban consumers, are not conscious about the raising, transportation and killing methods employed in animals to provide humans with food. Generally speaking, and with the exception of countries with a tradition of high-meat consumption, such as Uruguay, Brazil and Argentina, South American consumers have little detailed knowledge of meat quality. They know little about different cuts and qualities, and even about cooking methods, leaving price as the main factor influencing their choice at the time of buying meat. The traditional consumption habits in South America show a marked preference for fresh meats (refrigerated), and distrust towards frozen, vacuum-packed meats or otherwise processed meats. This characteristic, as well as the fact that supermarkets like to have a long shelf life for fresh meats, reduces the potential there is to increase the consumption of meat that has been transported from distant slaughterhouses, as an alternative to long-distance transport of the live animals. Although there are no studies in this respect, it is perceived that the acceptance of processed meats by consumers has in general increased in the last 10 years. Probably this is due to improvements in the technologies used for the processing of the meat, quality assurance systems and also increased consumer knowledge on the relationship between processing the meat and eating quality.

In most countries, farmers, slaughterhouses, livestock markets, supermarkets and consumers (national consumer services) have their own independent association. The livestock transporters are usually not associated. Interestingly, in some countries there are also associations and institutions where all stakeholders jointly take part; this is promoting improvements in animal welfare as well as in meat quality, through quality assurance schemes in the meat chain (for instance the Instituto Nacional de Carnes (INAC) in Uruguay, Instituto de Promoción de Carne Vacuna in Argentina (IPCVA), Corporación de la Carne in Chile (CORPCARNE), Agrocadenas in Colombia).

Stakeholders in some South American countries seem prone to improve quality in order to keep external markets, and it is important to use this strategy for improving animal welfare and also product quality.

Live Animal Transportation

General aspects

Although records on carcass weight, fatness, bruising, downgrading, pH and other quantitative and qualitative measurements can be found in export slaughterhouses, this information is usually private and there is little published scientific evidence available on the characteristics of live animal transportation in South America in general or on its consequences on animal welfare and meat quality.

Most of the information available on transportation of live animals deals with cattle. In Uruguay and Chile there have been studies undertaken in order to identify the critical points of transport for cattle and sheep (Castro and Robaina, 2003; Huertas and Gil, 2003; Bianchi and Garibotto, 2004; Gallo et al., 2004; Cáraves et al., 2006; Strappini et al., 2006) and also recently in Argentina (IPCVA, 2007). Most studies refer to the economic impact and losses in product quality and quantity due to transport, mishandling and structural deficiencies that can cause injuries to the animals rather than strictly to ethical aspects of animal welfare. Dark cutting and bruises are some of the main problems encountered due to bad handling, loading and also to inadequate infrastructure and climatic conditions during the journey (bad roads, extreme weather conditions). Some descriptive data about the handling and transport of cattle in Perú were found (Battifora and Adama, 2000) and data on bruises were also collected in Brazil (Teseimazides, 2006). Worryingly, no studies giving figures on deaths of animals due to transport were found, except for pigs and poultry in Brazil (Paranhos da Costa et al., 2006) and Chile (Becerra, 2006). Some studies on pigs and poultry have also been carried out in Brazil (Branco, 2004; Dalla Costa et al., 2005a,b,c; Costa, 2006), and some on horses for slaughter in Chile (Gallo et al., 2004; Werner and Gallo, 2006). For the other species of farm animals commonly produced in South American countries, such as water buffaloes, goats and camelids, information about transport and slaughter is lacking.

Several South American countries export live animals but these are mainly pets and animals for breeding, sports or show purposes (Gallo, 2006b); in these cases animals are usually handled very carefully considering high welfare standards. The international trade deals mainly with the meat and only in exceptional cases there are live exports for slaughter at destination; in these cases it is clients from the countries of destination who buy the animals at origin and then transport them.

Occasional cases of live exportation of horses from Uruguay and Argentina, occur once or twice a year (for instance a shipment of 1390 horses in April 2006, Ponce, 2006). The horses are bought in Argentina by the clients and then sent by ship on a 17-day journey to Italy. Although the horses are initially bought for sport purposes, according to the Association for the Defense of Animal Rights in Argentina, the horses are eventually slaughtered in Italy. Occasional shipments of sheep from Uruguay to Jordan and Lebanon also occur. The exportation of live cattle from Brazil is recent (Paranhos da Costa et al., 2006); some years ago farmers started to export cattle from the states of Rio Grande do Sul and Pará, sending them to Lebanon. Usually they are loaded in ships (from 2,000 to 10,000 animals/ship). The decision about this is strictly based on economic figures, the number of live exported animals increasing during a crisis. In 2006, Brazil was on course to export 70,000 head of cattle from southern Brazil, while investment in a huge new port in Pará, as a northern export hub, looked set to further increase this new trade route.

There are also some unofficial markets with illegal live imports between bordering countries, in the cases of Ecuador, Venezuela, Perú and Colombia (Gallego, 2006), and also between Brazil, Paraguay and Argentina (Ponce, 2006).

The main reason for transporting large numbers of animals in most countries in South America is within country for sale at livestock markets or directly to slaughterhouses; therefore, problems arise mainly from this type of transport. In some cases long-distance transport is still predominant because of the large extent of some countries (Argentina, Brazil, Colombia and Chile), while in other cases (Uruguay) it is not. Short transport is commonly between 1 and 12 h (300–500 km) but long transport can be occasionally up to 60 h (600–1500 km). When animals are sold by the farmers directly to slaughterhouse companies, this simplifies and shortens the transport journeys for animals, and this is the commonest case in the larger beef-producing countries and those exporting beef, such as Argentina, Brazil and Uruguay. Nevertheless, in most countries cattle farmers sell through dealers; for those who are only breeders or only fatteners, transport of younger animals also occurs from farms to cattle auction markets and between dealers (weaned calves and other cattle for fattening). In most countries, slaughter for consumption is still performed close to the consumption areas (main cities) rather than close to production areas, although the trend is towards selling directly to the slaughterhouses.

The great majority of the animals destined for slaughter are transported by road, in trucks; there are only a few circumstances where a small proportion of animals also have to travel by ferry (Chile) or on boats (some journeys in Amazonian countries). In most South American countries there are paved carriageways in good condition leading to the main cities, but there are also many unpaved or stone roads, often in bad condition; this is especially the case of side roads serving the farms. The roads' conditions depend on the geographic localization, and there are large variations within South America in general and also within countries, due to the varying climate (in tropical countries the heat and humidity are a problem, in which cases animals are transported mainly by night) and geography (including mountainous and winding roads). In general, travelling is slow because of the geography and the nature of the roads, so there is no close relationship between distance travelled and journey duration.

Usually, during road transport, cattle in South America do not have access to food and water inside the load compartment, and it is not common to unload the animals during the trip, even when travelling long distances. Bad practices used to get the animals to move, especially when loading and unloading, are observed to be commonplace; there is a tendency to use aggressive strategies to drive animals and inappropriate aids (sticks, goads, shouting and sometimes even unsuitable handling practices, as described by the OIE, such as pulling sheep by the fleece and twisting of tails). High stocking densities occur, and overloading trucks when transporting live animals has been observed to be a problem in the region, but no precise figures on actual stocking densities used are available, except for the case of Chile (455 kg/m^2 by Gallo et al., 2005) and Uruguay (450 kg/m^2 by Bianchi and Garibotto, 2004). Livestock transport drivers as well as animal handlers who help in the process of loading and unloading the animals lack training (Strappini et al., 2006).

With the exception of some countries (Paraguay, Brazil, Uruguay, Argentina) it is not a common practice to separate cattle in smaller groups within the truck. During the journey the truck drivers inspect the animals regularly, but do not have support to solve any problems that arise with injured or dead animals. When cattle falls occur during transit, the driver or another person in charge of the load will prod or hit the fallen animals to make them stand up. But no written evidence of this or any other driving events is kept by the drivers, because journey logs are not compulsory. In many cases truck drivers drive long shifts. If there are official inspection points en route, these are mainly concerned with ownership, traceability and with the sanitary conditions of the animals and the vehicles, but do not actually take into consideration the journey duration and welfare of the animals. It would be interesting if official veterinary services could perform studies themselves or make this information available to researchers, as this would promote enforcement of transport regulations in South American countries.

Vehicles used for cattle transport are articulated or non-articulated trucks, single-decked, not roofed, usually of metallic structure (the larger and better ones) or wooden structure (the smaller ones) and have no facilities for watering or feeding the animals.

Vehicles used for the transport of sheep and pigs are multideck (2–3 decks). Small trucks and pickups transporting livestock for short distances to auction markets and small slaughterhouses are also common. Trucks with video cameras to monitor animals in transit are occasionally used in Brazil (Paranhos da Costa *et al.*, 2006) and are being used experimentally in Chile for registering animal behaviour during transport (FONDECYT, 2005). Most vehicles use woodchips as bedding, a carpet of rubber covering the load compartment floor and a wooden or iron grid fixed on the floor, to prevent cattle from slipping. In countries like Argentina, Brazil, Chile, Paraguay and Uruguay, most trucks are reported to be in fair to good condition (Bianchi and Garibotto, 2004; Paranhos da Costa *et al.*, 2006; Ponce, 2006; Strappini *et al.*, 2006). There are also some structural deficiencies at farm and slaughterhouse level, mainly in terms of proper design of pens (corrals), races, crates, loading ramps and others.

In general, there is lack of organization and planning between farmers, transporters and slaughterhouses, in order to reduce to a minimum the transport and waiting times spent by the animals. Another problem mentioned is long waiting times before the animals are unloaded in the stockyards in some countries, as well as long lairage times once at the slaughterhouses (Delgado, 2006; Gallego, 2006; Gallo, 2006a; Paranhos da Costa *et al.*, 2006; Ponce, 2006).

Specific aspects of the transport of cattle in each country

The distances travelled by cattle to slaughterhouses within Argentina vary between 100 and 1200 km. Problems with cattle include the use of prods, dogs and whips, unsuitable systems of loading and unloading, prolonged journeys, lack of rest, water and food provision and a lack of trained workers. A recent report states that per animal slaughtered, the losses due to bruises, injection sites and other meat quality problems reached US$0.89 (IPCVA, 2007).

In order to supply internal and external markets, 45.4 million cattle are transported and slaughtered every year in Brazil. Most cattle are sold by farmers directly to slaughterhouses, and auction markets are common only for young cattle. The great majority of animals destined for slaughter are transported by road, in trucks. The Brazilian road network has more than 1.6 million km (Confederacão Nacional de Transportes, 2004) and the condition of the roads varies with geographical location; most of the roads are not paved and they are frequently in bad condition. This situation causes serious problems, mainly during the rain season, increasing the duration of travel and the numbers of broken trucks and road accidents. According to the truck drivers sometimes it is very difficult to continue the trip due to problems with mud and broken bridges.

An evaluation of vehicles arriving at 16 slaughterhouses in Brazil (Paranhos da Costa *et al.*, 2006) showed that most vehicles were in good condition; all of them had either iron or wooden structure, they were non-articulated rather than articulated trucks with three axles and 81% of them had a carpet of rubber covering the load compartment floor. The load capacity of the vehicles varied between 17 and 20 animals, considering animals of 400 kg live weight on average. However, there is a growing trend towards the use of articulated trucks, with one and two decks, given their high capacity of cattle transport (between 22 and 36 animals).

Although in Brazil distances travelled by animals can range from 20 to 2560 km, most commonly trips are up to 500 km, only in 35% of the trips the animals travelled more than 500 km (Paranhos da Costa *et al.*, 2006). In terms of journey duration, this is usually up to 8 h. However, because of road problems, in some cases the animals will be on the road for 60 h or more, without food and water. In general, the roads connecting the farms with main routes are stone roads often in bad condition, with obstacles, bumpy, and driving needs to be slow. There are no services along these roads (fuel or other facilities for either drivers or animals). The combination of the precarious condition of roads, long journeys and unfavourable weather conditions usually increases journey duration and the operational cost of the loads transported, as well as the risk of extreme stress to the animals during transport, sometimes resulting in death. Cattle from distant counties are bought for slaughter only when there are few cattle available in closer counties. The freight cost and the cattle value are usually important to determine if long-distance transport will be done or not.

Due to the size of Uruguay, journeys of more than 400 km are usually not undertaken by sheep or cattle, hence transport time rarely exceeds 8 h; only occasionally cattle are transported for up to 2 days (Ponce, 2006). It is a small homogenous country and this favours the different parties involved in the meat chain to reach agreements. Its geographical and climatic situation does not greatly vary and it is therefore possible to talk of one region with one set of standards. Meat export being one of the main activities in Uruguay, the meat processing industry is submitted to regular animal welfare control audits and studies have been done in order to establish the critical points during handling, transport and slaughter (Castro and Robaina, 2003; Huertas and Gil, 2003; Bianchi and Garibotto, 2004). These reports show that one of the critical points is the use of aggressive methods during driving of the animals, especially at stunning. Structural conditions of trucks and loading–unloading facilities were described as in good to fair condition and transport times as short (mean 331 km, range 200–600 km).

In Chile, due to the impact of transport of live animals for slaughter on animal welfare, meat quality and economic losses, studies funded by the National Commission for Science and Technology (FONDECYT, 1998, 2001, 2005) have been undertaken over the last 10 years in order to quantify the problem and propose appropriate solutions for the case of cattle and, recently, sheep. The results of the various studies carried out have been published (Gallo *et al.*, 1995, 2000, 2001, 2003a,b,c, 2005; Gallo, 2005; Tadich *et al.*, 2000, 2003, 2005; Amtmann *et al.*, 2006). Roads and vehicles have been improved, but there are still many unpaved bumpy roads connecting farms with the main roads. Although most common distances travelled by cattle vary between 100 and 800 km (3–12 h), there are still some long transport journeys (14 h and up to 48 h including 24 h ferry crossings, Aguayo and Gallo, 2005, 2006; Manríquez and Gallo, 2005) which involve 20% of the cattle slaughtered. Stocking densities used in cattle transport are high (455 kg/m^2) compared to EU standards, especially considering the frequent 12 h journeys (Gallo *et al.*, 2005). Within trucks, cattle are usually not segregated in different compartments, except in the case of the long trailers that carry 20–24 cattle, which are divided in two compartments. A survey of infrastructure and loading conditions on farms, of vehicles arriving with cattle at the main slaughterhouses and of handling before slaughter in these is being carried out in conjunction with the Association of Chilean slaughterplants (FIA, 2005; Cáraves et al., 2006; Strappini et al., 2006). Bruises and high-meat pH are the main problems as a result of long journeys in the trucks and in lairage (Gallo et al., 2003c). According to the results of concentrations of blood constituents related to stress, long journeys (especially over 24 h) are detrimental for animal welfare and there is no beneficial effect on the welfare of the animals of a long lairage time at the abattoir afterwards (Tadich et al., 2000, 2003, 2005).

According to Delgado (2006) in Colombia cattle have to travel on unpaved roads and then on main roads to get to the larger urban centres of consumption. Usually these animals are transported on wooden-bodied trucks with a capacity of 10–12 heads; in many cases this capacity is exceeded by as much as 20%. Journeys of more than 8 h are common, because production areas are far from consumption centres and often roads are closed or trucks break down (Ospina Restrepo, 1995). The presence of several middlemen delays transport and lairage, with journeys often reaching up to 24 h. This situation is reduced in modern slaughterhouses, as these buy directly from farmers and pay by yield. Despite the fact that Law 84 (República de Colombia, 1989) regulates the transport conditions, this is not always followed and it is observed that drivers tend to avoid stopping the vehicle for inspection of the animals on the journey. Current improvements include a guide for good handling practices in slaughterhouses (Ministerio del Medio Ambiente, 2002).

Due to the geography of Ecuador, livestock trade varies according to different regions (Gallego, 2006). The large Andean cities are mainly supplied with cattle from the coastal plains, which travel for over 12 h to the abattoirs. Some cattle bred in the sierra slopes, also slaughtered in the main cities, travel no more than 8 h. In general, the cattle are traded through cattle dealers (introducers) who gather animals from different farms. Usually cattle dealers also own the vehicles for livestock transport and load as many animals as possible, because their profit depends on the numbers of animals sold. In the case of farmers transporting their own animals directly to slaughterhouses, their main worry is just getting the animals to their destination alive. Due to the fact that there is not enough awareness among consumers about the transportation of animals and their welfare, the issue is not an influence on the market and not much attention is given to it. There is also legal and illegal import of live cattle from Perú to Ecuador. One main reason for this is that whenever cattle prices are low in Perú, this country sees the Ecuadorian market as a price stabilizer. In these cases animals are subjected to long journeys, sometimes 18–24 h to reach the border. It is not uncommon that some animals die before reaching the border. At the border the animals are moved into an Ecuadorian truck before continuing the journey for another 8 h until reaching the main cities. Currently there is a project which aims for the use of a single vehicle (Ecuadorian, Peruvian or Colombian) for the entire transportation of the cattle, but this project is strongly opposed by Ecuadorian transporters (Gallego, 2006).

Cattle transportation is not an organized industry in Bolivia. Although there is legislation (Servicio Nacional de Sanidad Agropecuaria, 2002, 2005) regarding the quality of the vehicles and a registry of vehicles, embarkations and wagons used for the transportation of live animals is being implemented, the legislation is so far only partially followed (about 50% are registered) and it is rare to find vehicles fulfilling all the requirements. However, an enquiry (Gallego, 2006) revealed that truck own-ers are seeing a more profitable business in the transportation of meat instead of live animals, which could lead to improvements in animal welfare.

In Bolivia most cattle travel for no more than 8 h, but cattle coming from Beni or Santa Cruz travel an average of 1000 km and at least for 16 h without resting, feeding and in some cases not even drinking. There are also cattle coming from Andean regions closer to the abattoirs, but because of the road conditions as well as of the difficult terrain the cattle are transported for at least 6 h to their final des-tination. Transportation time can be frequently extended to 2–3 days because of the 'bloqueos' (road blockings); this is a common way of protesting against unfavourable socio-economic conditions in the Andean region. As a result of these delays there are not only economic losses but also a great amount of suffering (hunger, thirst, unrest, stress, lesions and exposure to extreme climatic conditions) for the animals being transported (Gallego, 2006). In the Chaco region the trucks transport between 15 and 17 cattle, but in the Altiplano (highlands) the same vehi-cles load up to 22 animals, increasing the stocking density. According to Gallego (2006) the enforcement of the legislation is weak due to a lack of knowledge of the law and also to the lack of resources and support for the enforcing authorities. As a result, it is common for animals that have died on the journey to be transported on to the slaughterhouse and enter the food chain. For drivers of livestock vehicles another problem encountered is the feral temperament of the cattle as, in most cases, they have had little handling or contact with humans.

In Bolivia the trekking or 'arreo' of cattle is also common (Gallego, 2006); here cattle are driven from the farm where they were raised up to the place where they will be traded. This is practised in areas where roads do not exist or are not suit-able for trucks, and also in the case of small farmers who take only a few animals to the market. Cattle traders gather the cattle, farm by farm, along established routes and walk for long hours sometimes even days, leading the cattle to fairs and markets. One famous cattle fair for cattle trekking is held at the city of Punata, which markets from 1000 to 2500 animals in a single day.

According to Ponce (2006), due to the high temperatures in Paraguay the ani-mals are transported mainly by night. The trucks have compartments for smaller groups of cattle. Distance travelled is usually around 500 km, but it is common that animals do not go directly to the slaughterhouse and pass through livestock markets in Asunción; hence the journeys can sometimes be increased up to 10 h. Most routes are paved, although sometimes not in good condition, some run through urban areas and very often they are being repaired and have delays. The Chilean market has been a good opportunity for Paraguayan beef and it has encouraged improve-ments in quality assurance, including the transport and handling of cattle.

In Perú (Gallego, 2006) beef is supplied mainly by fattened dual-purpose ani-mals instead of specialized beef breeds. A bovine commercial meat chain is devel-oping but there are still some discrepancies between some of the components resulting in lack of integration. The fattening farms are located near the big cities to be near to the slaughterhouses. Generally, animals slaughtered for meat are transported for less than 8 h, but there are also some longer journeys (Battifora and Adama, 2000). Vehicles used for transportation of animals are designed for the purpose. The best of them are destined for transportation to the slaughterhouses ('camales') located in big cities. Those that transport animals from local primary production centres to local or regional slaughterhouses are inadequate.

In Venezuela, distances between cattle producing areas and urban centres range from 4 to 18 h and depend on the weather conditions and also on the maintenance of the roads. Once the cattle are loaded into the vehicles, the animals do not go out of them until they are unloaded at the abattoir. From time to time the driver stops to see if there is a fallen animal, and if it is so, it is forced to stand up even by the use of cruel methods such as puncturing the legs or breaking the tip of the tail (Gallego, 2006).

Specific aspects of the transport of pigs

Although 30.6 million pigs are transported and slaughtered every year in Brazil, there are only a few reports on actual studies on the handling of pigs during transport and slaughter (Dalla Costa et al., 2005a,b,c). Pig production occurs both in integrated and independent farms. Transport distance can be up to 400 km. For long-distance transport it is common to use articulated trucks, most of them have 2–3 decks, with a load compartment divided in pens; the regular number of adult pigs per pen is 9–10, regardless of their body weight. There are a few trucks equipped with nipples, for drinking, but none have ventilation systems. Loading and unloading facilities are often mediocre or bad, trucks are often overloaded. Sometimes pigs are not immediately unloaded after arrival at slaughterhouses, and waiting times can be long (2 h) (Paranhos da Costa et al., 2006).

In Colombia, only 50% of pig farms are integrated. According to Delgado (2006) there are no studies that have measured the impact of animal welfare on meat quality and the death of pigs during the transportation process clearly indicates a lack of animal welfare in the period before slaughter.

In Chile, 80–90% of the pigs are from large producers that are vertically integrated and are sent directly to the slaughterhouses. The journeys are short (2–4 h) and mortality ranges between 0.1% and 0.8%. Large multideck trucks are used and ambient and handling conditions are monitored carefully to reduce losses (Becerra, 2006); animal welfare standards during transport and handling have been improved and much more modern technology is applied compared to ruminants and horses.

In Argentina, there are welfare problems in pigs due to prolonged journeys, overcrowding in the vehicles and a lack of trained personnel for transporting the load (Ponce, 2006).

Generally, the integrated pig farms in South America are close to the slaughterhouses and the transport conditions like animal density, loading time and number of animals per truck are carefully coordinated by the industry in order to reduce losses and comply with export markets. Nevertheless, it has been observed by various authors (Delgado, 2006; Gallego, 2006; Gallo, 2006a; Paranhos da Costa et al., 2006; Ponce, 2006) that in pigs destined to small communal slaughterhouses in most South American countries there is a great deal of mistreatment during transport; the use of electric prods and other physical aggression (kicking and use of sticks) is common during loading and unloading procedures. Animals also face a high risk of accidents when loaded and unloaded into the second and

third deck of the trucks, some of them fall off the trucks or the loading ramps during these procedures. This situation is worse when trucks do not park properly and
ramps are inadequate, leaving gaps between the load compartment and the ramp,
some animals also fall into this space. Pigs produced in smaller farms are transported in non-specialized vehicles (adapted for cattle), small trucks and pickups;
these vehicles not only transport pigs but also all sorts of merchandise. Overcrowding
is commonly observed causing a high degree of stress to the animals. Furthermore,
pigs are sometimes also mixed with cattle or goats, and therefore can be trampled
on and injured. In Bolivia there are festivities, such as 'San Juan' and New Year
that promote the consumption of pork meat and during these seasons pigs have to
travel at least 1000 km and 20 h from farms to abattoirs located at Andean cities.
The lack of protection against climate exposure provokes skin lesions, heat stroke,
suffocation and stress.

Specific aspects of the transport of poultry

Considering that the risk of death and bruising is high in poultry, broiler producers
and the industry in South America are concerned about the transport of broilers.
In the main poultry producing countries, broiler farms are integrated, located close
to the slaughterhouses and the transport is done mainly during night time and
closely monitored.

The risk of welfare problems related to heat stress in poultry is very high in
many regions of Brazil, where more than 8 billion broilers are transported each
year. Here transport is done mainly in non-articulated trucks with three axles,
which are usually loaded with three columns of crates with 5–7 layers (depending
of the road types and conditions). The vehicles do not have any drinking or ventilation systems; because of this the broilers which are transported in the central col-
umn of the load compartment usually face a high risk of hyperthermia on hot days.
On the other hand, those birds inside the crates, on both sides of the trucks' load
compartment, are more susceptible to problems on cold days. Broken crates are a
common problem, increasing the risk of body injuries to the broilers (Paranhos da
Costa et al., 2006). Considering that the risk of welfare problems is higher when the
broiler and crate densities increase, poultry farmers and industries use lower stocking densities per crate during hot days, dropping from 0.51 to 0.43 kg broilers/m^2
(Branco, 2004); also on hot days it is common practice in large commercial
companies to spray water over the load compartment to reduce the air and body
temperature.

According to the Brazilian Union of Poultry Producers (Uniao Brasileira de
Avicultura, UBA 2006), the biggest barrier for expansion of the activity is the logistics of transport, reporting that roads in bad condition increase the transportation
time (up to 40%). According to Costa (2006), who evaluated 85 trucks with 308,944
birds in two slaughterplants in the state of Sao Paulo, in one plant 70% of the
broilers were transported for 130–170 km (mean journey time was 3 h 25 min) and
30% for 250 km (mean time 5 h 37 min) and in the other plant 50% were transported for 50–90 km (average time 1 h 37 min) and 50% for 130 km (average time
3 h). Under these conditions the mortality rate was 0.83 and 0.31 in the first and

second plant, respectively. Another indicator of welfare in poultry is the frequency of bruises and in the study by Costa (2006) the figures (18% up to 44%) indicated a deterioration of welfare with increase of the distance travelled.

The big poultry companies of Colombia usually have their own slaughter-house, which allows them a great deal of working autonomy. Small-size and medium-size poultry farmers do not have vertical support infrastructure, and there-fore, they have to use different agents and processes. They hire the transport equip-ment (small trucks) with the corresponding wooden or plastic crates, and also hire the workers for loading and unloading, drive to the destination place and arrive there in accordance with the slaughter schedule at the plant, a journey which can last between 4 and 5 h. A truck can contain between 200 and 280 wooden crates, so that the minimum load capacity will be around 7.7 t (the crates weigh 7.5 kg each). It is uncommon to use bigger trucks because whenever there are ventilation problems, those animals located in the centre of the vehicle tend to die from asphyxiation (Delgado, 2006).

In Argentina, the main welfare problems in poultry arise from the high density in cages or crates, the system for collection in the sheds and the prolonged waiting time at the slaughterhouse (Ponce, 2006). In regard to poultry in Chile, Becerra (2006) indicates that the journeys are short (average 1.5 h) and mortality ranges between 0.1% and 0.5%; large multideck trucks are used and ambient and hand-ling conditions are monitored carefully to reduce losses. As with pigs, welfare stan-dards during transport and handling have been improved and modern technology is applied.

Specific aspects of the transport of small ruminants

Very little information on small ruminant transport in South America is avail-able. In Chile most common distances travelled by over 80% of the sheep are short (up to 300 km, Taruman and Gallo, 2006). The space availability found by Taruman and Gallo (2006) for lambs was 0.16–0.22 m²/lamb (4.55–6.14 lambs/m²) and 7.5% bruised carcasses were found; bruises were mainly small in extension and affected only the subcutaneous tissue. Even though distances were rather short, a relationship between distance travelled and bruising was observed. On the other hand, around 5% of sheep are submitted to very long-distance trans-port (up to 48 h including maritime ferry crossing). Carter and Gallo (2006) found 12.5% and 41% of bruises in lambs transported for 12 h and for 46 h (including ferry crossing); although no differences were found between the two groups in terms of pH, the values registered are considered high (5.8–6.1). It is important to state that these studies did not consider handling of the lambs on the farm during loading or at the slaughterhouse during unloading and lairage, factors which should be taken into consideration in future studies, especially con-sidering that light bruising could be related to handling more than transport. In the south of Argentina sheep also have to travel sometimes over long distances in a cold climate that is very windy (Ponce, 2006). In Bolivia there are fairs for different species such as llamas, vicunas, guanacos, pigs, sheep and goats, and it is common to see these animals driven by trekking (Gallego, 2006).

Specific aspects of the transport of other species

Although in Argentina 200,000 horses are transported for slaughter within the country every year (Ponce, 2006), the meat being mostly destined for export, there is no mention of studies describing horse transport. In the case of horses for slaughter in Chile (around 50,000 per year), Gallo *et al.* (2004) found that distances travelled are usually short (90% travelled less than 300 km and 10% over 300 km), but improvements in terms of conditions during transport and handling are needed. Trucks used for the transport of horses for slaughter are the same as for the transport of cattle, and horses are transported loose within the truck. Stocking density used for horses was found to be 355 kg/m^2 and at inspection during arrival mortality was 0.25%, fallen horses 1.5%, lame horses 1%, fractured horses 0.5% and wounded horses 0.25%. The effects of transport and pre-slaughter handling of horses on the concentrations of blood variables related to stress were studied by Werner and Gallo (2006), who found for a transport journey of 1 h, significant increases in the concentrations of lactate, glucose, cortisol, packed cell volume and CK activity during loading, transport and unloading, but the highest peaks for most variables were found in the stunning box, before stunning and during bleeding. Currently studies are being carried out in Chile regarding the transport and pre-slaughter handling of salmon (Gatica *et al.*, 2004).

Legislation on Animal Welfare and Transport

In some South American countries there are laws aiming to avoid cruelty to animals, avoid unnecessary suffering and protect animals in general (República del Paraguay, 1953; República de Colombia, 1989; Cámara de Diputados, 1998; ARGENTINA, 1999a; URUGUAY, 2000; República del Perú, 2006). There is also legislation regarding the transport of animals for consumption in most countries but it deals mainly with sanitary requirements and public health issues (cleanliness of vehicles, animal health inspection ante-mortem and carcass/meat inspection post-mortem), rather than the welfare of animals (SENASA, 1974; Comisión del Acuerdo de Cartagena, 1983; República de Colombia, 1989; Ministerio de Agricultura, 1995; Servicio Ecuatoriano de Sanidad Agropecuaria, 1996, 2006; Ministerio de Agricultura, 1997; ARGENTINA, 1999b; SENACSA, 2000; Servicio Nacional de Sanidad Agropecuaria, 2002, 2005; Servicio Nacional de Seguridad Agropecuaria, 2004; República de Venezuela, 2006). In Brazil there is no specific legislation regulating the transport of farm animals, although governmental agencies and big slaughterhouse companies are aware of the need for improvement and most of them know OIE recommendations (Paranhos da Costa *et al.*, 2006). Nevertheless, in countries where there is legislation on the transportation of animals, this includes aspects related to the vehicle structure, the stocking density, inspections required, conditions for the journey and others, with similar recommendations to those of the OIE. There are bodies controlling the movement and transport of cattle, but again they deal mainly with ownership and taxes documentation, traceability and avoiding the transmission of diseases, rather than welfare of the animals. When there is a maximum journey time expressed in the regulations, this can be as high as 24 (Ministerio de Agricultura, 1997) or 36 h (SENACSA, 2000). Slaughterhouse regulations also indirectly take into account animal welfare, by including aspects such as appropriate unloading structures, lairage pens and races designed according to animal behaviour, water access for the animals, proper handling and stunning methods (República Federativa de Brasil, 1952; Comisión del Acuerdo de Cartagena, 1983; República de Venezuela, 1996; Servicio Ecuatoriano de Sanidad Agropecuaria,1996; República del Perú, 1998; Servicio Nacional de Sanidad Agropecuaria, 2001, 2002, 2004; Ministerio de Agricultura, 2004; URUGUAY, 2004). Again, the main problem seems to be that there is not much enforcement of the existing legislation and that compliance with regulations is commonly overlooked. Cultural problems, lack of motivation and education of transporters, authorities with scarce human resources, lack of trained personnel and lack of physical resources and infrastructure are some of the reasons making law implementation and enforcement difficult (see Chapter 5, this volume, for more on enforcement).

In South American countries, the institutions that deal with, and control, all matters related to animal health are under the Ministry of Agriculture (see Table 10.2).

Although these institutions have to deal mainly with phyto and animal health actions (protection of plant and livestock health, quarantine, food safety, public health, environmental risks), in most cases, they also have to deal with meat exports, and therefore animal welfare issues have also been handed to them for their development (Rojas *et al.*, 2005). At present these institutions are also in charge of the observance of animal welfare during production, transport and slaughter. In most cases this is an extra task which the same personnel are expected to handle, usually with no extra financial resources.

Measures are being taken by the government of many South American countries (Ministry of Agriculture, Official Veterinary Services), universities (educational programmes, research projects) and also by the private sector (meat industry associations) in order to disseminate animal welfare issues and create an awareness of the importance of good animal handling practices and animal welfare, not just to prevent economic losses but also on ethical grounds. Efforts are being made by the governments in order to disseminate the OIE recommendations for transport

Table 10.2. South American institutions responsible for animal health-related matters.

Argentina	Servicio Nacional de Sanidad Agropecuaria (SENASA)
Bolivia	Servicio Nacional de Sanidad Agropecuaria e Inocuidad Alimentaria (SENASAG)
Brazil	Serviço de Inspeção Federal (SIF)
Chile	Servicio Agricola y Ganadero (SAG)
Colombia	Instituto Colombiano Agropecuário (ICA)
Ecuador	Servicio Ecuatoriano de Sanidad Agropecuaria (SESA)
Paraguay	Servicio Nacional de Calidad y Salud Animal (SENACSA)
Peru	Servicio Nacional de Sanidad Agropecuaria (SENASA)
Uruguay	Dirección Nacional de Sanidad Ganadera (DNSG)
Venezuela	Servicio Autónomo de Sanidad Agropecuaria (SASA)

and humane slaughter, and to either produce or complement existing regulations with the standards of the World Organisation for Animal Health (OIE, 2005). In many cases it will be the consumers themselves who will demand ethical products, produced by taking animal welfare into account. In others it will be the supermarkets and fast food stores, as in many developed countries. At present quantity, quality and safety of the meat are much prioritized over animal welfare by consumers in most South American countries.

Although in most South American countries there is some kind of legislation indirectly intended to protect animal welfare in general, during transport and slaughter, it can be concluded that the main problem is not the lack of legislation, but that enforcement of it is lacking.

Education and Training

Animal welfare is an issue that, for ethical reasons, should be the responsibility of all human beings, but particularly so of veterinarians. However, it is noteworthy that, with the exception of only a few veterinary schools in South America (in Argentina, Brazil, Colombia and recently in Chile), the veterinary curriculum does not include any specific courses on this matter and the subject is only included collaterally within other courses (Gallo, 2006b). Hence, there is an urgent need for education and training in animal welfare in veterinarian and agricultural schools, which should include all levels: formal education (postgraduate studies) down to informal training (courses, workshops, seminars, etc.). Trained veterinarians are required for the technical introduction of animal welfare in the activities of public institutions, but also for the implementation of animal welfare on a daily basis in farm activities.

The governments in some countries (Argentina, Brazil, Chile, Paraguay, Uruguay) are conscious of the importance of animal welfare for trade and national market purposes, meat quality as well as for ethical reasons, and are working on it to improve animal welfare. These countries have recently created animal welfare committees, which include members of governmental institutions (usually the same ones that have delegates to OIE), academic staff and private stakeholders, and are an instance for the discussion of animal welfare subjects. These committees are working on the OIE guidelines for animal transport and slaughter (OIE, 2005), in order to adapt them to local conditions.

There are some groups of researchers working in the subject in Brazil (Universidad Nacional Estadual de Sao Paulo, meat industries, transporters and producers), Chile (Universidad Austral de Chile, together with the association of meat plants and the Fundación para la Innovación Agraria-FIA, producers and transporters), Uruguay (Universidad de la República, Instituto Nacional de Investigaciones Agrarias, Instituto Nacional de Carnes, rural associations and transporters) and Argentina (Instituto de Promoción de Carne Vacuna Argentina, Universidad de El Salvador) universities, ministerial institutions and stakeholders are working in line with the recommendations of the OIE in order to train people handling animals in relation to transportation and humane slaughter. Manuals, technical guides, CDs and other educational material on animal welfare and handling of animals on farms, markets, during transport and at slaughterhouses are being developed in Argentina (Giménez Zapiola, 2006), Chile (Ministerio de Agricultura, 2005) and Uruguay (Castro and Robaina, 2003). The issue is also being disseminated through meetings with producers, farmers, transporters and slaughterhouse operators.

Plants exporting meat undergo periodic audits to verify their competence to receive livestock that has been transported in a suitable way and slaughtered humanely; in most cases this is controlled by governmental bodies from the Ministry of Agriculture and/or Ministry of Health in the different countries.

However, there are other countries in South America (Perú, Bolivia, Venezuela, Ecuador) where the issue is practically unknown, and in most countries the farmers are not keen on animal welfare issues because they are seen as an imposition that will not return any economic benefits.

Discussion and Recommendations

Among people involved in the handling and transport of livestock in South America such as farmers, drivers, cattle market dealers, animal handlers and even veterinarians, there is insufficient knowledge about animal behaviour and about proper handling methods to avoid suffering of animals and poor meat quality. The evidence presented in this chapter strongly suggests that there is a need for training veterinarians and workers handling animals in the meat chain (farms, transport, slaughterhouses, etc.). This is in agreement with the recommendations arisen from the 18th Regional OIE Conference for the Americas held in December 2006. Animal welfare was one of the technical items discussed and all delegates agreed on some general recommendations, such as to promote animal welfare at all levels through training at veterinary schools, official veterinary services, producers, transporters and all those who handle animals (OIE, 2006). The results of training and/or regulatory actions so far, which are in many cases joint between academy, industry and government, have been positive, but they need to be expanded to a larger number of recipients. More veterinary and agriculture schools need to introduce the subject into the syllabus, which probably also requires training of lecturers (academic staff). In order to gain major progress, institutions which have groups of researchers that are already working together with the industry should be strengthened and stimulated to work in collaboration among them.

Infrastructure and sociocultural problems are common to most South American countries and differ from other regions. Hence, it is important that countries produce local research describing the current situation and that the critical points in terms of transport (journey lengths, conditions and others) are identified. Scientific evidence should be published in worldwide journals, and not only in local bulletins as is mostly the fact now. In order to enhance the publication of scientific papers, researchers should seek expertise from countries in more developed regions where appropriate. Furthermore, regional scientific evidence should be used as a basis to find solutions to local problems and also for the introduction of new legislation.

In most of the South American countries included in this study there is also an important role to be played by the government and veterinary services. Legislation in South American countries needs to be more effectively implemented. Although in

most countries there is legislation directly or indirectly related to animal transportation, the main problem detected is that there is not enough enforcement. To do so,
the authorities responsible for its enforcement should be provided with the necessary
resources. Also, the institutions in charge of animal welfare in each country should
introduce and implement OIE guidelines on animal transport and slaughter.

Generally speaking, there is within South America an increasing awareness
among consumers and young people about animal welfare, but the issue is by no
means as important as in more developed countries, because emphasis has to be
given to other priority needs (human welfare, health, poverty, education).
Nevertheless, at present there seems to be a good opportunity to set up technical
and educational projects to improve animal welfare during transport and slaughter,
in countries from the southern cone, that are trading meat (Argentina, Brazil,
Chile, Paraguay and Uruguay), because of the relationship between animal welfare, product quality and market demands from more developed countries. This
should be used as an advantage to more readily promote animal welfare.

Acknowledgements

The authors wish to acknowledge the collaboration of Dr Héctor Delgado, Dr Juan
F. Gallego, Dr Mateus Paranhos da Costa and collaborators and Mónica Ponce
who collected the information for South American countries, other than Chile,
included in this study.

References

ABCZ (2005) Associação Brasileira de Criadores de Zebu. Available at: www.abcz.org.br
Aguayo, L. and Gallo, C. (2005) Tiempos de viaje y densidades de carga usadas para bovines transporta-
    dos vía maritima y terrestre desde la región de Aysén a la zona centro-sur de Chile. XII Congreso
    Latinoamericano de Buiatría y VII Jornadas Chilenas de Buiatría, Valdivia, Chile, 15–18 de
    Noviembre.
Aguayo, L. and Gallo, C. (2006) Densidad de carga y comportamiento de bovinos transportados vía
    marítima desde Puerto Chacabuco a Puerto Montt, Chile. Submitted to XX Congreso
    Panamericano de Ciencias Veterinarias y 14° Congreso Chileno de Medicina Veterinaria,
    Santiago, Chile. 12–16 Nov.
Amtmann, V.A., Gallo, C., Van Schaik, G. and Tadich, N. (2006) Relaciones entre el manejo ante-
    mortem, variables sanguíneas indicadoras de estrés y pH de la canal en novillos. *Archivos de
    Medicina Veterinaria* 38(3), 259–264.
ANUALPEC (2005) *Anuário da Pecuária Brasileira*. Instituto FNP, São Paulo, Brazil, 340 pp.
ARGENTINA (1999a) Ley 14,346. Prohibición de malos tratos y actos de crueldad. Available at:
    http://infoleg.mecon.gov.ar
ARGENTINA (1999b) Resolución N°097/1999 sobre el transporte animal. Comisión de Agricultura
    y Ganadería.
Battifora Villa-Garcia, L.E. and Adama Pilla, J. (2000) *Análisis descriptivo del manejo del ganado bovino de
    carne desde su embarque en distintas provincias del Perú hasta su llegada y posterior proceso en Centros de Beneficio
    en Lima*. Estudio auspiciado por la FAO y Humane Society Internacional, con la Coordinación
    de la Asociación Amigos de los Animales, Perú.

Becerra, R. (2006) Animal transport: the situation in Chile (pork and poultry). Presentation at seminar: 'Animal welfare in Chile and the EU: experience and assessment in animal transport', held at Universidad Austral de Chile, Valdivia, Chile, 10 November.

Bianchi, G. and Garibotto, G. (2004) Bienestar animal: relevamiento de puntos críticos en Uruguay. *Serie Técnica N°37*, Instituto Nacional de Carnes (INAC) y Facultad de Agronomía, Paysandú, Uruguay, pp. 40.

Branco, J.A.D. (2004) Manejo pré-abate e perdas decorrentes do processamento do frango de corte. In: *Conferência Apinco de Ciência e Tecnologia Avícolas, 2004*. Santos. Anais . . . Campinas: FACTA, 2:129–142.

Camara de Diputados (1998) Proyecto de ley sobre protección de animales. *Diario Oficial* No 1977, 03/06/98

Cáraves, M., Gallo, C., Strappini, A., Aguayo, L, Allende, R., Chacón, F. and Briones, I. (2006) Evaluacion del bienestar animal de bovinos durante el manejo ante mortem en seis plantas fae-nadoras en Chile. Congreso XXXI de la Sociedad Chilena de Producción Animal, 18–20 de Octubre de 2006, INIA Quilamapu, Chillàn, Chile.

Carter, L. and Gallo, C. (2006) Efectos del transporte prolongado terrestre marítimo sobre pérdidas de peso vivo y algunas características de la canal en corderos. XX Congreso Panamericano de Ciencias Veterinarias y 14ª Congreso Chileno de Medicina Veterinaria, Santiago, Chile, 12 al 16 de Noviembre de 2006.

Castro, L.E. and Robaina, R.M. (2003) Manejo del ganado previo a la faena y su relación con la cali-dad de la carne. Serie de Divulgación N°1, Instituto Nacional de Carnes (INAC), Uruguay.

Comisión del Acuerdo de Cartagena (1983) *Norma y Programa Subregional Sobre Tecnologia, Higiene e Inspección Sanitaria del Comercio de Ganado Bovino para Beneficio, Mataderos y Comercio de Carne Bovina, Decisión 197* Comisión del Acuerdo de Cartagena, Lima, Perú.

Confederação Nacional de Transportes (2004) Available at: www.cnt.org.br

Costa, F.M.R. (2006) Influência das condições de pré-abate no bem estar de frangos de corte. 2006. 73p. Dissertação de Mestrado – Faculdade de Ciências Agrárias e Veterinárias, Universidade Estadual Paulista, Jaboticabal, Brazil.

Dalla Costa, O.A., Paranhos da Costa, M.J.R., Ludke, J.V., Faucitano, L., Peloso, J.V., Coldebella, A., Holdefer, C., Dalla Roza, D., Ventura, L.V. and Triques, N.J. (2005a) *Efeito do transporte e tempo de jejum durante o manejo preé abate sobre a qualidade da carne dos suinos, contuido estomacal e lesoes de úlcera esófago-gástrica*, Comunicado Técnico 405. EMBRAPA, Concordia SC, Brazil.

Dalla Costa, O.A., Paranhos da Costa, M.J.R., Ludke, J.V., Faucitano, L, Peloso, J.V., Coldebella, A., Holdefer, C., Dalla Roza, D., Ventura, L.V. and Triques, N.J. (2005b) *Efeito do manejo pre abate e da posicao do box dentro da carroceria sobre o perfil hormonal dos suinos*, Comunicado Técnico 406. EMBRAPA, Concordia SC, Brazil.

Dalla Costa, O.A., Paranhos da Costa, M.J.R., Ludke, J.V., Faucitano, L, Peloso, J.V., Coldebella, A., Holdefer, C., Dalla Roza, D., Ventura, L.V. and Triques, N.J. (2005c) *Efeito do época do ano, modelo da carroceria e posicao des animais no carroceria sobre qualidade da carne dos suinos*, Comunicado Técnico 407. EMBRAPA, Concordia SC, Brazil.

Delgado, H. (2006) Transport of animals in Colombia. Report for the World Society for the Protection of Animals.

FAO (2003) Production of meat and share in the world. 2003. Available at: http://fao.org/statistics/yearbook/vol_1_1/pdf/b02.pdf

FAO (2005) Production statistics, live animals. Available at: http://faostat. fao.org/site/568

FIA (2005) *Diagnóstico e implementación de estrategias de bienestar animal para incrementar la calidad de la carne de rumi-antes*, de responsabilidad de la Asociación de Plantas Faenadoras de Carne y Frigoríficos, colabora-dores Universidad Austral de Chile, U. de la Frontera y Pontificia U. Católica de Chile. Proyecto FIA – PI – C-2005-1-P-010, 2005 – 2009. Fundacion para la Innovacion Agraria, Ministerio de Agricultura, Santiago, Chile.

FONDECYT (1998) *Efectos del transporte y ayuno previo al sacrificio sobre la producción cuantitativa y cualitativa de carne de bovinos*. PROYECTO 1980062, 1998–2000. Fondo Nacional para el Desarrollo de la Ciencia y Tecnologia, Caracas, Venezuela.

FONDECYT (2001) *Efecto de diferentes condiciones de transporte, ayuno y manejo de bovinos previo al faenamiento, sobre el bienestar animal y la calidad de la carne.* PROYECTO 1010201, 2001–2004. Fondo Nacional para el Desarrollo de la Ciencia y Tecnologia, Caracas, Venezuela.

FONDECYT (2005) *Efectos del transporte prolongado de bovinos y ovinos sobre el bienestar animal y la calidad de la carne.* PROYECTO1050492, 2005–2008. Fondo Nacional para el Desarrollo de la Ciencia y Tecnologia, Caracas, Venezuela.

Gallego, J.F. (2006) Farm Animal Transportation in the Andean Region. Report for the World Society for the Protection of Animals.

Gallo, C. (2005) Transporte de ganado: situación nacional y recomendaciones internacionales. En: Gonzáles, G., Stuardo, L, Benavides, D. and Villalobos, P. (eds) *Actas del Seminario La Institucionalización del bienestar animal, un requisito para su desarrollo normativo, científico y productivo.* Comisión Europea, Gobierno de Chile (SAG) y Universidad de Talca, Santiago, Chile, pp. 83–99.

Gallo, C. (2006a) Chile Specific Status Report with Key Focus on Animal Transport. Report for the World Society for the Protection of Animals.

Gallo, C. (2006b) Animal Welfare in the Americas. *Compendium of Technical Items Presented to the International Committee or to Regional Commissions of the OIE.* OIE, Paris, France, pp. 151–158.

Gallo, C., Carmine, X., Correa, J. And Ernst, S. (1995) Análisis del tiempo de transporte y espera, destare y rendimiento de la canal de bovinos transportados desde Osorno a Santiago. En: *Resúmenes de la XX Reunión Anual de la Sociedad Chilena de Producción Animal (SOCHIPA).* Coquimbo, Chile. pp. 205–206.

Gallo, C., Pérez, S., Sanhueza, C. and Gasic, J. (2000) Efectos del tiempo de transporte de novillos previo al faenamiento sobre el comportamiento, las pérdidas de peso y algunas características de la canal. *Archivos de Medicina Veterinaria* 32, 157–170.

Gallo, C., Espinoza, M. and Gasic, J. (2001) Efectos del transporte por camión durante 36 horas con y sin período de descanso sobre el peso vivo y algunos aspectos de calidad de carne bovina. *Archivos de Medicina Veterinaria* 33, 43–53.

Gallo, C., Teuber, C., Cartes, M., Uribe, H. and Grandin, T. (2003a) Mejoras en la insensibilización de bovinos con pistola neumática de proyectil retenido tras cambios de equipamiento y capacitación del personal. *Archivos de Medicina Veterinaria* 35, 159–170.

Gallo, C., Altamirano, A. and Uribe, H. (2003b) Evaluación del bienestar animal durante el manejo de bovinos previo al faenamiento en una planta faenadora de carnes. *VI Jornadas Chilenas de Buiatría,* Pucón, Chile, 26–28 Noviembre.

Gallo, C., Lizondo, G. and Knowles, T.G. (2003c) Effects of journey and lairage time on steers transported to slaughter in Chile. *Veterinary Record* 152: 361–364.

Gallo, C., Caraves, M. and Villanueva, I. (2004) Antecedentes preliminares sobre bienestar en los equinos beneficiados en mataderos chilenos. Resúmenes Seminario. *Producción animal de calidad contemplando bienestar animal,* pp. 70–77.

Gallo, C., Warriss, P., Knowles, T., Negrón, R., Valdés, A. and Mencarini, I. (2005) Densidades de carga utilizadas para el transporte comercial de bovinos destinados a matadero en Chile. *Archivos de Medicina Veterinaria* 37, 155–159.

Gatica, C., Gallo, C. and Vasquez, S. (2004) Determinación de constituyentes sanguíneos durante el transporte de salmones a faena y su asociación a estrés. 13° Congreso Chileno de Medicina Veterinaria, 4–6 de Noviembre, Valdivia, Chile.

Giménez Zapiola, M. (2006) *Manual de buenas prácticas ganaderas. Cámara Argentina de Consignatarios de Ganado.* Lara Producciones Editoriales, Buenos Aires, Argentina, pp. 63.

Huertas, S.M. and Gil, A.D. (2003) Efecto del manejo prefaena en la calidad de las carcasas bovinas del Uruguay. In: *XXXI Jornadas Uruguayas de Buiatría.* 11–13 Junio 2003, Paysandú, Uruguay.

IPCVA (2007) *Evaluación de las prácticas ganaderas en bovinos que causan perjuicios económicos en plantas frigoríficas de la República Argentina (año 2005),* Cuadernillo Técnico N°3. Instituto de Promoción de la Carne Vacuna Argentina, Buenos Aires, Argentina, pp. 35.

Manríquez, P. and Gallo, C. (2005) Efecto del transporte marítimo y terrestre prolongado de novillos, sobre la presencia de contusiones en canales, el pH, glucógeno muscular y color de la carne. En: *Resúmenes de la XII Congreso Latinoamericano de Buiatría,* Universidad Austral de Chile, Valdivia, pp. 348–349.

Ministerio de Agricultura (1995) *Decreto 22 Sobre el Control del Transporte Animal.* República del Perú, Lima, Perú.

Ministerio de Agricultura (1997) Modifica Decreto n° 240, de 1993, que aprueba reglamento general del transporte de ganado y carne bovina. Decreto N° 484.. Reglamento General de Transporte de Ganado Bovino y Carne. *Diario Oficial* 05 de Abril de 1997.

Ministerio de Agricultura (2004) Reglamento sobre estructura y funcionamiento de mataderos, cámaras frigoríficas y plantas de desposte y fija equipamiento mínimo de tales establecimientos. Decreto N° 61. *Diario Oficial* 09 de Septiembre de 2004. Ministerio de Agricultura, Chile.

Ministerio de Agricultura (2005) Guía técnica de bucoaenas prácticas en bienestar animal para el manejo de bovinos en predios, ferias, medios de transporte y plantas faenadoras. Comisión Nacional de Buenas Prácticas Agrícolas, Ministerio de Agricultura, Chile.

Ministerio del Medio Ambiente (2002) *Guia Ambiental para las Plantas de Beneficio del Ganado,* Mayo 2002. Available at: www.ideam.gov.co/apc-aa/img_upload/ccf8a2325cc9292dc1cf8549cc72e8d8/guia_plantas_ganado.pdf

OIE (2005) *Terrestrial Animal Health Code,* Section 3.7 Animal Welfare. World Organization for Animal Health, Paris, France.

OIE (2006) Informe Final (Recomendación N° 1, Bienestar Animal en las Américas), de la 18° Conferencia de la Comisión Regional de la OIE para las Américas, Florianópolis, Brasil, 28 de Noviembre al 2 de Diciembre de 2006, pp. 52.

Ospina Restrepo, J.M. (1995) El Cebú No. 281 Dic. – Enero 1995, p.32.

Paranhos da Costa, M.J.R., Dalla Costa, O.A., Cruz Barbalho, P., Biagiott, D., Panin Ciocca, J.R., Naves, J.E., Quintillano, M.E., Ribeiro Gabriel, J.E., Naves, G. and Das Barbosa Silveira, I. (2006) The transport of farm animals in Brazil: First report. Report for the World Society for the Protection of Animals.

Ponce, M. (2006) Transport of animals in Argentina, Uruguay and Paraguay. Report for the World Society for the Protection of Animals.

República de Colombia (1989) *Ley 84: Estatuto Nacional de Protección Animal.* Republica de Colombia, Bogota, Colombia.

República del Paraguay (1953) *Decreto Ley N° 67/1953.* República del Paraguay, Asncion, Paraguay.

República del Perú (1998) Resolución Suprema 007–98-SA. Reglamento sobre salud y control de bebidas y alimentos. República del Perú, Lima, Perú.

República del Perú (2006) *LEY 27265. Ley de protección a los animales domésticos y a los animales silvestres mantenidos en cautiverio.* República del Perú, Lima, Perú.

República de Venezuela (1996) *Reglamento General de Alimentos.* República de Venezuela, Caracas, Venezuela. Available at: www.gobiernoenlinea.ve/legislacion-view/sharedfiles/reglamentogeneralalimentos.pdf

República de Venezuela (2006) Decreto 078 del 26 de Mayo de 2006. Creación de la Guía Única de Despacho para la Movilización de Animales, Productos y Subproductos de Origen Animal. *Gaceta Oficial de la Republica Bolivariana de Venezuela.* Caracas, 2006.

Repúbla Federativa de Brasil (1952) Regulamento da inspeção industrial e sanitária de produtos de origem animal – RIISPOA. Available at: www.fabricadoagricultor.pr.gov.br/arquivos/File/RISPOA.pdf

Rojas, H., Stuardo, L. and Benavdes, D. (2005) Políticas y prácticas de bienestar animal en los países de América: estudio preliminar. *Revue Scientific et Technique de Office International de Epizooties* 24, 549–565.

SENACSA (2000) *Resolución de SENACSA N°256/2000.* Sevicio Nacional de Calidad y Salud Animal, Asuncion, Paraguay.

SENASA (1974) *Resolución Suprema N° 480–74-AG Código sanitario para el transporte interno.* Servicio Nacional de Sanidad Agraria, República del Perú, Lima, Perú.

Servicio Ecuatoriano de Sanidad Agropecuaria (1996) *Reglamento a la ley sobre mataderos inspección, comercialización e industrialización de la carne.* Available at: www.sesa.gov.ec/leyes/matadero3.htm

Servicio Ecuatoriano de Sanidad Agropecuaria (2006) *Marco Jurídico para la Producción y Comercialización de Carne.* Servicio Nacional de Seguridad Agropecuaria (SESA). Ministerio de Agricultura y Ganadería del Ecuador, Ecuador.

Servicio Nacional de Sanidad Agropecuaria (2001) *Resoluciones Administrativas No. 087/01 y 088/01. Requerimientos sanitarios para el transporte de animales, infraestructura y clasificación de mataderos, procesamiento, conservación y transporte de carnes y sobre la inspección ante y postmortem en mataderos.* Ministerio de Asuntos Campsinos y Agropecuarios, República de Bolivia.

Servicio Nacional de Sanidad Agropecuaria (2002) *Resolución Administrativa N° 156/02. Requerimientos sanitarios para el transporte de aves, infraestructura y clasificación de mataderos de aves, procesamiento, conservación y transporte de carnes.* Ministerio de Asuntos Campsinos y Agropecuarios, República de Bolivia.

Servicio Nacional de Sanidad Agropecuaria (2004) *Resolución Administrativa N°013/04. Inspección periódica de mataderos y registro sanitario.* Ministerio de Asuntos Campsinos y Agropecuarios, República de Bolivia.

Servicio Nacional de Sanidad Agropecuaria (2005) *Registro de Vehículos, Embarcaciones y Vagones Utilizados para el Transporte de Animales Vivos.* Servicio Nacional de Sanidad Agropecuaria (SENASAG). Ministerio de Asuntos Campesinos y Agropecuarios, República de Bolivia.

Strappini, A., Gallo, C., Cáraves, M., Navarro, G., Barrientos, A., Chacon, F., Briones, I. and Allende, R. (2006) Relevamiento preliminar del transporte de ganado bovino en chile: vehículos y manejo de los animales durante la descarga. Congreso XXXI de la Sociedad Chilena de Producción Animal, 18–20 de Octubre de 2006, INIA Quilamapu, Chillàn, Chile.

Tadich, N., Gallo, C. and Alvarado, M. (2000) Efectos de 36 horas de transporte terrestre con y sin descanso sobre algunas variables indicadoras de estrés en bovinos. *Archivios de Medicina Veterinaria* 32, 171–184.

Tadich, N., Gallo, C., Knowles, T., Uribe, H. and Aranis, A. (2003) Efecto de dos densidades de carga usadas para el transporte de novillos, sobre algunos indicadores sanguíneos de estrés. XXVIII Reunión Anual de SOCHIPA, Talca, Chile, 15–17 Octubre.

Tadich, N., Gallo, C., Bustamante, H., Schwerter, M. and Van Schaik, G. (2005) Effects of transport and lairage time on some blood constituents of Friesian cross steers in Chile. *Livestock Production Science* 93, 223–233.

Taruman, J. and Gallo, C. (2006) Contusiones en canales ovinas y su relación con el transporte. Submitted to XX Congreso Panamericano de Ciencias Veterinarias y 14° Congreso Chileno de Medicina Veterinaria, Santiago, Chile. 12–16 Noviembre.

Teseimazides, S.P. (2006) Efeitos do transporte rodoviário sobre a incidência de hematomas e variações de pH em carcaças bovinas. Dissertação de Mestrado – Faculdade de Ciências Agrárias e Veterinárias, Universidade Estadual Paulista, Jaboicabal-SP.

UBA (2006) União Brasileira de Avicultura. Available at: www.uba.org.br URUGUAY (2000) Decreto 82/2000 sobre protección animal.

URUGUAY (2004) Resolución N° 21 de la Dirección Nacional de Sanidad Animal. Voluntary Code of Practice.

Werner, M. and Gallo, C. (2006) Efectos del transporte y manejo pre-sacrificio sobre las concentraciones de algunos constituyentes sanguíneos relacionados con estrés en equinos. XX Congreso Panamericano de Ciencias Veterinarias y 14ª Congreso Chileno de Medicina Veterinaria, Santiago, Chile, 12 al 16 de Noviembre de 2006.

World Atlas (2007) World Atlas. Available at: www.worldatlas.com/webimage/countrys/samerica/lgcolor/bocolor.htm

The World Bank (2006) A simple table from the *World Development Indicators*, 2006. Available at: http://go.worldbank.org/RVW6YTLQH0

11 Asia

P.J. Li,[1] A. Rahman,[2] P.D.B. Brooke[3] and L.M. Collins[3]

[1]Social Sciences Department, University of Huston-Downtown, Huston, USA;
[2]Commonwealth Veterinary Association, Jayanagar, Bangalore, India;
[3]Compassion in World Farming, River Court, Mill Lane, Godalming, Surrey, UK

Abstract

This chapter brings together information gathered from a variety of sources in key countries across Asia (China, India, Pakistan, Bangladesh, Sri Lanka and the region of Taiwan), chosen to highlight a range of contexts in which live transport takes place and to examine the extent of welfare concerns regarding live animal transport and opportunities for change.

Meat consumption and live animal transport are increasing across Asia. In China, factors that may have tended to lead to a large and increasing live animal transport industry include:

- The large size of the country;
- Increasing urbanization;
- Increasing consumption of meat, especially among urban populations;
- Long distances between some significant areas of production and consumption;
- A preference for freshly killed meat; and
- Reduction in localized production.

Animals are transported from all over China to key centres such as Beijing, Shanghai and Hong Kong involving journeys of up to 2500 km. Animals are transported live for sale at wet markets or for slaughter in modern slaughterhouses such as at Shenzhen north of Hong Kong. China also has a developing live export trade to Muslim countries such as Malaysia, Jordan and Kuwait.

The same basic principles apply to live transport elsewhere in Asia. Distances travelled in Taiwan are shorter, but the same processes have occurred as the population has urbanized, become wealthier and meat consumption has increased. Most consumption is in the north including the capital Taipei; most production is in the centre and south and there is a demand for 'warm', freshly killed meat. On the other hand, imports and exports of live animals have been low since the outbreak of foot-and-mouth disease in 1997.

Live animal transport in India involves fewer animals than in China. Less meat is consumed, but consumption is increasing rapidly as the economy develops. An issue is the transportation, often long distance, of worn-out draught oxen and dairy cattle. Most Indian states have banned cattle slaughter for religious and moral reasons, so many of these animals are transported to Kerala and West Bengal or exported to Pakistan and Bangladesh.

©WSPA 2008. *Long Distance Transport and Welfare of Farm Animals*
(eds M.C. Appleby *et al.*)

In Bangladesh, animals from all over the country and those from India are transported to markets in the capital Dhaka.

Countries across Asia are developing legislation to protect the welfare of animals. Indian legislation includes rules which guide transport of poultry, pigs and cattle. For religious reasons, some Indian states have passed laws to prohibit transport of animals, especially cattle, for slaughter. China has not yet passed specific animal welfare legislation, but its Animal Husbandry Law (AHL) includes measures for conditions during transport, and there is also legislation relating to disease control during transport. In Taiwan, the Animal Protection Act includes legislation about transport, and Animal Transport Management Regulations were enacted in 2005, partly derived from World Organisation for Animal Health (OIE) standards.

Much needs to be done, both to pass legislation in line with the new OIE standards and to enforce that which is on the statute book. Research across Asia shows that, in addition to long journeys, animals are often subjected to long periods without food or water, rough handling and overcrowding. They are often carried in vehicles in which they cannot be protected from extremes of weather. There are many cases of actual abuse listed in this chapter, such as forced watering in China and the use of sharp sticks to goad animals in India. A top priority must be to reduce transport times. As journey length increases, animal welfare tends to decrease. Reducing journey times requires the development of local and regional slaughterhouses with high health and welfare standards together with an infrastructure of refrigerated transport. Markets for frozen and refrigerated meats need to be developed as higher living standards are attained.

Consumers need to be educated to demand high welfare food. The food industry needs to respond by demanding high standards of its suppliers, and governments by ensuring good legislation and enforcement. All those involved in animal transport and marketing need to be trained to a high professional standard and then treated and respected accordingly.

Introduction

Traditionally, most animals in Asia were raised and slaughtered locally. As countries have urbanized, and trading has developed, the distances that animals have travelled to market or to slaughter have increased.

The welfare of animals during transport depends on:

- The number of animals involved;
- The treatment they receive;
- The length and duration of the journeys;
- The time spent at market;
- The number of times they have to be loaded and unloaded;
- Levels of thermal stress, thirst and hunger;
- The quality of the vehicles and the roads;
- Design of loading and unloading facilities;
- Quality of lairages;
- Provision of food and water; and
- The skills and attitudes of the traders and hauliers.

This chapter aims to detail the current cultural, legal and trade factors in each country which impact on the long-distance live transport of animals for slaughter; the extent and welfare consequences of live animal transportation for slaughter; and potential opportunities for reducing or eliminating the negative effects of long-distance transportation for animal welfare.

China

Background

China ('China' here refers to the People's Republic of China including Hong Kong and Macau – the region of Taiwan is elaborated upon in a separate section 'Taiwan' of the chapter) covers 9.6 million km² – 6.5% of the world land area. The human population of over 1.3 billion – representing one-fifth of the total world population – is unevenly distributed on the mainland – the south-west, north-west and north-east regions, though large, are sparsely populated. The country's east, south-east and central parts hold more than 60% of the population. Urban residents consume more animal protein than their rural counterparts. As the urban community grows, the demand for farm animal products simultaneously increases. China's geographical location permits extensive international exchanges and a variety of possible modes of transport for these exchanges (Figs 11.1 and 11.2). It

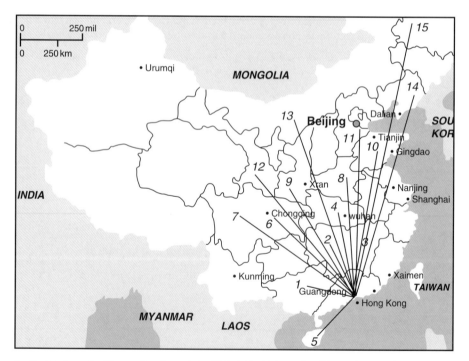

1 = Guangdong; 2 = Hunan; 3 = Jiangxi; 4 = Hubei; 5 = Hainan; 6 = Chongqing; 7 = Sichuan; 8 = Henan; 9 = Shanxi; 10 = Shandong; 11 = Hebei and Beijing; 12 = Gansu; 13 = Inner Mongolia; 14 = Liaoning; 15 = Heilongjiang.

Fig. 11.1. Transport routes to the Pearl River Delta Region, China.

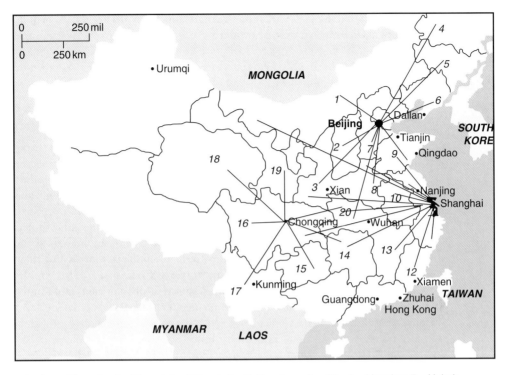

1 = Inner Mongolia; 2 = Shanxi; 3 = Sh'anxi; 4 = Heilongjiang; 5 = Jilin; 6 = Liaoning; 7 = Hebei;
8 = Henan; 9 = Shandong; 10 = Anhui, Jiangus; 11 = Zhejiang; 12 = Fujian; 13 = Jiangxi; 14 = Hunan;
15 = Guizhou; 16 = Sichuan; 17 = Yunnan; 18 = Qinghai; 19 = Gansu.

Fig. 11.2. Live transport to Beijing, Shanghai and Chongqing regions, China.

shares common borders with 14 countries – for instance, road links with Vietnam, Pakistan, North Korea, Russia and Mongolia. The shipping lanes connecting mainland China with Taiwan, Japan, South Korea and North America are among the busiest commercial routes in the world.

With its large population, China is the world's biggest market for meat and dairy products (Xu, 2005). Since 1990, the Chinese animal husbandry industry has grown by approximately 9% year-on-year. In 2005, China produced 77.5 million t of meat (in 2005, China produced 50 million t of pork, 10 million t of chicken, 7.2 million t of beef and 4.3 million t of mutton), representing 29.3% of the world's total (FAOSTAT, 2006) (Table 11.1.2A). The very large growth of China's livestock production underlies the scale of its live transport industry. China's live transport industry overshadows that of any other country. With the modernization of the Chinese livestock industry and the increasing utilization of modern slaughter technology, live transport is set to increase in quantity and extent.

The Chinese economy has increased by 9.4% per annum in the last quarter century. Livestock farming accounted for 34% of Chinese agricultural output in 2004 (China Animal Husbandry Industry Association, 2005b) though in some provinces and municipalities such as Beijing, Henan, Sichuan and Hebei, livestock accounted for more than 40% (China Animal Husbandry Industry Association, 2005b). Among the top ten animal producing provinces, only one (Xinjiang) is a traditional livestock farming region. Livestock farming households are no longer producing solely for local rural markets. Increasing concentration of production tends to lead to the shipment of large numbers of animals over long distances to remote large-scale slaughterhouses.

China's rapid urbanization and enhanced environmental awareness are influencing governments to phase out animal farming from the immediate suburbs of the nation's major metropolitan regions. Shanghai's animal farms are being relocated to Jiangsu and Anhui. Shenzhen, near Hong Kong, has begun to source live farm animals from nearby Jiangxi and Hunan. Beijing is moving egg producing farms to its remote suburbs. This trend is set to spread to the medium and smaller urban centres. China has pursued an active strategy to evolve its existing farm animal stock by importing large numbers of breeding stock into the country. Simultaneously, it seeks to re-enter the world markets for live farm animals as the Middle Eastern countries are trying to diversify their supplying sources. All these developments have a direct impact on live transport in China.

China's animal farming industry is still primarily a peasant household-based production system, although this is likely to change with time. Ninety-five per cent of China's pig farms in 2003 were small-scale operations maintained by individual peasant families (China Animal Husbandry Industry Association, 2005a, pp. 225–226). China's dairy cow farming has grown most rapidly, yet the vast majority of farms (85%) are peasant-owned, small operations with an average of six dairy cows (China Animal Husbandry Industry Association, 2005a, pp. 227–228). Despite the fact that China's animal farming industry remains traditional in operation and scale, it has undergone considerable growth in the last quarter century.

In 2005, frozen meat, eggs and dairy exports from China were valued at US$5.2 billion. In comparison, live farm animal export earned US$326 million (China Animal Husbandry Industry Association, 2005a, p. 262). However, the live export industry is expanding its markets – since 2002, China has exported live animals to Malaysia (Anonymous, 2006), Jordan and Kuwait (Chaoyuan Xinfa Rep., Mongolia, 2006, personal communication).

The rapid growth of China's animal husbandry industry is correlated with the shift in eating habits of the Chinese nation. In mainland China, per capita meat consumption doubled to 47 kg between 1988 and 1998 and further increased to 60 kg in 2005 (China Animal Husbandry Industry Association, 2001; China Animal Husbandry Industry Information Network, 2006). City-dwellers consume much more than their rural counterparts – in Hong Kong, meat consumption per person reached 124 kg in 2000, whereas mainland Chinese peasants consumed 19 kg of meat on average (Yonggang, 2005). This suggests that livestock raised on peasant farms are mostly sold (rather than consumed by the household) and shipped to slaughterhouses or markets to supply urban residents. Overall increases in consumption will trigger greater volumes of live transport.

Animals are transported live to a number of destinations – wet markets (farmers' markets where assorted terrestrial and aquatic animals are sold live or are slaughtered in front of the customers), slaughterhouses and for national and international trade. China's wet markets are popular across the nation. Chinese consumers prefer meat from freshly slaughtered animals. Farm animals are shipped to these markets by a variety of vehicles, over varying distances, as discussed later in the chapter.

Slaughterhouses are the second major destination of China's live transport. For disease control and prevention, the Chinese authorities have in recent years called for centralizing animal slaughtering operations with modern slaughtering facilities. By the end of 2006, more than 30,000 slaughterhouses across the country had been designated as centralizing pig slaughtering facilities (*Nanjing Daily Correspondent*, 2007). The distance travelled to these centralized slaughtering facilities can be considerable. For example, Shenzhen's Ming Ji Food Processing slaughters chickens transported from Jilin, Shandong, Fujian and Henan in north-east, north and east China. Live chickens travel up to 2500 km from Jilin (near North Korea) to Shenzhen (just north of Hong Kong) (China State Quality Inspection and Quarantine Administration, 2006).

China's traditional slaughtering practices, where butchers slaughtered animals in the farmers' backyards, are being phased out. Although potentially providing for better welfare at slaughter, a major drawback of the new centralized slaughterhouses is the need for animals to travel long distances from where they were reared to reach the slaughterhouses. As the major metropolitan centres are closing animal farms or relocating them to nearby provinces, this also extends the journey time to slaughter.

In every Chinese province, farm animals are traded in wholesale markets before transport to fattening, breeding or slaughter facilities. Such markets may be used as centres for buying and selling animals by farmers from surrounding provinces – for example, Zhangbei Livestock Trading Centre, in Hebei province, is the biggest livestock trading venue in north China, attracting traders from 18 provinces (Chinese Cultivation Network, 2003).

All animals intended for the wet market are transported there live. Although many are slaughtered at the market, in front of the customers, some animals are sold live and are transported away from the market also. These animals could therefore endure two long journeys with just a few hours rest in between.

Context: legal and regulatory issues

Since the early 1980s, China has accelerated policy making and legislation in animal-related issue areas. For example, with the endorsement of the State Council, the Agriculture Ministry issued in 1980 a decision on accelerating the development of the nation's livestock farming. This decision reiterated the Communist Party Central Committee's decision to encourage diversified agricultural production (Agriculture Ministry, 1980). Government regulatory actions culminated in the enactment of the Animal Husbandry Law (AHL) in December 2005. The outbreak of SARS in 2003 in particular was an issue that motivated the Chinese authorities to deal more proactively with issues affecting animal and human health, including the animal farming industry. While animal farming-related policies, regulations, government decisions and laws are designed to ensure normal production and safeguard economic interests of the parties concerned, considerations of farm animal welfare are seen in a limited number of articles in some of these policy documents. No animal farming-related laws or regulations are completely devoted to animal welfare.

The Animal Husbandry Law (AHL) 2005

The AHL is the latest legislative accomplishment related to agriculture and the livestock industry in particular. The AHL was adopted on 29 December 2005. It took effect on 1 July 2006. Article 53 is devoted to live transport, stating that

> Livestock transport should conform to laws, administrative rules, and requirements regarding disease prevention set by the relevant State Council veterinary offices. Measures must be taken to ensure livestock safety and to provide the necessary space and drinking conditions for livestock in transport.

The AHL is similar to most other Chinese laws relating to non-human species, in that it does not lay down specific regulations to protect the welfare of animals in farming, trade and transport. The AHL is also potentially open to interpretation – for example, there are no stipulations as to what exactly 'necessary space and drinking conditions should be. No time limits are imposed on live transport under particular weather conditions or for specific farm animal species. The AHL is first and primarily a law serving the nation's economic interests, even though the Chinese government states that it understands the adverse affect of poor welfare on the economy and on human health.

Rules regarding railroad transport

The Rules Regarding Railroad Transport of Fresh Meat Products and Live Animals, 1 October 1994 (hereafter the Rules), enacted by the Communications Ministry, contain more detailed requirements regarding live transport (Railway Ministry, 1994). The Rules define fresh meat products and live animals as those goods in transport that call for special attention to prevent them from becoming rotten, decayed, diseased and deceased. In view of seasonality, fluctuating transport amounts and demands on speedy transport, the Rules call on the rail transport businesses to follow four priority principles. These are to give priority: (i) to fresh and live goods in transport planning; (ii) to purchasing and loading of fresh and live goods; (iii) to fresh and live goods arrival and delivery; and (iv) to train car assignments for animals and fresh goods. The Rules also suggest that express cargo trains or non-stop cargo trains should be planned and operated in sections where fresh and live products concentrate.

The Rules state that no live animals can be loaded from the departing station without a quarantine certificate. This requirement ensures that live animals are healthy and fit for the planned journey. The cars for shipping live animals must be clean and free from contamination by poisonous materials. They can be either covered or uncovered, but they must have windows. Cars for shipping cattle cannot have iron sheet flooring. Although the cars can have multiple layers, the number of animals permitted on each layer depends on the season, weather conditions, transport distance, size of the animals and the type of car used. One or two experienced persons should be on board for the journey and are responsible for providing feed and drinking water to the animals. They are also responsible for changing the water, keeping the animals cool and acting as security guards. Designated water stations along the railroad must provide water free of charge to cargo trains with live animals. No animal wastes can be discharged along the railroad except at designated stations and infected or dead animals must be handled in accordance with the rules laid down by the government's disease prevention departments. Shipping documents should be clearly labelled 'live animals' to draw the attention of those handling the shipment. No articles, however, address how the livestock should be handled during loading and unloading.

The rules are also designed to reduce journey times through the 'four priority' principles for live transport. This means that live animals receive priority consideration in transport planning, loading, shipping and unloading. Most noteworthy is the requirement that 'live animals' should be written or marked conspicuously on all transport documents to catch the attention of the transport processing personnel to ensure priority attention. Exporting firms are required to notify the quarantine and inspection agency of the departure city 24 h before loading of the livestock to ensure that the inspection and quarantine persons are ready. The Rules are designed to reduce mortality during transport. This has obvious economic benefit, but is also helpful to welfare.

Live transport in China is regulated by a number of other relevant laws and regulations, including:

- Quarantine Law Governing the Import and Export of Animals and Plants (1991);
- Railroad Law (1990);
- Regulations on Cargo Transport by Automobiles (2004);
- Administrative Measures of Inspection and Quarantine of Live Poultry Transported to Hong Kong and Macau (2001);
- Administrative Measures on Animal Quarantine (2002);
- Administrative Regulations on Pig Slaughtering (1997);
- Railroad Cargo Transport Regulations (1994);
- Administrative Regulations on Cargo Transportation on the Waterways (1997); and
- Regulations on International Ocean-Going Cargo Transport (2001).

China and World Organisation for Animal Health (OIE) transport guidelines

China is a member of the OIE. In 2004, the State Council approved the creation of the Bureau of Veterinary Service (BOVS) in the Ministry of Agriculture. Earlier in 2000, the Ministry of Agriculture had already drafted a surveillance programme on BSE by following the OIE's Animal Health Code.

China follows the UN Food and Agriculture Organization (FAO), World Health Organization (WHO) and OIE recommendations on disease reporting and notification (Ministry of Agriculture, 2006). The Chinese authorities are aware of OIE live transport guidelines, though this knowledge is not necessarily shared with transporters, trading firms and other stakeholders. Live transport is regulated by administrative regulations issued by the General Administration of Quality Supervision, Inspection and Quarantine (AQSIQ), the Communications Ministry, the China Civil Aviation Administration and other ministerial-level agencies.

The gap between Chinese standards and the new OIE animal welfare standards is considerable. It is not clear what proportion of Chinese animal production, transport and slaughter goes essentially unregulated. OIE standards state unequivocally that animal welfare should be the 'paramount consideration' in live transport. Chinese rules regarding live transport are concerned with fulfilling business transactions or the safe delivery of the goods. Furthermore, Chinese standards tend to be general and broad, in contrast with OIE's detailed, comprehensive and operationally focused standards. For example, OIE standards of live transport by land identify eight parties responsible for the welfare of livestock in transport and the detailed responsibilities of each party, as well as identifying the areas of competence that animal handlers and care providers should possess through training and experience. Chinese standards address the need for care providers on the journey to provide feed and water, though do not specify a required level of competence.

Disease control during live transport

Disease control has, in recent years, received more attention from the Chinese government. In 1997, China's national legislature enacted the Animal Disease Control and Prevention Law. The Agriculture Ministry issued a Work Target for Animal Disease Control calling on agricultural departments and animal disease control offices in the provinces to strengthen inspection of live transport vehicles. Certification of healthy, live animals prior to transportation was established to prevent sick livestock from leaving their place of origin. In 1998, the State Council issued evaluation measures regarding the management of animal disease control and prevention targets. These measures were designed for the evaluation of compliance with regulations by village communities, backyard and commercial farms, wholesale markets and slaughterhouses.

Every province has in place road checkpoints for disease control purposes. In 1999, for example, Beijing's road checkpoints inspected 118,500 trucks with live animals. Some 4324 sick animals were prevented from entering the city's slaughterhouses (China Animal Husbandry Industry Association, 2001, p. 277). In 2003, animal disease control and prevention continued to receive much attention. Provincial road checkpoints strengthened inspection of livestock in transit. Hubei had 4880 animal health inspectors who conducted inspection of more than 35 million livestock. More than 119,000 diseased livestock were stopped from entering the food market (China Animal Husbandry Industry Association, 2005a, p. 143). The percentage of the inspected consignments out of the total live transport volume in these provinces was not known. How the inspections were conducted was also hard to establish. Yet, the number of the inspections suggested enhanced government awareness of the need to strengthen farm animal disease prevention. Hubei conducted training for 7000 inspectors who work on the various road checkpoints. Most Chinese provinces adopted a more proactive approach to animal disease control as a result of the outbreak of SARS (China Animal Husbandry Industry Association, 2005a, pp. 142–144).

Context: journey routes and trade issues

Imports

The majority of animals shipped live into China are for breeding purposes. For example, in 2003, it imported a total of US$596 million worth of live farm animals from North America, Oceania and West Europe.

Exports

In general, export of farm animal products overshadows export of live animals. China's live animal export is focused towards South-east Asia, Mongolia and the Middle East.

SOUTH-EAST ASIA.   In recent years, China has resumed live animal export to Muslim countries following an 8-year suspension imposed as a consequence of the poor quality of the livestock. In 2002–2003, 3000 cattle were shipped to Malaysia by sea. In 2006, China agreed to increase its live cattle export to Malaysia. Cattle from Fujian, Inner Mongolia, north and north-east China will be exported to Malaysia. Cattle may also be exported from Xinjiang, Hebei and Shandong in north China (Anonymous, 2006). Xinjiang's livestock travel 3000 km by rail to reach the nearest port.

MONGOLIA.   In September 2006, China's Inner Mongolia shipped 83 adult Holstein cows by truck to Mongolia. This was the first ever live cattle export to Mongolia from China. These dairy cows were purchased for milk production purposes (Fengjun and Haifeng, 2006). The cows exited China from a Chinese city bordering Mongolia. For ocean-going live transport of consignments sourced from north and north-east China, Qinghuangdao, Jinzhou and Tianjin are the designated exit points (Communications Ministry Official, Beijng, 2006, personal communication).

MIDDLE EAST.   In the early 1990s, China began exporting large numbers of live cattle and sheep to the Middle East by sea. In 1996, trade was suspended due to severe health and mortality problems found among some animals that were force-fed rocks, concrete and other weight-adding materials by some Chinese farmers. In 2004, live exports resumed. Journeys can be as far as from Inner Mongolia to Jordan – a trip that requires a 500 km rail journey followed by 20 days at sea. Export volumes are expected to increase to half a million head by 2007 (Chaoyuan Xinfa Rep., Mongolia, 2006, personal communication).

Chinese domestic markets and routes

THE GREATER PEARL RIVER DELTA REGION.   The Greater Pearl River Delta Region includes China's most developed urban centres. Hong Kong dominates this region both in industrial production and farm animal consumption, despite having reduced livestock farming. More than 95% of the live animals slaughtered for the Hong Kong market are transported by truck and special cargo trains from inland provinces. This can include provinces in north-east China and Inner Mongolia, which involve travelling distances in excess of 3500 km. Sichuan is Hong Kong's main pork supplier. Hong Kong's live animals are supplied by the following provinces in order of increasing distance: Guangdong, Jiangxi, Hainan, Hunan, Guangxi, Hubei, Henan, Shandong, Shanxi, Chongqing, Sichuan, Hebei, Beijing, Gansu, Inner Mongolia, Liaoning and Heilongjiang (Fig. 11.1, Table 11.1) (Anonymous, Jiangxi Province, 2006, personal communication). Guangzhou has the second highest meat consumption after Hong Kong. As different estimates put it, 90% of pigs slaughtered daily there are supplied from outside the city and other provinces.

In addition to Hong Kong and Guangzhou and their combined population of some 13 million people, the region's other major cities (Shenzhen, Zhuhai, Macau and Dongguan) and their sprawling suburbs have attracted transport businesses of all sorts. The entire province of Guangdong (including the cities of Guangzhou and Hong Kong) has a combined population of 180 million. Yet, Guangdong province (excluding Guangzhou and Hong Kong) could only satisfy 60% of its pork consumption. In contrast, 70% of Guangzhou's pork and 95% of Hong Kong's pork

Table 11.1. Destination of live transport and distance from livestock suppliers, China.

Destination	Farm animal species	Transport vehicle	Supplier province	Longest distance
Beijing	Sheep, cattle, broilers	Trucks	Inner Mongolia, Hebei, Liaoning, Jilin, Shandong, Henan, Hubei, Heilongjiang	1500 km
Shanghai	Sheep, cattle	Trains	Shandong, Liaoning, Inner Mongolia, Henan, Anhui, Hebei	2000 km
Shanghai	Pigs, broilers	Trucks	Jiangxi, Zheijiang, Fujian, Anhui, Jiangsu, Henan, Hubei, Hebei, Shandong	2000 km
Hong Kong, Macau, Guangzhou	Sheep, goats, cattle, pigs, ducks, geese, broilers	Trains, trucks, cargo plane, ocean-going vessels	Jiangxi, Hunan, Hainan, Guangdong, Guangxi, Yunnan, Sichuan, Chongqin, Hainan, Shanxi, Inner Mongolia, Anhui, Henan, Liaoning, Heilongjiang	Over 2500 km

have to come from the nearby areas and inland provinces. Guangdong is arguably China's busiest live transport destination (Anonymous, 2005).

THE GREATER SHANGHAI REGION.   The Greater Shanghai Region covers China's second largest manufacturing centre, with a population of more than 30 million. Inner Mongolia is the most distant supplier, a journey likely to be well over 1000 km. Jiangsu is the closest (Fig. 11.2, Table 11.1). In 2005, Shanghai accelerated efforts to phase out livestock farming to nearby Anhui and Jiangsu provinces. This suggests that in the future, more livestock may be transported greater distances. The majority of transport is conducted by truck. Broilers transported by truck come from as far away as Hebei province in north China (a median distance of 1085 km), whereas sheep and cattle come from Inner Mongolia (Communications Ministry Official, Beijng, 2006, personal communication) (median distance of 1770 km). Large numbers of pigs come from Shandong (median distance of 620 km), Jiangxi (685 km), Hubei (920 km), Anhui (360 km) and Zhejiang (280 km). Like the Greater Pearl River Delta Region, the Shanghai Region is undergoing rapid urbanization. Phasing out livestock farming or moving it to the very periphery of the urban centres is the trend in the region. In recent years, the Greater Shanghai Region has increased its beef and mutton consumption. Cattle, sheep and goats are being transported to the region from distant northern provinces. To ensure that livestock in transit are healthy, there are 23 checkpoints in operation at main entrance points to Shanghai.

THE BEIJING–TIANJIN AREA.   The Beijing–Tianjin metropolitan area has a total of 20 million residents. Both cities have well-developed peripheral animal farming industries, though they are still destinations for live transport. Live transport is mostly conducted by truck. As Beijing and Tianjin are the main meat processing centres supplying north China and other regions, slaughterhouses in both cities purchase livestock from other provinces (Table 11.1). For example, Hebei's live animal sales to Beijing accounted for 55.7% of its total sales in 2002 (China Animal Husbandry Industry Association, 2004a,b, p. 18). In 2003, the total number of pigs slaughtered in Beijing was 4.66 million, whereas the total number of pigs in stock in local farms at the end of the year was 2.48 million – a deficit of 2.18 million pigs that was to be made up with pigs from outside the city. Chongqing has 2.5 million residents and its location by the side of the Yangtze River makes it an important transit depot for industrial and agricultural produce. A significant number of live animals are shipped to Chongqing by boats, trucks and railway and from the city to nearby provinces by cargo boats downstream along the Yangtze River and by other vehicles (Table 11.1).

Methods and welfare consequences

Several factors impact animal welfare in transport. China faces the universal challenges as well as problems unique to its current stage of development.

Distance

The welfare of transported animals decreases as the length and duration of the journey increases (SCAHAW, 2002). Because of the increasing geographical specialization of livestock farming, livestock are shipped over extensive distances. For example, a truck company shipping live chickens from north-east China to Beijing covers more than 2000 km. Table 11.1 summarizes the longest distances live transport covers between China's main urban centres and their supplier provinces. Often, trucks loaded with up to 3000 birds are on the road for 18 h or more without stopping for more than 2–3 min. Since the only ventilation is provided by the movement of the vehicle, transporters cannot stop for too long without a risk of birds suffocating or suffering heat stress. In a chicken slaughterhouse in Henan province, between 2% and 5% of the broilers are dead on arrival (research carried out by a regional investigator in June 2005).

Cattle transport from north China's Zhangbei livestock trading centre to Guangdong in South China is particularly stressful for the animals. After the journey (in excess of 2500 km), the animals, who are usually provided with water but no food, have lost up to 10% of their former body weight (from an interview with beef cattle traders in Zhangbei by a local investigator, 16 July 2006).

Extent of training and poor practice

Transporters have very varied levels of experience of live animal transport. In China, short to intermediate-distance live transport is conducted by small transport firms. Except for transport firms in livestock trading centres, most transport firms are general purpose transporters. Consequently, drivers have at best rudimentary knowledge or, at worst, no knowledge of animal needs.

Many of the general purpose transporters and even some specialized livestock transporters are known to handle animals very roughly during transport. At north China's biggest livestock trading centre at Zhangbei, researchers have witnessed and photographed the unloading practice, which appears to be typical, of forcing cattle to jump down from trucks without using ramps. To unload the cattle, the transporters harass them by making loud noises and using hard objects as goads. However, major transporters such as China Southern and China Railway Special Cargo Corporation have standard livestock handling procedures in place which take into consideration some basic needs of live animals.

One of the worst practices that has come to light is the forced watering or force-feeding of slaughter animals with mud and stones in order to increase their slaughter weight. This practice used to be most common among long-distance live transport dealers who resorted to such cruel methods rather than good care and adequate food and water to maintain livestock weight. Force-feeding led to exports being banned by the Middle East in the mid-1990s (Anonymous Official, Communications Ministry, Beijing, 2006, personal communication). Forced watering remains a big problem in China's live transport industry. According to an interview with livestock dealers who transport live cattle from north China to the south, cattle are forced to drink water immediately prior to arrival at the slaughterhouses to recoup some of the weight lost during transport. Despite the state's specific rules against forced watering of livestock (Article 12, State Regulation on Pig Slaughter, 1998), the practice continues and is a target of government regulatory action.

Transportation methods

There is no doubt that the methods of transport used can have a major impact on animal welfare. Most of China's live transport is conducted by road. No exact statistics are available on the number of livestock transported by trucks, but according to interviewees, trucks are the main vehicle used to transport pigs, cattle and sheep from northern provinces such as Shanxi to Guangzhou where the animals are slaughtered, processed and shipped to Hong Kong, Macau and nearby cities in the Zhujiang Delta region (Manager, Special Cargo Transport Corporation, Beijing, 2006, personal communication). Compared with animals bound for the Hong Kong and Macau markets, those shipped to Guangzhou are loaded on trucks, trains and other vehicles. Death from exhaustion, lack of space and water, and exposure to severe weather conditions is common. This explains the continued existence of illegal trade in dead animals at some of Guangzhou's slaughterhouses.

Small transport firms often operate with little consideration to the space needs of the animals, let alone other needs such as water, feed and protection from adverse weather or climatic extremes (local researcher, Jiangsi province, 2006, personal communication). Pigs in long-distance transport are often loaded in trucks with two to three tiers with those on the upper tier most exposed to the weather conditions. Overloading is also common, with the objective of cutting transport costs.

China's export of livestock to the Middle East, for example, is currently conducted by foreign ocean-going vessels chartered by the importing firms. An official from the Communications Ministry confirmed that special care is usually taken to ensure space allowance, veterinary service, daily inspection and provision of water and feed during the journey by sea (Official, Communications Ministry, Beijing, 2006, personal communication). Yet, similar consideration is lacking in the transport of livestock in China's internal waters. Fatalities can increase sharply when emergencies or unscheduled delays occur. For example, in 2003, routine maintenance of the Three Gorges ship locks led to a blockage of transport on the Yangtze River and caused the death of pigs and other livestock on transport boats (Wang, 2004).

Rail transport of livestock is becoming less popular as a consequence of booming road transport. As of July 2006, the most popular rail transport links carried live animals between Guangzhou and Henan, Hubei and Shandong. Despite government regulations against using boxcars for transporting live animals (as they have particularly poor air circulation), they are believed to be still in use.

Impact of policies on animal welfare

Rather than working to improve both economic growth and animal welfare simultaneously, some may consider China's agricultural policies to be more concerned with economics than welfare protection or environmental conservation per se. The drive to increase livestock production has led to overuse of grassland, which led to desertification and frequent sand storms in north and north-west China. China's phenomenal growth in livestock farming has led to its increasing intensification. Short-term growth rates and profits are some of the determinant factors that have underlain the drive to raise or transport more livestock.

Impact of welfare developments on live transport

An improvement in transport conditions could have a direct impact on the welfare of animals during live transport. According to a local investigator, most of the small- and medium-sized transporters in a southern Chinese province are family-owned operations. The majority of China's transporters are individually owned small operations. These transporters are not motivated to comply with many of the welfare recommendations and widespread non-compliance is likely. Some transporters feel that complying with animal welfare regulations will force them out of business as a consequence of the additional expense.

A typical way of cutting cost and maximizing profit from each truckload of live animals is by overloading. A 5-tonne truck often carries 80–100 pigs, severely exceeding the holding capacity of the vehicle. The practice of overloading occurs nationwide. In 2005, the Chinese government began to enforce measures to prevent overloading. To some transporters, reducing loading to permissible tonnage results in losses of 30–40% of the revenue. Similarly, improving conditions of livestock during rail transport is likely to further raise the already high transport fees (50% more than the transport fees of all other goods). Although improvements are necessary, they will not come without increased costs in some areas. However, compliance with health and welfare standards is becoming increasingly important for both domestic and export markets.

One important condition to ensure animal welfare in transport is the use of qualified handlers to load and unload the animals and experienced care providers to accompany the livestock in transport. In the experience of one of the chapter authors, many of the welfare problems experienced during transport are the result of a lack of training among animal handlers and transport workers.

Prospects of welfare improvements in live transport

The Chinese government states that it is aware of the negative impact of poor animal welfare on the economy and on human health (although as stated previously, current laws seem to favour economy over welfare). Secondly, the AHL and relevant regulations address in more detail the issue of live transport. New articles are added to relevant state regulations to ensure the satisfaction of some most basic welfare conditions such as provision of water to transported animals free of charge. If the OIE transport guidelines are used as the benchmark, existing Chinese laws and regulations currently fall short of the required standards. However, as a member of the OIE, China will be expected to refer to the OIE guidelines for future law and regulation making. Most important of all, measures to reduce journey times are urgently required.

As a major and expanding participant in the global economy, China's livestock and export industries are set to continue to expand. However, it is imperative that China adapts to the increased world demand for farm animals that are humanely raised, transported and slaughtered.

Discussion

China has the world's largest animal farming industry and this industry continues to grow, which is likely to lead to concomitant increases in industry influence on government policy making. There has been a convergence of interests between China's developmental state and business sector. Instead of being pressured to make policies favouring industry development, the Chinese authorities have, in the last quarter century, anticipated industry demands and actively adjusted state policy to meet those demands. Underlying this convergence of interests is the reformist government's determination to push for economic growth. As a result, economic considerations occupy important positions in government and industry calculations of risks and benefits. Other policy issues such as animal welfare and environmental protection are often given lower priority, although exports to the EU will require that these be given greater precedence.

China's geography and demographics pose further challenges to potential welfare improvements for farm animals in China. Long-distance live transport routes criss-cross the entire Chinese mainland. A truckload of beef cattle from Zhangbei near Inner Mongolia covers more than 2500 km in distance to Guangdong. Live farm animals sold to the Middle East are often reared in China's remote inland provinces more than 1000 km from the ports they will ship out from. Farm animal production and live transport will continue to be stimulated by China's burgeoning population – most of which is clustered in the more industrial central and eastern areas – and increasing urbanization. The scale of China's farm animal production and transport portends greater challenges to welfare improvements.

Some very practical steps in state and country planning could fundamentally, and positively, influence the welfare of farmed animals in relation to long-distance transport. For example, the provision of regional slaughterhouses, processing and refrigerated transport could drastically reduce the use of long-distance transportation, improving welfare while also boosting local economies in more rural provinces.

China is an active member of the global community. Its participation in, and with, world organizations, particularly those that aim to improve animal welfare should be encouraged. China's implementation of globally agreed standards for animal care has positive impacts on the health and well-being of both farm animals and humans. The Chinese authorities should be encouraged to make further positive moves on the front of animal welfare legislation. It is important for Chinese domestic and international animal welfare advocacy groups to drive home the message that welfare improvement should not be ignored at a time of rapid economic growth. Legislating and enforcing welfare laws do not need to have negative impacts on China's animal husbandry industry. On the contrary, it could help eliminate factors that have hindered China's export of farm animal products to international markets.

Training of transport personnel will also prove central to improving welfare where animals are transported to slaughter. Increasing awareness of animal welfare and the importance of its consideration in areas such as planning and animal handling will have benefits not only for the animals, but also for issues such as disease management. The OIE provides clear operational guidelines on animal welfare during transportation (OIE, 2006). These solutions will be similarly valuable in improving welfare in other Asian countries.

Taiwan

Background

Taiwan is an island in East Asia, situated just off the coast of mainland China. Taiwan is also commonly used to refer to the territories administered by the Republic of China (ROC), which governs the island of Taiwan, Lanyu (Orchid Island) and Green Island in the Pacific Ocean, the Pescadores in the Taiwan Strait and Kinmen and Matsu islands off mainland Fujian. Taiwan is bound to the east by the Pacific Ocean, to the west by the Taiwan Strait, to the north by the East China Sea and to the south by the South China Sea. The island is 394 km (245 mi) long and 14 km (89 mi) wide. It has a land area of 36,179 km² and 1,566 km of coastline (including the Penghu Islands). It lies on major air and sea transportation routes in the region.

Topographically, Taiwan is steep, mountainous terrain, covered by tropical and subtropical vegetation. Taiwan has an oceanic and subtropical monsoon climate, with average daily temperatures ranging from 28°C in July to 14°C in January, and annual rainfall around 2500 mm.

In 2004, Taiwan had a population of nearly 23 million, about 70% of which live in cities. It is one of the most densely populated areas of the world with a population density of 626 persons per km² (Taiwan Government Information Office, 2006). Languages spoken include Mandarin Chinese and Holo Taiwanese. Principal religions are Taoism, Buddhism and Yi Guan Dao.

Although traditionally an agrarian society, Taiwan's agricultural sector has decreased from 35% of the gross domestic product (GDP) in 1952 to less than 2% in 2004. As Taiwan has developed both economically and in the population's quality of life, the demand for animal protein in the daily diet has also increased (Council of Agriculture, undated). Livestock production has increased with the introduction of intensive technology. Animals are mainly reared indoors under intensive conditions and are fed imported feed grains rather than locally produced crops and by-products. Due to high land prices (particularly in the north) and rising environmental awareness, most of the agricultural industry is focused in the more rural areas of middle and southern Taiwan.

Around 400 million farm animals are reared and slaughtered each year. On average, during the period 1995–2004, this total was made up of over 10 million pigs, around 372 million chickens, 35 million ducks, 7 million geese, 170,000 goats and 28,000 cattle (Council of Agriculture, 2005). Meat consumption mainly revolves around pork, beef, mutton and poultry, with pork being most popular. In 2004, pork accounted for 51.3% of meat consumption, followed by poultry with 42.5%.

Live transport

Since the outbreak of food-and-mouth disease (FMD) in 1997, the live imports and exports into, and out of, Taiwan have declined to very low levels. Now, live imports and exports are mostly for breeding purposes, rather than slaughter or further fattening. In the domestic market, however, live transport continues to be de rigeur. As most of the population is in the north of the country and the majority of the agricultural sector is situated in the south, livestock tends to be transported live from south to north (Wen-Fu, 1996). Pigs and poultry are the most prevalent livestock in Taiwan. Live animal markets are popular with the effect that time from first loading on farm to slaughter is protracted. For example, the time spent in transit from loading on the farm to final slaughter of pigs can be as long as 48 h (Lymbery, 2001). Prolonged transit time increases stress (Liao, 2000).

There is some demand for warm meat, which typically comes from animals sold at live auction or from animals slaughtered during the night, for sale the following morning (Liao, 2001). Generally, the older generations and those with large families tend to be more likely to buy warm meat; by contrast, more highly educated consumers and those with higher incomes are more likely to buy chilled or frozen meat (Lin *et al.*, 2003).

Context: legal and regulatory issues

The main law-making body in Taiwan is the Legislative Yuan. The Council of Agriculture (COA) deals with national agricultural issues. Within the COA, the Department of Livestock Industry deals with on-farm matters and animal transport and the Bureau of Animal and Plant Health Inspection and Quarantine (BAPHIQ) is responsible for animals at slaughter.

Taiwan joined the OIE in 1954. The Taiwanese government is aware of OIE's transport guidelines (Council of Agriculture, 2006). There are two livestock-relevant laws in Taiwan – the Animal Protection Act (APA) and the Animal Industry Act (AIA) – both of which were enacted in 1998. The AIA is more specific and as such is adhered to first, while the APA governs matters not specified within the AIA. The AIA covers farm registration and administration; the regulation of breeding flocks and stocks; the supply and market; and regulation of livestock and poultry slaughter. The APA covers animal transport in addition to the general protection of animals; the use of animals in science; and the management of pets (Ministry of Justice, undated).

Article 9 of the APA states: '[W]hile carrying an animal, the guardian shall take good care of its food, water, excrement, environment, and safety. Furthermore, it shall be prevented from being frightened or hurt. The carrying vehicles, the carrying ways and the other carrying measures to be complied with shall be determined by the central competent authority.'

Based on the delegation of the APA, the COA called together a group of academics to draft a regulation in 2001. The resultant Animal Transport Management Regulations (ATMR) were enacted and came into effect in 2005 (Zheng, 2005). Taiwan joined the OIE in 1954, and the ATMR were derived to a certain extent from OIE standards. The ATMR covers welfare issues such as loading, unloading, space allowance and vehicle specifications, in addition to regulating water, feed and rest provision for animals; and making record-keeping and the completion of appropriate training courses on animal transportation a legal requirement.

With regards enforcement of the ATMR, a transition period of 1 year has been set up for improvements and 2 years for the completion of training. If transporters fail to comply after this deadline, they will be penalized. Enforcement after the deadline will need to be looked at by the COA.

According to Article 31 of the APA, a fine of NT$2000 (US$60) to NT$10,000 (US$300) shall be imposed on those who violate transport laws by using means of transport not permitted by the central competent authority. If improvements are not made, a penalty will be imposed according to the frequency of violation. As the APA is what is classed as an 'administrative law' (rather than criminal or civil law), the penalty for violation of the regulations tends to be financial or confiscatory.

Meat production

Taiwan's pig industry is the foremost agricultural sector. In 2004, the total value of pig production was about NT$65 billion (US$1.9 billion), accounting for 51% in animal production and 17% in total agricultural output. In 2004, six counties including Changhua, Yunlin, Chiayi, Tainan, Kaohsiung and Pingtung accounted for 83% of pig supply, 84% of cattle supply and 87% of goat supply.

In 2003, national pig sales reached 7.6 million, with 23% from Yunlin County, 19% from Pingtung County, 15% from Changhua County, 14% from Tainan County, 10% from Kaohsiung County and 5% from Chiayi County. Hence, southern Taiwan accounted for 87% of the annual sales volume at wholesale pig markets. In Taipei County (in northern Taiwan), annual sales at pig markets have reached 960,000 animals. The average number of pigs sold daily in Taipei County is in excess of 3000. In Taoyuan County (also in northern Taiwan), 770,000 pigs are sold annually, with an average of over 2500 a day (Wu, 2005).

The poultry sector is the second largest in scale. In 2004, there were approximately 350 million chickens. The total value of the poultry industry in 2004 was NT$49 billion (US$1.5 billion), comprising NT$30 billion (US$900 million) of broilers, NT$13 billion (US$400 million) of eggs and NT$6.8 billion (US$200 million) of domestic waterfowl.

Live import and export

Before the 1970s, Hong Kong was the top destination for Taiwan's live pigs. In the 1980s and early 1990s, owing to increasing demand for breeding pigs, there was live export to Malaysia and Japan as well as Hong Kong (Wen-Fu, 1996).

However, the outbreak of FMD in 1997 dealt the pig industry a serious blow. Since then, government policy priorities have shifted to the domestic market and have reduced livestock exports to a low level to effectively balance Taiwan's production surpluses (Council of Agriculture, undated).

Domestic live transport

Pigs

Currently, nearly every city or county in Taiwan has a wholesale pig market (Wu, 2005). Most transactions at these markets take place by auction. In the mornings, trucks of pigs are transported to holding areas at markets. Transactions take place in the afternoon via computer. That night, the pigs are either slaughtered at the market or are transported to other slaughter sites before being shipped to retail markets.

Transport costs account for 50–60% of the total market cost (Lee and Du, 1997). Transport costs cover vehicle depreciation, interest, tax, warehouse costs, auto insurance, salaries, fuel, maintenance and highway tolls (Liu *et al.*, 1998).

VEHICLES. There are mainly four types of vehicles used in pig transport in Taiwan. Two-tiered trucks (generally carrying 100 pigs) are more popular in northern Taiwan and are used mainly for long-distance transport. Similarly, one-and-a-half-tiered trucks (carrying 75 pigs) are also used for transport over longer distances. By contrast, single-decked trucks (carrying 50 pigs) are used mainly for shorter journeys and small trucks (carrying 35 pigs) are used solely for short distances (Liu *et al.*, 1998).

DISTANCE AND JOURNEY TIME. On average, the distance of domestic pig transport is 80 km. The distance from farms to Taipei County market is the longest (230 km), followed by Miaoli County (220 km), Taichung County (180 km), Taoyuan County (170 km) and Hsinchu County (140 km). The average journey time of domestic pig transport is approximately 1.5 h. The journey time from farms to Taipei County market is the longest (3.5 h), followed by Miaoli County (3 h 10 min) and Taichung County (2.5 h) (Chou *et al.*, 1995a). The primary routes taken for pig transport are depicted in Fig. 11.1.

Poultry

As with pigs, because poultry production is mainly located in central and southern Taiwan, while consumption is concentrated in the northern area, poultry are transported from the south to the north. Two kinds of broiler chickens are common in Taiwan, standard commercial white broilers and buffy-brown-coloured broilers. The standard white broilers are mostly transported to slaughterhouses while the more traditional breeds of coloured chicken are more likely to be transported to general dealers, wholesale markets and distributing centres (Kuo, 1997).

In 2003, of 190,000,000 white broilers, 88% were slaughtered in legal or illegal slaughterhouses and then transported and traded as refrigerated carcasses (70–75%) or frozen carcasses (25–30%) (Lin, 2004). Of 156,000,000 coloured broilers, only 7–10% were slaughtered in slaughterhouses; 20% were traded as refrigerated or frozen carcasses, while 80% were transported and traded live (Lin, 2004).

There are three wholesale poultry markets, located in Taipei, Taichung and Kaohsiung (Chen, 2004). Taipei market is in the far north of Taiwan, Taichung is in central Taiwan and Kaohsiung is situated in the south. Taking Taipei as an example, in the 'Huan-nan' poultry market, chickens are transported mostly from central and southern Taiwan to the market starting from eleven o'clock in the evening and then auctioned from one to six o'clock in the morning, while ducks are auctioned from ten o'clock in the morning to four o'clock in the afternoon (Markets Administration Office, undated).

Methods and welfare consequences

Loading and unloading

Animal welfare is first compromised in the trucks as animals are loaded and unloaded. It is reported from field investigations that trucks are usually overloaded, pigs are often wedged against the door and sides of the trucks, workers are often rough when moving pigs on and off the trucks and electric goads and sticks are commonly used to hit and kick animals.

Transport conditions

Transporting trucks typically have no roof or shelter from the sun. On hot days, transporters have to sprinkle water on pigs in transit to prevent heatstroke (Wu, 2005).

The main facilities that trucks for pig transport are equipped with include a railing (82%), sprinkler (79%), adjustable rear door (23%) and a roof (4%). Railings are used to protect pigs against being crushed when the vehicle is braking. During the journey, while most transporters use a water sprinkling system (72%), rest stops are less common (32%) (Chou *et al.*, 1995a).

Pigs are usually given no food for at least 8h before transit. By the time they are auctioned or slaughtered, they will typically have been starved for 16–36h. In some cases, pigs are kept from water for 24h before being slaughtered (Kao, 2000). Although ATMR does not cover feed and water pre-journey, this practice could violate Article 5 of APA stating that 'the feeder shall provide adequate food and water and sufficient space for activities for the animal'.

Holding facilities

Pigs are often left on trucks for some time before being unloaded. At poultry markets, most birds are held in stackable crowded wire crates. Packed crates were often stacked on top of each other (EAST, 2006). The holding pens are sometimes very crowded. Pigs are rarely provided with clean water in troughs for drinking. Overcrowding at markets is a welfare issue as serious as overcrowding during transport.

Rough handling is commonplace at markets. Animals are sometimes jabbed and poked with sticks and often are deprived of water throughout their time at market. Injured animals are sometimes given no veterinary care at markets. They can be left for hours before being taken off to slaughter. Pigs often vocalize during handling and prodding. According to a local investigation, 90% of pigs in almost all slaughterhouses in Taiwan vocalized during handling and prodding (EAST, 2000).

Mortality and injuries

According to studies on the mortality of pigs during transport, the proportion ranged from 0.25% to 0.3%. In terms of regions, mortality is highest in northern Taiwan (up to 0.45%), followed by central Taiwan (0.27%) and southern Taiwan (0.21%), while there is no mortality reported in eastern Taiwan (Chou *et al.*, 1995a). However, it should be noted that these are rather low mortality rates and may not realistically represent true mortality.

The mortality and injury rate of pigs carried by trucks with single decks is higher than that with two tiers; also in hotter seasons the rate is higher than in other seasons. These trends are obvious both in Taiwan as a whole and in Taipei County specifically (Chou *et al.*, 1995b). Opinion polls carried out among transporters suggested that the main cause of accidents with pigs is breed-related (45%), followed by human fault (driver's emotion, handling, overuse of electric goads, use of emergency brakes) (42%), overloading (36%), weather (29%), disease (14%) and journey time (12%), among other factors (Chou *et al.*, 1995a).

The research revealed that there is positive relationship between distance/ speed and accidental death of pigs. As the distance or speed increases, pig mortality also increases. On the other hand, including rest stops reduced mortality (Chou *et al.*, 1995a).

Discussion

Since the outbreak of FMD in 1997, the live import into, and export from, Taiwan has declined to a low level. The government policy priorities have been shifted to the domestic market, reducing livestock export to the role of balancing production surpluses. Furthermore, live import and export are mostly for breeding purposes, rather than for slaughter or further fattening.

Most domestic live transport takes place from the agricultural south to the urban north. The preference for wholesale auctions necessitates the live transport of pigs, goats and poultry. In the long term, the focus has to be on substituting carcass transport and auction for live transport and auction. With the rise in the popularity of supermarkets and the increasing availability of refrigeration equipment, the likelihood of achieving this long-term goal seems feasible.

Indian Subcontinent

Background

The Indian subcontinent includes Bangladesh, India, Pakistan and Sri Lanka. It has a predominantly agrarian population, dependent on crops and livestock for their livelihood. The region is one of the world's most densely populated areas, with more than a billion people in India alone. Tradition, cultural and religious variations (e.g. Muslim, Hindu and Christian) across the region also impact on the structure and practice of animal farming and hence on live transport.

Animal production systems on the subcontinent vary greatly, depending on climate, soil and socio-economic factors. Traditionally, small farms produce milk, eggs and meat. The predominance of cattle for milk, bullocks for ploughing, sheep and goats for meat and wool, and poultry for eggs and meat is common to all the countries. Besides milk and meat, however, livestock are used as sources of draught power, capital, credit, hides and skins, fuel, fertilizer and for religious purposes. The Indian subcontinent production systems are characterized by small, scavenging herds, which are provided with only basic housing, feed or health care.

Backyard goat, chicken and duck farming systems are less capital-intensive than larger enterprises. On small farms, chickens and ducks feed mostly by scavenging and foraging household waste, fruit by-products, roots, tubers, small amounts of grains, grain by-products and anything edible found in the immediate environment.

Many animals in the subcontinent are dual-purpose and are used for meat at the end of their breeding or milk-producing lives. It is therefore important to consider the whole animal system, not just the meat industry (Ranjhan, 1999). As such, most of the animals undergoing live transport will be aged, spent livestock. They may be weakened and more at risk of injury or death on the longest or hardest journeys.

India

With a population of 1050 million, India is the second most populated country after China and is the 11th largest economy in the world. The livestock sector is one of the most important components of agriculture in India. The output from livestock and fisheries sectors together was approximately US$31,000 million during 1998–1999, accounting for 27% of the total agriculture output (Ranjhan, 2006). Financial contributions of the meat sector rose from US$356 million in 1980–1981 to US$890 million in 1986–1987.

India has 28% of the world's large ruminants, 26% of its small ruminants and is the number one producer of milk and the fifth largest producer of eggs in the world today. Seventy-six percent of milk and 40% of eggs are produced by small farmers and landless labourers. This group also owns over 90% of small ruminants and almost the entire pig sector is in the hands of the socio-economically poor (Rao, 1997). The use of animal power alone provides employment to nearly 20 million people (Ramaswamy, 2000). The country's over 160 million ha (400 million acres) of farmland are ploughed with some 40 million traditional ploughs pulled by bullocks (Ramaswamy, 2000).

According to the 2003 census, India has 185 million cattle (25 million are crossbred, 160 million are indigenous), 98 million buffaloes, 61.5 million sheep, 124 million goats, 14 million pigs, 1 million camels and 489 million poultry (Rahman, 2004). In total, these animals produce 5 million t of meat and meat products, valued at US$130 million (Ranjhan, 1999). The per capita consumption of meat, milk and eggs is 5.2, 42.1 and 1.5 kg per year, respectively (Rahman, 2005).

Pakistan

Pakistan has a population of 150 million and a current growth rate of 3%. Population densities vary from Punjab (240 people per km^2) to Balochistan (19 people per km^2).

Agriculture is the largest sector of the economy, contributing 24% of the GDP. It accounts for half of the employed labour force and is the largest source of foreign exchange earnings. Its growth rate over the last five decades has remained at around 4% per annum (Ministry of Finance, 2006).

Livestock is an important sector of agriculture in Pakistan. It accounts for 39% of agricultural value added and about 9.4% of the GDP. Demand for livestock products exceeds supply and milk powder, baby foods and live sheep are imported.

Pakistan has a large livestock population. There are an estimated 26 million buffaloes, 24 million cattle, 25 million sheep, 57 million goats, 1 million camels and 692 million commercial poultry. There are also 81 million rural poultry. Buffaloes are kept mainly in the northern and southern irrigated plains, and cattle are raised throughout the country. More than 50% of sheep are reared in the western and northern dry mountains and the western dry plateau. Goats are raised in all regions, but larger herds are common in areas with forage and grazing (Afzal, 1997).

Pakistan is the fifth largest milk producer in the world, producing 28 million t of milk per year. The daily consumption of milk in Lahore is 2–3 million l and that of Karachi is 4 million l. During the last two decades, processed milk has increased its share in the milk market of Lahore to 4% (The Dawn, undated).

Demand is often greater than can be supplied, with the result that meat prices can be high. In order to reduce meat prices, Pakistan imports nearly 5000 animals daily. A recent governmental decision to import cattle from India was made in order to improve the situation (Dinker Vashisht, 2006).

Bangladesh

The livestock system in Bangladesh is predominantly comprised of smallholdings, with few large-scale farms. Traditionally, livestock farming involved keeping animals in proportion to the 'free' crop residues and family labour available in the household, and converting these into food, fuel and draught power. By this method, households are virtually self-contained production systems with no purchased inputs and few marketed outputs. This age-old system has undergone rapid changes in recent times.

Bangladesh's livestock growth rate is almost 5% and the contribution of livestock to the national economy is 3%. The animal population in Bangladesh is estimated to consist of 23 million cattle, 2 million sheep, 19 million goats and 221 million poultry. The average number of bovines, sheep/goats and poultry per rural household is 1.3, 0.8 and 7.1, respectively. The density of large ruminants is much higher than in many other countries in South-east Asia (Department of Livestock Services Activities, 2006). With limited resources and a growing human population, the supply of milk, meat and eggs is outstripped by demand, leading to increased pressure for imported animals and their products.

Sri Lanka

Sri Lanka has a population of 18.5 million, the majority of which are Buddhist (69%), Hindu (15%), Christian (8%) or Muslim (7%). Sri Lanka is primarily an agricultural country, with 42% of employed women and 35% of employed men engaged in agriculture and allied sectors.

Specific policies have been developed to encourage the livestock sector, with special programmes for both poultry and dairy development. Livestock are typically considered as a potential source of supplementary income, particularly for small farmers with limited land. Sri Lankan farming households maintain livestock for draught power and as a source of cash income. In 2005, Sri Lanka had 1.2 million cattle, 307,750 buffaloes, 405,250 goats, 85,020 pigs and 11.6 million poultry (FAOSTAT, 2006).

Context: legal and regulatory issues

India

India has a number of official bodies whose remit is concerned with animal welfare. For example, the Animal Welfare Board of India (AWBI) was set up in 1962 in accordance with Section 4(1) of the Prevention of Cruelty to Animals (PCA) Act, 1960. In 2000, in his role as Prime Minister of India, Atal Behari Vajpayee directed the state governments to enforce India's animal protection laws, particularly with regard to animal transport and slaughter. Mr Maran, former Minister of Commerce and Industry, publicly urged Indian state governments to set up committees to ensure enforcement of the PCA Act, including unannounced inspections in places where cattle are sold and loaded on to trucks (PETA, 2005).

The PCA Act of 1960 has been amended by the central government from time to time to formulate rules to enforce animal welfare issues. For example, the Transport of Animals (Amendment) Rules (2001) is a set of guidelines on transport of poultry by rail, road and air, and transport of pigs by rail or road. Rule 55(k) states that as far as possible, cattle should be moved at night only (when it is cooler). Rule 55(l) states that if possible, cattle should be unloaded, given feed and water, rested and if lactating, milked during daylight hours.

Rule 96 authorizes the Chairman of the AWBI or their representative to exercise the powers under the Rules to check certificates and inspect any agency transporting animals. Without the appropriate certificate, they are in breach of the PCA Act, 1960. Transporting agencies are obliged to help the authorized representative to conduct inspections.

For religious reasons, some state governments have enacted separate laws to prohibit transport of animals (particularly cattle) for slaughter. The law requires that permits for animal transport are given only for genuine agricultural purposes and not for animals meant for slaughter. The AWBI also has identified representatives in each state that are authorized to issue permits.

In addition to the transport rules mentioned above, the PCA Act 1960 has formulated the Prevention of Cruelty to Animals (Slaughterhouse) Rules 2001, to prevent cruelty at slaughter. A recent move to ban cow slaughter entirely is currently a matter of controversy in some states.

Three million cattle and buffaloes as well as three million sheep and goats are brought into Kerala, a state where cow slaughter is currently legal, every year for slaughter and re-export to other states. About 300,000 t of beef (cattle and buffaloes) and 30,000 t of mutton are produced in 50 municipal and hundreds of illegal slaughterhouses (PETA, 2005).

The state of Gujarat has implemented a total ban on cattle (cows, bulls and bullocks) slaughter. The Supreme Court of India upheld this decision, a move which may have consequent repercussions throughout the other states.

Pakistan and Bangladesh

There is an Animal Cruelty Act in existence in Bangladesh. Pakistan's Prevention of Cruelty to Animals Act, 1890 has since been amended and renamed 'West Pakistan Prevention of Cruelty to Animals Rules, 1961'.

Sri Lanka

Sri Lanka implemented the 'Prevention of Cruelty to Animal Ordinance' in 1907. This has been amended many times to include current issues of animal welfare. There is growing public concern for animal welfare and environmental issues. There are 15 enactments associated with animals and their welfare in Sri Lanka. Of these, four have been enacted since 1948 (Weeraratna, 2003). The primary animal welfare legislation in effect in Sri Lanka comprises: (i) The Prevention of Cruelty to Animals Ordinance, No.13 of 1907; and (ii) Animals Act, No.29 of 1958, which prohibits the slaughter of cows and female buffaloes under certain conditions and cow and female buffalo calves.

To secure the rights and welfare of animals in the country, the Law Commission of Sri Lanka has drafted an Animal Welfare Act. This proposes a wide range of well-being and responsible-care measures for animals while envisaging rigorous legal action against violators of the Act (Jayatilleke, 2004). The main objective of the draft Act is to replace the Prevention of Cruelty to Animals Ordinance, No. 13 of 1907 and to bring the law governing animal welfare in the country in line with modern legislation. The draft Act includes many features to help strengthen law enforcement and envisages an enhanced penalty system compared to existing legislation. According to the draft Act, the Authority is required to create awareness of the objects and provisions of this Act among Government agencies, Provincial Councils, local authorities, schools and other educational institutions and the general public and foster kindness and compassion in society.

Methods and welfare consequences

India

ILLEGAL TRANSPORT PRACTICES.   The Animal Welfare Board of India alongside a number of animal welfare organizations have been gathering evidence of widespread abuse of cattle, buffaloes, sheep and goats used for meat and leather in various parts of the country. Their findings show high levels of overcrowding leading to severe injuries and fatalities, with animals gouged by horns or crushed. Despite directives from the state government and the public assurances made by top Indian government officials over several years, the evidence collected by these groups suggests that there has been little improvement in the treatment of animals during transport or at slaughter.

As the slaughter of cows is prohibited in most states, most slaughtering takes place in Kerala and West Bengal, where there is no ban. Animals are transported to the slaughterhouses of Kerala from the southern and western states and to the slaughterhouses of West Bengal from the northern and eastern states. On the trucks, many cattle are held with their pierced noses and tails tied to the top of the trucks. Many of the animals suffer injury as a consequence of this. Animals travel in this state for what can be more than 8 h. Often, wooden planks are laid down in the truck to create a second tier on to which more animals can be loaded. To prevent defaecation, some animals are starved before embarking on the journey (PETA, 2005). This is adverse to the welfare of the animals. Rather than being transported over long distances, they should be slaughtered as close to the farm as possible.

Following slaughter, the meat is usually sent back to its state of origin or is exported and the raw hides are taken to centres such as Erode in Tamil Nadu and Kolkata in West Bengal. However, it is thought that slaughter takes place even in states where slaughter is illegal.

It has been estimated that nearly 1.5 million cattle are brought into Kerala every year and of these, just 0.6 million pass officially through government check-points. The rest are smuggled in through other transport routes from Tamil Nadu and Karnataka. Tamil Nadu is the largest livestock holding state in India. The illegal route begins at Gundalpet in Karnataka, entering Tamil Nadu at Madhuamali in the Nilgiris. The animals transported along this route enter Kerala through the border at Gudloor. The second transport point is at the Waliyar post near Coimbatore in Tamil Nadu, which is an entry point into Kerala (PETA, 2005).

Illegally transported animals do not have valid health certificates to indicate that they are free of contagious disease and fit for transport. Though there are a number of checkpoints in the forested areas of this route to prevent poaching, vehicles are known to pass through without being checked, by paying bribes to the person operating the checkpoint.

TREKKING. More than 170 million livestock are trekked for hundreds of miles to slaughterhouses all over the country. Small ruminants accustomed to pasture grazing are driven by this mode and they are allowed to graze or browse on the side of the road (PETA, 2005).

Spent cows and bullocks are purchased by middlemen directly from villagers or village markets and are walked to the nearest town, tethered in twos and fours. The middlemen employ casual labour to walk the animals along busy roads and highways until they reach the slaughterhouse. On the way, the animals may be subjected to whipping, goading by inserting sharp instruments into the vagina or anus, or by applying irritants such as tobacco or chilli power in the orifices and eyes to make the animals move faster.

TRUCKS. Animals are transported in trucks for distances ranging from 500 to 1500 km. With an average speed of 60 km/h (which is likely to be unattainable in poor weather or road conditions), the longest journeys could take in excess of 25 h. Animals are bought by middlemen from the village, or the villagers themselves join together to hire a truck, tempo (auto rickshaw), three-wheeler or tractor. Vehicles are frequently open, without any cover or are covered by tarpaulin. The animals are tied together with ropes attached to the roof or side of the truck and are loaded to maximum capacity. Sometimes a temporary two-tier system is employed to increase the number of animals that can be transported. The animals are loaded on to trucks by goading or twisting or biting of the tails. Once inside the trucks, the animals are unable to move for the entire duration of the journey, which can often exceed 24 h. Often, the animals receive neither water nor feed en route. Unloading at the slaughterhouse is performed by employing the same procedures as at loading.

TRAIN.   Animals, especially dairy animals, are transported between railway stations in wagons designed to hold a maximum of 10 horned cattle or 15 calves or 8 cattle with calves in broad gauge trains. There are strict rules to which the consignor must adhere regarding loading and positioning of animals within the wagon and their feed and water provision. The station master has to obtain a declaration from the owner that the animals are not meant for slaughter. Often this method of transport is misused and cattle from states where there is a total ban on cow slaughter are sent to other states for slaughter using false documentation. Some Indian NGOs monitor animals at railway loading yards and it is thought that their actions have resulted in a reduction of illegal rail transport of animals for slaughter (Rahman, 2006, personal communication).

OTHER MEANS (ON BICYCLES, CARTS, ETC.).   There are many other ways by which animals are transported. In states where rivers have to be crossed, ferries are used. Most often these are small boats mostly meant for commuter traffic but which are also used for animals. Of all the animals transported for slaughter, poultry present some of the most concerning welfare problems. Birds are transported in overcrowded cages inside small trucks and tempos for 2–6 h under harsh environmental conditions (CUPA/WSPA, 2001). The birds are handled roughly while loading and unloading from the cages. Due to cramped conditions in the metal cages, birds have bruises especially in the breast region. Birds are sometimes transported on bicycles and scooters with their legs tied, hanging upside down from the handlebars, with up to 60 other birds.

Pakistan

According to the Pakistan Slaughterhouse Act of 1983, the killing of animals outside slaughterhouses is prohibited. The majority of slaughterhouses are now located in densely populated areas. Slaughtering, carcass dressing and by-product handling are all done in the same space (Khushk and Memon, 2006).

There is considerable illegal live animal transport between India and Pakistan. The border of Rajasthan state in India in the region of Monabao and Khokrapar in Pakistan is particularly well used for illegal transport. Animals are also illegally transported by truck from Pakistan to Afghanistan and Iran. However, the government has recently decided to legally import cattle from India as a way of combating illegal trade and its effects on meat prices in Pakistan (Dinker Vashisht, 2006). However, a number of welfare issues arise as a consequence. One issue is that due to the absence of quarantine facilities at Wagah border, the animals procured from Amritsar (India) are transported nearly 500 km to Delhi, where they are inspected before travelling back to Wagah and into Pakistan. The entire duration of this journey can take more than 6 days.

Bangladesh

According to the Bengal Act I of 1920, The Cruelty to Animals Act (now the Society for Prevention of Cruelty to Animals Ordinance) has made provisions for proper transport of animals prior to slaughter. However, the Act is poorly implemented.

Illegal transport from India (West Bengal) into Bangladesh is extensive. The main crossing points are the border towns of Benapool in India and Kustia, Jessore and Rajsahi in Bangladesh and across Aricha Ghat ferry crossing into Dhaka and other towns. Animals can be transported for over 14 h in cramped conditions on trucks from the Indian borders all the way to Dhaka.

Sri Lanka

The transport system for all species of animals is similar to that described in India. Animals are transported by road or by trekking. According to observations made over the last decade, mortality during transport was 0.01–0.05% for pigs, 3–6% for goats, 3–4% for cattle and 2–4% for broilers. These figures are based on a journey lasting a maximum of 6 h for cattle and goats and 3 h for pigs and broilers (Dharmawardena, 2003). There is limited illegal transport from India into Sri Lanka. When it does occur, it is most frequently in the north-eastern parts of the country.

Discussion

Meat production on the Indian subcontinent predominantly involves poultry, goats, sheep, cattle and buffaloes (see Table 11.1.1A). Livestock production for domestic consumption, especially meat and meat products, is dependent on a diversity of socio-economic and religious factors, which influence the dynamics of the trade in meat and live farmed animals for slaughter. The majority of the population of India are Hindus. Among meat-eating Hindus, beef is taboo. For all except the poorest people in society, beef is not eaten. Further, many Indian states have a ban on cow slaughter, with the exception of West Bengal and Kerala. However, the predominantly Muslim-populated countries of Pakistan and Bangladesh have no such aversion to eating beef. This results in large volume movement of live animals across the borders, especially from India into Bangladesh, where demand for beef far outstrips domestic supply. Generally, Indian Muslims do not eat beef unless they cannot afford other types of meat, but still wish to include meat in their diet. In India, as a consequence of the low demand for beef, it is one of the cheapest meats. In Pakistan and Bangladesh, where there is a great demand for beef, it is as expensive as any other meat and there are no social or religious taboos in eating it. In Sri Lanka, chicken is the most popular meat.

As a general rule, beef products available in these countries on the Indian subcontinent come from spent dairy animals and aged bullocks. Similarly, buffaloes are sent for slaughter once their use as draught or dairy animals has expired – except for commercially reared buffalo for exports.

Small ruminants (sheep and goats) are reared by small and marginal farmers on free range systems. Once these animals reach maturity, they are either transported in trucks or trekked to the slaughterhouses. Meat from sheep and goats is the most preferred red meat across the whole subcontinent. There is high demand for this meat both in the domestic market, in addition to international markets in the Middle East.

Poultry production has become an increasingly intensive industry. Spent hens are transported to resale outlets which have opened in the suburbs of every city.

Poultry meat is the most commonly available as well as one of the most affordable. Hindus prefer to eat chicken when dining out, reportedly because chicken is easily identifiable. In some restaurants, beef products are sold as lamb in cost-cutting exercises.

The consumption of pork is restricted to Christians and tribal Hindus especially in the north-eastern states of India. There is virtually no pork production or consumption in Pakistan and Bangladesh. In Sri Lanka, with large Buddhist and Hindu communities, poultry forms the bulk of meat production (see Table 11.1.1A).

Transport of animals in the subcontinent is mainly by trekking, road and rail. Cattle, sheep and goats are made to walk long distances to economize on the cost of transport by poor farmers or middlemen who wish to maximize profits. On trucks, animals are loaded to or above capacity, driven for long distances often without feed and water and then slaughtered inhumanely – especially if the slaughter is being carried out in an illegal slaughterhouse.

Legislation exists, especially in India, to regulate animal welfare during transport. However, enforcement can be difficult. Corruption is reported to occur at many levels in the subcontinent and contributes to the lack of implementation of regulations governing live transport. A number of NGOs and animal welfare organizations, including the Animal Welfare Board of India, are lobbying to ensure that cruelty during transport does not occur. Voluntary workers for these organizations have, at great risk to their personal safety, intercepted trucks and animals on the move, confiscated them and have charged the responsible individuals with the help of police. It is reported that, in India, a number of volunteers have been killed trying to monitor and control this trade (Rahman, 2006, personal communication).

Some Indian states back up laws banning cattle slaughter with measures to prohibit transport for slaughter across their borders. Unfortunately, these have again proved difficult to enforce. As a result, much cattle transport operates outside the law with predictable consequences for welfare. Serious abuse has been documented (PETA, 2005).

There is a need for the countries of the Indian subcontinent to cooperate to prevent animal smuggling. Cooperation is also required to ensure that the legal trade in live animals is operated as humanely as possible and is replaced, once practicality allows, with a trade in meat.

Animal welfare issues and implementation measures

With the growing economy and human population of many of the countries surveyed in the Indian subcontinent, there is likely to be a continuing rise in the demand for meat and meat products. Increases in animal transport are likely to follow. It is imperative that long-term measures to protect these animals are initiated and, importantly, are implemented.

A top priority is to reduce journey times to slaughter. The construction of small local slaughterhouses close to the farms and villages where the animals are reared would reduce journey time. Carcasses can be transported to markets for sale to those who have access to refrigeration. The development of an infrastructure for refrigerated transport is also required.

Measures are also needed to make such transport as does occur more comfortable for animals. Legislation is required to incorporate OIE guidelines. Strategies are needed to develop capabilities of enforcement. Efforts are needed to improve the design of trucks, lairages and loading and unloading facilities. Training programmes are required for all involved in animal transport.

A particular challenge will be to develop and enforce legislation that ensures that aged draught and dairy cattle in India are protected from illegal trafficking. There is an additional challenge to ensure that those animals which no longer have economic value, but which it is illegal to slaughter, can be provided for.

Throughout the Indian subcontinent there is a tradition of concern for the lives and welfare of animals. There is a clear role for animal welfare societies to build on this tradition, to educate the public and to put pressure on politicians to develop and enforce animal welfare legislation. More state and national boards of animal welfare could be set up to increase the likelihood that animal welfare regulations are given high priority and are implemented.

International understanding and concern for the welfare of animals is on the increase. In 2005, the OIE adopted its first global guidelines for animal welfare, specifically in the areas of land transport, sea transport, slaughter of animals for human consumption and killing of animals for disease control. The passing of the guidelines by all 167 member countries, some of which did not have national animal protection legislation of their own, signalled that animal welfare was no longer a concern only of certain nations, but had become an issue for official attention at a global level.

Conclusion

Improving welfare during transport throughout Asia depends on:

- Legislation and regulatory codes, appropriate to each species and drafted with sufficient precision and detail, to ensure that welfare needs during transport are met;
- Measures to ensure that animals are slaughtered as close as possible to the farms where they are reared;
- Effective enforcement of welfare regulations;
- Education and training which ensures that animal transport professionals have the necessary attitudes and skills to ensure the welfare of the animals in their charge; and
- Increased concern and understanding among the general public of these issues leading in turn to pressures on governments and the food industry to ensure high welfare standards.

All the countries covered in this chapter are members of the OIE. As mentioned earlier in this chapter, the OIE has produced separate guidelines for the transportation of animals by sea, land and air (CIWF, 2006; OIE, 2006). These form a basis for developing detailed regulations regarding the treatment of animals in transit. These guidelines were unanimously adopted by the OIE's 167 member countries.

Countries in the Indian subcontinent have a long history of producing legislation governing animal welfare going back to the days of colonial rule, if not before.

Some of these address the issue of transport. China has also recently passed legislation and/or published regulations which attempt to address the welfare of animals including that during transport. There is now an urgent requirement to update this legislation in line with the requirements of the OIE Guidelines to which all of the countries covered in this chapter have signed up. Effective enforcement structures and processes are then required to see that the regulations are adhered to.

However, the welfare of animals during transport throughout Asia depends primarily on the length of the journey. However well the animals are transported, fatigue and exhaustion are likely to become an increasing problem as the journey continues. Any issues of ill-treatment are compounded the longer the period of that ill-treatment.

It is vital to take steps to reduce the distances which animals need to travel. As more consumers gain access to refrigeration equipment, this can be achieved by setting up well-designed and managed regional and local slaughterhouses close to the production areas. Refrigerated transport with an accompanying infrastructure will be required to take the meat to its markets. Any demand for fresh meat needs to be met from local sources. Legislation is required to limit journey times.

Animal transporters are key animal welfare workers and their professional role should be appreciated. Proper training programmes are required, with recognized professional qualifications, to ensure that animals do not suffer through ignorance or a lack of respect. Animal welfare expertise among those who design and maintain vehicles, ramps and holding facilities is also crucial.

Animal welfare groups, the media and educators of all sorts have an important role to play in developing public concern for the welfare of animals during transport. Pressure from the public will in turn help legislators, the regulatory authorities and the food industry itself to play their part in ensuring that animals do not suffer during transport.

There is a danger that, as economic liberalization leads to the availability of cheaper meat, the welfare of both farmers and animals could suffer. The concerns of global society, as reflected in the unanimous agreement to the OIE Guidelines, suggest there is a better way forward.

The development of humane farming systems for both local and international high welfare markets is likely in future to offer opportunities for farmers throughout Asia. High welfare production gives animals a good life followed by humane transport and slaughter. The final journey should be short and comfortable.

Acknowledgements

The author wishes to thank Peter Li, Frank Chen, Abdul Rahman, Abe Augulto, and Swarg Yudisthira for researching and compiling the regional reports used to inform this chapter.

References

Abdul Rahman, S. (2004) Animal welfare: a developing country perspective. *Proceedings of the Global Conference on Animal Welfare: An OIE Initiative.* OIE, Paris, France, pp. 101–112.

Abdul Rahman, S. (2005) Animal resource development as an alternate source of income generation. *Thought* 8, 8–27.

Afzal, M. (1997) Pakistan country paper. *Proceedings of a Consultation on Setting Livestock Research Priorities in West Asia and North Africa (WANA) Region.* International Centre for Agricultural Research in the Dry Areas, Aleppo, Syria.

Agriculture Ministry (1980) Agriculture Ministry Report on Accelerating Farm Animal Production. Available at: http://vip.chinalawinfo.com/Newlaw2002/SLC/SLC.asp?Db=chl&Gid=687

Anonymous (2006) China's live cattle and sheep given permission to enter the Malaysian market. Available at: www.saincom.com/2006003.htm

Chen, S.-C. (2004) A study on price determination process of chicken industry in Taiwan (*trans.*). MSc Thesis, National Taiwan University, Republic of China (Taiwan).

China Animal Husbandry Industry Association (2001) *China Animal Husbandry Industry Yearbook: 2000.* China Agriculture Press, Beijing, China.

China Animal Husbandry Industry Association (2004a) The share of animal husbandry industry in GDP. Available at: www.caaa.cn/show/newsarticle.php?ID 858

China Animal Husbandry Industry Association (2004b) *China Animal Husbandry Industry Yearbook: 2003.* China Agriculture Press, Beijing, China.

China Animal Husbandry Industry Association (2005a) *China Animal Husbandry Industry Yearbook: 2004.* China Agriculture Press, Beijing, China.

China Animal Husbandry Industry Association (2005b) The share of animal husbandry industry in GDP. Available at: www.caaa.cn/show/newsarticle.php?ID=858

China Animal Husbandry Industry Information Network (2006) China's total meat output. Available at: www.caaa.cn/show/newsarticle.php?ID=1045

Chinese Cultivation Network (2003) North China Domestic Animal Transaction Center (*trans.*). Available at: www.chinabreed.com/news/cover/2003/06/2003061824828.shtml

China State Quality Inspection and Quarantine Administration (2006) List of the export-oriented poultry slaughtering facilities and their supplier farms: up to 12 July 2006. Available at: www.caaa.cn/show/newsarticle.php?ID=84374

Chou, J.-G., *et al.* (1995a) *Studies on the Relationship Between Various Transport Facilities with Exhaust and Death of Pig Transporting* (*trans.*). Council of Agriculture, Taipei, Taiwan.

Chou, J.-G., *et al.* (1995b) Surveys of accidental death of pigs during transport in winter *(trans.)*. *Taiwan Agriculture* 31, 55–72.

Council of Agriculture (2005) *Annual Report 2004.* Council of Agriculture, Taipei, Taiwan.

Council of Agriculture (2006) Animal Production in Taiwan. Available at: http://eng.coa.gov.tw/./content.php?catid=9158&hot_new=8877

Council of Agriculture (undated) Animal Production in Taiwan. Available at: http://eng.coa.gov.tw/./content.php?catid=9158&hot_new=8877

CIWF (2006) *Animal Welfare During Land Transportation – A Brief Guide.* Compassion in World Farming Trust, Petersfield, UK.

CUPA/WSPA (2001) *Humane Standards in Poultry Rearing and Transport.* The World Society for the Protection of Animals, London.

Department of Livestock Services Activities (2006) *At A Glance 2004–2005.* Government of Bangladesh, Bangladesh.

Dharmawardena, I.V.P. (2003) Status of animal welfare in Sri Lanka. *Proceedings of the 2nd International Seminar on Animal Welfare, 14–16 February.* Commonwealth Veterinary Association and World Society for the Protection of Animals, Bangalore, India, pp. 18–21.

Dinker Vashisht (2006) Pakistan looks to India for livestock. *The Economic Times,* 15 April 2006.

EAST (2000) *Investigation of Governmental Slaughterhouses in Taiwan.* Environment & Animal Society of Taiwan, Taiwan.

EAST (2006) Online Footage. Available at: http://my.so-net.net.tw/hicnh2/chicken-2d.wmv

FAOSTAT (2006) UN Food and Agriculture Organization statistics. Available at: http://faostat.fao. org/site/569/DesktopDefault.aspx?PageID=569

Fengjun, X. and Haifeng, H. (2006) China exported its first group of dairy cows to Mongolia. *Xinhua Network*, 30 September 2006.

Jayatilleke, C. (2004) New chapter opens in animal welfare. *Daily News*, 23 October 2004.

Kao, C.-T. (2000) Report on studies of reaction in behaviour and physiology of pigs after recovering from 24 hour transport for 6 hours in holding pens *(trans.)*. *Today's Pig Magazine* 21, 72–78.

Khushk, A.M. and Memon, A. (2006) Unhygienic slaughter houses. *The Dawn*, 27 March 2006.

Kuo, C.-Y. (1997) A study on chicken marketing and establishing chicken carcass exchange center in Taipei city. MSc thesis, National Chung Hsing University, Republic of China (Taiwan).

Lee, T.-R. and Du, Y.-T. (1997) A study of optimal weight of transshipment points and optimal number of wholesale markets after Taiwan carcass meat transportation system established. *Taiwanese Agricultural Economic Review* 3, 69–95.

Liao, C.-Y. (2000) Effects of ideas of pork consumption on animal welfare in Taiwan *(trans.)*. *Animals' Voice of Taiwan* 23, 20–21.

Liao, C.-Y. (2001) *Improving Pig Production Quality-Importance of Animal Welfare (trans.)*. Animal Technology Institute, Taiwan.

Lin, C.-Y *et al.* (2003) The analysis of household purchasing behaviours of pork in the Taipei area. *Taiwanese Agricultural Economic Review* 9, 43–62.

Lin, H.-C. (2004) Outlook on the slaughter of live poultry in traditional markets *(trans.)*. *Public Forum of Animal Protection* 2004, 211–214.

Liu, Y.-C, *et al.* (1998) Cost-analysis of transport ways in pig transportation. *Journal of National Chiayi Institute of Technology* 57, 97–117.

Lymbery, P. (2001) *Farm Animal Welfare in Taiwan, A Situation Report*. The World Society for the Protection of Animals, London.

Markets Administration Office (undated) Taipei City. Available at: www.tcma.gov.tw/market/home/samarket.asp?market_no=7

Ministry of Agriculture (2006) *Animal Health in China: 2004–2005*. The Ministry of Agriculture of the People's Republic of China, Beijing, China.

Ministry of Finance (2006) *Economic Survey 2004–05*. Government of Pakistan, Pakistan.

Ministry of Food Processing Industries (2006) Sectoral Review. Available at: http://mofpi.nic.in

Ministry of Justice (undated) Laws and Regulations Database of the Republic of China. Available at: http://law.moj.gov.tw/Eng/Fnews/FnQuery.asp

Nanjing Daily Correspondent (2007) Pigs slaughtered by machine reached 52% of the total slaughtered in Nanjing. Available at: http://nj.blogbar.cn/default.php?op=ViewArticle&articleId=525

OIE (2006) Terrestrial Animal Health Code, Section 3.7 – Animal Welfare. Available at: www.oie. int/eng/normes/mcode/en_titre_3.7.htm

PETA (2005) *Investigative Report – Assessment of Animal Welfare, Worker Welfare and Environmental Impact. Investigate Report on Transport and Slaughter Conditions for Animals Used for Meat and Leather in Tamil Nadu, India*. People for the Ethical Treatment of Animals, Norfolk, Virginia.

Raha, S.K. (2000) Development of livestock sector: issues and evidences. In: Mondal, M.S. (ed.) *Changing Rural Economy of Bangladesh*. Bangladesh Economic Association, Dhaka, Bangladesh.

Railway Ministry (1994) The Rules Regarding Railroad Transport of Fresh Meat Products and Live Animals. Available at: http://vip.chinalawinfo.com/Newlaw2002/SLC/SLC.asp?Db=chl&Gid=28084

Ranjhan, S.K. (1999) Production of meat from buffaloes. *National Seminar for Sustainable Development of Buffaloes for Milk, Draft and Meat*, NDRI, Karnal. 14–16 October 1999.

Ranjhan, S.K. (2006) Farmer-industry linkage in meat production: a symbiotic relationship. Paper presented at the International livestock Research Institute – Food and Agriculture Organisation Workshop, Nairobi, East Africa, 13–15 February 2006.

Ramaswamy, N.S. (2000) Report on the study of the feasibility of economics of rural abattoir scheme to Bangalore City, April 2000.

Rao, K.C. (1997) Livestock in India – Instrumental of social justice. Paper presented at Dr V.S. Alwar Memorial Oration Lecture, Chennai, 1997.

SCAHAW (2002) The welfare of animals during transport (details for horses, pigs, sheep and cattle). Available at: http://ec.europa.eu/food/fs/sc/scah/out71_en.pdf

Taiwan Government Information Office (2006) Taiwan Yearbook 2005. Available at: www.gio.gov.tw/taiwan-website/5-gp/yearbook/p015.html

The Dawn (undated) All about livestock – dairy industry in pakistan. Available at: Pakistan.com

Xu, B. (2005) China is replacing the US as the world's biggest consumer nation. Available at: www.voanews.com/chinese/archive/2005-02/w2005-02-16-voa71.cfm

Wang, W. (2004) The negative impact of the shiplocks of the three gorges dam on water transport on the Yangtze. *Observer*, 3 June 2004.

Weeraratna, S. (2003) Animal rights and law enforcement. *Buddhist News Network*, 30 July 2003.

Wen-Fu, H. (1996) *Pig Industry & Production and Marketing of Pork in Taiwan (trans.)*. Taiwan Meat Development Foundation, Pingtung, Taiwan.

Wu, M.-M. (2005) How to commercialize pig markets: a discussion of the residents' protest against the Taichung City wholesale pig market. Available at: www.taiwanthinktank.org/ttt/servlet/OpenBlock?Template=Article&category_id=18&article_id=342&lan=en

Yonggang, G. (2005) Per capita meat consumption in China shall reach 65 kg. *Southern Rural News*, 23 July 2005.

Zheng, Z.-J. (2005) Brief introduction to animal transport management regulation *(trans.)*, *Agriculture Policy & Review*. Available at: http://bulletin.coa.gov.tw/view.php?catid=9674

Appendix 11.1

Table 11.1.1A. Number of animals slaughtered for meat in 2005. (From FAOSTAT, 2007.)

Country	Buffalo ('000s)	Cattle ('000s)	Chicken ('000s)	Duck ('000s)	Goat ('000s)	Pig ('000s)	Rabbit ('000s)	Sheep ('000s)	Total (millions)
China	3,600	50,300	7,300,000	1,800,000	150,000	673,000	343,000	161,000	10,481
India	10,000	14,500	2,000,000	50,000	47,500	14,200		19,900	2,156
Pakistan	4,400	2,470	366,000	3,600	21,800			10,000	408
Bangladesh	47	2,570	144,000	14,000	19,600			429	180
Sri Lanka	35	215	67,000	16	75	30		8	67

Table 11.1.2A. China's meat production by livestock as a percentage of world total. (From FAOSTAT, 2006.)

Category	1978	1985	1990	1995	2000	2005
Meat total	8.7	13.6	16.9	23.3	26.9	29.3
Pork	19.2	29.3	34.4	41.7	46.0	48.9
Chicken	5.4	5.3	7.5	13.0	15.3	14.5
Mutton/lamb	3.2	4.8	7.8	12.5	18.9	33.3
Beef/veal	0.5	0.8	2.1	6.1	8.8	11.3
Poultry meat	6.7	6.5	9.1	15.8	18.2	18.1

12 Australia and New Zealand

M.W. FISHER[1] AND B.S. JONES[2]

[1]Kotare Bioethics, PO Box 2484, Stortford Lodge, Hastings, New Zealand;
[2]RSPCA Australia, PO Box 265, Deakin West, ACT Australia

Abstract

The transport of animals to slaughter in Australia and New Zealand is complex with differences in geography, animals, distances travelled, slaughter destinations, animal welfare guidelines and regulations and people's expectations. However, three main themes are apparent, two of which raise significant issues for animal welfare.

First, the majority of livestock within each country are transported to slaughter over varying, but not long, distances. Such transport is regarded as necessary, and both countries have government endorsed codes or guidelines describing the conditions (e.g. handling, vehicle design and operation, environmental factors, unavailability of food and water) under which livestock should be transported. There also exist various industry guidelines and quality assurance programmes which serve to reinforce government standards. In New Zealand, codes of welfare have legal status, and failure to adhere to the minimum standards can be used to support a prosecution. In contrast, in Australia there are, as yet, no enforceable standards for the land transport of livestock for slaughter. Reported instances of poor animal welfare associated with this type of transport in either of the two countries are rare, but there is little publicly available information on journey outcomes.

Second, in Australia some livestock and feral animals may be transported long, sometimes extremely long, distances for slaughter. These distances suggest the animals experience less than ideal conditions, particularly when the journey exposes animals to significant changes in environmental conditions. While long journeys can compromise animal welfare, the responses of these animals to such journeys and the conditions they experience have not been adequately documented. Of particular concern is the transport of livestock from remote areas. Often unused to human contact and handling, they may suffer from the stress of mustering, confinement and long-distance transport.

Finally, animals are also exported by sea from Australia (and potentially New Zealand) for slaughter overseas. In view of the conditions they inevitably endure during transport; the inherent mortality demonstrating that some are unable to tolerate those conditions; the equivocal benefits to the exporting country; and the alternative of supplying chilled product, it is concluded that the long-distance export of livestock for slaughter should not take place.

Acronyms

AQIS   Australian Quarantine Inspection Service
DAFF   Department of Agriculture, Fisheries and Forestry (Australia)
MAF   Ministry of Agriculture and Forestry (New Zealand)
RSPCA   Royal Society for the Prevention of Cruelty to Animals
ASEL   Australian Standards for the Export of Livestock
OIE   World Organisation for Animal Health

Introduction

Livestock production in Oceania is dominated by two countries: Australia and New Zealand (see Rahman *et al.*, 2005, for a description of general animal welfare in Oceania). The 5.2 million t of meat they produced in 2004 represented 92% of the region's total production. Most of the remainder was produced in Papua New Guinea (7%) and Fiji (0.5%) (FAO, 2005).

Poultry, sheep, cattle, pigs, goats and deer are the most commonly farmed species (Table 12.1). With the exception of poultry and pigs, most of which are housed, livestock typically are pasture-farmed, often in relatively extensive environments. There are also smaller numbers of rabbits, ostriches, horses, possums, emus, buffaloes and camels slaughtered for meat.

Livestock are slaughtered for both domestic consumption and export. Australia is the largest exporter of red meat in the world. New Zealand is responsible for half of the world's sheep meat exports. Both countries derive much of their export earnings from agriculture (Australia 20.7% and New Zealand 48.1% in 2003; OECD, 2005).

All livestock are transported by road to slaughter (or export port), except for some cattle transported by rail in Australia. While some sheep, cattle and goats can be transported long distances to slaughter, pigs and poultry are generally not. The journey from farm to abattoir can vary from less than 50 km to more than 3000 km in Australia (MacArthur Agribusiness, 2001; Hassall and Associates, 2006). In New Zealand, two-thirds of journeys may be less than 100 km (Sanson, 2005), but others can be up to 400–500 km (Graafhuis and Devine, 1994).

In addition to the slaughter of livestock within each country, a proportion of live animals are also exported for slaughter. In 2005, Australia exported 4.2 million sheep and 573,000 cattle (Hassall and Associates, 2006). Virtually all the sheep were for slaughter, but some 9% of the cattle exported in 2003–2006 were for dairying (LiveCorp, Australia, 2006, personal communication), including 75,000 to China in 2004. The largest markets for Australian sheep exports were Kuwait, Saudi Arabia and Jordan,

Table 12.1. The total number (millions) of animals (excluding live exports) slaughtered per year within Australia (2005–2006) and New Zealand (2004). (From Rural Industries Research and Development Corporation, Australian Bureau of Statistics; Statistics New Zealand, Meat & Wool Economic Service, Poultry Industry Association of New Zealand.)

	Poultry	Sheep	Cattle	Pigs	Goats	Deer
Australia	437.9	30.5	7.58	5.4	1.1	0.04
New Zealand	87.5	28.3	5.2	0.8	0.1	0.4
Total	525.4	58.8	12.8	6.2	1.2	0.4

with Indonesia the key export cattle market (Hassall and Associates, 2006). Animals exported for slaughter are transported by sea in vessels carrying up to 100,000 sheep and/or several thousand cattle. They are assembled for quarantine, health inspections and varying levels of preconditioning to ship-like diets and feeding regimes, prior to loading for export (see Chapter 6, this volume). Distances travelled within Australia prior to shipping vary from as little as 100 km to more than 2000 km.

In 2004, New Zealand exported 62,000 cattle mostly to Asia including 60,000 dairy heifers to China (Hini, 2005). While in the past New Zealand also shipped animals for slaughter, cattle could not be exported for slaughter to countries which did not practise pre-slaughter stunning. The last shipment of sheep was in 2003 (MAF, 2007).

Clearly, in countries so dependent on agriculture, the transport of livestock to slaughter is important. Both New Zealand and Australia have legislative provisions, regulations and detailed codes or guidelines designed to protect the welfare of transported animals, but these differ in the standards imposed and their enforceability. These provisions are supported by veterinary involvement and some scientific research. Nevertheless, there are significant concerns over some practices and conditions, namely the export of livestock for slaughter and, in the case of Australia, the lack of adequate regulation of livestock transport standards. The transport of livestock and feral animals from remote areas in Australia for slaughter is of particular concern, since in these situations, the handling, long distances, availability of food and water and thermal conditions to which these animals are exposed can severely compromise their welfare.

The remainder of this chapter describes: (i) livestock industries and the regulations and factors affecting them in relation to transport; (ii) welfare of the animals and the conditions affecting them; finishing with (iii) a discussion of these results. Although there are many similarities between Australia and New Zealand, there are several important differences. Among the most significant is the geography and climates of the two countries. This is reflected in the distances travelled and conditions experienced by animals being transported. The other main difference is in the complex system of government in Australia which has both federal and separate state and territory governments – whereas New Zealand has just a national government. Where these differences have an impact on livestock transport practices and outcomes, this chapter presents information on the two countries separately.

Background

Key stakeholders

The primary stakeholders are the farmers or producers and livestock transporters, processors and exporters. There are also a number of secondary stakeholders including regulators, animal welfare interest groups, animal scientists and researchers, retailers and consumers. Individuals and organizations of both types of stakeholders may also be represented by collectives or associations.

Livestock farmers and producers
There are about 130,000 farms in Australia, with the beef cattle industry being the largest with around 28% of all farms (Australian Bureau of Statistics, 2006). Sheep

and cattle are farmed in most parts of the country although there are some regional variations in their distribution. Dairy cattle tend to be restricted to southern and coastal areas, while the majority of beef cattle are reared in Queensland (40% of the herd) and New South Wales (22%) (Australian Bureau of Statistics, 2005) (see Fig. 12.1 for the boundaries of each state/territory). The farming of cattle in the cooler temperate south of the continent tends to be relatively intensive, with mainly European breeds (*Bos taurus*), particularly Hereford and Angus. In northern Australia, farms (properties) are extremely large, with reduced contact between animals and humans. In these regions, beef breeds are dominated by *Bos indicus*, such as the Brahman and Brahman-derived breeds, able to tolerate high temperature and humidity. Most pigs are farmed in grain-producing areas of New South Wales, Queensland and Victoria, with poultry mainly farmed in New South Wales and Victoria.

New Zealand's 50,000 livestock farms are spread throughout the country, over 32,000 in the North Island and 17,000 in the South Island and most are sheep

Fig. 12.1. Location of export abattoirs processing sheep and cattle in Australia. The map shows all functional abattoirs owned by the 'Top 25' largest processors and all additional AUS-MEAT-accredited export abattoirs not included in the 'Top 25' (Meat and Livestock Australia, 2005; AUS-MEAT, 2005). The abattoirs shown are responsible for processing more than 80% of the sheep and cattle slaughtered each year.

and/or beef (56%) or dairy (28%). There are some regional differences with more dairying in the North Island and more deer farming in the South Island, for example. While many operations are small and family-owned, there are a number of large producers. For example, three poultry meat producers supply approximately 98% of the market (Cooper-Blanks, 1999).

Livestock transporters and processors

All livestock are transported by road within Australia, except for some long-haul rail transport of cattle in Queensland (Access Economics and Maunsell Australia, 2002). Large consignments of animals (especially sheep and cattle) are usually transported in heavy truck and trailer units and road trains capable of carrying up to twice the number of livestock as more conventional units. Smaller consignments are transported in units varying from double-decker open-top trucks to small single-deck trucks. Similarly, in New Zealand most livestock are transported to slaughter by road, with a small proportion moved by a combination of road and coastal ship between the North and South Islands (see Chapter 7, this volume, for details of road transport).

Most abattoirs in Australia are located within regions of Victoria, New South Wales, Queensland and Western Australia (see Fig. 12.1). The centralization of the meat processing industry and associated reduction in the number of abattoirs and meat processors, as well as competition between companies, has meant that the average distances from farm to slaughter have increased significantly in some regions (Industry Commission, 1994; QAF Meat Industries, 2004; Productivity Commission, 2005). New Zealand has a significant number of animal processing facilities operating throughout both the North and South Islands. Loyalty to particular meat processors, competition between meat processing companies and occasional droughts resulting in larger than normal numbers of animals being slaughtered, mean livestock are not necessarily always transported to the nearest plant. Inter-island transport of livestock is usually undertaken for reasons other than for slaughter, such as the expansion of dairy farming into the South Island (Fisher *et al.*, 1999). However, on occasions livestock are shipped between the North and South Islands, a sailing time of 3 h, when meat processors compete for livestock outside of their normal procurement catchment. The two islands have separate payment schedules with slightly different returns for farmers (Countrywide, 2006).

Livestock exporters

Livestock are exported from up to 23 Australian ports to 44 overseas ports in 29 countries, mostly by 12 companies with the involvement of around 25 other businesses – e.g. transporters, fodder suppliers, feedlot operators, agents, shearers (Hassall and Associates, 2000, 2006). Most sheep exported are adult Merino wethers and lambs (Norris and Norman, 2005) and the majority of cattle are steers with an increasing proportion being *B. indicus* breeds (Bindon and Jones, 2001).

Although the export of sheep has been in existence since the late 1800s, the more regular trade began in the 1970s, with more than 3.8 million sheep exported in 2003–2004, or 12% of the national turn-off (Hassall and Associates, 2006). Most (71% over the period 2003–2005) originate from Western Australia, representing 62% of the annual turn-off (animals leaving farms) of sheep in that region (Hassall and Associates, 2006).

Cattle exports began in the 1970s, initially for breeding purposes, but expanded in the 1980s with the development of feedlots in Indonesia and the Philippines. In 2003–2004 over 680,000 cattle were exported, most originating from the Northern Territory, Western Australia and Queensland. In the Northern Territory, the export trade accounted for 47% of the region's cattle leaving farms for slaughter or live export in 2004 (Hassall and Associates, 2006), there being no other significant market providing a similar financial return and no functioning slaughter facilities within the region.

Animal welfare research

Research into the welfare of livestock transported to slaughter is funded and undertaken by a variety of groups in Australia and New Zealand.

In Australia, transport research is funded by industry groups and is largely export-related (Meat and Livestock Australia, 2007). More recently, the effects of land transport conditions on the physiology and behaviour of livestock have been the focus of research. Examples of the topics investigated (some are yet to be completed) and the associated funding agency include:

- Developing animal welfare indicators on ship and in pre-export assembly depots (Livecorp/MLA);
- The physiology of heat stress and use of electrolytes, including responses of cattle to prolonged and continuous heat and humidity relevant to live export by sea (Livecorp/MLA; Beatty et al., 2006);
- Updating biological assumptions in the current heat stress risk management tool (Livecorp/MLA);
- The death rate and causes of death in cattle exported by sea (Norris et al., 2003); and
- Animal welfare outcomes of livestock road transport practices for cattle and sheep (MLA).

Examples of relevant animal welfare research conducted in New Zealand include:

- The effects of stock crate design, stocking density, stationary periods and temperature and humidity on conditions for sheep during road transport (Fisher et al., 2002, 2004);
- Specifications on pens, rails and troughs for live sheep exports (Waghorn et al., 1995);
- Monitoring conditions on stock trucks (Fisher et al., 2002);
- The welfare aspects of stock unloading (Johnson and Gregory, 1998); and
- Responses of red deer to conditions during road transport (Wass et al., 1997).

Regulators

The Australian Government Department of Agriculture, Fisheries and Forestry (DAFF) is responsible for Australia's agricultural sector at a national level, including the regulation of livestock exports. The Australian Quarantine Inspection Service (AQIS) is the agency within DAFF focused specifically on the regulation of imports and exports. DAFF also administers the Australian Animal Welfare Strategy, developed to coordinate improving the welfare of animals in Australia.

State and territory governments are responsible for approving and implementing animal welfare legislation and codes of practice relating to the transport of livestock within Australia.

The Ministry of Agriculture and Forestry (MAF) administers the New Zealand Animal Welfare Act. Biosecurity New Zealand, a division of MAF, has Animal Welfare and Compliance and Enforcement Groups. The Animal Welfare Group is responsible for policies and standards for the humane treatment of animals. Working closely with the National Animal Welfare Advisory Committee (NAWAC), the group focuses on resolving animal welfare problems, identifying research priorities, liaising with New Zealand and international agencies involved in animal welfare, supporting New Zealand's policy in relation to animal welfare and international trade and ensuring all complaints of cruelty are investigated in accordance with the requirements of the Animal Welfare Act 1999. The Compliance and Enforcement Group employs experienced auditors and law enforcement officers (investigators) to audit, investigate and respond to serious breaches of legislation, as well as initiate prosecutions. Since 1993, MAF (see NAWAC, 2005) has also commissioned operational research in areas of animal welfare where there is a lack of information.

Industry organizations

Many industry stakeholders are represented by national organizations which promote the interests of their member bodies:

- The Australian Red Meat Advisory Council promotes the interests of the sheep meat, cattle, feedlot, meat processing and live export industry organizations in Australia.
- Meat and Livestock Australia and Meat & Wool New Zealand provide research and development, market information and marketing support to the main livestock industries, funded by producer levies. Producers are also represented by the National Farmers' Federation (Australia) and Federated Farmers (New Zealand).
- Other sectors of the industry are represented by individual national producer groups such as Australian Pork Limited; the Australian Chicken Meat Federation; Dairy Australia; and the Central Australian Camel Industry Association. In New Zealand, these include the Poultry Industry Association of New Zealand; Deer Industry New Zealand; and Dairy Insight (New Zealand).
- The Australian Meat Industry Council and the Meat Industry Association of New Zealand represent processors, retailers and others involved in the post-farm gate meat industry.
- The Australian Livestock Transporters Association and the Road Transport Forum New Zealand represent the interests of transporters.
- In Australia, livestock exporters are represented by Australian Live Exporters Council and LiveCorp, the latter responsible for research and development, industry standards, accreditation of livestock exporters and the provision of training.

Some of these groups have initiated or contributed to animal welfare guidelines relevant to the transport of livestock to slaughter such as:

- *Is It Fit to Load? A National Guide to the Selection of Animals Fit to Transport* (Meat and Livestock Australia, 2006b);

- *Managing your Bulls before Slaughter* (Meat New Zealand, 1999);
- *Animal Welfare – Stunning and Transporting* (Meat New Zealand, 1998); and
- *Wetting Cattle to Alleviate Heat Stress on Ships* (Meat and Livestock Australia, 2003).

Legislation

Animal welfare legislation in Australia and New Zealand aims to prevent cruelty and promote the welfare of animals. In Australia, state and territory governments have primary responsibility for animal welfare, and each has its own Animal Welfare, Prevention of Cruelty to Animals or Animal Care and Protection Act. In contrast, New Zealand has a single, national Animal Welfare Act. Legislation is complemented with codes and guidelines providing a mixture of legislative standards, recommended best practices and information/guidance on the welfare of animals during transport.

Australia has specific Model Codes for the land transport of cattle, pigs, poultry and horses. These provide guidance on key management issues such as loading conditions, maximum stocking densities, maximum journey times, animals that should not be transported and maximum food and water deprivation times (see Table 12.2). In all cases, the maximum recommended journey times are significantly longer than those required in New Zealand (see Table 12.3). For other species, the general husbandry code provides some limited guidance on transport conditions. Most states and territories have incorporated the codes under their animal welfare legislation, but compliance is not compulsory, although it can be used as defence against prosecution. In reality, the format of the codes is not compatible with a compliance tool, as much of the wording is in the form of guidance rather than setting clear standards. A process is currently under way to develop nationally accepted standards and guidelines for the land transport of animals, based on the Model Codes, with the aim of reaching greater consistency in regulation and application across the livestock industries (Animal Health Australia, Australia, 2006, personal communication).

New Zealand codes of welfare are an integral part of the framework and philosophy of the Animal Welfare Act. They detail appropriate behaviour, establish minimum standards and promote best practice for those responsible for the welfare of animals. Codes of welfare have legal status and failure to adhere to the minimum standards can be used as evidence to support a prosecution. The codes are developed through a process of consultation with stakeholders, including animal welfare groups and the public. They are intended to be reviewed regularly. The transport code has sections dealing with: the responsibilities of owners, drivers and vessel masters; minimizing stress; the transport of livestock by road, sea, rail and air; the transport of pregnant and lactating animals; and general guidelines for loading and the provision of food and water (see Table 12.3).

The regulation of the Australian livestock export industry is the responsibility of the Department of Agriculture, Fisheries and Forestry. Within DAFF, AQIS administers the legislation covering the livestock export industry: the Australian Meat and Live-stock Industry Act 1997 and the Export Control Act 1982. In addition, Biosecurity Australia has responsibility for assessing the health requirements of importing countries that need to be certified by the Australian Government.

Table 12.2. A summary of some of the recommendations for the land transport of sheep, cattle, pigs and poultry detailed in the Australian Model Codes of Practice for the Welfare of Animals. Note that allowable journey times include loading and unloading, and the ranges of maximum times without food and water relate to favourable conditions. (Available at: www. publish.csiro.au)

Feed, water and allowable journey times	Maximum stocking density	Animals that should not be transported
Sheep[a]		
Mature stock, 36h without food or water; extended to 48h if followed by 24h rest; young, 24–36h; ewes more than 4 months pregnant, 8h.	20kg live weight: 0.17m² per head; 60kg: 0.29m² per head.	Late-pregnant; weak, sick, or injured.
Cattle		
Mature stock, 36–48h; lactating dairy cows and calves, 24h; late-pregnant cows, 8h.	250kg live weight: 0.77m² per head; 650kg: 1.63m² per head; 5% fewer cattle if horned.	Late-pregnant and recently calved cows; unfit and injured, weak or diseased animals unless on veterinary advice.
Pigs		
After 24h there should be a rest period of 12–24h. All animals should be fed and watered at least once (preferably twice) in each 24h period, or more frequently if young.	50kg live weight: 0.22m² per head; 200kg: 0.61m² per head; with an additional 10% when ambient temperature exceeds 25°C.	Suffering animals.
Poultry		
24h (except day-old chicks) unless food and water available.	Day-old chicks: 400–475 per m²; birds ≤1–1.6kg: 40 per m²; birds >5kg: 100cm² per kg.	Sick, weak or injured birds.

[a]Recommendations for sheep are taken from the *1983 Model Code of Practice for the Road Transport of Livestock* Model Code of Practice, in draft at time of writing, for the road transport of livestock as there is no sheep-specific code.

The export of live animals from Australia comes under the Australian Standards for the Export of Livestock (ASEL) (DAFF, 2006). It covers sourcing and on-farm preparation of livestock; land transport of livestock; management of livestock in registered premises (holding and assembling points); vessel preparation and loading; on-board management of livestock (see Table 12.4); and the air transport of livestock. While the ASEL have provided the first regulated standards for livestock exports, a number of serious shortcomings have been identified in both their approach and content. Criticisms include that many sections of the standards lie outside the jurisdictional powers of AQIS and can only be enforced by state

Table 12.3. A summary of some of requirements for the land transport of sheep, cattle, pigs and poultry detailed in the *Code of Recommendations and Minimum Standards for the Welfare of Animals Transported within New Zealand*.

Feed, water and allowable journey times	Maximum stocking density	Animals that should not be transported
All animals		
		Animals with broken legs (unless treated by a veterinarian); animals unable to stand or bear weight on all limbs and unable to withstand the journey without suffering pain or distress.
Sheep		
12 h without water and 24 h without food but more frequently depending on animals and conditions.	Depending on live weight 0.14–0.31 m² per head.	Sheep within 3 days of being shorn.
Cattle		
12 h without water and 24 h without food but more frequently depending on animals and conditions; calves 12 h.	Depending on live weight: bobby (young) calves 0.16–0.4 m² per head; young 0.36–0.73 m² per head; adult 0.86–1.59 m² per head.	Recently dehorned, wild or unaccustomed to handling until they have quietened down; unfit, ill or weak calves; and calves of low birth weight or with severe physical or painful congenital defects.
Pigs		
Should not exceed 14 h with a limit of 8 h without water and 24 h without food, but more frequently depending on animals and conditions.	Depending on live weight 0.31–0.73 m² per head, but 20% more when temperatures exceed 25°C.	None in very hot, humid weather (an ambient temperature of 28°C or more).
Poultry		
Except day-old chicks, 12 h without water; travel must be completed in 24 h (birds) or 72 h (chicks).	21–25 cm² per chick; 175–105 cm² per kg bird depending on live weight.	Unhealthy or unvigorous day-old chicks; turkeys on hot days.

or territory legislation; some standards require actions that are not measurable or enforceable; and some do not set a sufficiently high benchmark for animal welfare (RSPCA Australia, 2006).

The export of live animals for slaughter from New Zealand is covered by the *Code of Recommendations and Minimum Standards for the Sea Transport of Sheep from*

Table 12.4. A summary of some standards for the on-board management of livestock described in the *Australian Standards for the Export of Livestock*. (From Commonwealth of Australia, 2006.)

Guiding principle	Required outcomes
On-board facilities, management and husbandry must be adequate to maintain the health and welfare of livestock throughout the sea voyage.	• The voyage is completed safely. • Adequate livestock services are maintained throughout the voyage. • On-board care and management of the livestock is adequate to maintain their health and welfare throughout the voyage. • Statutory reporting requirements are met, both during and after the voyage.

Standards

An accredited stock person and a veterinarian where required must be part of each voyage and, with the vessel's master and crew, maintain the health and welfare of the livestock. All personnel responsible for animals must be experienced and skilled.

Sick or injured animals must be treated, transferred to a hospital pen or killed without delay where euthanasia is necessary.

Livestock must have been loaded according to the loading plan, and feed and water must be offered within 12 h of loading, preferably as soon as possible.

All animals must have access to water; there must be a contingency plan for providing feed and water in the event of failure of automatic systems; feed and water must be provided during discharge of livestock.

Livestock must be regularly inspected to ensure their health and welfare is maintained; stocking density must be checked and adjusted as required; ventilation must be monitored to ensure adequate animal thermoregulation; faeces and litter must be disposed of with regard to animal health and welfare.

Veterinary medicines must be stored and used according to veterinary and manufacturers' instructions and records of their use kept.

Any bedding material must be maintained to ensure animal health and welfare.

Contingency plans must be prepared for mechanical breakdowns, feed and water shortages, disease outbreaks, extreme weather conditions or rejection of the shipment.

Australian Government agencies must be notified of excessive mortality rates, their causes and the location of the vessel, its destination and arrival time.

Daily animal health and welfare reports must be provided for voyages over 10 days.

An end of voyage report on the health and welfare of the livestock must be completed within 5 days of the completion of discharge.

New Zealand. It details standards for preassembly, assembly, preparation, shipping and discharge phases.

Finally, both countries are actively involved in the development of the international animal welfare guidelines, including land and sea transport, undertaken by the World Organisation for Animal Health (OIE).

Auditing and assurance programmes

There is a range of voluntary quality assurance programmes developed by many industry groups covering the major livestock production industries. The animal welfare content of these programmes is generally based on relevant codes and guidelines, and are seen as a means of ensuring compliance. Some of these programmes include animal welfare as a mandatory section, while in others this is optional. Their aim is, among other things, to provide guidelines for those responsible for animals, to ensure compliance with regulations and to assure customers and the public that participating farms or companies have high standards of animal welfare, hygiene and product quality. They include:

Australia
- CattleCare – cattle production for slaughter and live export (Australia, www.ausmeat.com.au);
- National Feedlot Accreditation Scheme – feedlot cattle (Australia, www.ausmeat.com.au);
- FlockCare – sheep production including for slaughter and live export (Australia, www.ausmeat.com.au);
- Australian Pork Industry Quality Programme – pig production including for slaughter (www.apl.au.com);
- Australian Deer Industry Quality Assurance Programme – deer production for slaughter and live export (www.diaa.org);
- TruckCare – livestock transport including for slaughter and live export (Australia, www.alta.org.au); and
- National Saleyards Quality Assurance Programme – saleyard livestock and management in Australia (www.ausmeat.com.au).

New Zealand
- Pocket Guide for Stock Truck Drivers for the Welfare of Sheep, Goats and Cattle Transported by Truck Within New Zealand (www.biosecurity.govt.nz);
- Livestock Handling Best Practice Guide – for New Zealand transport operators (www.rtfnz.co.nz);
- Deer Transport Quality Assurance Programme (New Zealand, www.deernz.org);
- Best Practice for Safe Handling and Transport of Ostriches (Morley, 2002) – for producers and transport operators (New Zealand);
- CARTA Stock Crate Code for Transportation of Livestock Quality Assurance Programme (Road Transport Forum New Zealand, 1999) – for the transportation of livestock by road; and
- AFFCO's Select Beef and Select Lamb – a New Zealand meat processing company's programme for farmers and transport operators (www.affco.co.nz).

There are few animal welfare reporting requirements during transport in either country. Requirements for the assessment of animals on arrival at an abattoir differ. In Australia, only export abattoirs are required to have an AQIS veterinary officer present to assess whether animals are fit for slaughter and to check the status of animals on arrival where there might be dead or injured stock (Meat and Livestock

Australia, 2006a). In New Zealand, the welfare status of animals is assessed on arrival at the abattoir to ensure that they are fit for slaughter (Petersen *et al.*, 1991).

Social, economic and political influences and livestock

Australia

The Australian Government supports the livestock industries, including the meat industry, through promotion of the 'clean and green' Australian environment, freedom from debilitating livestock diseases, traceability of products and quality assurance measures, stringent food safety standards and growth in niche industries such as halal and kosher production (Agribusiness Australia, 2005).

The Australian Government has continuously supported the live export trade since its inception. Where market weaknesses or public pressure have highlighted potential problems, the Government, along with the live export industry, has taken steps to address them through initiatives such as the Action Plan for the Livestock Export Industry in October 2002 (Agribusiness Australia, 2005). This included formal risk management processes to address animal disease, heat stress, animal husbandry techniques, weather conditions and other risk factors identified by ongoing research and development.

One of the more contentious arguments for the continuation of live animal exports for slaughter relates to the economic benefits to Australia. A report commissioned by the meat and live animal export industries estimated around 25 different business types are involved in animal exports, each generating added value. With live cattle, producers, transporters and fodder suppliers are the major recipients with the industry generating an estimated AUS$30/head. Sheep producers, feedlot operators, agents, shearers, fodder suppliers and transporters bring an additional AUS$5/head (Hassall and Associates, 2006). It has been estimated that the livestock export industry contributes AUS$1.8 billion to Australia's gross domestic product, over 12,000 jobs to the national economy and wages and salaries totalling AUS$0.99 billion (Hassall and Associates, 2000). The 2003 Keniry report (Keniry *et al.*, 2003) described the livestock export industry as providing 'a valuable alternative market for Australia's livestock producers and is particularly important to the economies of the sheep growing areas of Western Australia and the cattle regions of northern Australia'.

In direct contrast to these conclusions, a report commissioned by the Australian meat processing industry concluded that the live export trade could be costing the country around AUS$1.5 billion in lost gross domestic product (GDP), AUS$270 million in household income and 10,500 lost jobs (Heilbron and Larkin, 2000). Furthermore, the report states that profitability of the live export trade is supported by market distortions and incentives that are created by government and industry policies. It has also been argued that growth in live exports, supported by governments, has contributed to a decline in the meat-processing sector in some regions in Australia (Ministerial Taskforce Report, 2003). According to Heilbron and Larkin (2000), if it were not for these factors, the rising demand for meat in importing countries would have been met by exports of chilled and frozen meat.

It was also noted that there was a risk of adverse incidents in livestock export affecting the wider Australian meat export industry. Examples are the *Cormo Express* (see Box 12.2) and the inhumane treatment of cattle at Bassatin abattoir and sheep during the Eid-al-Adha festival in Egypt, as well as in other Middle East countries including Kuwait, Oman, Bahrain and Qatar (Sidholm, 2003; Animals Australia, 2006a,b). In the case of Egypt, the response of the Australian Government was to temporarily suspend exports until agreement had been reached that Egypt would comply with OIE transport guidelines and also meet specific standards regarding the treatment of cattle (DAFF, 2006). However, subsequent reports of non-compliance with these requirements have not resulted in any further suspension of trade or change in government policy (Animals Australia, 2006b).

A major problem here is that among countries that are members of the World Trade Organization (WTO), national governments are limited in restrictions that they can impose on trade with other members. Thus, while New Zealand can take responsibility for its own animals – as it did by imposing an age limit on lambs to be transported – it suggests that it cannot impose conditions on trade based on what will happen to animals after they are sold to another WTO country. MAF has stated that as a member of the WTO New Zealand 'cannot prohibit the export of animals to other WTO member countries on the basis of management procedures in the country of importation' (MAF, 2007).

However, companies can act where governments cannot. Thus, livestock export interests can assist to enhance the treatment of animals in recipient countries. For example, Meat and Livestock Australia and LiveCorp have run joint development activities in several feedlots and abattoirs in the Middle East (Brown, 2004; Meat and Livestock Australia, 2004). These include stock handling and processing in Egypt; electrical pre-slaughter stunning in Jordan; and assistance with the building, modification and management of feedlots in the United Arab Emirates, Jordan, Egypt, Kuwait, Saudi Arabia, Bahrain, Qatar, Oman and Eritrea. A similar strategy is planned for Asian/South-east Asian importing countries. This approach has been criticized by animal welfare organizations as having little impact on welfare outcomes or cultural practices, with some facilities remaining unused (Lightfoot, 2003). It is also argued that such programmes act to prolong the continuation of live exports to countries with demonstrably poor welfare standards, instead of encouraging moves towards a chilled meat trade.

New Zealand

Agriculture is fundamental to New Zealand's economy and, given the importance of exports, the influence of the international trading environment is significant. Livestock numbers are largely determined by world market prices for farm products. Uniquely, among the developed countries, New Zealand farmers are almost totally exposed to world market forces with no subsidies and having to compete with subsidized production from other countries. Distortions in the world trade in agriculture (high tariffs, export subsidies and domestic support for farming in many countries) limit the benefits New Zealand derives from its competitiveness in the farming sector.

Meat products and live animals

While many countries consume the majority of meat they produce, Australia and New Zealand export most of the meat they produce, with the exception of poultry and pork. Around 65% of Australia's beef is exported, along with 35% of its lamb and 65% of its mutton production (Meat and Livestock Australia, 2000). Similarly, New Zealand exports 88% of its beef and 95% of its sheep meat.

In contrast to the value of livestock slaughtered for export as meat, the Australian livestock export industry is small. Annual earnings from live exports (most of which are for slaughter) are around AUS$531 million for cattle, sheep AUS$279 million and goats AUS$4.3 million, accounting for 4.6% of the livestock industry's total value in 2004–2005 (Australian Bureau of Agricultural and Resource Economics, 2006; Department of Agriculture and Food Western Australia, 2006). Despite its lower relative value, historically there has been strong political support for the live export trade as it is considered to provide an alternative market to meat exports.

Markets for livestock and meat products can be competitive or serve very different regions. With sheep, there is considerable overlap between the two markets, especially in the Middle East. In contrast, the main destinations for live cattle are Indonesia and Malaysia, yet most beef exports end up in the USA and Japan (Australian Bureau of Statistics, 2005; Norris and Norman, 2006).

The demand for live sheep exports to the Middle East has traditionally been attributed to two reasons: a cultural (religious) requirement for freshly killed sheep and a lack of refrigeration and modern food distribution systems (Ministerial Taskforce Report, 2003). However, there is a growing demand for chilled and frozen sheep meat in the region, as refrigeration and distribution systems expand (Roger Fletcher, Australia, 2006, personal communication).

Further enhancing the export of meat, rather than the export of live animals for slaughter, is dependent on a complex of factors. They include efforts in expanding the markets for chilled and frozen meat, and factors such as climate, land use and exchange rates affecting supply and demand. Inevitably, if left to market forces, there will be periods when export of livestock for slaughter will be commercially attractive. The alternative is some form of political or cultural intervention restricting the practice.

If livestock exports of sheep and cattle were to cease, a proportion of the revenue would be recovered through cross-substitution from the carcass trade. The impact would be felt most by the northern cattle industry and the Western Australia sheep industry, with a fall in prices likely together with increased transport costs. This could, in part, be avoided by the construction of slaughtering facilities in the Northern Territory/Kimberley regions (Hassall and Associates, 2006).

Ultimately, what is required to bring an end to live exports from Australia is the political will to support a shift in the export market away from live animals, together with the development of slaughter and processing facilities in the areas most reliant on the current trade.

Conditions and Welfare of Animals Being Transported

Introduction

With the exception of mortality on sea voyages, there is relatively little quantitative information available to accurately assess animal welfare or the conditions that animals experience during transport for slaughter. In the absence of such information, it is necessary to infer welfare from the conditions expected to influence welfare, especially the distance travelled.

Land transport – Australia

Distances travelled

The distances (many hundreds or thousands of kilometres) and climate (average maximum shade air temperatures exceeding 35°C over much of Australia's interior) particularly in northern Australia have the potential to create significant welfare problems (Petherick, 2005).

There are no estimates of the average distances sheep travel to slaughter. Export abattoirs are located throughout Australia's sheep farming regions albeit mainly near the coast. The country's largest meat processor draws stock from close to the abattoir to as far away as 700–985 km. Sheep grown inland must travel longer distances. In addition, when prices are higher in other regions, sheep may be transported further than the nearest available abattoir to reach a higher price. Sheep are regularly transported from Western Australia to eastern states for this reason, a distance of over 4000 km (Hassall and Associates, 2006).

Cattle can also be transported long distances, with 940 km described as typical by road from feedlot to an abattoir (MacArthur Agribusiness, 2001). While most cattle moved by rail in Queensland originate within the state, some are transported by road 1400 km from the Northern Territory, or 1635 km from Alice Springs, to spelling (resting) yards prior to rail transport. Depending on prices, cattle from the north may also be transported across to Queensland (more than 3000 km) or down to Perth (3300 km). From the south-west of Western Australia, cattle are also transported to South Australia (2700 km), or further east to Victoria and New South Wales (RSPCA, Western Australia, 2006, unpublished data).

Pigs are usually transported shorter distances as farming is concentrated in a smaller region. The 20 largest pig abattoirs are located close to major pig production areas (QAF Meat Industries, 2004). For example, Australia's largest pig producer has its own abattoir at its main production site, with its furthest farm 405 km away (QAF Meat Industries, 2006).

Chicken grower farms are generally located within 100 km of the processing plant (Australian Chicken Meat Industry Association, 2005). Layer hens are usually loaded by hand into transport crates. Journeys for layer hens can often be longer as fewer abattoirs slaughter these lower-value birds.

Animal welfare and conditions during transport

There is very little published research on the effects of land transport on livestock under Australian conditions. Unpublished, industry-funded research on the long-distance transport of cattle indicated that cattle can 'cope' with long journeys when transported under good conditions. The study examined the effects of transport on mature cattle under controlled conditions for journey times from 6h up to 48h. It found that loading and initial transport had the greatest effect on stress, and that cattle recovered weight loss by 72h after transport (Meat and Livestock Australia, 2007). Cattle in northern Australia are generally thought to be tolerant of heat stress during transportation, largely as a consequence of them being better adapted to warmer environments (Petherick, 2005). The main challenge affecting *B. indicus* cattle subject to 48h transportation appears to be dehydration (Parker *et al.*, 2003). However, there is very little reporting of actual journey outcomes to allow comparison with experimental results. One exception is rail transport: Queensland Rail reported that in 2003 they moved 423,280 cattle with a total of 48 deaths (0.011%) (NCCAW, 2004). It should be noted that the use of mortality as an indicator of animal welfare during transport does not reflect that animals can suffer in many ways that do not result in death (Petherick, 2005).

Other reports detailing breaches of welfare standards suggest that compliance with standards cannot be assumed and that the welfare of some animals is unacceptable during transport. The effectiveness of the Australian Standards for the Export of Livestock has been questioned by observations (Animals' Angels, 2005, 2007) of livestock loaded for export in breach of those standards. They included:

- Sheep presented for loading with cancer, scabby mouth, pink eye, shearing cuts or long horns;
- Sheep unable to rise, dead sheep, sheep with trapped limbs on transport vehicles;
- Horned and unhorned sheep penned together on transport vehicles;
- Sheep not having sufficient space between transport vehicle decks to stand in a normal position without contacting overhead structures;
- Sheep transport vehicles being overloaded;
- Inappropriate handling of sheep (dragging sheep by their wool or ears, throwing sheep along races, kicking sheep); and
- Inappropriate use of electric goads.

Land transport – New Zealand

Distances travelled

Distances travelled to slaughter in New Zealand are shorter than in Australia, due mainly to the geography of the country and the distribution of meat processors. Average distances for sheep, cattle and deer are of the order of 88km (median 58km, range 3–372km) with most (66.7%) of the distances up to 100km (Sanson, 2005); 72km (range 24–424km; Jago *et al.*, 2002); or up to 400–500km (Graafhuis and Devine, 1994). The average maximum estimated distance travelled by bobby calves to slaughter in one region of New Zealand was 150km (Stafford *et al.*, 2001).

Animal welfare and conditions during transport

The effects of heat stress on welfare during transport have been investigated in a number of studies, apparently in part in response to some deaths during transport between the North and South Islands (Anonymous, 2001). There was little evidence of a rise in core body temperature, in ewes transported by road up to 14 h. The work was carried out over two summers, and the relatively low temperatures experienced (between approximately 15°C and 30°C), the design of the stock crate and transport company practices meant heat stress was negligible (Reid, 2004; Reid *et al.*, 2005). In a separate experiment, lack of airflow in stationary vehicles saw increases in the temperature–humidity index that could be detrimental to welfare (Fisher *et al.*, 2004). Those conditions might be alleviated by stock crate design and lowering the stock density (Fisher *et al.*, 2002).

Voluntary compliance with transport standards has been described by MAF as very good. Furthermore, responses to incidents usually involve transport operators responding in training and disciplining their own staff, with low levels of recidivist offending. MAF has investigated some 30 transport-related complaints during the last 2 years, mainly related to overcrowding, or vehicles remaining stationary for long periods. Minimum standards were seldom broken and the cases resolved by reminding transport operators of the standards and their responsibilities and obligations. Excluding two complaints still under investigation, there have been no prosecutions during the last 2 years (MAF, New Zealand, 2006, personal communication).

Dairy or bobby calves can be transported for slaughter at a minimum of 4 days of age and may be more vulnerable to the physical demands encountered during transport (Hargreaves *et al.*, 1993). In one study, only 27 of 7169 calves presented for slaughter after up to 8 h of transport and lairage (75–300 km travelled) were deemed in an unacceptable state necessitating euthanasia soon after unloading. A further group of 306 calves were classified as marginal but most calves were deemed acceptable. The authors concluded that provided the minimum standards are met, calves satisfactorily tolerate transport and pre-slaughter lairage (Stafford *et al.*, 2001).

A survey of nearly 10,000 beef and sheep at nine processing plants indicated that the distances travelled (up to 400 km for sheep and 500 km for cattle) had no significant effect on the ultimate pH of the meat (a measure of stress). This result does not mean that transport effects on meat quality are not important just that the distance travelled is not a major issue in New Zealand (Graafhuis and Devine, 1994).

Livestock exports for slaughter

The export of livestock for slaughter involves a number of phases. It begins with the selection of stock, road (or sometimes rail) transport to an assembly farm, spelling yard or feedlot and once there, preparation for export and road transportation to the port of departure (Norris, 2005). Table 12.5 shows some typical distances for sheep and cattle transported within Australia prior to loading for export. In one study, the journey from sheep farm to feedlot varied from 10 to 850 km, taking between 30 min and 25 h with food unavailable for 2–43 h (Norris *et al.*, 1989). The major cause of sheep mortality in these feedlots and on the subsequent ship journey is salmonellosis (Richards *et al.*, 1989; Moore, 2002b; see Chapter 6, this volume).

Table 12.5. Examples of the distances travelled for some typical sheep and cattle road journeys in preparation for export from Australia. (From Hassall and Associates, 2000; MacArthur Agribusiness, 2001; AQIS, 2002b; DAFF, 2002.)

Origin	Stage 1	Stage 2
Sheep		
Sheep property (farm) in Western Australia	180 km from the farm to a quarantine feedlot	35 km to the port and loaded on to the ship
Merino property in Victoria	85 km to a feedlot to adapt to pelleted feed	4 km to the port and loaded on to the ship
Sheep property in New South Wales	80 km from the farm to a saleyard and combined with sheep from other properties	1174 km to a feedlot then 4 km to a wharf for loading on to the ship
Cattle		
Cattle property in Victoria	26 km directly from the farm to the wharf, for loading on to the ship	
Cattle company operating five stations (farms) in the Northern Territory	450 km directly to the port for loading on to the ship	
Cattle property in Queensland	2100 km from the farm to an export yard for pre-export quarantine and spelling (resting)	90 km to the port for loading on to the ship
Cattle property in New South Wales	115 km from a farm to a saleyard and combined with cattle from other properties	1174 km to a feedlot for 24 h spelling, prior to loading on to the ship
Cattle property in Queensland	180 km to assembly and spelling yards	452–1720 km to the port depending on the season and region
Extensive cattle properties in Western Australia	620 km to assembly yards	60 km to the port stockyards for spelling, quarantine and preparation for loading

The live sheep trade from Australia is based around three main export ports (Fremantle in Western Australia, Portland in Victoria and Port Adelaide in South Australia), cattle are exported from 15 ports and goats from three. Loading can take between 1 and 5 days, depending on the size of the vessel and the number of animals.

In 2005, there were approximately 48 sheep voyages to the Middle East averaging 71,000 sheep each. Of these, 11 voyages had sheep loaded onboard ship at more than one Australian port. Sixteen shipments of live sheep also went to South-east Asia (these were likely to have been combined with cattle shipments). In the same year, there were 38 voyages of cattle to the Middle East and northern Africa, 168 to South-east Asia, 36 to north-eastern Asia and nine to Mexico, averaging 2240 head each.

Sea voyages to the Middle East can last 11–25 days, although ships stopping at more than one port can take up to 32 days to reach their final destination (Norris and Richards, 1989; AQIS, 2002a). Journey times from Darwin and ports in the far north of Queensland are considerably less (3–10 days). Voyages to Mexico (mainly cattle and, occasionally, sheep) take 19–26 days from Fremantle and 14–34 days from Portland (Norris and Norman, 2002). The duration of sea voyages also depends on prevailing weather and sea conditions and the requirements of the importing country. Vessels can encounter hazardous seas (AMSA, 1999) and extreme changes of temperature and humidity, particularly during long journeys from the southern to the northern hemisphere. For example, animals travelling from a relatively cool Australian winter can encounter midsummer air temperatures in Middle East waters of around 40°C.

Two disastrous examples of livestock exports are shown in Boxes 12.1 and 12.2. The first describes the 32-day maiden voyage of the *MV Becrux* carrying sheep and cattle from two ports in Australia to four ports in the Middle East (Box 12.1). This voyage was investigated by the Australian Quarantine and Inspection Service due to the high mortality suffered after the ship left Fremantle: 865 cattle and 1437

Box 12.1. The maiden voyage of the *MV Becrux* carrying sheep and cattle from Australia to the Middle East between 14 June and 11 July 2002 (AQIS, 2002a; Moore, 2002a).

The *MV Becrux* is a dual-cargo vessel capable of carrying both sheep and cattle. Its maiden voyage involved transporting livestock from two ports in Australia (Portland and Fremantle) to four ports in the Middle East (Dammam, Fujiarah, Doha and Muscat).

- On 9 June 2002, the *Becrux* completed a preloading inspection in Fremantle (Western Australia). It then proceeded to Portland (Victoria) to load its first consignment of livestock.
- On 14 June, 1734 cattle and 46,055 sheep were loaded at Portland. The vessel left for Fremantle.
- On 18 June, a further 243 cattle and 17,358 sheep were loaded at Fremantle. The vessel then left Australian waters for the Middle East. During the voyage, the ship encountered extreme weather conditions in the Arabian Gulf with very high temperatures (45°C) and humidity.
- By 1 July, 200 cattle were reported dead. 23,658 sheep were discharged at the port of Dammam in Saudi Arabia. Saudi authorities refused to allow the unloading of the cattle consignment.
- By 6 July, a total of 380 cattle and 1248 sheep had died. The *Becrux* arrived in Fujiarah, United Arab Emirates, where 1337 cattle and 3000 sheep were discharged. While at Fujiarah, a further 234 cattle died on the vessel and approximately 200 died at the feedlot after unloading.
- On 8 July, 23,217 sheep were discharged at the port of Doha, in Qatar.
- On 11 July, 44 cattle and 12,122 sheep were discharged at the port of Muscat, Oman. The remaining 51 cattle were declared moribund and were destroyed.

The overall mortality for the journey was 1437 sheep and 865 cattle.

Box 12.2. The voyage of the *Cormo Express* carrying sheep from Australia to Saudi Arabia and eventually to Eritrea from 6 August to 24 October 2003 (Keniry *et al.*, 2003; Wright and Muzzatti, 2005; with additional information from Animals Australia, RSPCA Australia, Livecorp and DAFF web sites).

- On 6 August 2003 the *Cormo Express* left Fremantle (Western Australia) for Jeddah (Saudi Arabia) carrying 57,937 sheep. Built in 1979 as a 'roll-on-roll-off' ferry, the *Cormo Express* had been converted to an 11-deck animal transporter in 1989.
- En route to Jeddah, 1.1% of the cargo (637 sheep) died.
- On arrival in the heat of Jeddah on 21 August, the shipment was rejected by Saudi authorities who claimed 6% of the animals suffered from 'scabby mouth', a condition resembling cold sores in humans but which cannot be transmitted to humans by eating the meat of affected animals. Australia strongly denied the claim and was supported by independent veterinary opinion.
- With no solution to the impasse in sight, Australia suspended live animal exports to Saudi Arabia on 28 August. By this stage on-board mortality had risen to 3%.
- The importer ordered the sheep back to Australia. With limited fodder on board and the prospect of the sheep starving to death, the Australian Government bought the entire consignment of sheep for AUS$4.5 million.
- The *Cormo Express* travelled the region, stopping occasionally to take on feed, and eventually docked in Kuwait on 2 October, where it spent 14 days re-provisioning. Mortality had climbed to 8.7% on arrival in Kuwait and 9.4% on departure. Throughout this period Australian authorities unsuccessfully negotiated with more than 30 countries in several regions of the world to take the sheep.
- Alternatives, including returning the sheep to Australia or slaughtering them at sea were considered. However, Australia's disease-free status precluded their return and slaughter at sea ultimately was considered a task of extreme proportions and environmental risk.
- The East African nation of Eritrea finally accepted the remaining sheep as a gift of 'food aid' from the Australian Government, together with 3000 t of additional feed and AUS$1 million to assist with the associated costs.
- After almost 11 weeks at sea in extreme temperatures and humidity, 9.8% or some 5692 animals had died, mostly from heat stress and exhaustion while waiting in ports.
- Unloading commenced on 24 October and took several days. Once on land, sheep were transported by road to a large refurbished holding facility at Ghathelal, 58 km from Massawa. The Australian industry body LiveCorp advised that the last of the *Cormo Express* sheep had been slaughtered in Eritrea on 16 January 2004.
- After this voyage the *Cormo Express* was renamed *Merino Express*.

---

sheep died as a result of the journey. The AQIS report found several examples of non-compliance with export regulations during the pre-export process. Cattle mortality during the voyage was attributed to extreme heat stress suffered by the *Bos taurus* breeds, whereas sheep deaths were due to enteric disease (salmonellosis) and subsequent inanition (failure to eat) (AQIS, 2002b).

The second example relates to a shipment of sheep on board the *Cormo Express* rejected by Saudi Arabia. In 1989–1990, Saudi Arabia rejected 11 Australian sheep shipments, the animals remaining on board until an alternative country was found. One shipment remained at sea for 16 weeks, and the average mortality rate on these shipments was 6%. Shipments between Australia and Saudi Arabia were subsequently suspended for almost 10 years. In August 2003, a shipment of Australian sheep on the *Cormo Express* was rejected by Saudi Arabia, resulting in arguably the most significant incident in the history of live exports from Australia, with the voyage lasting from 6 August until 24 October and the death of 5692 sheep (Box 12.2). The trade was again suspended, but recommenced in July 2005 following the signing of a Memorandum of Understanding between Australia and Saudi Arabia to avoid rejected sheep remaining at sea. Australia subsequently instituted new standards for the selection, preparation and shipment of Australian animals (the ASEL, Table 12.4), in an attempt to reduce on-board mortality.

Livestock export mortality

Information on mortality rates during export and unloading is reported and regularly published by government with industry cooperation (Norris and Norman, 2006).

The main causes of sheep deaths during sea transport from Australia are inanition (failure to eat resulting in exhaustion and starvation) and salmonellosis, with a 0.95% overall death rate (38,234 sheep) for all destinations in 2005 (Norris and Norman, 2006). The main causes of death in cattle are heat stroke, trauma and respiratory disease (Norris *et al.*, 2003) with an overall death rate of 0.14% for all destinations in 2005 (Norris and Norman, 2006).

The highest overall cattle death rate was to the Middle East/North Africa (0.34%), followed by Mexico (0.26%), with the lowest rate (0.09%) to South-east and North-east Asia (Norris and Norman, 2006). These differences in death rates reflect both the size of consignments and the length of the journey. Voyages to South-east Asia were mainly small shipments on short voyages; voyages to the Middle East involved larger single shipments and longer journeys.

Death rates of sheep from Western Australia to the Middle East have steadily declined from a peak of over 3% in 1992 to less than 1% in 2003 (Table 12.6). This trend is most probably due to the export of fewer older animals, faster ships and shorter pre-embarkation periods, better facilities and management and newer ships with better ventilation systems (Norris, 2005).

Table 12.6. Comparison of average Australian livestock export mortalities (%) in 1996 and 2005. (From RSPCA Australia, 1998; Norris and Norman, 2006.)

	1996	2005
Sheep	2.79	0.95
Cattle	0.05–0.63	0.14
Goats	no data	0.77

However, mortality rates are a very blunt instrument when it comes to assessing animal welfare. Those animals that die on an export journey are at the most extreme end of a scale of suffering: animals can suffer during the export journey in many ways that do not result in death (Petherick, 2005; Phillips, 2005).

New Zealand livestock export

Initial attempts to export live sheep in the early 1970s were strongly opposed by animal welfare groups and trade unions, and the government initially banned the export of sheep for slaughter. However, a market for live sheep, a deteriorating economic situation for New Zealand sheep farmers and a political climate of deregulation, resulted in the first shipment to Mexico in December 1985. By the end of 1990, nearly 4 million sheep had been shipped for slaughter, mainly to the Middle East though the trade became more focused on Saudi Arabia (Black *et al.*, 1991).

A 1990 New Zealand shipment on the *Cormo Express* suffered significant mortalities (12%) and a subsequent shipment on the *Mawashi Al Gasseem* was rejected on the basis of, albeit questionable, health grounds. Consequently, New Zealand temporarily suspended live sheep shipments to Saudi Arabia. Trade recommenced in 1991, reaching a peak of over 1 million sheep being exported in the mid-1990s. Since 1991, the Code of Recommendations and Minimum Standards for the Sea Transport of Sheep from New Zealand has been in place, setting and maintaining animal welfare standards and reducing mortality to less than 1%. A comprehensive reporting system was also instigated to review each shipment and adopt improvements leading to improved animal welfare. MAF also commissioned research on pneumonia, an issue with very young animals, resulting in the setting of a minimum age limit based on animal health and welfare grounds (the sea transport of lambs, except for animals of the Awassi breed, was stopped in 1994 due to repeated shipboard epidemics of pneumonia in sheep less than 9 months of age). By 2004, due to a combination of economic factors, falling New Zealand sheep numbers and the higher age limit, shipments dwindled to one a year (approximately 40,000 sheep). Deaths on these shipments were consistently below 0.8% (MAF, 2004).

Smothering (animals, mainly lambs, piling up over one another, especially in corners and suffocating) and pneumonia were the two major causes of death in the early shipments with a lower prevalence of inanition (Black, 1997). While a high proportion of sheep gained weight and arrived in the Middle East in good condition, there was much cause for concern for the health and welfare of the animals. This was evident in significant death rates (average 2%, range 0.5–19%), animals losing weight and arriving dirty and in ill-health and behavioural evidence of heat stress on most voyages. Risk factors included increasing age and fatness, unpalatable concentrated food and increasing duration of the voyage (Black *et al.*, 1991).

Heat stress has since been implicated in pneumonia in lambs since openmouthed panting, a cooling mechanism, may predispose the animals to the disease. On two voyages of sheep to the Middle East (1993 and 1995), nearly all of the lamb deaths were due to pneumonia (total mortality rates were 4.3% and 2.4%, respectively, Black *et al.*, 2004). Typically, there was a high prevalence (2–6%) of pneumonia in lambs transported to the Middle East, but a low prevalence (<0.3%) in sheep over 1-year old similarly transported (Black and Duganzich, 1995).

Discussion

The transport of animals to slaughter in Australia and New Zealand is complex with differences in geography, animals, distances travelled, slaughter destinations and animal welfare guidelines and regulations. However, three main themes are apparent, two of which raise significant issues for animal welfare.

First, the majority of livestock within each country are routinely transported to slaughter over varying, but not long, distances with little evidence of compromises to their welfare. Such transport is regarded as necessary and both countries have government-endorsed codes or guidelines dealing with this transport. In New Zealand, codes of practice have legal status, and failure to adhere to the minimum standards can be used to support a prosecution. In contrast, in Australia there are no enforceable standards for the land transport of livestock for slaughter.

Second, some livestock and feral animals are transported long to extremely long distances in Australia and these distances suggest the animals experience less than ideal conditions, particularly when the journey exposes animals to significant changes in environmental conditions.

Finally, the live export of animals by sea from Australia (and potentially New Zealand) where consistent deaths, and occasional disasters, indicate significant animal welfare issues are an inherent feature of this practice.

Transport to slaughter within Australia

The majority of meat processing facilities in Australia are located within the livestock production areas in the south-east, with smaller numbers in the south of Western Australia and along the coast of Queensland. Thus, while most animals are within reasonable range of an abattoir, long-distance transport for slaughter is unavoidable in some regions and distances travelled may be up to 3000 km. There has been a steady decline in the number of abattoirs in Australia in the last 30 years, and this centralization of the industry has meant that average distances from farm to slaughter have increased significantly. In addition, differences in meat prices between states can result in longer journeys as producers send their livestock interstate to chase better prices.

The welfare of livestock transported in Australia is covered by a system of voluntary codes of practice and industry quality assurance programmes. These describe the conditions (e.g. handling, vehicle design and operation, environmental factors, food and water deprivation periods) under which livestock should be transported to slaughter. Recommendations may vary from state to state and there are no means of ensuring compliance, particularly when transport journeys cross state boundaries. A process is currently under way to develop nationally accepted standards and guidelines for land transport, based on the recommendations in existing voluntary codes, with the aim of reaching greater consistency in regulation and application across the livestock industries. The intention is to regulate the standards through state and territory legislation. Development of the standards is being carried out in consultation with industry, government and animal welfare groups.

The lack of reported data on individual journeys is an impediment to assessing the welfare problems associated with transport or monitoring compliance with current standards. Concerns have been raised over some specific issues, such as inadequate resting of animals during transport, journeys exceeding maximum food and water deprivation times, the transport of poultry for slaughter and the legal difficulties in dealing with animal welfare problems with interstate transport (RSPCA Inspectors, Australia, 2006, personal communication).

Transport to slaughter within New Zealand

The transport of livestock to slaughter in New Zealand is presently unavoidable, as modern processing plants tend to be located in areas with greater access to labour and other resources. New Zealand's livestock are mostly transported less than 100 km but sometimes up to 400–500 km.

New Zealand has extensive requirements describing the conditions under which livestock are transported to slaughter in order to ensure their welfare. Important instruments in ensuring the care of animals in New Zealand are the codes of animal welfare. These are based on enforceable standards and are drawn up by animal welfare regulatory bodies with industry and public involvement (note that the New Zealand transport code of welfare is being revised as of 2007). In New Zealand there are few concerns about adequacy of the standards or the extent of compliance with them.

Long-distance transport to slaughter within Australia

It is generally accepted that prolonged journeys are stressful, necessitating either limits on their duration (i.e. slaughtering livestock as close to the farm as possible) or specific strategies (e.g. vehicle design, resting periods) to allow animals to travel extended distances without suffering (Webster, 2005). Therefore, it is relevant to differentiate between the journeys most livestock undertake in Australia and New Zealand, and the long journeys, up to 2000–3000 km, undertaken in parts of Australia.

It is clear that long-distance livestock transport in Australia is not a simple issue. The sheer scale of the country, its variety of geographical and climatic regions, the breadth and diversity of livestock production and the distances travelled between farms and slaughter destinations, all make for a very complex picture. It is further complicated by the Australian system of federal, state and territory governments, which means that for every activity there can be up to nine different regulatory systems.

The lack of reported data on the outcomes of actual journeys makes it difficult to gauge the extent of any welfare issues. By necessity then, poor animal welfare has had to be inferred from the length of such journeys, a reasonable stance given the weight of scientific evidence of the stress of transport over comparatively shorter journeys (Knowles *et al.*, 1995; Hall *et al.*, 1999; see Chapters 1, 4 and 7, this volume), but acknowledging that this may not necessarily be true in all cases. Conclusions on good or poor outcomes require evidence of what the animals experience (their physiological and behavioural responses), and factors such as the class and body condition of the animals, stocking rates, temperature and humidity, ability of individuals to rest and availability of feed and water. Unfortunately, these are largely unknown as there is very little scientific research on the effects of transport on animal welfare under Australian conditions.

An aspect which deserves critical examination is the transport of livestock from remote areas and of feral animals. Often unused to human contact and handling, they may suffer unreasonably from confinement and long-distance transport. Although there is a Code of Practice for Feral Livestock Animals (which has been under revision for more than 5 years) it does not provide any useful guidance on stocking densities, maximum journey times, withholding times after capture and before transport, rest periods or restrictions on transport. The new transport standards and guidelines (currently under development) must take account of the specific problems encountered with feral animals, such as the need for minimum rest periods after mustering (gathering) and prior to loading and the exclusion of vulnerable classes of animals from transport. The potential to eliminate the need for transport of feral animals, where humane slaughter in the field is an alternative, should also be explored.

Export of livestock from Australia for slaughter in other countries

The export of livestock for slaughter consists mainly of livestock shipped from Australia to either the Middle East or South-east Asia (in the past New Zealand has also been involved, see the following section on New Zealand). Primary guidance for this practice is given in the Australian Standards for the Export of Livestock, but the ability of government agencies to effectively enforce these standards from the farm to wharf has been strongly questioned, as state and territory government officers are not trained or empowered to enforce the standards (RSPCA Australia, 2006). While there has been a progressive reduction in mortality rates during sea voyages, mortality is unavoidable in the process (over 38,000 sheep died in 2005). The voyages remain long and arduous and incidents such as the *Cormo Express* emphasize the adverse consequences of the practice.

Many of those not dying may nevertheless be suffering. Furthermore, the conditions under which the animals are transported (e.g. heat, diet, confinement) are unacceptable. The responses of cattle to these conditions has been described in an experiment mimicking the continuous and prolonged high temperature and humidity such as might be experienced during live export (Beatty *et al.*, 2006). In this work, the animals' heat-loss mechanisms could not fully compensate for the excessive heat load and the *B. taurus* cattle showed clinical signs of heat stress (open-mouth panting, drooling, reluctance or inability to rise, increased licking of coat and general dullness including neurological signs with staring and glazed eyes). Clinical signs of heat stress were not, however, observed in *B. indicus* animals. While changes in indicators of animal welfare do not necessarily mean animals are suffering, or that society believes that their treatment is unreasonable and unnecessary, the magnitude of these changes suggests that the conditions to which some animals may be exposed are intolerable and unacceptable. The mortality associated with the live export industry further indicates that some animals are unable to tolerate those conditions.

The treatment of livestock once they arrive at their export destination must also be included in any consideration of the live export trade. Inhumane handling and slaughter of Australian cattle at Bassatin abattoir and of sheep during the Eid-al-Adha festival in Egypt have both been documented in the recent past. Exports have been permitted to continue despite demonstrated breaches of an agreement between Egypt and Australia that unloading and slaughter would meet OIE standards. The morality of allowing animals to be exported to countries where it is likely that they will be treated inhumanely must be questioned (Petherick, 2005).

New Zealand

New Zealand has also exported animals for slaughter in other countries, most notably in the Middle East. Here, at least in the early days of the trade, there were significant animal welfare issues. However, they were identified and researched, a code for the Sea Transport of Sheep from New Zealand developed and restrictions placed on the transport of the more vulnerable animals. No animals have been exported for slaughter for some time and the New Zealand Government is currently reviewing the export of livestock. It should be noted that resumption of the trade could be prompted by unfavourable economic conditions for farming in New Zealand. The trade has been a contentious issue within New Zealand (both because of the welfare of the animals during shipment and because of their treatment in the country of destination) and is likely to resurface should it resume.

Conclusion

The transport of livestock to slaughter in Australia and New Zealand is characterized by: (i) transport over short to moderate distances within each country; (ii) long-distance transport within Australia; and (iii) long-distance travel of animals by sea for slaughter in other countries.

In the first case, the aim should be to ensure that all journeys are conducted with adequate consideration for the welfare of transported animals. At present, while New Zealand has community-endorsed enforceable minimum standards for the land transport of animals, Australia does not. Australia must replace its voluntary system of codes of practice with regulated standards and guidelines (a process currently under way) and ensure that these standards are reflective of good animal welfare practice.

Australia is a vast country and thus very long distance transport of livestock occurs, often exposing animals to significant changes in environmental conditions. Distances travelled have increased due to the centralization of the meat processing industry, resulting in a significant decline in the number and geographical spread of abattoirs. In some cases journeys are unnecessarily long, such as where animals are transported interstate in order to fetch higher prices at slaughter. The mustering, loading and transport of feral animals and livestock from remote areas raises particular problems. Standards for transport in Australia must take into account the specific issues raised by very long journeys and the transport of animals unused to human contact. Further research is sorely needed to better understand the effects of transport on animal welfare under Australian conditions.

It is difficult to justify the live export of animals for slaughter in other countries. However, while animal welfare is essentially about what the animal experiences, making decisions concerning welfare is clearly complex. For example, the OIE regards animal welfare as a multifaceted public policy issue which includes important scientific, ethical, economic and political dimensions (Petrini and Wilson, 2005). As such, methods of avoiding livestock exports for slaughter need to be investigated in both exporting and importing countries within the context of the economic, social and cultural factors which support the practice or alternatives. This needs to include consideration of the expansion of meat processing facilities into regions that are not currently serviced by the industry. However, the welfare compromise to livestock during such exports is not complex – their welfare is at risk and it would clearly be better for the animals if these journeys did not take place.

References

Access Economics and Maunsell Australia (2002) *Benchmarking Technology on the Australian Waterfront – Implications for Agricultural Exports.* Rural Industries Research and Development Corporation, Barton, Australia.

Agribusiness Australia (2005) *Australian Meat Industry Capability Report.* Invest Australia, Canberra, Australia.

AMSA (1999) Investigation into Excessive Livestock Mortality on the MV Kalymnian Express. Available at: wwwlive.amsa.gov.au/Shipping_Safety/Codes_Manuals_and_Reports/Livestock_investigations/Investigation_into_excessive_Livestock_Mortality.asp

Animals' Angels (2005) Interim Report on the MV Maysora 14–15 October 2005. Available from Animals' Angels.

Animals' Angels (2007) Height and density. Lack of headroom for sheep in transport and density of sheep in pens in transport. Available from Animals' Angels.

Animals Australia (2006a) Middle East Investigation Report. Available at: www.animalsaustralia.org/default2.asp?idL1=1272&idL2=1865&idL3=1910

Animals Australia (2006b) Egypt Investigation Report, December 2006. Available from Animals Australia.

Anonymous (2001) Lamb deaths crackdown. *The Dominion Post* 20 February, p1. Montreal, Canada.

AUS-MEAT (2005) List of AUS-MEAT Accredited Abattoirs. Available at: www.ausmeat.com.au

Australian Bureau of Agricultural and Resource Economics (2006) *Australian Commodity Statistics 2006.* ABARE (Australian Bureau of Agricultural and Resource Economics), Canberra, Australia.

Australian Bureau of Statistics (2005) *Yearbook Australia 2005* – Agriculture – Livestock. Australian Bureau of Statistics, Canberra, Australia.

Australian Bureau of Statistics (2006) *Agricultural Survey Australia, 2004–05.* Australian Bureau of Statistics, Canberra, Australia.

Australian Chicken Meat Industry Association (2005) Available at: www.chicken.org.au

AQIS (2002a) *Investigation into High Livestock Mortality MV Becrux Voyage of 8 June to 14 July 2002.* Australian Quarantine Inspection Service, Department of Agriculture, Fisheries and Forestry, Canberra, Australia.

AQIS (2002b) *MV Becrux V1: Investigation into the Causes of High Mortalities in Cattle and Sheep.* Australian Quarantine Inspection Service, Department of Agriculture, Fisheries and Forestry, Canberra, Australia.

Beatty, D.T., Barnes, A., Taylor, E., Pethick, D., McCarthy, M. and Maloney, S.K. (2006) Physiological responses of *Bos taurus* and *Bos indicus* cattle to prolonged, continuous heat and humidty. *Journal of Animal Science* 84, 972–985.

Bindon, B.M. and Jones, N.M. (2001) Cattle supply, production systems and markets for Australian beef. *Australian Journal of Experimental Agriculture* 41, 861–877.

Black, H. (1997) Sea water in the treatment of inanition in sheep. *New Zealand Veterinary Journal* 45, 122.

Black, H. and Duganzich, D. (1995) A field evaluation of the efficacy of two vaccines against ovine pneumonic pasteurellosis. *New Zealand Veterinary Journal* 43, 60–63.

Black, H., Matthews, L.R. and Bremner, K.J. (1991) The welfare of sheep during sea transport. *Proceedings of the New Zealand Society of Animal Production* 51, 41–42.

Black, H., Alley, M.R. and Goodwin-Ray, K.A. (2004) Heat stress as a manageable risk factor to mitigate pneumonia in lambs. *New Zealand Veterinary Journal* 53, 91–92.

Brown, N. (2004) Developments on animal welfare in Egypt. *Australian Veterinary Journal* 82, 240.

Commonwealth of Australia (2006) *Australian Standards for the Export of Livestock. Version 2.* Department of Agriculture, Fisheries and Forestry, Canberra, Australia.

Cooper-Blanks, B. (1999) *The New Zealand Poultry Meat Industry.* Poultry Industry Association of New Zealand, Auckland, New Zealand.

Countrywide (2006) This Weeks Meat Schedules. Available at: www.country-wide.co.nz/a-man/articles/meat_schedules.pdf

DAFF (2002) *MV Novantes Voyage 83, 8–19 February 2002. AQIS Investigation into Cattle Mortalities.* Department of Agriculture, Fisheries and Forestry, Australian Quarantine Inspection Service, Canberra, Australia.

DAFF (2006) Frequently Asked Questions: Live Export Trade – Egypt. Available at: www.daff.gov.au/__data/assets/pdf_file/0003/157422/faq_egypt_14feb07.pdf

Department of Agriculture and Food Western Australia (2006) Australian Trade Statistics Database. Available at: http://www.agric.wa.gov.au/servlet/page?_pageid=639&_dad=portal30&_schema= PORTAL30

FAO (2005) Meat Production: Total Meat Production. World Resources Institute. Available at: http://earthtrends.wri.org/searchable_db/index.php?theme=8&variable_ID=190&action=select_countries

Fisher, A.D., Pearce, P.V. and Matthews, L.R. (1999) The effects of long haul transport on pregnant, non-lactating dairy cows. *New Zealand Veterinary Journal* 47, 161–166.

Fisher, A.D., Stewart, M., Tacon, J. and Matthews, L.R. (2002) The effects of stock crate design and stocking density on environmental conditions for lambs on transport vehicles. *New Zealand Veterinary Journal* 53, 6–9.

Fisher, A.D., Stewart, M., Duganzich, D.M., Tacon, J. and Matthews, L.R. (2004) The effects of stationary periods and external temperature and humidity on thermal stress conditions within sheep transport vehicles. *New Zealand Veterinary Journal* 53, 6–9.

Graafhuis, A.E. and Devine, C.E. (1994) Incidence of high pH beef and lamb II: results of an ultimate pH survey of beef and sheep plants in New Zealand. *Proceedings of the 28th Meat Industry Research Conference.* Auckland, pp. 133–141.

Hall, S.J.G., Broom, D.M., Goode, J.A., Lloyd, D.M., Parrott, R.F. and Rodway, R.G. (1999) Physiological responses of sheep during long road journeys involving ferry crossings. *Animal Science* 69, 19–27.

Hargreaves, A.L., Matthews, L.R. and Verkerk, G.A. (1993) Is the welfare of dairy cows at risk from current farm practices? *Proceedings of the New Zealand Society of Animal Production* 53, 179–182.

Hassall and Associates (2000) *Economic Contribution of the Livestock Export Industry: Report for LiveCorp and Meat and Livestock Australia.* LiveCorp, North Sydney, Australia.

Hassall and Associates (2006) *The Live Export Industry: Value, Outlook and Contribution to the Economy.* Meat & Livestock Australia, North Sydney, Australia.

Heilbron, S.G. and Larkin, T. (2000) *Impact of the Live Animal Sector on the Australian Meat Processing Industry.* Australian Meat Processor Corporation, Sydney, Australia.

Hini, C. (2005) Reports from biosecurity New Zealand – international animal trade. *Surveillance* 32, 4–7.

Industry Commission (1994) *Meat Processing Volume 1: Report. Report No. 38 20 April 1994.* Australian Government Publishing Service, Melbourne, Australia.

Jago, J.G., Matthews, L.R., Hargreaves, A.L. and van Eeken, F. (2002) Preslaughter handling of red deer: implications for welfare and carcass quality. *Deer Branch New Zealand Veterinary Association* 19, 27–39.

Johnson, K.G. and Gregory, N.G. (1998) *Welfare Aspects of Stock Unloading.* MAF, Wellington, Australia.

Keniry, J., Bond, M., Caple, I., Gosse, L. and Rogers, M. (2003) *Livestock Export Review: Final Report.* Department of Agriculture, Fisheries and Forestry, Canberra, Australia.

Knowles, T.G., Brown, S.N. and Warriss, P.D. (1995) Effect of feeding, watering and resting intervals on lambs transported by road and ferry to France. *The Veterinary Record* 139, 335–339.

Lightfoot, J.S. (2003) Welfare of cattle transported from Australia to Egypt. *Australian Veterinary Journal* 81, 432–433.

MAF (2004) Live sheep exports – what's happening? *Primary Source: News from MAF* 25, 3.

MAF (2007) Why Are Live Sheep and Cattle Shipments Still Allowed? Available at: http://www.biosecurity.govt.nz/faq/term/913#Why-are-live-sheep-and-cattle-shipments-still-allowed

MacArthur Agribusiness (2001). *Economic and Social/Community Impacts of the Live Cattle and Processed Beef Export Supply Chains in Queensland.* Queensland Department of Primary Industries, Queensland, Australia.

Meat and Livestock Australia (2000) *Fast Facts: Australia's Beef Industry and Fast Facts: Australia's Sheepmeat Industry.* Meat and Livestock Austalia, Sydney, Australia.

Meat and Livestock Australia (2003) Wetting cattle to alleviate heat stress on ships. *Tips and Tools, Animal Health and Welfare.* Meat and Livestock Australia, Sydney, Australia.

Meat and Livestock Australia (2004) Animal Handling Workshops to Begin in Kuwait. Media release 29 March 2004. Available at: www.mla.com.au/TopicHierarchy/News/MediaReleases/Default. htm?currentYear=2004

Meat and Livestock Australia (2005) Feedback: top 25 processors. *Meat and Livestock Industry Journal Supplement.* October, 2005.

Meat and Livestock Australia (2006a) Paddock to Plate. Available at: www.mla.com.au

Meat and Livestock Australia (2006b) *Is It Fit to Load? A National Guide to the Selection of Animals Fit to Transport.* Meat and Livestock Australia, Sydney, Australia.

Meat and Livestock Australia (2007) Animal Welfare Project. Available at: http://www.mla.com.au/ TopicHierarchy/ResearchAndDevelopment/On-farm/Animal+welfare+project.htm

Meat New Zealand (1998) *Animal Welfare – Stunning and Transport: R & D Brief 25.* Meat and Wool New Zealand, Wellington, New Zealand.

Meat New Zealand (1999) *Managing Your Bulls Before Slaughter: R & D Brief 44.* Meat and Wool New Zealand, Wellington, New Zealand.

Ministerial Taskforce Report (2003) Cattle and Sheep Meat Processing in Western Australia. A report to the Minister for Agriculture, Western Australia.

Moore, S. (2002a) Investigation of Cattle Deaths During Voyage 1 of the *MV Becrux.* Report prepared for Meat and Livestock Australia and LiveCorp, September 2002.

Moore, S. (2002b) Salmonellosis Control and Best-Practice in Live Sheep Export Feedlots. Final Report prepared for MLA and LiveCorp, October 2002.

Morley, L. (2002) Best Practice for Safe Handling and Transport of Ostriches. Available at: www. ostrich-association.co.nz

NCCAW (2004) National Consultative Committee on Animal Welfare 33rd Meeting 6–7 April 2004. Draft Summary of Proceedings, pp. 11.

NAWAC (2005) *National Animal Welfare Advisory Committee Annual Report 2004.* Ministry of Agriculture and Forestry, Wellington, New Zealand.

Norris, R.T. (2005) Transport of animals by sea. *Animal Welfare: Global Issues, Trends and Challenges, Revue Scientifique et Technique Office International des Épizooties* 24, 673–681.

Norris, R.T. and Norman, G.J. (2002) *National Livestock Exports Mortality Summary 2002.* Meat & Livestock Australia, Sydney, Australia.

Norris, R.T. and Norman, G.J. (2005) *National Livestock Exports Mortality Summary 2004.* Meat & Livestock Australia, Sydney, Australia.

Norris, R.T. and Norman, G.J. (2006) *National Livestock Exports Mortality Summary 2005.* Published by Meat & Livestock Australia, Sydney, Australia.

Norris, R.T. and Richards, R.B. (1989) Deaths in sheep exported by sea from Western Australia – analysis of ship Master's reports. *Australian Veterinary Journal* 66, 97–102.

Norris, R.T., Richards, R.B. and Dunlop, R.H. (1989) Pre-embarkation risk factors for sheep deaths during export by sea from Western Australia. *Australian Veterinary Journal* 66, 309–314.

Norris, R.T., Richards, R.B., Creeper, J.H., Jubb, T.F., Madin, B. and Kerr, J.W. (2003) Cattle deaths during sea transport from Australia. *Australian Veterinary Journal* 81, 156–161.

OECD (2005) OECD in figures. Available at: http://ocde.p4.siteinternet.com/publications/doifiles/ 012005061T010.xls

Parker, A.J., Hamlin, G.P., Coleman, C.J. and Fitzpatrick, L.A. (2003) Quantitative analysis of acid-base balance in *Bos indicus* steers subjected to transportation of long duration. *Journal of Animal Science* 81, 1434–1439.

Petersen, G.V., Madie, P. and Blackmore, D.K. (1991) Veterinary aspects of meat quality. *FCE Publication* 138.

Petherick, J.C. (2005) Animal welfare issues associated with extensive livestock production: the northern Australian beef cattle industry. *Applied Animal Behaviour Science* 92, 211–234.

Petrini, A. and Wilson, D. (2005) Philosophy, policy and procedures of the world organisation for animal health for the development of standards in animal welfare. *Revue Scientifique et Technique Office International des Épizooties* 24, 665–671.

Phillips, C.J.C. (2005) Ethical perspectives of the Australian live export trade. *Australian Veterinary Journal* 83, 558–562.

Productivity Commission (2005) *Australian Pigmeat Industry Productivity Commission Inquiry Report No. 35.* Commonwealth of Australia, Canberra, Australia.

QAF Meat Industries (2004) Submission to the Productivity Commission Inquiry in the Australian Pigmeat Industry. Available at: www.pc.gov.au/inquiry/pigmeat/subs/sub029.pdf

QAF Meat Industries (2006) Available at: www.qafmeats.com.au

Rahman, S.A., Walker, L. and Ricketts, W. (2005) Global perspectives on animal welfare: Asia, the far east, and oceania. *Revue Scientifique et Technique Office International des Épizooties* 24, 597–612.

Reid, T.C. (2004) Transport and Heat Stress. MAF Operational Research Report. MAF, Wellington, New Zealand.

Reid, T.C., Urquhart, R.A., Stewart, M. and Matthews, L. (2005) Effects of Heat Stress During Transport. MAF Operational Research Report. MAF, Wellington, New Zealand.

Richards, R.B., Norris, R.T., Dunlop, R.H. and McQuade, N.C. (1989) Causes of death in sheep exported live by sea. *Australian Veterinary Journal* 66, 33–38.

Road Transport Forum New Zealand (1999) *CARTA Stock Crate Code for Transportation of Livestock Quality Assurance Programme.* Road Transport Forum New Zealand, Wellington, New Zealand.

RSPCA Australia (1998) *Report on the Livestock Export Trade.* RSPCA Australia, Canberra, Australia, 29pp.

RSPCA Australia (2006) Comments on Draft Version 2 Australian Standards for the Export of Livestock – available on request from RSPCA Australia.

Sanson, R.L. (2005) A survey to investigate movements off sheep and cattle farms in New Zealand, with reference to the potential transmission of foot-and-mouth disease. *New Zealand Veterinary Journal* 53, 223–233.

Sidholm, P.M. (2003) Welfare of cattle transported from Australia to Egypt. *Australian Veterinary Journal* 81, 364–365.

Stafford, K.J., Mellor, D.J., Todd, S.E., Gregory, N.G., Bruce, R.A. and Ward, R.N. (2001) The physical state and plasma biochemical profile of young calves on arrival at a slaughter plant. *New Zealand Veterinary Journal* 49, 142–149.

Waghorn, G.C., Davis, G.B. and Harcombe, M.J. (1995) Specification of rail spacing and trough heights to prevent escape and enable good access to feed by sheep during sea shipments from New Zealand. *New Zealand Veterinary Journal* 43, 219–224.

Wass, J.R., Ingram, J.R. and Matthews, L.R. (1997) Physiological responses of red deer (Cervus elaphus) to conditions experienced during transport. *Physiology & Behaviour* 61, 931–938.

Webster, J. (2005) *Animal Welfare. Limping Towards Eden.* Blackwell, Oxford, UK.

Wright, W. and Muzzatti, S.L. (2005) Not in my port: the 'death ship' of sheep and crimes of agri-food globalization. *Agriculture and Human Values* 24, 133–145.

13 Europe

S. CORSON[1] AND L. ANDERSON[2]

[1]Positive Pet Behaviour, London, UK, positivepetbehaviour@googlemail.com;
[2]ANNEX Consultancy, Edinburgh, UK

Abstract

Millions of animals are transported long distance (over 8h) in Europe each year and there has long been concern that this poses serious welfare issues for livestock. These journeys, sometimes in adverse conditions, regularly involve transporting animals thousands of kilometres to slaughter. Routes identified as causing particular problems for animal welfare include those where: horses are transported from Central and Eastern Europe to Southern Italy; pigs are transported from the Netherlands to Southern Italy; and where sheep are imported into Greece.

Even where legislation governing the welfare of transported animals is adhered to, animals may endure long journeys in cramped conditions and the stress of being handled, mixed, loaded and unloaded. They may also be exposed to regional variations including differences in climate and infrastructure which can further compromise their welfare. Furthermore, inconsistencies in enforcement of transport legislation within the EU and/or ineffective penalties for non-compliance mean that legislation is all too often ignored with serious welfare implications for the transported animals. Reports of welfare infringements en route are not uncommon and include: overstocking; illegal route plans; inadequate road vehicles; and sick, injured and dead animals.

Animal welfare organizations in Europe believe that the long-distance transport of live animals over 8h should be ended completely and that transportation of animals over these distances should be in carcass form. However, the issue of the live export trade in Europe is complex and multifaceted, influenced by many factors from the economy to the religions and cultures prevalent in the region. Legislative initiatives and action plans to be implemented within the next few years will offer opportunities to improve the welfare of transported livestock and alleviate, if not eradicate, the suffering they can endure on their, sometimes exceptionally long, journeys to slaughter.

Acronyms

CEEC	Central and Eastern European Countries
DG SANCO	Directorate General Health and Consumer Protection
EEA	European Economic Area
EEC	European Economic Community
EU	European Union

EUROSTAT	Statistical Office of the European Communities
FAO	United Nations Food and Agriculture Office
FAOSTAT	Statistical Office of the United Nations Food and Agriculture Office
FAWC	Farm Animal Welfare Council, UK
FVE	Federation of European Veterinarians
FVO	Food and Veterinary Office
ILPH	International League for the Protection of Horses
NGO	Non-governmental organization
OIE	Office International des Épizooties (World Organisation for Animal Health)
PMAF	Protection Mondiale des Animaux de Ferme, France
RSPCA	Royal Society for the Prevention of Cruelty to Animals
USDA	United States Department of Agriculture
WSPA	World Society for the Protection of Animals

Introduction

Approximately 356 million farmed animals (not including poultry) are transported around Europe each year (Caporale *et al.*, 2005). The reasons for transport of live pigs, cattle, sheep, horses and goats are: for breeding; for transhumance (seasonal movement of animals for pasture); for further fattening; and for slaughter. Many of these journeys are long distance, i.e. over 8 h, which in Europe is acknowledged as the source of significant animal welfare concerns by consumers, animal welfare organizations, industry, farmers and governments; many people argue that it should be more effectively addressed by legislation.

Any long-distance journeys of livestock unused to such travel, including those made to slaughter, may expose the animals to unfamiliar conditions such as noise, movement, vibrations, heat or cold, lack of food and water, strange surroundings and different stock which could put them under stress even under good travelling conditions (RSPCA, 1999; European Commission, 2000; Stevenson, 2000; Eurogroup for Animal Welfare, 2002; RSPCA, 2003; Pickett, 2005; see Chapters 6 and 7, this volume). Although statistics for long journeys within states are not easily obtained, journeys that cross borders are routinely recorded and it is known that journeys of several thousand kilometres are everyday occurrences. On such journeys, European law requires that animals be unloaded for rest at set intervals to meet the animals' most basic requirements for food, water and rest. Unfortunately, scientific evidence shows that loading and unloading are stressful to animals and, since journeys to slaughter can be extremely long, whether for economic reasons or for access to an appropriate slaughterhouse, the animals may have to endure the same stressful process several times during transportation.

Although relevant European legislation sets down minimum standards for transport, long journeys themselves continue to be legal. Problems of insufficient enforcement of legal requirements have been identified for many years, monitored by animal welfare groups, formally reported by European veterinary inspectorates and recognized by the European Commission as an area requiring urgent and detailed attention (see Chapter 5, this volume). While lack of resources and manpower constrain the authorities that are tasked with enforcing the law, maximum journey times could well be exceeded undetected.

A shortage of slaughterhouses within reach is one of the factors cited as making reform impossible. While this may be true in some areas, it is not a logical alternative simply to continue to permit animals to travel for up to 90 h to be slaughtered at the end. Animal welfare organizations and farmers' unions alike have called on governments to address the shortage of slaughterhouses, taking the view that this could be done if they were committed to reducing animal suffering caused by transport, and that it would be economically attractive for governments to do this.

A solution proposed by animal welfare organizations has been to limit journey times to slaughter to a single journey of 8 h or less. Although the European Parliament has endorsed this in principle, it has been opposed by an industry dependent on exporting live animals to distant markets and resistant, for a variety of reasons, to changing over to a carcass trade. Writing in a (OIE) publication (OIE, 2005), industry representatives considered that improvements in long-haul transport vehicles and road systems had greatly improved conditions for animals travelling long distances in addition to reducing journey times. They stated that mixing of stock, loading and unloading and human intervention were the most stressful aspects of transport. While these are stressful elements of travel, animal welfare organizations would be unlikely to concur with the authors' conclusion that journey times and stocking densities were less relevant as a result.

In the following chapter, the first section will contain a short overview of the livestock industry within the EU: identifying the prevalence and extent of livestock transportation and the species involved as well as legislative, economic, cultural and religious factors influencing such live transportation. Subsequent sections will detail: the general conditions of animals during transport under legislative requirements; breaches of those requirements; and situations outside the scope of existing law. The final section discusses the findings of the report and draws attention to problem routes and particular areas of good and poor practice within the region and identifies opportunities for improving welfare in the transport of farmed animals during the next 5–10 years.

An Overview of the Livestock Industry in Europe

The European region

While the focus of this chapter is the EU – the 27 current member states of the EU – it also refers to states outside the EU, and in some cases on the European regional boundary.

Europe is a relatively small continent, but is composed of striking geographical contrasts. In mountainous areas, livestock production is one of the few means of subsistence, whereas those areas with broader and flatter plains are suitable for both arable and livestock production. Most live transport of animals in Europe is by road, but sea journeys are still significant.

Production and consumption of meat

The production, consumption and demand for meat are what drive the long-distance transport of live animals within Europe. The meat sector in the EU consists of farmers, farmer cooperatives, slaughterhouses and companies involved in various aspects of processing and distribution and marketing of meat. Certain areas specialize in one or more types of meat production. For example, pig rearing is concentrated in regions of Belgium, France, Germany, the Netherlands and Spain, while most sheep rearing (54%) takes place in Spain and the UK. In many areas, such as Finland and Sweden and the mountainous terrain of Austria and Italy, livestock farming has been seen as a vital economic activity where few alternatives to animal husbandry exist (Directorate General for Agriculture, 2004). The predominance of France in cattle rearing is accounted for by an agriculture where the breeding of dairy cows (for a major cheese-making industry) is significant (Lebegge, 2004).

Poultry rearing for meat and eggs is widespread throughout the EU. For 2005, the number of poultry slaughtered were: France 993 million; UK 906 million; Germany 581 million; Spain 564 million; and Italy 438 million. A large majority of slaughtered poultry is chickens (FAO, 2006).

Horses are not generally reared for consumption but are slaughtered at the end of their working lives, so they are not always classed as livestock. Precise figures tend to be available for registered, rather than non-registered, fattening or slaughter animals. It is estimated that there are around 6 million horses in Europe. Horses are naturally more prevalent where they continue to be used in agriculture (e.g. Eastern European countries such as Poland, Romania, Ukraine, Belarus). However, there is not a particular tradition of eating horse meat in these countries and the export market has existed for many years, principally to Italy where horse meat and horse meat products are very popular.

Meat consumption

The USA and Europe constitute the two biggest consumers of meat on a worldwide level. Per capita consumption in recent years is shown in Table 13.1 below.

The European Commission (2004) further reported that the EU consumed approximately 35 million t of the various meat types per year – around 92 kg per head, plus around 5 kg of edible offal.

The needs of a particular country may not, however, be satisfied by its indigenous production. Surpluses and deficits combine with national tastes and traditions to drive transactions between states. As has already been said, Italy imports live horses and horse meat from Eastern European countries where it is not commonly eaten; the UK is the largest producer of sheep meat, but not a high consumer, whereas mutton is popular in Greece.

Meat is a significant part of the EU economy. Altogether, the four main meat types – beef and veal, pig meat, poultry meat and sheep meat/goat meat – accounted for a quarter of the total value of agricultural production in 2003 (Directorate General for Agriculture, 2006).

Consumption of meat is projected to increase steadily year by year, as in Table 13.1. Pig meat, with a share of about 50%, was by far the most preferred meat by EU consumers, followed by poultry, recording a share of around 27%, which had in turn overtaken beef and veal since 1996 (Directorate General for Agriculture, 2005).

Imports and exports

Most meat is traded within the EU, and imported and exported to/from the EU, in carcass or prepared (product) form.

Table 13.1. Per capita consumption of meat, Europe and the USA, in kg.

	2001	2002	2003	2004	2005
Europe (From European Commission Agriculture and Rural Development. Available at: www.ec.europa.eu/agriculture/index_en.htm.)					
Pig meat	43.1	43.4	43.8	43.4	–
Beef/veal	18.3	19.9	20.2	17.9	–
Poultry meat	23.0	23.4	23.4	23.0	–
Sheep meat/goat meat	3.4	3.5	3.5	2.9	–
Equine meat	0.4	0.4	0.3	0.3	–
Other	2.7	2.5	2.5	2.5	–
Total	90.8	93.1	93.7	90.0	–
USA (From National Agricultural Statistics Service, 2007.)					
Pork	22.9	23.4	23.5	23.3	22.6
Beef	30.1	30.8	29.5	30.0	29.7
Veal	0.3	0.3	0.3	0.2	0.2
Broilers	34.9	36.6	37.1	38.3	38.9
Turkeys	8.0	8.0	7.9	7.7	7.6
Lamb/mutton	0.5	0.5	0.5	0.5	0.5
Total	97.1	100.4	99.5	100.6	100.2

USA total includes other chicken.

Lebegge (2004) calculated the extent of imports and exports of both live animals and meat in 2001, throughout and with countries outside the EU. Although the statistics do not take into account journey times, they are a useful indication of the proportion of live animals transported in relation to total trade in meat. Out of the total 27,295,000 t transported, 3,568,000 t (13.1%) was in the form of live animals. Transactions in live animals (in EU and with non-EU countries) amounted to 19.8% of the total bovine trade; 10.3% of the pig trade; 15.9% of ovines; 13.3% of poultry; and 46% of equines.

Stevenson (2004a) also calculated the proportions of live trade and meat imports and exports for a number of states for 2002–2003. He found, for example, that live sheep occupied only 2.6% of UK exports, with the remaining 97.4% being in meat form. France, on the other hand, exported 43.4% of its sheep live, and 56.6% as meat. Hardly any UK sheep imports were live (0.02%), whereas Italy imported a high proportion of 45.8% live and 54.2% as meat. For pigs, in 2003, only the Netherlands exported a significant proportion live (28.8%), with exports from the other main exporting countries all under 10%.

The fact that live transport occupies a low proportion of the overall trade is a strong argument for eradicating it altogether for slaughter animals. With regard to equines, which are transported in smaller numbers than other species – probably due to the shortage of slaughterhouses with capacity to handle horses in some countries – the high proportion of the trade occupied by this method may make it less susceptible to alteration. There have, however, been more recent changes in the equine market from Romania (the second largest exporter to the EU before it joined in 2007), driven by animal health controls, and other countries are also showing some downturn.

Traffic flow of live animals

Within the EU, journeys across member states are considered a major part of the welfare problems of live animal transport and the traffic flow tables, found in Appendix 13.1, show the vast flow of live animals between these states. Although these statistics do not indicate which journeys exceed 8h, clearly many will give the distances involved, for example, in the transportation of 5025 horses from Lithuania to Italy (Appendix 13.1), the shortest distance between the two countries is over a 1000 km (Meridian World Data, 2007).

Notes to traffic flow tables in Appendix 13.1

After poultry, piglets were the animals that travelled in the greatest numbers in 2005, the main flows were from the Netherlands and Denmark, countries where the focus was on pig breeding, to Germany and Spain, which specialized in fattening. The main flows of lambs were from Hungary, Romania, France and the Netherlands, with large numbers flowing into Italy. While the sheep trade was dominated by lambs, there were also considerable flows of sheep (over 1 year) from Hungary and Spain to Italy, France and Greece, where mutton is consumed more readily.

These figures show there is considerable two-way traffic in the same species and age of animals between states. There may in some cases be valid logistical reasons for this – such as the greater ease of access to a slaughterhouse just across a neighbouring border – and even within the categories there will be differences of age and purpose that mean there is not direct duplication.

Legal and regulatory issues

Farmed animal protection through legislation

The flow of live animals travelling in the EU is regulated by the EU's comprehensive legislation on animal welfare and nominated officials and inspectors are charged with enforcing this legislation throughout the transportation process.

The first European animal welfare legislation was passed in 1974, and dealt with the protection of animals at the time of slaughter; it was subsequently updated with Directive 93/119/EEC (European Council, 1993). Since then a considerable body of EU legislation related to the treatment of animals has been created.

A Protocol on Protection and Welfare of Animals was annexed to the EC Treaty in 1997, recognizing that animals are sentient beings. It stated that full regard should be paid to animal welfare concerns when formulating or implementing policies relating to agriculture, transport, research and the internal market. At the same time the legislative or administrative provisions and customs of member states relating in particular to religious rites, cultural traditions and religious heritage were to be respected (Directorate General for Health and Consumer Protection, 2006a).

General minimum standards for the protection of farm animals were set out in Directive 98/58/EC (European Council, 1998a). These rules reflect the 'Five Freedoms', i.e. freedom from hunger and thirst; freedom from discomfort; freedom from pain, injury and disease; freedom to express normal behaviour; and freedom from fear and distress (see Chapter 1, this volume). Additional specific

Directives have been provided for calves, pigs and laying hens, and a Council Directive on the welfare of broiler chickens has been proposed.

The first EU Directive on the protection of animals during transport was adopted in 1977; the second Directive 91/628/EEC was adopted in 1991 (European Council, 1991a) and amended in 1995 by Council Directive 95/29/ EC (European Council, 1995), further provisions were added in 1998 by Council Regulation EC/411/98 (European Council, 1998b). Council Regulation 1/2005, adopted in December 2004 (European Council, 2005), implemented on 5 January 2007 has replaced Directive 91/628/EEC and Regulation EC/411/98.

As a Regulation, binding in its entirety, the new law encourages greater harmonization between member states and will, it is hoped, deliver more consistent enforcement. For example, there will be uniform documentation and additional rules relating to long-distance international journeys. Article 1 also states that the Regulation shall not be an obstacle to any stricter national measure aimed at improving the welfare of animals during transport within the relevant member state, or transport from it by sea. The content of Directives and Regulations will be discussed in further detail in the section 'Conditions and Welfare of Animals'.

Guidelines or Codes of Practice

It has been observed (Broom, 2005) that while laws can have a significant impact on the ways in which people manage animals so too can codes of practice. Some of the most effective, as stated by Broom (2005), are retailer codes of practice, since retailers need to protect their reputation by enforcing adherence to their own codes.

The European Food Safety Inspection Service (EFSIS) is an independent inspection and certification service, providing retailers, manufacturers, farmers and caterers with inspection and certification of their operations. Those who draw up specifications include major multiple retailers, grocery independents, contract caterers, multiple catering outlets, fast food chains and other companies operating in the food chain. EFSIS also audits various industry quality assurance schemes, covering an extensive range of food products.

EFSIS has offices in a number of European cities. In the UK, its Assured British Meat transport standard covers identification and traceability; driver training; driving, loading/unloading vehicles; stocking densities; provision of bedding; fitness of animals; cleanliness; ventilation; and vehicle operating procedures.

Stevenson (2004b) noted that voluntary quality assurance schemes might require meat labelled as home-produced to be derived from animals born, fattened and slaughtered in the home country. This represents an improvement on the 'cosmetic' approach whereby many animals designated as for further fattening in the country of origin are retained only for the minimum period required to qualify as home-reared, before being slaughtered. One of the complicating factors in assessing the transport problem is that many further fattening animals are to all intents and purposes destined for slaughter, and treated no differently from slaughter animals. Mandatory labelling of country of origin, rearing and slaughter – as is currently required for beef – for all meat products would increase consumers' knowledge of their food and allow for informed purchasing decisions.

Welfare Quality

Welfare Quality is an EU scientific project on animal welfare and sustainable agriculture throughout Europe. It aims to develop on-farm monitoring, product information systems and practical strategies to improve animal welfare (Welfare Quality, 2004).

A report by Welfare Quality in 2006 into research into quality assurance labelling found that labels are the main form of communication between quality assurance schemes and shoppers but, for a variety of reasons, do not convey to consumers a clear picture as to what welfare standards the products have been produced.

They also found that most labelling schemes were concerned with quality rather than animal welfare, with a few exceptions such as the RSPCA Freedom Food Scheme in the UK, which is the only dedicated quality assurance for animal welfare in Europe. However, organic production in a number of member states is regulated so that the maximum travel time is 8 h.

OIE Guidelines

The OIE identified animal welfare as a priority in its 2001–2005 Strategic Plan. A permanent working group on animal welfare was established to provide guidance on the development of science-based standards and guidelines including, as a priority, the topics of transportation by land and sea. It was understood that change would take time and that the needs of all OIE member countries would have to be addressed (Petrini and Wilson, 2005) so that in many countries there is likely to be little progress in implementing the guidelines in the near future.

For the purposes of this report, OIE delegates (chief veterinary officers) from 43 member countries were asked to comment on their countries' progress towards disseminating and implementing the transport guidelines. Of the seven European countries that responded Switzerland observed that its legislation was directly influenced by the Council of Europe draft Code of Conduct for the international transport of animals by road and that: 'The OIE guidelines for terrestrial transport were taken over in large parts from the mentioned Code. Therefore, the OIE guideline will have an indirect influence on our legislation.' Turkey responded that there was no progress towards disseminating and implementing the OIE Guidelines. Germany welcomed the compilation of OIE Guidelines, saying that the rules ought to find international application, and that the standards would contribute to raising the level of animal protection worldwide (E. Konigs, 2006, personal communication).

At present OIE Guidelines are not sufficient to protect animal welfare and if they are to be useful it will be necessary for them to become more prescriptive and also to be enforced effectively, which is not currently the case.

Socio-economic factors

Economics

Being generally viewed as freight or a commodity, live animals are subject to a constant drive for reduced costs. The import/export trade seeks to purchase animals at the lowest possible price: thus, CEEC, with their large production and lower labour costs, have traditionally been a source of horses, lambs and cattle for export to the rest of the EU.

Following purchase, the importer aims to get the product to the consumer for as low a price as possible and the suffering of the animals has not been a consideration in this approach: in economic terms, it is disregarded as costing nothing. However, in fact, diminished welfare is often associated with decreased economic value, as discussed in Chapter 4, this volume.

It has often been suggested that the cost of transporting meat or meat products in refrigerated vehicles inhibits the conversion of the live slaughter trade into an entirely carcass trade. Stevenson (2004a), however, calculated in a hypothetical example (see Box 13.1) that the transport of carcasses was often cheaper and that the main factor affecting profitability was not the transport cost, but the ultimate sale, with live animals fetching a better price.

Other studies have similarly found transporting live animals to be the most economic option, but only given unacceptable transport conditions. Leckie's 2002 report on the economics of the live horse and horse meat trade from Hungary to Italy also showed that transport costs were cheaper for carcasses. However, she concluded that it was cheaper to transport live horses for slaughter than it was to transport carcasses when all costs were taken into account so long as the horses were transported under the suboptimal conditions that were commonly in use at that time. When the conditions of live animal transport were raised and enforced properly in accordance with EU rules the economic balance changed abruptly and it became cheaper to transport meat rather than live horses.

A variety of factors combine to maintain demand within the market for a certain amount of cross-border traffic. These include different preferences in methods used for carcass-cutting (even though there is little impediment to slaughterhouses using the method preferred by their customers, regardless of country), and the exploitation of the so-called fifth-quarter (offal) which is not included in the prepared carcass.

Trade follows the availability of higher prices, and this may be complicated by the marketing of imported animals to the consumer as domestic animals, so long as they have been killed in the importing country. Improved labelling could address this, and indeed detailed labelling was introduced for beef and veal by Regulation (EC) 1760/2000 (European Council, 2000). It would be desirable to see pig and sheep meat subject to a similar regime.

Box 13.1. Cost comparison of transportation of live and carcass-form animals. (From Stevenson, 2004a.)

The total cost of exporting 400 live lambs from Wales to France, including purchase of the lambs and transport to France, would have been £18,416, of which approximately £2300 related to transport costs. According to the exchange rate in operation at the time (November 2004), the sale price would have been £22,904.75.

The total cost of exporting 400 Welsh lambs in carcass form, including purchase of the lambs, slaughter in England and transport to France, was calculated as £20,382, of which £730 related to transport costs. Their sale value was set at £17,353.22 at the exchange rate in operation at the time.

The main factor affecting profitability, therefore, was not the transport cost, but the ultimate sale, with the live lambs fetching a better price.

In order to take advantage of any economic and welfare benefits of transporting meat, rather than live animals, there must be slaughter facilities near to the places of rearing. Both farmers' unions and animal welfare groups in Europe regularly state that under-provision of slaughterhouses is an obstacle. To this needs to be added the requirement for refrigerated warehouses to guarantee hygiene standards during transfer of carcasses.

Slaughterhouse provision today is much more concentrated than 30 years ago, when most countries had a network of large and small, public and private facilities. Following EU Directive 91/497/EC on slaughterhouses, which came into force in 1995 (European Council, 1991b), small slaughterhouses that were unable or unwilling to meet the required standards of hygiene and animal welfare closed their doors instead of upgrading. In France, the number of slaughterhouses declined by one-third between 1990 and 2000, to 339 (Agreste, 2004). In the UK, the number of slaughterhouses fell from 1890 in 1971/72 to 480 in 2007 (FSA, 2007). This increased travelling times within a member state such as the UK and increased the motivation for crossing borders within mainland Europe to reach a slaughterhouse with the appropriate facilities. However, the availability of slaughterhouses is not the only factor involved in the decision to transport live animals long distances.

Conditions and Welfare of Animals

Although legislation provides minimum standards for welfare during transportation these are, in any event, considered inadequate by many groups including animal welfare scientists and government regulators and they do not take into account the regional differences in infrastructure, topography and climate throughout the EU. Furthermore, there are often breaches of legislation due to inadequate enforcement and/or penalties, which have lead to serious welfare contraventions (see Chapter 5, this volume).

Legal conditions

Prior to commencing their journeys to slaughter or further fattening destinations, export animals can spend many hours in holding facilities. The maximum length of time animals should spend at assembly centres and markets, which are the most common places of departure for long-distance transport, has not been determined by law. As a result, competent authorities continue to apply national provisions, and animals spend anything from 1 to 24 h at assembly centres before departure, depending on the member state (European Commission Food and Veterinary Office, 2004a). The time spent is related to the time between veterinary inspection for export (under different legislation) and the time of departure.

During transit to slaughter, the second Directive amended in 1995 by Council Directive 95/29/EC outlines the general conditions in which different animals are allowed to travel. It allows 'standard' vehicles to transport animals for a maximum of 8 h. 'Improved' vehicles (carrying bedding and appropriate feed, with adaptable ventilation, moveable compartments and connection to a water supply) are permitted a maximum journey time of:

Unweaned animals	9 h travel, 1 h rest[1], 9 h travel
Pigs	24 h travel (access to water at all times)
Adult cattle and sheep	14 h travel, 1 h rest, 14 h travel
Horses	24 h travel, during which water must be provided every 8 h

After the first phase, all livestock must be unloaded, rested, fed and watered for a minimum of 24 h, before the journey can continue on another round.

Further provisions for 'improved' vehicles such as forced ventilation and direct access for attendants in order to provide appropriate care were added in 1998 by Council Regulation EC/411/98 on additional animal protection standards for road vehicles used for the carriage of livestock on journeys exceeding 8 h.

Compared to the previous legislation, the newly adopted Council Regulation (EC) No 1/2005 implemented in January 2007 is wider in scope, makes the responsibilities of different operators more clear and places a new emphasis on training. It provides for better harmonization with regard to authorization, certificates and forms, and permits only approved lorries to transport animals on journeys of more than 8 h. In order to reduce the regulatory burden, member states will allow a large degree of self-certification, so problems will probably only emerge where breaches of the Regulation are observed. Space allowances and travel times have not been revised, and this is seen as a major deficiency by the European Commission (see Chapter 5, this volume). However, a requirement for vehicles to carry satellite navigation equipment, to be introduced in phases, will improve monitoring of some key elements of the Regulation.

Transport conditions under the European Convention on the protection of animals during international transport

For the 46 member countries of the Council of Europe, the Convention on the Protection of Animals during International Transport applies (Official Journal of the European Union, 2004). As well as laying down general conditions for transporting animals, such as the design of the transport, the health of the animals, the humane handling of the animals and veterinary controls and certificates, the Convention sets out more detailed rules for different types of transport.

For rail transport, the Convention requires that wagons be marked to indicate the presence of live animals; specifies loading and tethering techniques; requires jolting of wagons to be avoided; and provides that every opportunity must be taken to check animals.

For road transport, the special conditions also require clear marking of vehicles; appropriate driving standard; the carriage of suitable loading/unloading equipment; and that every opportunity be taken to check animals.

For transport by water (except roll-on roll-off vessels), the provisions are much more extensive, and include prior inspection of vessels by competent authorities and supervision of loading and unloading by an authorized veterinarian. Animals

[1] Unloaded, fed, watered, rested, reloaded.

are not to be transported on open decks unless in suitable containers or structures that give protection from sea water. Suitable gangways, ramps, walkways and accommodation must be provided and adequately lit so that animals can see where they are going. Sufficient water, feed and bedding must be carried; provision must be made for isolating sick or injured animals and for humane destruction.

For road or rail vehicles on roll-on roll-off vessels, provision is required for adequate ventilation, inspection and watering or feeding of animals, including emergency watering and feeding if necessary. Vehicles and containers must be well secured and if transported on the open deck, protection from sea water must be given. Provision for humane destruction must be available for sick and injured animals.

For transport by air, no animal may be transported unless air quality, temperature and pressure can be maintained within appropriate ranges; the commander of the aircraft must know the number, species and location of animals on board; and they must be loaded as close as possible to the time of departure. Drugs may only be administered by a veterinarian if a specific problem occurs, although sedation or euthanasia must be available. The animals' attendant must be able to communicate with the crew.

However, as mentioned previously, concern was expressed at the time of revision that, due to the narrow definition of 'international transport', the scope of the Convention is limited only to livestock journeys which include passage to, from or across a non-EU country. Thus, the many journeys between and across EU member states, some of these being thousands of kilometres, may not be subject to the more stringent conditions mentioned above designed to protect the welfare of animals in transit.

Monitoring of animal welfare during transportation

Monitoring, to ensure compliance with EU legislation, is facilitated by:

Control points (staging posts)
Control points (previously called staging posts), for resting animals, are subject to EU approval (Council Regulation (EC) 1255/97 – Council of Europe, 1997) and in March 2006 there were no approved points in Austria, Cyprus, Denmark, Estonia, Finland, Greece, Ireland, Luxembourg, Lithuania, Latvia, Malta, Portugal, Sweden or Slovakia. France, with 46 staging posts, had considerably more than any other member state. Control points are authorized to take specific numbers of animals and specific species, according to the facilities available. For horses – a species for which unloading at regular intervals is a crucial welfare factor, as they will not drink water on board the transport – there were 10 staging posts in Italy, eight in France, four in Germany, four in Poland and one in Hungary.

Border inspection posts
Animals and animal products entering the EU and the EEA, incorporating Iceland, Norway and Liechtenstein, are subject to veterinary health checks at border inspection posts. These are designated, equipped and staffed to perform veterinary

border controls as laid down in the legislation. The purpose of the controls is to ensure that only live animals and animal products complying with the EEA legislation are imported into the EEA. It is the obligation of the EEA states to set up and run the border inspection posts.

Border inspection posts are critically important in monitoring the welfare of imported animals. It is on entry to the EU that animals become – beyond any question – subject to EU welfare requirements. However, the Commission has also taken the view that:

> ... it is beyond doubt that member states should take measures to prevent animals' suffering on their territory, even if that suffering had its origin in events taking place in third countries. member states' competent authorities normally have the means to establish the conditions under which the animals were transported prior to arrival at Community frontiers, including the documentation accompanying consignments notably veterinary and customs papers as well as the tachograph disc of vehicles. This evidence may be relevant to determining the welfare of animals.
>
> (European Commission, 2000)

There are around 300 border inspection posts around the EEA: these may be seaports, airports or road crossings, registered to handle specific species, such as ungulates, equidae or other animals.

Border inspection posts are under the control of state veterinary officials but health checks may be devolved on to local authority-appointed veterinarians, subject to guidance and training by the central authority. These official veterinary surgeons oversee veterinary checks and issue the Common Veterinary Entry Document or if necessary a rejection document, and are responsible for ensuring that TRACES messages (see section on 'TRACES') are monitored and generated. Veterinary checks cover documentation, identity and physical condition (including animal welfare).

When a consignment (a term that includes live animals) is rejected, the permitted options are re-dispatch or destruction. The person responsible for the consignment must pay the costs of storing or disposing of it.

With regard to intra-community trade, the conditions for trade are harmonized between the member states. In an open market, animals are perceived as any other commodity and can be transported over long distances. All live animals must be inspected at the point of departure and travel with a health certificate validated by an official veterinarian specifying that the animals fulfil the basic animal health requirements as set out in the relevant Council Directives. Further random checks on the animals may also be carried out at the final destination or en route.

For imports (i.e. the introduction of animals into the member states from third countries outside the EU), animal health requirements are set out in specific Commission Decisions and Regulations. Health certificates, signed by an official veterinarian of the competent authority of the exporting third country, must accompany all animal imports to guarantee that the conditions for import into the EU have been met. On arrival in the EU, the animals and the accompanying certificates must be verified and checked by EU official veterinarians at a designated border inspection post. Further checks on the animals may also be carried out at the final destination and also en route (Directorate General for Health and Consumer Protection, 2006).

FVO (a directorate of the Health and Consumer Protection Directorate General) inspection results can also contribute to the development of new legislation. The number of animal welfare inspections is low – in 2004, only 6% of FVO missions concerned animal welfare, 2% specifically addressing welfare in transport – but their effect is incremental and can contribute to special measures and general legislative review.

Where serious, but 'non-urgent' problems are found (and it is assumed that animal welfare will come into that category) inspection reports may be used by the European Commission in deciding to start infringement proceedings against a member state. The member states can also, through the Standing Committee on the Food Chain and Animal Health, withdraw export authorization from a third country.

TRACES

In April 2004, the EU introduced a new computerized system for monitoring interstate movement of live animals. The Trade Control and Expert System (TRACES) is a single central database to track the movement of animals and certain types of products both within the EU and from outside the EU.

TRACES was designed to be used directly by economic operators under the control of the competent veterinary authorities, so that relevant information can easily be shared with customs authorities.

Member states that were not ready by April 2004 were allowed to continue to use the old system (ANIMO) until 31 December 2004. Austria, Italy, France, Belgium, Luxembourg and Finland joined the system from the starting date, while other member states joined in the months following this date.

For trade within the EU, a dealer registered with TRACES can fill in all details of the consignment online, sending this electronic form to the relevant competent authority for the final destination. The electronic form is controlled and if the animals comply with the relevant requirements, the form is validated by the local veterinary unit or an official private veterinarian (Business Objects, 2004). As soon as validation is given, TRACES sends the information to the competent authority at the destination, to the central competent authority in states that will be crossed in transit and to all staging points, so that controls can be made en route and at the final destination.

For products or animals imported from outside the EU, if registered in TRACES, the agent at the border inspection post will be able to fill in part I of the Common Veterinary Entry Document describing the details of the consignment. After controlling the products, the veterinary authority at the post will give or refuse authorization. If authorized, the Document is sent to the competent authority at the destination. If the consignment is rejected, all border inspection posts within the EU will be informed via TRACES (Adapted from European Commission, 2004).

Theoretically, TRACES data should record the movement of every animal entering or travelling through the EU, and the purpose of its journey. For the purpose of this chapter, TRACES data for 2005 (the first year of full operation) were requested from the European Commission, and were still awaited at the time of writing.

Welfare of animals

Even where European legislation is complied with and monitored correctly, the welfare of animals during transportation has been shown to be compromised for a variety of reasons. However, minimum standards are often not adhered to, through a lack of enforcement and effective penalties, and serious breaches of welfare have been observed. Additionally, regional variations in climatic conditions and infrastructure within the EU are outside the scope of legislation and have been known to cause serious problems when animals are exported long distances and exposed to such variations.

Welfare under legal conditions

Journeys to slaughter can be extremely long, and journey times were not revised in the new legislation. Journeys of several thousand kilometres are not uncommon in the EU, such as those undertaken in the transportation of sheep and horses from CEEC to countries bordering the Mediterranean in southern Europe; as well pigs transported from the Netherlands to Italy (discussed further in the section 'Problem Routes').

The closure of many small slaughterhouses, which were unable to meet the criteria for operation under current legislation, tended to increase travelling times within member states. Crossing borders within mainland Europe is sometimes necessary to reach the nearest slaughterhouse with appropriate facilities and further delays may be experienced at border controls.

Under the new legislation, space allowances were not revised and, currently, permit stocking densities that make it impossible for all animals to lie down at once or to gain access to on-board drinkers.

Contraventions of legislation affecting welfare

Non-compliance with live transport legislation, according to NGO commentators, could be endemic and a Commission report from December 2000 (European Commission, 2000) found, in a number of member states, inadequate road vehicles; illegal route plans and non-compliance with travelling times; negligence and poor handling of animals; insufficient ventilation on road vehicles; overloading; and difficulties in checking the approval of transporters for animal transport.

The FVO monitors the compliance of states with EU legislation on food safety, animal health, animal welfare and plant health. Inspections are carried out in member states, accession countries, candidate countries and other third countries, and recommendations are made accordingly to the competent authorities (central and local government departments in charge of the relevant remit). Enforcement in member states may be by police, state veterinary authorities, local authority animal health inspectors or a combination of authorities.

Lack of enforcement of legal requirements is unquestionably one of the principal factors in the animal transport problem and this has been recognized at Commission level for many years. See Chapter 5, this volume, for a discussion of enforcement of transport regulations in the EU.

Intensive long-term monitoring of individual consignments and routes by animal protection NGOs has revealed disturbing examples of poor practice, non-compliance and animal suffering (Boxes 13.2, 13.3, 13.4, 13.5 and 13.6).

Box 13.2. Calves from Hungary to Italy.

Unweaned calves are often transported from north-eastern Hungary to southern Italy. On this route there is no single staging point equipped to feed and water unweaned calves. However, transporters' route plans will have watering marked, stamped and shown as paid for, even though it has not actually taken place. This would suggest that all transporters transporting calves on this route are in violation of current EU legislation. As observed by the local animal welfare group, 'it is hard to imagine watering 250 calves, loaded in three levels of transporter, in one hour' (U. Markelj, 2006, personal communication).

Box 13.3. Lack of enforcement on southern routes. (Available at: Animals' Angels, 2005.)

In summer 2005, Animals' Angels – a welfare organization with the long-term aim of abolishing the long-distance transport of farm animals – carried out a 14-day monitoring programme of the motorways of southern Europe, covering routes from Spain through France and Italy to Greece. The investigation revealed that pigs being transported from the Netherlands had died on route and been eaten by others in the vehicle; a horse that had died on a journey originating in Spain; as well as seriously injured and sick sheep. The investigators noted that conditions included routine breaching of EU transport regulations.

Box 13.4. Example of overstocking of pigs. (Available at: www.animals-angels.de.)

In February 2006, an Italian transport truck was observed overloaded with pigs en route from the Netherlands to Sicily. The watering system on the first loading deck of the truck did not work. Near Fréjus in France, the welfare organization monitoring the truck arranged for it to be stopped by the police and then weighed. Since it was found to be 2.7 t heavier than the maximum permitted, the police confiscated the truck and, in cooperation with the veterinary authority, arranged for the pigs to be taken to the nearest slaughterhouse. The transport company was fined and had to bear the high costs of emergency slaughter as well as collection and transportation of the slaughtered animals in refrigerated trucks to Sicily.

In this case, the cooperation of the police, the veterinary authority in Chambéry, the authorities in Savoy and the director of the slaughterhouse in Chambéry had prevented a long journey, without water, in overcrowded conditions.

Box 13.5. Trade with Spain. (Available at: www.animals-angels-de.)

Despite reprimands, complaints and correspondence, by the animal welfare organizations monitoring the transports, there was no significant improvement in the conditions on sheep transports from Burgos to Italy and Greece. Transports continued to be overloaded and sick and dead animals were found on board.

Evidence of the violations was reported to the chief veterinarian of the province concerned in an attempt to ensure that the competent government veterinarians within his area carried out their responsibilities strictly in compliance with regulatory requirements.

> **Box 13.6.** Authorization of illegal transports. (Available at: www.animals-angels.de.)
>
> In August 2005, on the highway near Frosinone in Italy, a check of the documents of a transporter carrying pigs and bulls from Spain to Sicily, revealed that no unloading was planned during the entire journey despite the fact that the permissible transport time would be exceeded by at least 12 h. The transport was fined by the local police and the truck unloaded at the next suitable location. The transport had been authorized by Spanish official veterinarians in violation of the law.
>
> Also in August 2005, in the port of Civitavecchia, Italy, an overloaded transporter was observed with horses arriving by ferry from Barcelona. The transporter had already exceeded the permissible transport time of 24 h and the final destination was a further 10 h away. Police required the horses to be unloaded and found two which were injured and unfit for transport; a foal standing between adult horses; and three stallions not separated from the mares. This transport had also been authorized by Spanish official veterinarians despite breaching regulations on journey times and stocking methods, as well as transporting injured animals.

Regional variations affecting welfare

In transit across countries within the EU, animals encounter variations in infrastructure, including the conditions of roads. Generally conditions in Eastern European countries are poor, with few or no motorways, in comparison with those in Western Europe where, for example, the French motorway network is well suited to animal transportation. Animals transported on a motorway will clearly travel more quickly and comfortably than those travelling on poor or badly maintained roads.

Different standards of training and competence within the member states can lead to ignorance of the correct procedures to be followed which, in turn, can result in the unnecessary suffering of animals.

Geographic and climatic variations

The mountainous geography in Europe is relevant to the subject of animal transport in that it poses physical obstacles to traffic, which can mean extended detours or long delays at bottlenecks such as mountain tunnels.

The European region is divided by significant bodies of water, the most notable in terms of animal traffic being the Mediterranean Sea, which has provided a route for trade between states, including the Middle East, for centuries. Journeys from the islands of Greece, Italy and Scotland to the mainland are routine, while international journeys include ferry transport from Spain to Italy, Italy to Greece and France to Lebanon.

While there has been concern about the welfare of animals transported by sea for many years, sea voyages within Europe are typically shorter than those in the rest of the world. For instance, large numbers of cattle and sheep have been traditionally transported by sea from the UK and the Republic of Ireland, mostly for slaughter in mainland Europe; animals travel between Italy and Spain by ferry; countries with significant island areas, such as Greece and Scotland, inevitably see many movements by sea for a variety of purposes. Derogations are available from EU legislation for outlying areas of the Community, such as islands.

Box 13.7. Winter conditions. (Availbale at: www.pmaf.org.)

An incident was reported in January 2006 when a large number of live animals, some of which were in transit to southern Italy, were trapped overnight near the Fréjus tunnel (in southern France), in temperatures of −15°C. The vehicles containing pigs, horses, sheep and cattle had no food or water for the animals even though some had already travelled for over 20 h, in contravention of EU law requiring sustenance to be carried on journeys of more than 8 h.

Longer journeys are involved in the export trade from southern and eastern Europe to the Middle East. Even then, however, these voyages are shorter than those involved in the trade between Australia and the Middle East (the world's largest importer of live animals for slaughter). None the less, some of the worst breaches of animal welfare have taken place at ports and on ferries.

Animals undertaking journeys that are frequently in excess of 2000 km, and sometimes up to 7000 km, within the European region inevitably cross different climate zones, and in different seasons are exposed to extremes of heat and cold.

Injury and morbidity

Even in optimum situations where live transportation is carried out in accordance with legislative requirements, some animals will be injured and/or die as a result of unforeseen but unavoidable problems such as traffic tailbacks or excessive heat or cold (Box 13.7, 13.8).

Clearly where animals are transported in non-optimum situations, often in illegal conditions, the injury and morbidity rates can be much higher. Advocates for Animals (www.advocatesforanimals.org) reported seeing welfare contraventions, including animals giving birth on board (it is illegal both to transport a pregnant animal in the final 10% of a gestation period, and to carry newborn animals); and animals' limbs being trapped between the transporter decks when they were raised and lowered, sometimes resulting in limbs being severed.

Box 13.8. Summer conditions. (Available at: Animals' Angels, 2005.)

An animal welfare group monitoring the transport of 400 piglets and 70 adult pigs from Best in the Netherlands to Salerno in Italy (1800 km), in the summer of 2005, found that, after scheduled stops for unloading of the piglets in Italy, 27 had died as a result of heat and overstocking. Following a 24 h rest period, the adult pigs were reloaded for travel to Sardinia but, by mid-afternoon, the transporter was in a traffic tailback lasting over 2 h in 40°C heat.

There was no water on the vehicle and by 18.00 h the driver counted six dead pigs. By the time the transporter was stopped by the police, at midnight, only 27 pigs were still alive.

Problem Routes

While the percentage of animals transported live long distance is statistically small in comparison to the carcass trade, this section reveals that many thousands of animals are still transported long distances across Europe and the welfare issues arising from such long-distance transportation. The transport routes most frequently giving rise to concern include the following.

Horses transported from Central and Eastern European countries to Italy

While the number of horses being transported long distance for slaughter is lower than for other species their welfare needs are so specific and reports of animals suffering so frequent that the trade has long been criticized. A large proportion of horses transported to Italy originate from CEEC where they are still widely used in agriculture. These journeys inevitably involve many kilometres, for example, even the shortest route between Poland and Italy is over 600 km, whereas a journey from Warsaw to Palermo (Sicily) is in excess of 1600 km (Meridian World Data, 2007).

In Poland, transport conditions have long caused concern at both official and NGO level, with transports being rejected during the 1980s due to the condition of the animals arriving at border crossings (Ratajczyk, 1994 as cited in Wilson, 1994).

A Polish Supreme Control Chamber (NIK) audit of the first 2 years of the Animal Protection Act (up to September 1999) (NIK, 2000) highlighted slow progress towards introducing necessary subordinate legislation; incomplete public administration of animal welfare measures; insufficient sanitary and veterinary supervision over the places where animals were gathered, kept, purchased and slaughtered; and concerns over the legality of trade in animals.

A later report (BIP, 2005) described the period in mid-2004 during which the process of harmonizing Polish law with EU regulations was taking place. Among other problems, the report recorded defects identified 'in transit' in 50% of vehicles transporting animals. There were cases of unrecorded animal transportation, vehicles in unfit or unsanitary condition, overstocking and unauthorized drivers. The NIK considered that central Veterinary Inspectorate and local veterinary supervision was inadequate, as was that of the Road Transport Inspectorate.

There is a considerable body of evidence from NGO monitors such as Animals' Angels and the ILPH for the length of journeys undertaken by slaughter horses and the suffering involved. An Animals' Angels mission of 1999 (Wardle, 2002) trailed a consignment of 20 horses handled by a Polish company. In this instance the horses originated in Kaunas, Lithuania, were transited through Poland, Czech Republic, Slovakia, Hungary and Slovenia (none of these countries were yet in the EU) and crossed into Italy at Gorizia. Their final destination was Sardinia. The journey time of 95 h included an 8 h ferry crossing, and the distance covered was 2530 km (1520 mi). On the way, the horses were not rested or fed as they would have been in countries subject to EU law; they were mixed so that fighting occurred, causing injury; and beaten by both the driver and workers at one of the rest stops. None of the effects of this treatment was observed during the cursory veterinary check at Gorizia, where they entered EU territory.

Romania has been the other principal source of horses exported live to EU states such as Italy, this in the past has been the subject of some animal welfare concerns (Box 13.8). ILPH has instituted a 5-year welfare project in the country and both the ILPH and the UK Department of Environment, Food and Rural Affairs provide advice to, and work with, the Romanian National Sanitary and Veterinary Authority.

As a full member of the EU, Romania is now subject to EU legislation including the new Regulation No 1/2005 concerning animal welfare during transport. Training is also under way. Animal transport was the subject of the first quarterly meeting of the 2006 National Training Programme for officials in charge of animal welfare. Training sessions in Resita, Oradea, Suceava and Ploiesti covered the report of a recent DG SANCO Mission, as well as practical issues raised by FVO inspectors. The presentation of standards for the vehicles, means of authorization and registration, and completion of route plans were also covered.

Although steps are being taken to implement change in CEEC to comply with EU legislation on animal welfare, it will take some time for the changes to be widely disseminated and fully understood in these regions. In addition, while these live imports from Eastern Europe have declined, welfare concerns have been growing over increasing imports of live horses from Spain to Italy in recent years. In 2005, this new trade route had grown to 10,136 horses, with the lack of enforcement of animal welfare regulations in Spain resulting in excessive livestock densities and journeys of over 30 h without rest at a staging point, water or feed.

Sheep exported to Greece

Greece has the highest per capita consumption of sheep meat in Europe and is a major importer of sheep. It is generally reported that local slaughter is preferred so that the meat can be marketed as Greek – a practice common in a number of the importing countries. Altogether, 94% of sheep imports to Greece in 2003 came from Romania, the shortest distance between these two countries being approximately 400 km (Meridian World Data, 2007). Romanian exports of sheep to Greece were shown as 765,129 (41% of the Romanian sheep export).

The main entry point for lorries originating in Romania is Promahonas on the Greek–Bulgarian border. A 2-day investigation (www.animals-angels-de) at Promahonas in April 2005 observed ten transporters crossing the border over the period, of which seven were detained by border authorities for legal breaches. Four Greek-registered and one Romanian-registered trucks were overloaded, while another two did not have the correct documentation. One of the lorries contained 880 lambs, 130 more than the permitted capacity.

Animals' Angels also reported seeing overloaded trucks being waved through the border with Former Yugoslav Republic of Macedonia (FYROM) in April 2005. The border post had only one veterinarian, and queues of up to 4 h were forming.

NGO reports are substantiated by a series of inspection reports from the EU FVO, including one from October 2004 (European Commission Food and Veterinary Office, 2004b) and a follow-up report published in June 2006 (European

Commission Food and Veterinary Office, 2006). In 2004, the FVO noted that enforcement of transport legislation was the responsibility of the Greek prefectures, which have a degree of autonomy in this respect, so that the Central Competent Authority (the Greek Ministry for Rural Development and Food) had been restricted in the action it could take to improve matters. Among other things, the mission report identified one prefecture where transporters' statements regarding their normal trade routes revealed that these would breach the permitted journey times under the relevant Directive, but the local authority had apparently not noticed this; nor had it asked importers from Romania even to make the required statement. Two other prefectures had not carried out this authorization exercise. Records of vehicle checks were often incomplete or not submitted by the authorities, and while some were checking long-distance transports, others were not. Illegal movements also took place; in one prefecture, Karditsa, ANIMO messages were not sent; while in another, Larisa, sheep certified for immediate slaughter had been sent to farms and slaughtered in different slaughterhouses over a period of months.

In response to criticism in the 2004 mission report, the authority indicated that it intended to recruit almost 400 veterinarians to be attached to the prefectures, as well as other measures such as improved training and legislative reform.

The 2006 mission to evaluate the Greek government's guarantees found little evidence of recruitment and improved practices or progress towards improving staffing levels. Illegal movements were still taking place within an ineffective control regime. Local competent authorities continued to give 'no priority' to transport issues (European Commission Food and Veterinary Office, 2006).

Transport checks were deemed insufficient and where deficiencies were identified, they were not followed up by written warnings or sanctions, and were not reported to the central competent authority as required by EU legislation. Out of seven prefectures where the local authorities were supposed to have drawn up programmes for veterinary checks, only one had done so and had focused on slaughterhouse checks rather than movement checks. At the port of Piraeus, where animals passed through en route between Cyprus and Greece, the competent authority said that controls on transiting consignments of live animals were not carried out because the ferry timetables did not coincide with the veterinarians' working hours. At the port of Patras, checks did not pick up problems such as insufficient headspace or even the fact that recumbent sheep were being trampled on in their compartment. At Promahonas BIP, animals were not unloaded for checking, and small ruminants with broken limbs, even though noticed, were allowed to continue to their destination.

Documentary checks were also insufficient, with one authority failing to check any route plans, while others did not address excessive journey times or other deficiencies.

The FVO requested that Greece take steps fully to comply with the requirements of its 2004 report, and noted that the central authority proposed to increase its own control and audit of the necessary measures. However, in 2007 Greece was referred to the European Court of Justice for persistent non-compliance (see Chapter 5, this volume).

Pigs transported from the Netherlands to slaughterhouses in southern Italy

Europe's largest long-distance trade is the export each year of around 2 million pigs from the Netherlands to Spain and Italy. Around 1.6 million of these are young pigs being sent for further fattening, while about 350,000 are older pigs being exported for slaughter (Stevenson, 2004a). The Netherlands' largest market for slaughter pigs is Germany, where high consumption of pig meat coupled with overcapacity in German slaughterhouses leads to higher prices. However, Italy also pays high prices for slaughter pigs, and it is this that leads to much longer journeys (e.g. pigs travelling from the Netherlands to Salerno, as in Box 13.8, had to endure a journey of 1800 km), often with serious consequences for animal welfare.

Discussion

Alternative to long-distance transportation

One solution to the suffering caused by long journeys to slaughter is to limit journey time to the first available slaughterhouse. Although it is often assumed that transport of carcasses is not an economic option due to the added costs of refrigerated warehouses and lorries, as discussed in the 'Economics' section, Stevenson (2004a) calculated that the transport of carcasses was often cheaper than live animals. However, the fact that destination countries can frequently charge higher prices for meat which is described as 'domestic', which only necessitates animals being slaughtered in that country, in itself, creates a demand for live exportation of farmed animals. This suggests that changes in legislation are needed to prevent such labelling. Since live transport occupies a low proportion of the overall trade in meat products there is a strong argument for totally eradicating long final journeys for slaughter animals.

Good welfare practices in Europe

While the long-distance transport of live animals is still a reality, steps have been taken by some member states to improve their welfare. Ironically, some of the states that take the most efficient approach to monitoring transport regulation compliance are those that lie beyond the main exchanges such as Sweden. Switzerland, although geographically at the heart of Europe, is not a member of the EU and requires veterinary checks at border stations and causes transit transports to be carried by rail. The effect of these measures is that no slaughter animals are transported through Switzerland (M. Rissi, 2006, personal communication).

Some countries have adopted more welfare-friendly practices and schemes, as well as the RSPCA Freedom Food Scheme in the UK, Swedish Meats also guarantee higher animal welfare standards; and Peter's Farm in the Netherlands and Thierry Scheitzer in France have created their own welfare-friendlier production system and labels (www.welfarequality). Such schemes can have an effect on the welfare of animals during transport, for example, organic production in a number of member states (e.g. Denmark and UK) is regulated so the maximum travel time is 8 h.

Future legislation

Many different factions connected with the long-distance transportation of animals, including NGOs, some government regulators and scientists, agree that live transport of slaughter animals should be banned. Unless and until that occurs, there are future opportunities that may be built on to improve animal welfare with support from NGO campaign work. Member states could not agree on a reduction of existing travel times and stocking densities in road vehicles during negotiations on Regulation 1/2005. The Commission therefore agreed that these two issues would be the subject of a further proposal to be presented no later than 4 years after the entry into force of the Regulation. This provides a legislative window fairly soon, and it will be valuable to present as much evidence from current transport as possible, before the end of 2009.

European Action Plan

In January 2006, the European Commission published an Action Plan for 2006–2010, outlining measures to improve the protection and welfare of animals and to influence consumer behaviour in that respect. The Action Plan aims to ensure that animal welfare will be addressed in the most effective manner possible in all EU sectors and through EU relations with third countries.

Although the Action Plan does not designate live transport as an area for action, specific actions to improve the protection and welfare of animals, some of them already under way, include the Preparation of a Commission proposal on the use of vehicle satellite navigation systems in the context of the protection of animals during transport (including technical support from the Commission's Joint Research Centre) and the development of a system for the real time monitoring of animal welfare conditions during transport for 2007.

Reporting on the Action Plan, the European Parliament Committee on International Trade called on the Commission 'to propose measures to enable taxes or tolls to be imposed in view of limiting the unnecessarily lengthy transport of live animals, or to actively encourage such charges'.

Labelling

The European Commission's EUROBAROMETER (2005) survey showed that 74% of consumers believed that they could improve animal welfare through their shopping choices; and over half of those surveyed were willing to pay more for welfare-friendly food products. However, consumers found such products difficult to identify.

These results indicate that consumers would welcome the introduction of an EU animal welfare label, as proposed in the Action Plan. Mandatory egg-labelling to indicate production methods has generally been considered a success.

Labelling of beef and veal, for traceability rather than animal welfare, is now current in the EU. Beef and veal labelling regulations in force since 2002 require that all beef products must be labelled with a traceability code at all stages of the supply chain within all EU member states. The regulations also require the country of origin of the product to be clearly indicated on the label. To ensure that the 'origin' is meaningful, four pieces of information must be provided:

- The country where the animal was born;
- The country or countries where the animal was raised;
- The country where the animal was slaughtered, and the official approval number of the abattoir; and
- The country or countries where the meat was cut, and the official approval number of the cutting plant.

Supermarkets and retail schemes

Major retailers such as supermarkets could be extremely influential in setting standards for meat sold in their outlets. However, the application of welfare standards is not uniform. The UK FAWC reported in June 2006 that 'Recent research under the EU Welfare Quality Programme indicates that animal welfare standards of livestock products in EU supermarkets are at the minimum regulatory level because supermarkets are mostly involved in price competition and only some retailers compete specifically on welfare quality traits' (FAWC, 2006).

The Carrefour chain, for example, with around 7000 stores worldwide, professes a variety of corporate social responsibility values, but as far as food is concerned, its interest in animals appears to concentrate on food safety. Other retailers, such as the UK supermarket chains Waitrose and Tesco, make a virtue of promoting animal welfare-friendly products. Waitrose informs its customers; 'We have created a dedicated supply chain where the meat goes directly from known farms to specially approved abattoirs. All Waitrose beef comes from animals born, reared and slaughtered in the UK. All processing of meat takes place in the UK.'

The link that is being made between animal welfare and food safety and food quality indicates that some retailers, at least, believe that shoppers are interested in paying for products with higher welfare credentials.

However, support can be patchy. There is still some resistance in producer and transporter circles to serious consideration of animal welfare: it is seen as an encumbrance, a cost burden. Campaigns such as the UK Farmers Weekly Food Miles campaign seek to persuade consumers that locally produced food is preferable to imported food. The campaign suggests that welfare standards in other countries might be lower than in the UK, but it ignores the welfare concerns inherent in travel.

Training of enforcers

Enforcement has been shown to be a problem and lack of training and information in the police services of various countries is an important part of this particular issue. Animals' Angels (Germany), PMAF (France), Lega Anti-Vivisezione (Italy), RSPCA International and Interessensgemeinschafdt für tierschonende Transporte und Schlachtung (Switzerland) have all worked successfully on training projects with enforcers such as police forces. There has been a high level of interest in these courses and an appreciation of the practical approach to training, such as the provision of user-friendly reference materials. Such courses tend to concentrate on enforcement by agencies such as the police. However, as FVO reports and some of the cases discussed in this chapter illustrate, there is also a need for better enforcement by the representatives of certain competent authorities, such as official veterinary surgeons.

Satellite navigation

The requirement under Regulation 1/2005 for vehicles to be equipped with satellite navigation systems is potentially far-reaching. While there are limitations – for example, a satellite system will not be able to confirm whether animals have been unloaded for their compulsory period of rest off the vehicle – on the most basic level, falsification of route plans will be made more difficult once these systems are in place. The standards recommended for on-board units will cover more than geographical position: the units will also record the vehicle speed, interior temperature and the number of times the animal compartment is opened and closed. Data will also be collected on the category and number of animals loaded; the transporter's name and authorization number; place, date and time of departure; place, date and time of arrival; and the number of animals dead during and after the journey, with the reason for their death. Information of this nature will be invaluable in providing evidence for the number of animals that die en route, and which transporters experience the greatest mortality – important information for enabling improvements to be made.

The primary aim of satellite navigation systems is enforcement. However, there may be difficulties in processing the volume of data that would be generated, and in handling this effectively to target the traders and journeys where there is possible welfare abuse.

It should be noted, however, that the requirement for satellite navigation systems only affects new vehicles from January 2007, while other vehicles will have until January 2009 to comply. No particular system has been recommended by the European Commission, although this may emerge following a study to be completed by January 2008. The intention is to lay down guidelines that would allow the vast majority of Global Positioning System (GPS) equipment suppliers to provide their equipment for use in livestock vehicles.

Conclusion and Recommendations

It might be concluded from this chapter that a series of transfers described by EU consumer protection authorities as 'from farm to fork' sometimes also involve the transportation of animals over extremely long distances, in overstocked, unventilated vehicles without adequate food, water or rest on their final journey to slaughter.

As long as specialist production areas and markets remain remote from one another – in other words, for the foreseeable future – meat will continue to be transported long distances in Europe although that does not mean that it must inevitably travel 'on the hoof' as opposed to 'on the hook'. Further evidence of the slim savings, if any, to be made in transporting live, could gain greater producer support as it is not necessarily the case that refrigerated transport is more expensive.

However, in general, industry will continue with its current practice of live long-distance transportation unless consumer demand dictates otherwise. Therefore, consumers need to be educated about the true nature of meat quality and the added value of welfare-friendly products. They should know where their meat comes from, making the labelling proposals discussed in the previous section a crucial piece of the pattern. Key messages for the 21st century would be that locally slaughtered does not necessarily mean locally reared, and that superior quality is to be derived from animals slaughtered near to their home farms, so long as their carcasses are transported in conditions of excellent hygiene within a secure cold chain. The practical issue of slaughterhouse availability also needs to be addressed, with government support for small rural slaughterhouses which otherwise might not be viable. In some situations mobile slaughterhouses could be valuable.

Educating the public to the welfare issues arising from the long-distance transportation of live animals is more likely to be effective in the more advanced economies of Western Europe than in the accession countries where consumer choice and control is less feasible. Economies such as Romania, with massive dependence on exports of live cattle, sheep and horses, would need practical assistance if they were to adopt different patterns of marketing. In this respect, accession to the EU offers great benefits, both for citizens and animals.

Past failures in enforcement must not be allowed to continue. Implementation of Regulation 1/2005 in all EU member states must be monitored by the Commission, perhaps with an increase in FVO missions between 2007 and 2011. Stringent enforcement of standards, coupled with maximum fines for breaches of legislation (and ensuring that fines are actually paid, rather than ignored) could have been a deterrent in the past, but was not. Infringement proceedings may be necessary against any member states that unreasonably fail to implement the legislation.

In this context it is impossible not to single out Greece as a member state that has persistently failed to implement European law on the protection of animals at transport, as successive FVO reports have shown. The use of infringement proceedings would be welcome, not only to hasten an improvement in conditions for the animals, but also to ensure that FVO reports are acknowledged as robust and likely to lead to consequences.

Overall, although some transportation of live animals is inevitable, the exceptionally long routes mentioned in the chapter, such as horses transported from CEEC to Italy or pigs from the Netherlands to southern Italy, are not always a necessity nor are they justifiable in economic terms. For these reasons and since many infringements of the legislation have been reported along these routes with serious welfare implications for the animals, it has been proposed by many organizations and officials connected with the industry that the transportation of live animals past the first available slaughterhouse should cease.

References

Agreste (2004) in Lebegge (2004) Live animal transport in Europe: status, context . . . towards an alternative. Thesis, ISARA-Lyon, Lyon, France.

Animals' Angels (2005) *Animals' Angels Infobrief 03.2005.* Animals' Angels, Frankfurt, Germany.

BIP (2005) Information on the Results of the Inspection into the Operation of Supervision of the Handling and Slaughter of Animal with Particular Reference to Animal Wellbeing (In Polish). Available at: www.bip.nik.gov.pl

Broom, D. (2005) The effects of land transport on animal welfare. *Revue Scientifique et Technique Office International des Epizooties* 24, 683–691.

Business Objects (2004) Traces On-line Reporting, DG SANCO User Introductory Guide. Available at: http://forum.europa.eu.int/Public/irc/sanco/tracesinfo/library?l=/manuals/dwh/serintroductoryguide/_EN_0.4_&a=d

Caporale, V., Alessandra, B., Dalla Villa, P. and Del Papa, S. (2005) Global perspectives on animal welfare: Europe. *Revue Scientifique et Technique Office International des Epizooties* 24, 567–577.

Directorate General for Agriculture (2004) *The Meat Sector in the European Union.* European Commission, Brussels, Belgium.

Directorate General for Agriculture (2005) *Prospects of Agricultural Markets and Income 2005–2012.* European Commission, Brussels, Belgium.

Directorate General for Agriculture (2006) *Agriculture in the European Union: Statistical and Economic Information 2005.* European Commission, Brussels, Belgium.

Directorate General for Health and Consumer Protection (2006a) European Animal Welfare Conference 30 March 2006, Questions and Answers. Available at: http://europa.eu.int/comm/food/animal/welfare/conference_30032006_en.htm

Directorate General for Health and Consumer Protection (2006b) Trade and Import of Live Animals, Introduction. Available at: http://europa.eu.int/comm/food/animal/liveanimals/index_en.htm

EUROBAROMETER (2005) *Attitudes of Consumers Towards the Welfare of Farmed Animals.* European Commission, Brussels, Belgium.

Eurogroup for Animal Welfare (2002) *Summary of Suffering Part II: An Investigation into the Poor Enforcement of Directive 91/628/EEC as Amended by Directive 95/29/EC on the Welfare of Animals in Transport.* Eurogroup for Animal Welfare, Brussels, Belgium.

European Commission (2000) *Final Report from the Commission to the Council of European Parliament on the Experience Acquired by Member States Since the Implementation of Council Directive 95/29/EC Amending Directive 91/628/EEC Concerning the Protection of Animals During Transport.* European Commission, Brussels, Belgium.

European Commission (2004) EC Press Release TRACES: Commission adopts new system to manage animal movements and prevent the spread of animal diseases. Reference: IP/04/487 Brussels, 15 April 2004. Available at: http://europa.eu/rapid/pressReleasesAction.do?reference=IP/04/487&format=HTML&aged=0&language=EN&guiLanguage=en

European Commission Food and Veterinary Office (2004a) *Overview of a Series of Missions Carried Out in 2003 Concerning Animal Welfare During Transport and at the Time of Slaughter,* DG SANCO 8506/2004. European Commission, Brussels, Belgium.

European Commission Food and Veterinary Office (2004b) *Final Report of a Mission Carried Out in Greece from 4–8 October 2004 Concerning Animal Welfare During Transport and at the Time of Slaughter,* DG SANCO 7273/2004. European Commission, Brussels, Belgium.

European Commission Food and Veterinary Office (2006) *Final Report of a Mission Carried Out in Greece from 21 February–1 March 2006 Concerning Animal Welfare During Transport and at the Time of Slaughter,* DG SANCO 8042/2006. European Commission, Brussels, Belgium.

European Council (1991a) Council Directive 91/628/EEC of 19 November 1991 on the protection of animals during transport and amending directives 90/425/EEC and 91/496/EEC. *Official Journal of the European Union* L340, 17–27.

European Council (1991b) Council Directive 91/497/EC. *Official Journal of the European Union* L268, 69.

European Council (1993) Council Directive 93/119/EEC of 22 December 1993 on the protection of animals at the time of slaughter or killing. *Official Journal of the European Union* L340, 21.

European Council (1995) Council Directive 95/29/EC of 29 June 1995 amending Directive 90/628/EEC concerning the protection of animals during transport. *Official Journal of the European Union* L148, 52–63.

European Council (1997) Council Regulation (EC) No.1255/97 of 25 June 1997 concerning community criteria for staging points and amending the route plan referred to in the annex to directive 91/628/EEC. *Official Journal of the European Union* L174, 1–6.

European Council (1998a) Council Directive 98/58/EC of 20 July 1998 on the protection of animals kept for farming purposes. *Official Journal of the European Union* L221, 23–27.

European Council (1998b) Council Regulation (EC) No.411/98 on additional animal protection standards applicable to road vehicles used for the carriage of livestock over 8 hours. *Official Journal of the European Union* L52, 8–11.

European Council (2000) Council Regulation (EC) No. 1760/2000 of the European parliament and of the council of 17 July 2000 on establishing a system for the identification and registration of bovine animals and regarding the labelling of beef and beef products and repealing council regulation (EC) No. 820/97. *Official Journal of the European Union* L204, 1–10.

European Council (2005) Council Regulation (EC) No.1/2005 of 22 December 2004 on the protection of animals during transport and related operations and amending Directives 64/432/EEC and 93/119/EC and Regulation (EC) 1255/97. *Official Journal of the European Union* L3, 1–44.

Eurostat (2006) Eurostat Database. Available at: http://epp.eurostat.ec.europa.eu

FAWC (2006) *Report on Welfare Labelling.* Farm Animal Welfare Council, London.

FAO (2006) ProdSTAT: Live Animals. Available at: http://faostat.fao.org/site/568/default.aspx

FSA (2007) Approved Red, Poultry, and Game Meat Establishments. Available at: www.food.gov.uk/foodindustry/meat/meatplantsprems/meatpremlicence

Lebegge, V. (2004) Live animal transport in Europe: status, context . . . towards an alternative. MSc Thesis, ISARA-Lyon, Lyon, France.

Leckie, E.J. (2002) *ILPH Transportation Campaign Report 2002.* The International League for the Protection of Horses, Norwich, UK.

Meridian World Data (2007) Distance Calculation. Available at: www.meridianworlddata.com/Distance-Calculation.asp?guid=2CD58612-D718–42F8–9316-EB59C25A19F1

National Agricultural Statistics Service (2007) Statistical Highlights 2006, Livestock. Available at: www.nass.usda.gov/Publications/Statistical_Highlights/2006/livestock.pdf

NIK (2000) Audit Findings of the National Audit of Animal Protection. Supreme Chamber of Control (Najwższa Izba Kontroli), Poland. Available at: http://eurosai.nik.gov.pl/en/site/px_Animal_Protection_Poland.pdf

Official Journal of the European Union (2004) 2004/544/EC: *Council Decision of 21 June 2004 on the Signing of the European Convention for the Protection of Animals During International Transport (as Amended)* OJ L241 13.7.2004.

OIE (2005) Animal welfare: global issues, trends and challenges. *Revue Scientifique et Technique Office International des Epizooties* 24, 475–477

Ratajczyk, A. (1994) Cruelty to animals: beastly treatment. *The Warsaw Voice* January 1994.

Petrini, A. and Wilson, D. (2005) Philosophy, policy and procedures of the World Organization for Animal Health in the development of standards for animal welfare. *Revue Scientifique et Technique Office International des Epizooties* 24, 665–671

Pickett, H. (2005) *Stop the Bull Ship: The Subsidised Trade in Live Cattle from the European Union to the Middle East.* Compassion in World Farming, Petersfield, UK.

RSPCA (1999) *Dead on Arrival: The Transport of Live Horses in Europe.* Royal Society for the Prevention of Cruelty to Animals, Horsham, UK.

RSPCA (2003) *Standing Room Only: The Suffering of Farm Animals During Transport.* Royal Society for the Prevention of Cruelty to Animals, Horsham, UK.

Stevenson, P. (2000) *Live Exports: A Cruel and Archaic Trade That Must Be Ended.* Compassion in World Farming, Petersfield, UK.

Stevenson, P. (2004a) *The Economics of the Live Transport Trade, A Report for the RSPCA.* Royal Society for the Prevention of Cruelty to Animals, Horsham, UK.

Stevenson, P. (2004b) *The Economics of the Long Distance Transport of Live Farm Animals in Europe. Report for the RSPCA: UK.* Royal Society for the Prevention of Cruelty to Animals, Horsham, UK.

Wardle, T. (2002) *Journey to Death: The Live Export of Horses for Meat from Poland.* Viva! Bristol, UK.

Welfare Quality (2004) Welfare quality: science and society improving animal welfare in the food quality chain. Available at: www.welfarequality.net/everyone/26536

Wilson, T. (1994) Where the bison still roam; wildlife preservation and animal welfare issues in Poland and beyond, http://condor.depaul.edu/~rrotenbe/aeer/v17n2/Wilson.pdf

Appendix 13.1.

Traffic flows of animals

The following statistics have been collated from the Eurostat database (Eurostat, 2006) to show the principal flows of traffic in 2005 in specific species and types of animals, and the purpose of their journeys. They show the largest trades within each type, ranked by size of trade. Some of the major flows of animals designated for further fattening have been included as they, too, are often long-distance journeys sometimes with a short retention in the destination country before being slaughtered (Stevenson, 2004b). As mentioned in the section 'Notes to Traffic Flow Tables', although these statistics do not indicate which journeys exceed 8 h, clearly many will give the distances involved, for example, in the transportation of 5025 horses from Lithuania to Italy (below), the shortest distance between the two countries is over a 1000 km (Meridian World Data, 2007).

Figures have been rounded to the nearest thousand. Where discrepancies were found in the statistics provided by the individual member states, the higher figures were included below:

Cattle

Lightweight calves (80–160 kg) for slaughter

Germany to the Netherlands	20,000
Spain to France	13,000
Germany to Italy	11,000
The Netherlands to France	1,000
France to Italy	7,000
Hungary to Slovenia	7,000
Poland to Spain	6,000
Poland to the Netherlands	6,000

Fattened calves (160–300 kg) for slaughter

The Netherlands to Belgium	30,000
Belgium to the Netherlands	25,000
France to Italy	23,000
Germany to the Netherlands	12,000
Germany to Spain	7,000
Portugal to Spain	5,000
France to Spain	4,000

Heifers (over 300 kg) for slaughter

France to Italy	21,000
Spain to Italy	14,000
Ireland to Spain	4,000

Cows (over 300 kg) for slaughter (generally spent dairy cows)

Portugal to Spain	474,000
Germany to the Netherlands	12,000
Belgium to France	12,000
The Netherlands to Germany	8,000

Large slaughter cattle (male over 300 kg), e.g. bull calves

France to Italy	83,000
Czech Republic to Austria	12,000
The Netherlands to Germany	12,000
Spain to Italy	10,000
Germany to Austria	9,000
Spain to France	6,000

Pigs

Piglets under 50 kg for fattening

Denmark to Germany	2,535,000
The Netherlands to Germany	1,803,000
The Netherlands to Spain	675,000
The Netherlands to Belgium	311,000
The Netherlands to Italy	217,000
Germany to Spain	193,000
The Netherlands to Poland	154,000
The Netherlands to Hungary	96,000
Germany to Luxemburg	70,000
Spain to Portugal	63,000
Germany to Austria	61,000
The Netherlands to Romania	58,000
Ireland to Great Britain	57,000
Denmark to Italy	56,000
Germany to the Netherlands	54,000
Germany to Austria	50,000
Belgium to France	50,000
Germany to France	49,000
The Netherlands to Slovakia	47,000
The Netherlands to France	45,000
Germany to Poland	44,000
Germany to Italy	38,000
Denmark to Poland	36,000
Belgium to the Netherlands	34,000
Germany to Belgium	31,000
The Netherlands to Greece	31,000
Belgium to Spain	26,000
Germany to the Netherlands	21,000

Spain to France	20,000
The Netherlands to Czech Republic	17,000
Germany to Czech Republic	16,000
Spain to Italy	16,000
France to Belgium	15,000
Belgium to Spain	14,000
Denmark to Italy	14,000
Lithuania to Poland	14,000

Sows (more than 160 kg, not breeding sows – likely to be cull sows)

Denmark to Germany	155,000
The Netherlands to Germany	71,000
France to Germany	69,000

Slaughter pigs (over 50 kg, not breeding or productive sows)

The Netherlands to Germany	2,162,000
Spain to Portugal	784,000
Germany to Austria	644,000
Denmark to Germany	299,000
Belgium to the Netherlands	268,000
Denmark to the Netherlands	214,000
France to the Netherlands	195,000
The Netherlands to Hungary	186,000
The Netherlands to Belgium	160,000
The Netherlands to Italy	159,000
Germany to Poland	136,000
Germany to the Netherlands	133,000
Spain to Italy	128,000
Spain to France	121,000

Sheep

Lambs (under 1 year) for slaughter or fattening

Hungary to Italy	770,000
France to Spain	509,000
Romania to Italy	436,000
The Netherlands to France	202,000
France to Italy	178,000
Ireland to Great Britain	134,000
Poland to Italy	99,000
Slovakia to Italy	87,000
Spain to France	74,000
Portugal to Spain	66,000
Italy to Spain	52,000
Spain to Italy	51,000

Sheep (over 1 year), for purposes other than breeding (slaughter or fattening)

Hungary to Italy	94,000
Spain to France	88,000
Spain to Italy	80,000
France to Italy	56,000

Spain to Greece	53,000
France to Spain	42,000
The Netherlands to France	40,000

Goats

Live goats (excluding pure-bred for breeding)

The Netherlands to Spain	113,000
France to Italy	28,000
The Netherlands to France	26,000
Spain to Portugal	10,000

Poultry

The main trade was in broiler chickens and in ducks, geese, guinea fowl and turkeys. Poultry are transported both at 1-day old, from hatchery to fattening farm, and subsequently to the slaughterhouse. Poultry journeys (excluding day-old birds) are normally shorter in duration than quadrupeds, because they are transported in crates which make it impossible to fulfil the legal obligation to give water to animals travelling for over 8 h.

Horses for Slaughter

Poland to Italy	29,000
Spain to Italy	10,000
France to Italy	9,000
Ireland to Italy	5,000
Lithuania to Italy	5,000
Hungary to Italy	4,000
Poland to France	3,000
Poland to Lithuania	3,000
Bulgaria to Italy	3,000
Spain to France	2,000
The Netherlands to Belgium	2,000
France to Spain	2,000
Romania to Poland	1,000
Lithuania to Poland	1,000
Belgium to France	1,000
Germany to Italy	1,000
Belgium to Italy	1,000

14 Middle East

S. Abdul Rahman

Commonwealth Veterinary Association, Jayanagar, Bangalore, India

Abstract

The Middle East due to arid climatic conditions in most parts of the region is currently not able to produce enough animals to be self sufficient in meat. This fact, coupled with traditional patterns of slaughter and consumption, has led to a continued demand for large numbers of live animals, particularly sheep, to be imported into the region. Religious and cultural factors in the region have caused the demand for huge numbers of live ruminants especially sheep. The Middle East imports millions of live ruminants from countries such as Australia, New Zealand (the live trade from New Zealand was effectively ended in 2000 when, in response to higher than acceptable in-transit deaths, minimum age limits were placed on the live export of New Zealand sheep by the Ministry for Agriculture and Forestry – see Chapter 2, this volume), Brazil, China, parts of Europe, Djibouti, Eritrea, Ethiopia, Sudan, Somalia and Pakistan. They are also traded locally among Middle East countries.

In addition to long-distance transport by sea, the animals after being unloaded from ship at the ports are kept for varying periods of time in 'feedlots'. They are then transported for considerable distances from the port to the cities and then on to the markets from where they are sold to the butchers for local trade or bought by consumers straight from the feed-lots and then transported by road. Various forms of vehicles from a car, open truck or a pickup van to a covered conventional truck are used.

At present, specific laws on animal welfare do not exist in most of the countries of the region where inhumane transport and slaughter are of concern. In many countries it is common for animals to be transported in open trucks or crammed in the boot of the cars or overloaded in open trucks and then kept without proper shelter, food and water for long durations prior to being sold.

Although cultural traditions on humane treatment of animals may exist, for example in religious teachings, specific laws on animal welfare are lacking in this region. Education of the traders and butchers in the area of animal sentience and animal welfare should be a priority. For some communities, this could well be done by highlighting religious teachings on animal welfare, for example those in the Quran and the Hadiths. There should be a political will to bring about changes in the existing methods of transport of animals for slaughter and at slaughter. Strict enforcement of legislation, in countries where it exists, and introduction in countries where there is none should be a priority.

Acronyms

DAFF   Department of Agriculture, Fisheries and Forests
FAO   United Nations Food and Agriculture Office
GDP   Gross Domestic Product
NGO   Non-Governmental Organization
OIE   Office International des Épizooties (World Organisation for Animal Health)
WSPA   World Society for the Protection of Animals

Introduction

The Middle East refers to the lands around the southern and eastern shores of the Mediterranean Sea, extending from Morocco to the Arabian Peninsula and Iran. It includes Cyprus, the Asian part of Turkey, Syria, Lebanon, Israel, the West Bank and Gaza, Jordan, Iraq, Iran, the countries of the Arabian peninsula (Saudi Arabia, Yemen, Oman, United Arab Emirates, Qatar, Bahrain, Kuwait), Egypt and Libya (Fig. 14.1). Animal production is increasing in most of these countries quite rapidly as economies expand (Bourn, 2003).

Changing modes of livestock production

The Arabian Peninsula, eastern Jordan and south-eastern Syria are characterized by their aridity, sparse vegetation and minimal surface water, which are not conducive to arable farming and animal husbandry. Less extreme conditions prevail in western Jordan, northern and western Syria and the highlands of Yemen to the south-west and Oman to the south-east.

Significant changes in the modes of livestock production have taken place in recent decades, including increased availability and utilization of crop residues, widespread use of supplementary animal feeds, mechanized pastoralism and introduction of modern dairy and poultry production units. Seasonal and tribal movement patterns of traditional nomadism and transhumance have been transformed, especially in countries such as Jordan, Saudi Arabia and Syria. With transport to supply animal feeds and tank trailers to provide water, pastoral livestock production is not as dependent on rainfall and range conditions as it used to be (Bourn, 2003). Traditional seasonal patterns of movement to, and from, specific areas are no longer followed and have been replaced by more market-based and opportunistic movements to areas with seasonal crop residues and natural pasture, and where water and supplementary feed can be supplied (Gibbon, 2001).

Livestock resources and environmental constraints

An encyclopaedia of animal resources in Arab countries was compiled in the 1980s and published in Arabic by the Arab Centre for Studies of Arid Zones and Dry Lands in Damascus (Tleimat *et al.*, 1981). Since the publication of this report, data on the sub-national distribution of livestock in the Arabian Peninsula and neighbouring countries are generally sparse or absent.

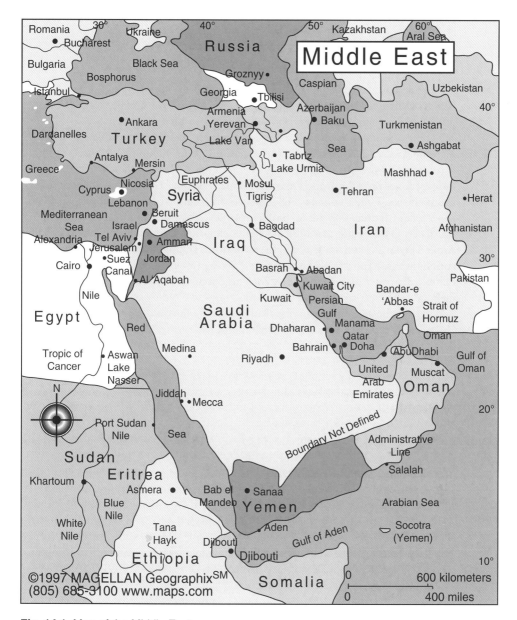

Fig. 14.1. Map of the Middle East.

The composition of livestock resources in the countries under consideration reflects the exigencies of prevailing environmental conditions. For example, chicken, sheep and goat are the most numerous livestock species, with the number of sheep generally exceeding that of goats; Syria and Saudi Arabia have the largest sheep populations; Saudi Arabia and Yemen have the most goats; Syria and Yemen have the largest cattle populations, although Saudi Arabia has the largest dairy farm of the world – Al Safi producing over 300,000l of milk per day (S.S. Hussain, Former Consultant, Livestock Services (Sheep Husbandry) Royal Farm, Saudi Arabia, 2006, personal communication). Saudi Arabia, United Arab Emirates and Yemen have the most camels.

While chicken, sheep, goats, cattle and camels are mostly reared in small numbers by individual farmers and Bedouin desert tribes, there are also a small number of large-scale commercial dairy and sheep farms in the region.

The production of meat from livestock reared locally is not sufficient to supply the domestic market. As a result, the region is heavily committed to import of both frozen and chilled meat. Livestock are also imported from all parts of the world ranging from Brazil, Europe Pakistan, China, Somalia, Ethiopia, Djibouti, etc. (see Chapters 8–12, this volume). Of the livestock imported for slaughter, and to a lesser extent for local breeding, import is by sea. Australia is the largest exporter of live sheep to the region. In 2005, Australia exported 4.2 million sheep and 573,000 cattle (Hassal & Associates, 2006), the largest market for sheep being Kuwait, Saudi Arabia and Jordan (Hassal & Associates, 2006).

As discussed above, local livestock resources are insufficient to supply local demand and both frozen and chilled meat and live animals are imported in large quantities. The expansion of refrigeration and the distribution system in the region has led to a growing demand for chilled and frozen meat. However, religious and cultural factors have necessitated the demand for a huge number of live ruminants, especially sheep.

Country Profiles

Saudi Arabia

Traditional agriculture and pastoral nomadism

In the past, the bulk of agricultural production in Saudi Arabia was concentrated in a few small areas. Produce was largely retained by local communities, although some surplus was sold to the cities. Nomads played a crucial role in this regard, shipping foods and other goods between the widely dispersed agricultural areas. Livestock rearing was shared between the sedentary communities and nomads, who also used it to supplement their livelihoods (Bourn, 2003).

Bedouin were not self-sufficient but needed some food and materials from agricultural settlements. Settled farming communities and traders needed the nomads to tend to their camels. Nomads would graze and breed animals belonging to sedentary farmers in return for portions of the farmers' produce.

Since the 1970s, the Government has undertaken a considerable massive restructuring of the agricultural sector to enhance food security through self-sufficiency and improve rural incomes. There has been a 25-fold increase in cultivation over the last three decades, from 150,000ha in the early 1970s to 3.7 million ha at the turn of the century – although arable land still amounts to less than 2% of the total. Agriculture's share of GDP has risen from 1.3% in 1970 to 9.4% in 1998, and 16% of the population are employed in the agricultural sector (Bourn, 2003). Sufficient surpluses are produced for export of some commodities, notably grain, milk and poultry both live and as meat. This increased production, however, has been very much dependent on exploitation of non-renewable water resources and extensive use of chemical fertilizers.

Profound changes in pastoral livestock production have taken place in Saudi Arabia over the last half-century (Abdalla *et al.*, 1998; Ahmad, 1998; Al-Eisa, 1998; Finan and Al Haratani, 1998). Traditional nomadism as a production system no longer exists in Saudi Arabia. Dependency on range forage as a basic feed resource has declined from 100% to less than 20%. Nomadic movements have been mechanized and operations commercialized (Abdalla *et al.*, 1998). A major shift has taken place from traditional camel-rearing to commercialized sheep-raising. Herd sizes have increased many-fold to suit the new economic conditions. Expansion in the sizes of production operations, in addition to other social changes, has resulted in a growing demand for foreign labour (Abdalla *et al.*, 1998).

Modern poultry production

Poultry consumption has steadily risen since 1975. The total 2001–2002 poultry production forecast was 410,000 t, which was about 55% of domestic consumption. Of this total, about 400,000 t was broiler meat. Per capita poultry consumption increased slightly from 32.5 kg in 1998–1999 to a forecast of 33.4 kg in 2001–2002.

Saudi Arabia is the largest importer of frozen poultry meat in West Asia. Expansion projects at two of Saudi Arabia's largest poultry producers in 1996 resulted in a 30% increase in poultry output, accompanied by a 26% increase in maize and soya meal exports to Saudi Arabia between 1996/97 and 2000/01 (Bourn, 2003).

Beef and veal production

Beef and veal production was estimated at 17,000 t for 2000/01, which was the same as the previous several years, but down from the all time high of 30,000 t in 1994/95. Consumption has been variable in the last 10 years, ranging from 62,000 t in 1990/91 to 98,000 t in 1993/94, but has been trending upwards. Consumption in 2000/01 was expected to surpass 83,000 t. Imports have also been variable but have trended upwards. In the past the European Union (EU) has provided more than half of Saudi meat imports, but import of EU beef and veal was banned following the outbreak of foot-and-mouth disease in 2001.

Demand for sheep

Saudi Arabia receives an annual influx of some 2–5 million pilgrims going on the Hajj to Mecca, which generates considerable extra demand for sheep, over and above baseline levels. About 11.5 million sheep are slaughtered annually in Saudi Arabia (RMF, 2006).

Consumption is particularly high during the Muslim festival of the Eid Al-Adha commemorating Abraham's sacrifice of a sheep. During this festival 3–5 million sheep are slaughtered and it is always not done in slaughterhouses but in the open without stunning.

Livestock imports

Saudi Arabia is by far the largest importer of live animals in the region and has regularly imported 3–5 million sheep and a million or so goats annually over the last decade. Historically, Somalia and Sudan have been major suppliers of sheep, goats and camels. Iraq, Jordan and Syria have also been important suppliers from the north, as have Oman and Yemen from the south. Australia is the major supplier of live sheep (Bourn, 2003).

Syria

Livestock resources

Livestock production is an important component of the agricultural sector and national economy, providing employment to approximately 20% of the workforce and is the main source of income and livelihood for Bedouin herders. The indigenous livestock breeds of Syria are well adapted to the arid conditions in the degraded steppes, the semi-arid rain-fed cultivated areas, such as Al-Djezera in the north-east and the irrigated Euphrates and Al-Ghab valleys and the Ghoota area near Damascus (Bahhady *et al.*, 1998).

SHEEP.   Sheep are the most important livestock resource, being found across most regions of the country. The fat-tailed Awassi is the main breed. It is famous for its meat and milk products and is known for its ability to tolerate heat, drought, cold and long treks. Awassi lambs grow fast and can reach 20-kg live weight at 2 months of age. If they are fed on concentrates, they can reach 40 kg, 5 months after birth. Ewes produce about 60 kg of milk after lambs have been weaned at 2–3 months old (Bahhady *et al.*, 1998).

GOATS.   Goats are still found in most parts of Syria, apart from the steppe where their numbers are low. There are two breeds, the mountain and the Shami (Damascus) goat. Mountain goats are found in the mountainous west and in the rain-fed and irrigated-cropped areas. They are dual-purpose animals, raised primarily for milk, and account for 83% of the goat population. The Shami goat, on the other hand, is distributed in the Ghoota area of Damascus and other irrigated regions. It is raised for milk and used as a milk-improver breed for crossing with the mountain goat (Bahhady *et al.*, 1998).

CATTLE.   There are several breeds of cattle in Syria, the most important being the Aksi that accounts for about half of the total. It is found along the Euphrates and Khaboor rivers and in the high-rainfall zone, and is raised for meat. Cattle of the exotic Friesian breed account for 29% of the total population. They are raised primarily for milk production. Shami (Damascus) cattle represent about 6% of the total. They are kept mainly for milk production and are found in the Ghoota area near Damascus. The rest of the cattle are a mixture of imported and local breeds (Bahhady *et al.*, 1998).

LIVESTOCK TRADE.   The Awassi breed of sheep is famous for its high-quality products, especially the quality and taste of its meat. For many years, Syria has supplied the Arabian Peninsula, Lebanon and Jordan with Awassi lambs and fattened kids because prices in these countries are higher due to the better incomes of many of their inhabitants. To compensate for the large numbers exported, Syria imported sheep from Turkey until 1993, but these have since been replaced by live sheep imports of sheep from Eastern Europe (Bourn, 2003).

Most of the imported and exported live animals are sheep, with exports fluctuating between 500,000 and 1,000,000 a year, while imports have declined over the last decade. Exports and imports are supervised either by the General Meat Company, or by allowing traders to export a limited number of sheep after obtaining a licence from the Ministry of Economics. Trade in sheep is in principle subject to certain rules, which include a requirement for traders to import twice as many sheep as they export.

Goats are not imported for two reasons. First, there are sufficient animals to meet the limited demand, and second, their meat, at least from adults, is not popular with consumers. As a result, goats are exported to neighbouring countries, but numbers have declined dramatically over the last decade. Some cattle, mostly pregnant Friesian heifers, are imported to be raised mainly for milk production. Cattle are not exported because their numbers are few. Sheep account for the major part of live animal export earnings (Bahhady *et al.*, 1998).

Israel

The State of Israel covers an area of approximately $20,000 \, km^2$ but only 20% of it is arable land. Israel's population has a relatively high standard of living compared to the region as a whole, with an annual GNP of nearly US$18,000 per capita. The society is mostly urban, with some 92% of the population living in cities. Although 8% of the populations live in rural areas, only 2.7% of the total national work force is engaged in agricultural production.

Most of Israel's agriculture is irrigated, making water the main limiting factor. Agricultural production in the desert takes advantage of some unique conditions: abundance of land, high temperatures and intensive radiation. The main water source is either saline water or recycled sewage water. The crops are winter vegetables and flowers in greenhouses, dates, grapes and olives irrigated with saline water. Animal husbandry consists mostly of dairy cattle, poultry and fish culture including the production of dairy cattle under methods to reduce heat stress.

During the last decade, productivity has grown by about 6–10% annually, while the percentage of those engaged in agriculture out of the total workforce in the economy dropped from 25% in the early years of the state to only 3% today.

Israel has a legal infrastructure regarding animal welfare. The Animal Welfare Act (Defence of Animals), 1994, states in article 2(a):

> 'A person should not torture an animal, act cruelly towards one or abuse one in any other way.'

The Supreme Court interpreted this phrase in the case of Let the Animals Live versus Hamat Gader (1996). According to the court's interpretation, the prohibition exists if three elements exist:

1. Any suffering, physical or psychological, to an animal;
2. A mental element: awareness to the suffering in the person causing it;

3. A legal element: the suffering is not justified. The justification of suffering is checked through the secondary checks used in examining constitutionality of statutes against the provisions of the basic laws on human rights: is the suffering for a worthy purpose? Does the suffering serve this purpose? Are there less severe means to achieve this purpose? Is there reasonable proportionality between the purpose and the degree of suffering?

Article 19 of the Act states:

> The Minister of Agriculture is in charge of the implementation of this act, and he may, with the approval of the Parliamentary Committee for Education and Culture, and with consideration for the needs of agriculture, make regulations to implement it and to achieve its goal, including regarding – (Ahmad, 1998) The transportation of animals.

The regulations should, thus, give concrete content to the prohibition in Article 2(a). They should reflect the normative balance (as accepted in Israel today) between worthy purposes and the cause of preventing animal suffering.

Lebanon

Livestock production methods in Lebanon

Animal production in Lebanon is not sufficient to meet local demand for animal products and is generally in low-yield systems. According to the FAO (2006), the whole Lebanese goat herd amounts to 435,000 goats. The herd is 95% Baladi, which is a mountainous and robust local breed (135l of milk per head per year) and 5% Damascus breed (250l). This herd's main feed is obtained by grazing which is limited by dryness and the amount of available land – resulting in low livestock production. In addition, the breeding of goats causes damage to the environment due to the uncontrolled grazing system which in turn reduces productivity.

New breeds of goats (Alpine and Saanen) and cattle (Holstein) have been imported. These breeds were found to adjust well to Lebanon's climate and have a higher milk yield (between 2 and 2.5 times more than local breeds). In addition to local and Damascus goats, 142 Alpine and Saanen goats were imported from Europe in 2003–2004 to introduce new and more productive breeds. As the population of imported goats increased through intensive breeding, goats were sold to farmers, which improved the Lebanese goat herd.

To meet the ever increasing demand for meat live transport of cattle and sheep, Lebanon has been importing cattle and sheep from other parts of the world, first from the EU until trade subsidies ended, and in more recent years from Brazil. Animal welfare issues are important as transport of these animals involves a 20-day sea journey, with long road transport both before and after, mostly in open tracks and vans without proper rest, food and water.

Jordan

Livestock resources

Sheep are by far the most numerous ruminant species, with more than 2 million followed by goats. The reduction in Jordan's small ruminant populations is believed to reflect the phasing out of government subsidies on imported grain that was widely used as a livestock feed during the 1990s (Environmental Research Group Oxford Limited, 1995).

Small ruminant production systems

There are three main systems of rearing sheep and goats in Jordan:

1. The 'traditional' system of the village Bedouin. Around 70% of sheep and goats are in this system. Livestock movements are restricted to short distances from the village. Barley and other crops are grown. Sheep and goats graze on crop residues after the harvest. Movements are seasonal and more extensive in drought years. Mainly found in northern and eastern Jordan (Badia area).

2. Smallholder 'village' systems in crop-livestock farming areas. Most farmers have less than 100 sheep and goats that graze on pasture and crop residues. More common in the north and west and some in the south of Jordan.

3. Fattening system, in which investors/small traders buy 100–150 sheep at weaning and fatten them. This is usually a seasonal trade associated with the Hajj and other festivals.

Transformation of traditional transhumance

As previously noted, traditional patterns of seasonal transhumance, documented in the classic literature on the Bedu in Jordan and neighbouring countries, have virtually disappeared in favour of a more opportunistic system using trucks for transport of animals, feed and water and telecommunications to exploit remote pastures (Blench, 1998).

Livestock imports and exports

Livestock markets and exports in Jordan are complex because many animals originating in other countries are sold through Jordan to Saudi Arabia and the Gulf States. All Jordan's small ruminant imports come from Australia, and exports of small ruminants are mainly to Saudi Arabia and, to a much lesser extent, Kuwait and Qatar. The biggest importing companies are Hijazi and Ghoshah which import and sell all live animals.

Re-export of animals form Jordan to neighbouring countries, particularly if they travelled long distances such as from Australia, increases stress and thus risk of illness or injury/death to animals.

Jordan Department of Statistics (JDS, 2006) estimates imports of live sheep from Australia for 2006 to be 341,762. Local production of meat in Jordan for the year 2005 was 5886 t of bovine meat, 7150 t sheep meat, 2801 t goat meat and 280 t camel meat (Mohammed Yacoub Saleh, Jordan, 2006, personal communication). Against this a total of 896,593 live sheep were imported from Australia (Animals Australia, 2006) and 98,145 goats from Syria. In addition, 25,722 cattle were also imported from Australia. China and Sudan are also exporting animals to Jordan. Jordan produces 110,000 t of chicken meat and 770 million eggs annually.

The typical cost of meat in town shops would be JOD 5.5–7.0 (Jordanian Dinar) per kg for local lamb, and JOD 3–4 per kg for Australian sheep. The local variety is thus more expensive than imported and thus makes live transport more economically viable.

Yemen

Livestock resources

According to World Bank estimates, livestock's contribution to agricultural GDP in 1999 was around US$300 million or 20% of agricultural GDP. Animal husbandry's contribution to employment, however, was considerably higher, with close to 80% of the rural population being involved in some form of animal production (Nizwa.Net, 2006).

Ten breeds of sheep and five breeds of goat are recognized. Cattle are mostly of the local Zebu type. Camels are used for milking, transport and ploughing, and are found mainly in coastal areas and in the eastern desert.

Livestock production systems

The principal livestock production systems in Yemen are:

1. Smallholder systems.

Animals are raised in confined space mainly found in the highlands. Almost all cattle and 20% of sheep in the highlands are raised in this system. Guarded grazing is common in agricultural areas. This applies mainly to small ruminants that are grazing (often as combined village flocks) under supervision in daytime and confined at night. Animal draught power is used for tilling (especially in the highlands, where animal traction is the only method, apart from manual labour, to till the small terraces) or for transport (mainly donkeys and camels) (The World Bank, 2000).

It is obvious that under the smallholder system when the animals are transported live for slaughter, cruelty in the form of overloading of trucks and long-distance travel without proper food and water will be a major factor which could compromise animal welfare.

2. Transhumant and nomadic systems.

These are mainly found in the drier areas where Bedu producers migrate their sizeable flocks (mainly sheep, goats and camels). Transport is mainly by trekking long distances.

3. Intensive production system.

These are mainly poultry farms, three large sheep farms (>3000 heads), six large intensive dairy farms (>100 cows), as well as some 60 urban dairies (each with approximately 20–50 cows) in, mainly Hodeidah and Aden city (The World Bank, 2000).

Most animals are raised in mixed arable/animal systems, where the linkages between crop production, feed, traction and manure are strong. 59% of Yemen's farmers are mixed arable/livestock farmers, 21% are arable farmers and 20% are livestock farmers.

Livestock trade

Considerable variation in Yemen's livestock trade has occurred over the last decade. Just over a million goats were imported in 1990, but numbers declined sharply thereafter, with no imports after 1994. There are no records of goat exports from Yemen. Sheep imports have also ranged rather erratically from a low of 41,000 in 1990 to a high of 657,000 in 1999. Sheep exports appear to have commenced in 1998 and rose to more than 400,000 in 2000. Annual cattle imports have been more consistent, ranging between 50,000 and 150,000 a year. It would appear that no cattle are exported from Yemen. Animal transport within the country is by trucks and trekking.

United Arab Emirates

The United Arab Emirates is a federation of seven Emirates on the Gulf coast. Vegetation is sparse and the general landscape is a flat sandy desert extending over 90% of the country.

Livestock resources

There has been a steady increase in dairy and poultry production since 2001. In the more traditionally managed, non-commercial sector, goats are the most numerous species of livestock, followed by sheep, camels and cattle.

The country imports substantial numbers of live animals from abroad, including Australia, whose sheep exports declined markedly from their peak of more than 1.7 million in 1996–1997 to 681,000 in 2001. Since then import of live sheep from Australia has been 466,421 in 2002, 225,313 in 2003, 196,095 in 2004 and 230,775 in 2005 (ABS, 2006).

Interestingly UAE, though an Islamic country where pork eating is taboo, imported 1266 t of pork and pork products in 2003 and also imported 3017 pigs as well. It is presumed that there is a demand for pork and pork products by the large expatriate community living there.

Oman

The Sultanate of Oman is an arid, oil-rich state with a varied topography with extensive plains, deserts, mountain ranges and valleys known as 'wadis'. With the exception of the mountains of the southern region which have a tropical monsoon climate, the rest of Oman is a subtropical desert with low irregular rainfall.

Livestock resources

The Sultanate is a leading livestock producer in the Gulf region. Goats are most numerous livestock species, followed by sheep, cattle and camels. The government's policy is to increase local production and reduce dependence on imports.

Model sheep production units have been set up, and the latest technology is being applied to improve fertility, lower death rates and increase growth rates. The Ministry of Agriculture and Fisheries supplies livestock breeders with concentrated feed, as well as fertilizers and seed for improved fodder production. Dairy development initiatives include providing dairy farmers with equipment and marketing assistance and demonstration of modern herd management techniques. An artificial insemination scheme has been introduced. Beef production and marketing is also being promoted in traditional cattle-breeding areas of the Jabali tribesmen on the coast-facing mountain slopes behind the coastal plain to meet increasing demand in northern Oman (Bourn, 2003).

Changing patterns of rural trade

Although some of this change is due to the natural growth of the rural economy, the most rapid and recent developments are a regional response to growth in the national economy resulting from the discovery of oil.

Food imports have made a particularly significant impact on the viability of the rural economy, although a much wider range of imported consumer goods are also now to be found in markets throughout Oman (Abdalla *et al.*, 1998).

Bahrain

Bahrain is an arid, oil-rich archipelago of 36 islands in the Arabian Gulf, with a total land area of 695 km^2 and by far the highest human population density in the region.

Livestock resources

Bahrain has relatively minor livestock resources, mainly, sheep, goats and cattle, with a few horses and camels and depends largely on imports for its supply of live animals and frozen meat, although some 13,000 heads of cattle are maintained on intensive dairy farms.

Kuwait

Kuwait is an arid, relatively small oil-rich state with limited water resources. Most of the territory is desert with a few oases and sparse vegetation.

Livestock resources

Small ruminant populations have increased progressively over the period from 1995 to 2000, numbering around 580,000 and 1,900,000 respectively and so also has the cattle population.

Livestock imports

Kuwait depends heavily on food imports, including live animals. Kuwait's imports of live sheep from Australian have steadily increased over the last decade to exceed 1.5 million in 2001, surpassed only by Saudi imports of 2.1 million.

Kuwait thus imports 2.5 times as many sheep from Australia than graze its arid plains. The biggest importing company is Almoushhi. Kuwait imports animals, mainly cattle, from Pakistan, Egypt and Tanzania but has stopped the imports of buffaloes from Pakistan. Fresh chilled meat is imported from Pakistan by air cargo. Dairy cattle are imported from Brazil, Australia and North America.

The cost of meat in Kuwait city varies depending upon from where the animal has originated. For example, beef from Pakistan is sold at KD 1 per kg, whereas beef from Egypt is sold at KD 1.5–2. Similarly while meat from Australian sheep costs KD 0.75, Arabian sheep meat sells at KD 2 (Mohammed Yacoub Saleh, Jordan, 2006, personal communication).

Qatar

Most of Qatar's peninsula is a flat, stony barren low-lying desert. Vegetation is found only in the north, where the country's irrigated farming areas are located.

Livestock resources
Qatar relies on imports for most of the local food requirements, although it does have some livestock resources of its own, mainly sheep and goats.

Livestock imports and exports
Qatar imports livestock from a variety of sources, including 300,000–400,000 live sheep annually from Australia.

Consumption of Meat and Animal Products in the Middle East

Patterns of food consumption are becoming increasingly similar throughout the world incorporating more meat and dairy products in daily dietary patterns. This trend is associated with increased international trade. Meat consumption in developing countries has risen from 10 kg per person per year in 1964–1966 to 26 kg in 1999 and is projected to rise further to 37 kg in 2030 (Slingenbergh *et al.*, 2002).

Human population increase and economic growth have raised demand for animal protein in developing countries and have resulted in a major expansion of international trade in livestock and livestock feeds. In most of the developing countries, better infrastructure such as electricity and refrigeration technology have paved the way for acceptance of chilled and frozen meat of every kind being available in huge supermarkets at affordable prices (Delgado *et al.*, 1999). However, in the Middle East this development has been limited so far and freshly slaughtered meat is still most commonly used. Religious and cultural factors have also made it difficult for a total replacement of freshly slaughtered meat.

Although largely unnoticed, a historic dietary shift is taking place in the Middle East. More eggs, chicken and meat are being eaten by an affluent middle class (S.S. Hussain, Saudi Arabia, 2006, personal communication). The production of meat in the Middle East has increased during the last two decades although it is still not sufficient to meet demand. Saudi Arabia produces the highest quantity with 641,100 t, followed by Israel, with 556,930 t and Syria with 367,364 t. To meet the increasing demand for meat, live animals, in addition to frozen meat, are imported into the region. The total value of imports of meat by Saudi Arabia alone was US$193 million in 2004. Very little export of meat occurs and is mostly from Saudi Arabia to neighbouring countries.

Very few vegetarians are found in the Middle East and these are mostly expatriate Asians, especially Hindus. This is especially reflected in the huge Asian workforce in the Middle East which has the means to afford meat on a regular daily basis which could not be afforded in their countries of origin. There is a preferential demand for fresh chilled meat, especially lambs from India, rather than frozen or freshly slaughtered lamb available in the local markets. To meet this large

demand for chilled meat, many meat exporting companies have started slaughtering sheep and goats in India and the meat is chilled and exported to the Middle
East. The fatty nature of the meat from both the imported Australian and the local
breeds is not preferred by the expatriate community as the meat tends to have a
pungent fatty smell. This meat is, however, preferred by the locals (S.S. Hussain,
Saudi Arabia, 2006, personal communication).

Cultural and Religious Factors Contributing to the Live Animal Trade

The predominant religions in the whole of the Middle East are Islam and Judaism
(in Israel) with an admixture of Christianity, mostly in Lebanon, Jordan, Syria and
Israel. However, expatriates belonging to all other religions such as Hindus, Christians,
Sikhs and Buddhists are found throughout the region employed in different profes-
sions and trades. To ensure that halal and kosher practices of meat production are
in place, some communities prefer freshly slaughtered meat to frozen meat.

In addition, the practice of slaughtering a live sheep and cooking it in honour
of a guest is a tradition in the region. The two important breeds of sheep, the Nadji
and Naeemi in the Middle East, especially bred, is a time honoured custom. This
breed which has the breed character of a fatty tail is the most relished meat though
expensive.

There are currently a number of rituals in Islam which require the sacrifice of
a live animal. One of the five pillars of Islam is the performance of Hajj, the pilgrimage which is obligatory for every Muslim man, women or child, provided they
have the means to do so, when they travel to Mecca and Medina and perform
various rituals spread over a period of 7 days. This festival is called Eid Al-Adha
and is performed on the 10th of Dulhajja, the month corresponding to the 10th
month in the Islamic lunar calendar.

An estimated 3.5 million people gather at Mecca during the week-long rituals
and on the 10th sacrifice a sheep as part of one of the rituals. However, on the
10th, 11th and 12th of Dulhajja sheep are slaughtered by Muslims all over the
world. At this festival, whether the person performs the Hajj in Mecca, or cele-
brates in their own house in any part of the world, it is mandatory as a part of the
ritual on every man, women and child that one sheep is sacrificed individually or
seven people join and collectively sacrifice a camel or a cow.

It is estimated that nearly 2–3 million sheep are slaughtered during the 3-day
period in Mecca alone. The slaughter at the time of Hajj is done in a huge modern
slaughterhouse in the suburb of Mina close to Mecca, employing thousands of
butchers who are especially brought from neighbouring countries. The animals are
examined by trained veterinarians and after slaughter the carcasses are chilled and
the meat sent by air cargo for distribution to the poor and hungry people of developing countries of the world. However, no stunning prior to slaughter is done. It
is obvious also that when such large numbers of sheep and goats have to be slaughtered within a specified time limit, animal welfare is bound to be compromised at
all levels from transportation to the time of slaughter.

A further two rituals requiring the slaughter of sheep as a sacrifice are those known as Aqiqa and Sadqa. The head-shaving ceremony of a newborn child is called Aqiqa. It is mandatory in Islam that the head of every child is shaved and the weight of the hair is valued in silver and the proceeds distributed to the poor. For a male child two, and female one, sheep have to be sacrificed and the meat consumed and also distributed among the relatives and poor. In the ritual of Sadqa, a sheep is sacrificed as a means of thanking God for averting a mishap or a bad event, such as recovery from illness, being saved from an accident, etc.

Other factors that result in the continuance of a large live animal trade are the continuing preferences for fresh meat rather than frozen meat, and for local Arabian sheep rather than sheep imported from outside the region.

It is also of interest to note that though Muslims and Jews are the two predominant religious communities in the Middle East, slaughter of pigs does occur in Lebanon and Israel. In addition, live imports of pigs do occur in UAE and pork is imported into UAE, Bahrain, Jordan, Lebanon, Syria and Oman.

Variation in Demand for Meat

A characteristic feature of the livestock trade in countries of the Arabian Peninsula is that demand for live sheep exceeds local national supply during religious festivals, especially Eid ul-Fitr and Eid ul-Adha.

The influx of over 3 million pilgrims during Hajj also generates considerable extra demand. Consequently, there are substantial variations in market prices and the number of animals traded. Indeed, it can be argued that much of the animal husbandry and sheep trade in the region is premised on the high prices commanded immediately before and during festival periods (S.S. Hussain, Saudi Arabia, 2006, personal communication).

Live Animal Trade in the Middle East

Many millions of live ruminants are imported into the Middle East each year from around the world, including Africa, Australia, East Asia and Europe, and also traded locally among Middle East countries. Analysis of global meat production and trade statistics indicates that Saudi Arabia is one of the world's largest importers of sheep and goat meat, and that Australia and New Zealand provide the bulk of supplies to the live sheep and sheep meat markets.

Livestock trade in the region is driven by demand from the Gulf States (Bahrain, Kuwait, Oman, Qatar, Saudi Arabia and UAE), which collectively import 71% of the recorded 11.3 million live sheep, goats and cattle imports. Imported animals are drawn in not only from as far afield as Australia and New Zealand, but also from Yemen, Pakistan, Sudan and the Horn of Africa, as well as Jordan and Syria, and almost certainly Iraq, Turkish Peninsula and from there into the Mediterranean Basin (Slingenbergh *et al.*, 2002).

Australia is the largest exporter of live sheep in the world. Exports to West Asia rose progressively during the early 1990s, declined during the late 1990s, and began to rise again during the first 2 years of the new millennium (MLA, 2006).

Key features evident from the patterns of livestock trade are:

- Imports far exceed exports for all countries and species, except for sheep exports from Syria and Yemen, and there is, therefore, a major net inflow of livestock into the region.
- Saudi Arabia was by far the largest importer of live animals, with 5.4 million sheep.
- Kuwait was the next largest sheep importer, followed by UAE, Jordan, Oman, Bahrain and Qatar.
- Syria was the largest exporter of sheep, followed by Yemen, Jordan and the UAE.
- In 2003, Oman was the main exporter of live goats with 387,000 recorded exports.
- Lebanon imported the highest number of cattle (219,675) in 2003.
- Qatar imported the highest number of camels (4568).
- Oman imported highest number of goats (850,310).
- Syria exported 3,600,000 sheep followed by Jordan with 178,029.

Livestock Trading and Marketing in the Middle East

Most animals in the Middle East are sold in city markets. Animals bred locally, including camels, cattle, sheep and goats are transported from the farms or villages to a central market which is close to a major town or city. The usual method of transport is by trekking where animals walk 3–10 km to reach the market. The market is the main focal point of all trade and from here the animals are either bought by the butchers who take them to the slaughterhouse for slaughter and subsequent retail trade at the butcher's shop or consumers themselves buy directly from the market and take them to their houses for slaughter.

Sheep, along with cattle and goats, are also raised by farmers in cropped areas, and are sold through local markets. Although there is no supporting statistical information, the highest percentage of sheep is said to pass through the fattening cooperatives before going back to the market to be sold for export or slaughter. The remainder are either taken directly to slaughterhouses or exported to neighbouring countries (e.g. from Syria to Jordan), travelling mostly by trucks.

Regarding long-distance transport of animals, in recent years, the Israeli market opened to free import of cattle and sheep destined for fattening and slaughter. The industry was totally privatized, and import barriers were lifted. To defend the local industry a subsidy is given to local breeding. According to the data of the Meat Council (1999 yearbook) the number of calves imported to Israel increased from 8680 in 1997 to 60,239 in 1999. The number of slaughter sheep imported increased from 7000 in 1998 to 89,144 in 1999. The data of the Veterinarian Services are different, and show an import of 52,416 cattle heads and 36,694 slaughter sheep in 1999.

The Council for the Production and Marketing of Cattle for Meat and Milk (The Meat Council) expects an increase in the scope of imports. The Council anticipates that within a number of years about 200,000 calves will be imported to Israel every year. Most of the imported animals are transported by lorries from breeding places in Australia to Australian ports, and from there by ships to Aqaba, Jordan. The duration of the sea journey is between 16 and 22 days. From the Aqaba port the animals are transported by trucks to the Arava border crossing and from there to quarantine in the Arava or to the Palestinian Authority territories. Some of the animals are for immediate slaughter and some for further fattening. A smaller part of the import originates in Eastern Europe, especially Poland and Czech Republic. The cattle are transported by trucks to Western Europe (Netherlands and sometimes Belgium) and from there to Ben-Gurion Airport. From the airport the cattle are transported in trucks to quarantine. In addition to that, there is relatively small import of cattle from Europe by sea (CAWA, 2001).

Large-scale trading is confined to big companies who buy live animals from overseas. There are also local sheep farms, such as Al Khalidiah in Saudi Arabia which has over 40,000 sheep and Al Watania with over 200,000 sheep. However, these do not have high enough production to meet the large demand for live sheep. Trading of livestock is either regular or seasonal. Some traders devote most of their time to buying and selling cattle and sheep and make it a means of their livelihood. For others it is a subsidiary income for a particular season while farming is the primary occupation.

Transport of Animals

The similarities of the existing methods of transport of animals for slaughter, the animal welfare concerns during transport, at feedlots, markets and at slaughterhouses prior to slaughter and lack of specific legislation to protect animal welfare are shared by most countries of the Middle East. The distance travelled and climate particularly in the desert regions of the Middle East can create welfare problems.

Transport of animals for slaughter is by various means, predominant being by sea where live animals are imported from other countries such as Australia, New Zealand, Brazil, China, parts of Europe and from the African countries of Djibouti, Eritrea, Ethiopia, Sudan and Somalia, and from Pakistan. Once the animals reach the port of destination they are unloaded from the ship and kept in areas close to the port which are designated as 'feedlots'. The animals are kept for varying periods of time ranging from 1 day to more than 2 weeks (S.S. Hussain, Saudi Arabia, 2006, personal communication) until they are transported by road to the nearest market or bought by individuals straight from the feedlots and then transported by road. Various forms of vehicles varying from a car, open truck, pickup van to a covered conventional truck are used.

In trucks, the animals are again cramped together and mostly the trucks are uncovered and travel in temperatures which reach more than 45°C in peak summer. In order to avoid loss due to this severe heat, animals are transported during the night (S.S. Hussain, Saudi Arabia, 2006, personal communication). Animals bred locally are made to walk to the nearest markets from nomadic Bedouin settlements.

In Israel, one can find a few provisions regarding animal transportation in the regulations under the Animal Diseases Order, 1945. The Animal Diseases Regulations (Registering, Marking and Transportation of Cattle), 1976, require a transport permit for any transport of cattle, in a form determined by the director of the Veterinary Services.

The Animal Diseases Regulations (Regulating the Movement of Animals in Israel), 1982, and the announcements issued under them put the transportation of animals in Israel under a licensing regulation. A permit is required for any transport of animals that crosses the borders of a local authority. Another permit is required for any vehicle used for animal transportation. The regulations and the announcements under them do not determine any criteria for the issuing of permits, but some criteria were laid by the means of internal guidelines.

Sea transport of live animals to the Middle East has always been a contentious issue with animal welfare organizations campaigning against this practice. In 2005, there were approximately 48 voyages to the Middle East from Australia; each voyage alone averaging 71,000 sheep. In the same year, there were 38 voyages of cattle to the Middle East and Northern Africa from Australia.

Sea voyages to the Middle East last 11–25 days but can be up to 35 days if ships stop at more than one port (Slingenbergh *et al.*, 2002). The sea voyage can also be hazardous to sheep when animals travelling from cool Australian winters can encounter high temperatures of 40°C and above in the Middle East. The two tragic examples of disaster of sea transport of live animals, the voyage of *MV Becrux* and *Cormo Express*, have been well documented (see Chapter 9, this volume).

Unfortunately, very long delays during the transport of cattle and sheep exist in Israel, especially for animals imported through the port of Aqaba in Jordan and from there to the Arava border crossing. Cattle unloaded at night reach the crossing only by the noon hours of the next day and the quarantine facility some hours later, all together about 12–20 h after unloading from the ship (CAWA, 2001).

Delays also occur at unloading of cattle and transferring them to Israel. Once the ship anchors it is around noon and by the time the Israeli government veterinarian inspects the animals and the animals are starting to unload it would be around 6:30 pm. Most often the animals are not immediately transferred to Israel but spend the nights in extreme overcrowding on lorries parked at the border crossing and after long delays are transported into Israel. Delays like that might happen also at the crossing points between Israel and the Palestinian Authority territories (CAWA, 2001).

Pigs are bred in Israel in Lahav (and are slaughtered there) and in Western Galilee. Pigs are slaughtered in Western Galilee near their breeding places, and at most they are transported to an abattoir in a nearby town. Thus, there is no obstacle from the point of view of needs of agriculture to shorten significantly the maximum duration for pig transportation. In order not to cause any inconvenience to the farmers, and taking into account that the duration of journey includes the duration of loading and unloading, the maximum journey time for pigs will be 4 h (CAWA, 2001).

Mortality of sheep during sea transport is mainly due to inanition and *Salmonella* infections, whereas in cattle, the main causes of death are heat stroke, trauma and respiratory disease (Norris, 2005). The highest overall cattle death rate is to the Middle East involving larger shipments and longer journey. Death rate of sheep from Australia to the Middle East has declined from a peak of over 3% in 1992 to less than 1% in 2003 probably due to export of younger animals, faster ships, shorter pre-embarkation periods, better facilities and management and newer ships with better ventilation system (Norris, 2005). Despite the reduced average death rate, higher mortalities do occur, sometimes over the rate of 2% for sheep and 1% of cattle to the Middle East which results in mandatory official investigation by Australian authorities.

Another cause of mortality is accidents and the probability for their happening grows the longer the journey is. Thus, for example, in 1999, 829 cattle suffocated on the *Temburong* when a power loss caused ventilation failure. In 1998, the *Charolais Express* carrying cattle to Israel hit heavy weather. Out of 1200 cattle, 346 died because of inadequate ventilation, 50 more died in Aqaba port, and 174 were rejected for being injured or ill. They were also rejected by Yemen and were ultimately disposed of at sea. In 1996, 67,488 sheep died in a fire upon the *Uniceb*, and in the same year, 1592 cattle died when the *Guerncy Express* sank at sea (Animals Australia, 2001).

Cattle imported from Ethiopia, Somalia and Djibouti are transported in small- to medium-sized ships which do not meet the welfare standards and most often there is no strict veterinary inspection. These vessels are overcrowded, resulting in large-scale deaths of animals during transport (S.S. Hussain, Saudi Arabia, 2006, personal communication).

Traditionally camels are traded for milk and meat and are transported by trucks between cities or trekked for long distances through the deserts. While being transported by trucks they are usually made to lie down in a recumbent position with two animals facing each other (Farid, 1980).

Animals generally are transported for purposes of slaughter or for fattening at another farm prior to slaughter especially for festivals. The animals are transported from markets or farms to the slaughterhouse by a wholesaler and after veterinary inspection they are slaughtered in a registered slaughterhouse and the dealers then sell the carcass to the retailer or butcher.

Welfare Consequences of Animal Transport for Slaughter

Various animal welfare organizations in recent years have gathered graphic evidence – photographic and otherwise – of widespread abuse of cattle, buffaloes, sheep and goats used for meat in various parts of the Middle East, as well as unhygienic and dangerous conditions prevailing in meat processing facilities. Problems associated with livestock trade in the region have been identified as far back as 1983 (Sultan Ahmed and Sultan Ali, 1983).

Among other acts of cruelty, this evidence shows that animals are regularly crammed on to trucks in such high numbers that many become severely injured, they are often gouged by horns and crushed and many of them die en route. Investigations in the Middle East countries by Animals Australia over a period of 3 years from 2003 to 2006 have significant documented and graphic information on the treatment/slaughter of animals, including Australian export animals, in the Middle East (Animals Australia, 2006).

There are a number of animal welfare organizations in the Middle East who are working to improve the welfare standards of transport of animals among other animal welfare issues. However, their task has become difficult, at least in some countries, due to absence of legislation to prosecute offenders who violate even the religious edicts on animal welfare.

A group of animal welfare organizations in the Middle East (comprising the Brooke Hospital for Animals, Dubai Kennels & Cattery, Protecting Animal Welfare Society, Kuwait, Society for the Protection of Animals Rights in Egypt, Society for the Protection of Animals and Nature Morocco, Society for the Protection of Animals Worldwide – SPANA, Sudanese Animal Care & Environment, The Center for Animal Lovers (CAL) and the Egyptian Society of Animal Friends) have joined together to form The Middle East Network for Animal Welfare (MENAW). This is being set up in Egypt to promote the exchange of news, ideas and lessons learned and to provide a networking forum among animal welfare societies in the Middle East.

OIE Animal Welfare Standards

The OIE's initiatives in animal welfare

Animal welfare was identified as a priority in the 2001–2005 OIE Strategic Plan. OIE member countries had decided that, as the international reference organization for animal health and zoonoses, the OIE was the organization best placed to provide international leadership on animal welfare by providing guidelines and recommendations to assist them in their international negotiations.

A permanent Working Group on Animal Welfare was then established and held its first meeting in October 2002. At this meeting, the Working Group saw as its primary task the development of policies and guiding principles to provide a sound foundation from which to elaborate draft recommendations and standards for the identified priorities. The Working Group's recommendations were adopted at the 71st General Session of the OIE in May 2003, and are included in the OIE Terrestrial Animal Health Code.

At its 73rd General Session in May 2005, the International Committee of OIE Member Countries adopted four animal welfare standards to be included in the OIE Terrestrial Animal Health Code. These standards, which were updated during the 74th General Session, are as follows:

- The transport of animals by land;
- The transport of animals by sea;
- The slaughter of animals; and
- The killing of animals for disease control purposes.

With the help of these guidelines, problems associated with livestock trade would be addressed, especially of animals in the region and recommendations made with existing guidelines and legislations to address the issues so that trade can continue while maintaining the highest standards of animal welfare.

All the countries of the Middle East are members of the OIE. However, the OIE Guidelines on Animal Welfare, especially with reference to land and sea transport of animals and slaughter of animals for human consumption, are yet to be followed in these countries. Except for commercial export-oriented meat processing plants managed in overseas countries by Middle East companies which strictly follow world class sanitary and phytosanitary (SPS) measures given in the OIE Guideline as well as being certified for HACCP (Hazard Analysis Critical Control Points), ISO-9001 and SGS, meeting the OIE norms, very few of the municipal or private slaughterhouses in some countries of the Middle East follow the above Guidelines.

While the identification and traceability of the animals from production source to the abattoir is fully maintained by commercial shipping and transport companies with the help of exporting countries such as Australia and New Zealand, which ensures that animals have been raised under disease-free conditions as per OIE Guidelines, none of the animals which are slaughtered from local sources for human consumption in the Middle East are subjected to any such measures.

Discussion

In the Middle East the production of livestock does not meet regional demand for sheep, goats and cattle. There is also a huge poultry industry especially in Saudi Arabia and Kuwait. Transport of animals in the Middle East is by trekking and by road. Live animals are imported for slaughter into the region by ships.

Unloading of these animals should be done as early as possible after the ship casts anchor. The transportation to the border crossing, especially in Israel, should be immediately and successively after the loading of each lorry. The border crossing should be open for the transferring of animals at any hour they arrive. The procedures in the crossing should be speedy, in a manner that will prevent any delay of the animals. If it is found that it is not possible to prevent all delays, appropriate facilities must be arranged that will enable the animals to stay in a shaded area (i.e. shaded, mechanically ventilated and with a cooling system).

It is obvious that due to the large-scale transport of animals both by sea and land, and to a lesser extent by air, animal welfare is often (or frequently) very likely to be compromised. Very few countries in the Middle East have rules and regulations comparable to the ones in the Western countries on specific issues of transport of animals, slaughter, etc. The religious rituals of kosher and halal slaughter are meant to be strictly adhered to in Israel and all Arab countries, however, there is much evidence to show that often these religious practices are not complied with. In the absence of specific regulations as to the manner in which the animals need to be transported, loading and unloading, the design of the vehicles, etc. animal welfare is likely to be compromised.

However, it is noteworthy to mention here that during the month of Hajj, when thousands of animals are slaughtered by every practising Muslim, it is mandatory in the countries of the Middle East that the ritual of slaughter be done only in registered abattoirs and not in open streets or houses.

A number of NGOs and animal welfare organizations have been working in the Middle East, but their priority has been more on pet animals rather than farm animals. In some countries such as Kuwait and Jordan, however, these NGOs have been very active in promoting animal welfare at feedlots, markets, butcher shops and at slaughterhouses to attempt to ensure that animals are not subjected to cruelty and ensure that legislations against cruelty to animals are implemented.

It is inevitable in the current conditions of the growing economy of the Middle East and with an increase in the demand for meat and meat products, that there will be large-scale movement of animals across the Middle East. Unless measures are introduced to have animal welfare legislation implemented in relation to the movement of live transport of animals, inhumane treatment of animals will continue. It is therefore imperative that long-term measures are initiated in an effort to avoid this ongoing cruelty.

Major stakeholders in the Middle East animal trade such as DAFF, Government of Australia and other allied agencies such as Meat and Livestock Australia and Live Corp have initiated efforts to improve animal welfare standards in the Middle East by organizing workshops and seminars and also setting up their own offices in some countries to oversee animal welfare standards both during transport and at slaughter. However, anecdotal evidence from investigations suggests that animal welfare standards are often not maintained despite best efforts. With an increase in international trade and concerns regarding awareness of animal welfare issues by trading partners, the countries of the region have recognized the need to address the problems. This was evident when a conference on animal welfare was held at Al Azhar University, Cairo, Egypt in 2004 on 'Welfare and Development of Animals Resources in Islamic Culture and Contemporary Systems'. With the help of international NGOs, such as Compassion in World Farming (CIWF), a conference on Farm Animal Slaughtering and Transportation was held in 2006. Workshops on animal welfare have also been held in Kuwait (MLA, 2004).

One of the measures which many animal welfare organizations all over the world are arguing for is the complete phasing out of the transport of live animals to the Middle East and instead export of frozen meat. It would have been an ideal situation even if it required establishment of slaughterhouses to perform halal and kosher slaughter in the countries exporting live sheep and cattle. However, some religious requirements, especially in Islam at least during the period of Hajj, prevent the slaughter of sheep in a place other than the holy sites of Mecca and neighbouring suburb of Mina.

However, the other countries of the Middle East which are importing live sheep could instead have frozen or chilled meat imported after ensuring that all religious rituals have been performed at the slaughterhouse in the exporting country, even by having their own supervising agency. This would also involve both economic development and changes to behaviour that would allow consumers to adopt the use of chilled or frozen meat rather than freshly slaughtered meat.

Thus, the issues are those which involve changing values, influenced by social, cultural and religious considerations. If countries such as Australia, New Zealand and the EU, which have relatively high standards of regulation, ceased to supply the Middle East with live animals, it would of course be essential to ensure that this market was not filled by imports from countries where regulation is weaker.

Establishment of animal welfare societies should be undertaken in those places where none exist, with the help of organizations, both at national as well as international level. Formation of state and national boards of animal welfare at the government level should be created to ensure implementation of rules and regulations of animal welfare which should be given high priority.

Strengthening of existing animal welfare organizations, especially in Kuwait and Jordan, with both financial as well as technical help for effective animal welfare work needs to be done urgently.

Most importantly there should be a political will to bring about animal welfare legislations to enforce high standards of animal welfare and prevent cruelty during transport. Constituting committees to recommend the governments on the issues involved would be the first step as has been done in Israel with the constitution of a Committee on the Animal Welfare Act and the recommendations of the committee regarding transportation of cattle, sheep, goats and pigs have been published (CAWA, 2001).

With a high potential import market of meat in the Middle East, interaction between the key stakeholders such as farmers, butchers, exporters and importers of the meat industry, animal welfare organizations, etc. should be facilitated by international animal welfare organizations such as WSPA and intergovernmental organizations such as OIE, etc.

In 2005, the OIE adopted its first global guidelines for animal welfare, specifically in the areas of land transport, sea transport, slaughter of animals for human consumption and killing of animals for disease control. The passing of the guidelines by 167 countries, some of which did not have national animal protection legislation of their own, signalled that animal welfare was no longer a concern only of certain (generally prosperous) nations, but had become an issue for official attention at a global level. It is envisaged that all the OIE member countries of the Middle East will eventually follow the OIE Guidelines of animal welfare in their respective countries which will strengthen their own existing religious doctrines of animal welfare.

Acknowledgements

Sincere thanks to Dr S.S. Hussain, Saudi Arabia; Dr S.A. Basheer, UAE; Dr Farida Molla Ahmad, Deputy Director Ministry of Agriculture, Kuwait; Ms Margaret Mcluskey and Ms Linnette Botha of Protecting Animal Welfare Society (PAWS) Kuwait; Dr Ibrahim Abdul Hamid Hammad, Capital Abattoir, Shuwaikah, Kuwait; Ms Margaret Ledger, Chairman, Humane Centre for Animal Welfare (HCAW) Jordan; Dr Mustafa Ghazi and Ms Chris Latear of SPANA in Jordan; Dr Abd El-Hafed Usili Veterinary Officer, Amman Abattoir; Dr Mohammad Yacoub Saleh, Head of Animal Production Department, Ministry of Agriculture, Amman, Jordan for their help in the collection of data in preparation of this chapter.

References

Abdalla, S.H., Hajooj, A. and Simir, A. (1998) Economic analysis of nomadic livestock operations in northern Saudi Arabia. In: Squires, V.R. and Sidahmed, A.E. (eds) *Drylands: Sustainable Use of Rangelands into the Twenty-first Century.* IFAD Series: Technical Reports. International Fund for Agricultural Development, Rome, pp. 375–383.

ABS (2006) Australian Bureau of Statistics. Available at: www.abs.gov.au/ausstats/abs@.nsf/web+pages/statistics?opendocument

Ahmad, Y. (1998) The socio-economics of pastoralism: a commentary on changing techniques and strategies for livestock management. In: Squires, V.R. and Sidahmed, A.E. (eds) *Drylands: Sustainable Use of Rangelands into the Twenty-first Century*. IFAD Series: Technical Reports. International Fund for Agricultural Development, Rome, pp. 329–344.

Al-Eisa, A. (1998) Changes in factors affecting Bedouin movement for grazing. In: Squires, V.R. and Sidahmed, A.E. (eds) *Drylands: Sustainable Use of Rangelands into the Twenty-first Century*. IFAD Series: Technical Reports. International Fund for Agricultural Development, Rome, pp. 369–373.

Animals Australia (2001) The Death Files 1981–2001: a Report from the Files of Animals Australia. Internal working document, Animals Australia, Melbourne.

Animals Australia (2006) Middle East Investigation Report. Available at: http://www.animalsaustralia.org

Bahhady, F., Thomson, E.F. and Boulad, M. (1998) Red meat production, marketing and trade in Syria. *Conference Proceedings: Filière des Viandes Rouges dans les Pays Méditerranéens* (trans Red meat industry in the Mediterranean countries). Tunis (Tunisia), 20–23 April 1997. CIHEAM-IAMZ, pp. 89–102.

Blench, R.M.B. (1998) Rangeland degradation and socio-economic changes among the Bedu of Jordan: results of the 1995 IFAD survey In: Squires, V.R. and Sidahmed, A.E. (eds) *Drylands: Sustainable Use of Rangelands into the Twenty-first Century*. IFAD Series: Technical Reports. International Fund for Agricultural Development, Rome, pp. 397–423.

Bourn, D. (2003) *Livestock Dynamics in the Arabian Peninsula: A Regional Review of National Livestock Resources and International Livestock Trade*. Environmental Research Group Oxford Limited, Oxford, UK.

CAWA (2001) The Committee on the Animal Welfare Act (Defense of Animals) (Transportation of Animals). *The Committee's Recommendations Regarding the Transportation of Cattle, Sheep, Goats and Pigs*. Chairperson Mr Yossi Wolfson. March 2001. Tel Aviv, Israel.

Delgado, C., Rosengrant, M., Steinfeld, H., Ehui, S. and Courbois, C. (1999) *Livestock to 2020. The Next Food Revolution*. International Food Policy Research Institute, Washington, DC, 72 pp.

Environmental Research Group Oxford Limited (1995) Formulation report for the pastoral resource assessment and monitoring component of the range rehabilitation and development programme. Oxford and Rome: Report to the Ministry of Agriculture, Hashemite Kingdom of Jordan, by Environmental Research Group Oxford Limited on behalf of the International Fund for Agricultural Development, Rome.

FAO (2006) Global livestock production and health atlas. Available at: www.fao.org/ag/aga/glipha/index.jsp

Farid, M.F.A. (1980) *The Camel in Arab Nations*. Arab Centre for the Studies of Arid Zones and Dry Land, Damascus, Syria.

Finan, T.J. and Al Haratani, E.R. (1998) Modern Bedouins: the transformation of traditional nomad society in the Al-Taysiyah region of Saudi Arabia. In: Squires, V.R. and Sidahmed, A.E. (eds) *Drylands: Sustainable Use of Rangelands into the Twenty-first Century*. IFAD Series: Technical Reports. International Fund for Agricultural Development, Rome.

Gibbon, D. (2001) *Global Farming Systems Study: Challenges and Priorities to 2030. Regional Analysis of the Middle East and North Africa*. Food and Agriculture Organisation of the United Nations, Rome, Italy.

Hassal & Associates (2006) *The Live Export Industry: Values, Outlook and Contribution to the Economy*. Meat and Livestock Australia, North Sydney, Australia.

JDS (2006) Jordan Department of Statistics. Available at: http://jotiis.dos.gov.jo:7001/JoTIIS/ControllerServlet

Meat Council (1999) Yearbook.

MLA (2004) Animal Husbandry Workshop, Kuwait. Available from Meat and Livestock Australia, North Sydney, Australia.

MLA (2006) Meat and Livestock Australia. Available at: www.meatlivestockaustralia.com

Nizwa.Net (2006) Agriculture in the Sultanate of Oman. Available at: www.nizwa.net/agr/agriculture.html

Norris, R.T. (2005) Transport of animals by sea. In: *Animal Welfare: Global Issues, Trends and Challenges.* Office International des Epizooties, Paris, France, pp. 673–681.

RMF (2006) Rene Moward Foundation. Available at: www.rmf.org.lb

Slingenbergh, J., Hendricks, G. and Wint, W. (2002) Will the livestock revolution in the developing world succeed? *Agriworld Vision* 2, 31–33.

Sultan Ahmed and Sultan Ali, K. (1983) *Problems Associated with International Trade and Movement of Livestock in the Region.* Office International des Epizooties, Paris, France.

The World Bank (2000) *Republic of Yemen, Comprehensive Development Review: Agriculture, Livestock and Fisheries.* Rural Development, Water and Environment Group (MNSRE), Middle East and Africa Region, The World Bank, Washington, DC.

Tleimat, F.M., Farid, M.F.A., Abbas, H.M.H., Al-Mufarreh, M.B. and Awa, O.A. (1981) Encyclopaedia of Animal Resources in the Arab Countries. Arab Centre for the Studies of Arid Zones and Dry Lands, Damascus, Syria. Available at: http://www.agr.gc.ca/mad-dam/e/bulletine/v14e/v14n11e.htm

Appendix: Introduction to the Guidelines for Animal Welfare

Article 3.7.1.1.

Guiding principles for animal welfare

1. That there is a critical relationship between animal health and animal welfare.

2. That the internationally recognised 'five freedoms' (freedom from hunger, thirst and malnutrition; freedom from fear and distress; freedom from physical and thermal discomfort; freedom from pain, injury and disease; and freedom to express normal patterns of behaviour) provide valuable guidance in animal welfare.

3. That the internationally recognised 'three Rs' (reduction in numbers of animals, refinement of experimental methods and replacement of animals with non-animal techniques) provide valuable guidance for the use of animals in science.

4. That the scientific assessment of animal welfare involves diverse elements which need to be considered together, and that selecting and weighing these elements often involves value-based assumptions which should be made as explicit as possible.

5. That the use of animals in agriculture and science, and for companionship, recreation and entertainment, makes a major contribution to the well-being of people.

6. That the use of animals carries with it an ethical responsibility to ensure the welfare of such animals to the greatest extent practicable.

7. That improvements in farm animal welfare can often improve productivity and food safety, and hence lead to economic benefits.

8. That equivalent outcomes based on performance criteria, rather than identical systems based on design criteria, be the basis for comparison of animal welfare standards and guidelines.

Article 3.7.1.2.

Scientific basis for guidelines

1. Welfare is a broad term which includes the many elements that contribute to an animal's quality of life, including those referred to in the 'five freedoms' listed above.

2. The scientific assessment of animal welfare has progressed rapidly in recent years and forms the basis of these guidelines.

3. Some measures of animal welfare involve assessing the degree of impaired functioning associated with injury, disease, and malnutrition. Other measures provide information on animals' needs and affective states such as hunger, pain and fear, often by measuring the strength of animals' preferences, motivations and aversions. Others assess the physiological, behavioural and immunological changes or effects that animals show in response to various challenges.

4. Such measures can lead to criteria and indicators that help to evaluate how different methods of managing animals influence their welfare.

Guidelines for the Transport of Animals by Sea

PREAMBLE: These guidelines apply to the following live domesticated animals: cattle, buffaloes, deer, camelids, sheep, goats, pigs and equines. They may also be applicable to other domesticated animals.

Article 3.7.2.1.

The amount of time animals spend on a *journey* should be kept to the minimum.

Article 3.7.2.2.

I. Animal behaviour

Animal handlers should be experienced and competent in handling and moving farm livestock and understand the behaviour patterns of animals and the underlying principles necessary to carry out their tasks.

The behaviour of individual animals or groups of animals will vary depending on their breed, sex, temperament and age and the way in which they have been reared and handled. Despite these differences, the following behaviour patterns, which are always present to some degree in domestic animals, should be taken into consideration in handling and moving the animals.

Most domestic livestock are kept in herds and follow a leader by instinct.

Animals which are likely to be hostile to each other in a group situation should not be mixed.

The desire of some animals to control their personal space should be taken into account in designing *loading* and *unloading* facilities, transport *vessels* and *containers*.

Domestic animals will try to escape if any person approaches closer than a certain distance. This critical distance, which defines the flight zone, varies among species and individuals of the same species, and depends upon previous contact with humans. Animals reared in close proximity to humans (i.e. tame) have a smaller flight zone, whereas those kept in free range or extensive systems may have flight zones which may vary from one metre to many metres. *Animal handlers* should avoid sudden penetration of the flight zone which may cause a panic reaction which could lead to aggression or attempted escape.

Animal handlers should use the point of balance at the animal's shoulder to move animals, adopting a position behind the point of balance to move an animal forward and in front of the point of balance to move it backward.

An example of a flight zone (cattle)

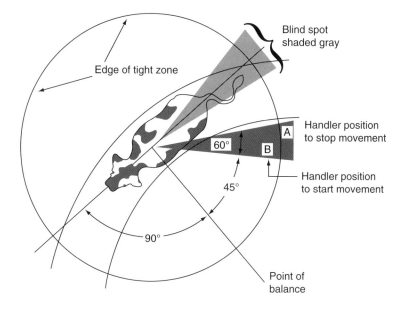

Domestic animals have a wide-angle vision but only have a limited forward binocular vision and poor perception of depth. This means that they can detect objects and movements beside and behind them, but can only judge distances directly ahead.

Domestic animals can hear over a greater range of frequencies than humans and are more sensitive to higher frequencies. They tend to be alarmed by constant loud noises and by sudden noises, which may cause them to panic. Sensitivity to such noises should also be taken into account when handling animals.

2. Distractions and their removal

Design of new *loading* and *unloading* facilities or modification of existing facilities should aim to minimise the potential for distractions that may cause approaching animals to stop, baulk or turn back. Below are examples of common distractions and methods for eliminating them:

Handler movement pattern to move cattle forward

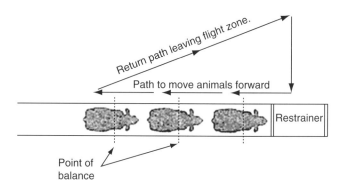

a. reflections on shiny metal or wet floors – move a lamp or change lighting;

b. dark entrances – illuminate with indirect lighting which does not shine directly into the eyes of approaching animals;

c. animals seeing moving people or equipment up ahead – install solid sides on chutes and races or install shields;

d. dead ends – avoid if possible by curving the passage, or make an illusory passage;

e. chains or other loose objects hanging in chutes or on fences – remove them;

f. uneven floors or a sudden drop in floor levels – avoid uneven floor surfaces or install a solid false floor to provide an illusion of a solid and continuous walking surface;

g. sounds of air hissing from pneumatic equipment – install silencers or use hydraulic equipment or vent high pressure to the external environment using flexible hosing;

h. clanging and banging of metal objects – install rubber stops on gates and other devices to reduce metal to metal contact;

i. air currents from fans or air curtains blowing into the face of animals – redirect or reposition equipment. Dead ends – avoid if possible by curving the passage, or make an illusory passage.

Article 3.7.2.3.

Responsibilities

Once the decision to transport the animals by sea has been made, the welfare of the animals during their *journey* is the paramount consideration and is the joint responsibility of all people involved. The individual responsibilities of persons involved will be described in more detail in this Article. These guidelines may also be applied to the transport of animals by water within a country.

The management of animals at post-discharge facilities is outside the scope of this Appendix.

1. General considerations

a. Exporters, importers, owners of animals, business or buying/selling agents, shipping companies, masters of *vessels* and managers of facilities are jointly responsible for the general health of the animals and their fitness for the *journey*, and for their overall welfare during the *journey*, regardless of whether duties are subcontracted to other parties during transport.

b. Exporters, shipping companies, business or buying/selling agents, and masters of *vessels* are jointly responsible for planning the *journey* to ensure the care of the animals, including:

   i. choosing appropriate *vessels* and ensuring that *animal handlers* are available to care for the animals;

   ii. developing and keeping up-to-date contingency plans to address emergencies (including adverse weather conditions) and minimise stress during transport;

   iii. correct *loading* of the ship, provision of appropriate food, water, ventilation and protection from adverse weather, regular inspections during the *journey* and for appropriate responses to problems arising;

   iv. disposal of carcasses according to international law.

c. To carry out the above mentioned responsibilities, the parties involved should be competent regarding transport regulations, equipment usage, and the humane handling and care of animals.

2. Specific considerations

a. The responsibilities of the exporters include:

   i. the organisation, carrying out and completion of the *journey*, regardless of whether duties are subcontracted to other parties during transport;

ii. ensuring that equipment and medication are provided as appropriate for the species and the *journey*;

iii. securing the presence of the appropriate number of *animal handlers* competent for the species being transported;

iv. ensuring compliance of the animals with any required veterinary certification, and their fitness to travel;

v. in case of animals for export, ensuring compliance with any requirements of the *importing* and *exporting countries*.

b. The responsibilities of the importers include:

(under study)

c. The responsibilities of the owners of the animals include the selection of animals that are fit to travel based on veterinary recommendations.

d. The responsibilities of the business or buying/selling agent include:

i. selection of animals that are fit to travel based on veterinary recommendations;

ii. availability of suitable facilities for the assembly, *loading*, transport, *unloading* and holding of animals at the start and at the end of the *journey*, and for emergencies.

e. The responsibilities of shipping companies include:

(under study)

f. The responsibilities of masters of *vessels* include the provision of suitable premises for animals on the *vessel*.

g. The responsibilities of managers of facilities during *loading* include:

i. providing suitable premises for *loading* the animals;

ii. providing an appropriate number of *animal handlers* to load the animals with minimum stress and the avoidance of injury;

iii. minimising the opportunities for disease transmission while the animals are in the facilities;

iv. providing appropriate facilities for emergencies;

v. providing facilities, *veterinarians* or *animal handlers* capable of *killing* animals humanely when required.

h. The responsibilities of managers of facilities during *unloading* include:

i. providing suitable facilities for *unloading* the animals onto transport *vehicles* for immediate movement or securely holding the animals in lairage, with shelter, water and feed, when required, for transit;

ii. providing *animal handlers* to unload the animals with minimum stress and injury;

iii. minimising the opportunities for disease transmission while the animals are in the facilities;

iv. providing appropriate facilities for emergencies;

v. providing facilities, and *veterinarians* or *animal handlers* capable of *killing* animals humanely when required.

i. The responsibilities of the *animal handlers* include humane handling and care of the animals, especially during *loading* and *unloading*.

j. The responsibilities of the *Competent Authority* of the *exporting country* include:

i. establishing minimum standards for animal welfare, including requirements for inspection of animals before and during their travel, and for certification and record keeping;

ii. approving facilities, *containers*, *vehicles* and *vessels* for the holding and transport of animals;

iii. setting competence standards for *animal handlers* and managers of facilities;

iv. implementation of the standards, including through accreditation of/interaction with other organisations and *Competent Authorities*;

v. monitor and evaluate health and welfare performance, including the use of any veterinary medications.

k. The responsibilities of the *Competent Authority* of the *importing country* include:

i. establishing minimum standards for animal welfare, including requirements for inspection of animals after their travel, and for certification and record keeping;

ii. approve facilities, *containers*, *vehicles* and *vessels* for the holding and transport of animals;

iii. setting competence standards for *animal handlers* and managers of facilities;

iv. implementation of the standards, including through accreditation of/interaction with other organisations and *Competent Authorities*;

v. ensuring that the *exporting country* is aware of the required standards for the *vessel* transporting the animals;

vi. monitor and evaluate health and welfare performance, including the use of any veterinary medications;

vii. give animal consignments priority to allow import procedures to be completed without unnecessary delay.

l. The responsibilities of *veterinarians* or in the absence of a *veterinarian*, the *animal handlers* travelling on the *vessel* with the animals include:

i. humane handling and treatment of animals during the *journey*, including in emergencies, such as humane *killing* of the animals;

ii. possess ability to report and act independently;

iii. meet daily with the master of the *vessel* to obtain up-to-date information on animal health and welfare status.

m. The receiving *Competent Authority* should report back to the sending *Competent Authority* on significant animal welfare problems which occurred during the *journey*.

Article 3.7.2.4.

Competence

1. All people responsible for animals during *journeys* should be competent to carry out the relevant responsibilities listed in Article 3.7.2.3. Competence in areas other than animal welfare would need to be addressed separately. Competence may be gained through formal training and/or practical experience.

2. The assessment of competence of *animal handlers* should at a minimum address knowledge, and ability to apply that knowledge, in the following areas:

a. planning a *journey*, including appropriate *space allowance*, feed, water and ventilation requirements;

b. responsibilities for the welfare of animals during the *journey*, including *loading* and *unloading*;

c. sources of advice and assistance;

d. animal behaviour, general signs of disease, and indicators of poor animal welfare such as stress, pain and fatigue, and their alleviation;

e. assessment of fitness to travel; if fitness to travel is in doubt, the animal should be examined by a *veterinarian*;

f. relevant authorities and applicable transport regulations, and associated documentation requirements;

g. general disease prevention procedures, including cleaning and *disinfection*;

h. appropriate methods of animal handling during transport and associated activities such as assembling, *loading* and *unloading*;

i. methods of inspecting animals, managing situations frequently encountered during transport such as adverse weather conditions, and dealing with emergencies, including euthanasia;

j. species-specific aspects and age-specific aspects of animal handling and care, including feeding, watering and inspection; and

k. maintaining a *journey* log and other records.

3. Assessment of competence for exporters should at a minimum address knowledge, and ability to apply that knowledge, in the following areas:

a. planning a *journey*, including appropriate *space allowances*, and feed, water and ventilation requirements;

b. relevant authorities and applicable transport regulations, and associated documentation requirements;

c. appropriate methods of animal handling during transport and associated activities such as cleaning and *disinfection*, assembling, *loading* and *unloading*;

d. species-specific aspects of animal handling and care, including appropriate equipment and medication;

e. sources of advice and assistance;

f. appropriate record keeping; and

g. managing situations frequently encountered during transport, such as adverse weather conditions, and dealing with emergencies.

Article 3.7.2.5.

Planning the journey

1. General considerations

a. Adequate planning is a key factor affecting the welfare of animals during a *journey*.

b. Before the *journey* starts, plans should be made in relation to:

i. preparation of animals for the *journey*;

ii. type of transport *vessel* required;

iii. route, taking into account distance, expected weather and sea conditions;

iv. nature and duration of *journey*;

v. daily care and management of the animals, including the appropriate number of *animal handlers*, to help ensure the health and welfare of all the animals;

vi. avoiding the mixing of animals from different sources in a single pen group;

vii. provision of appropriate equipment and medication for the numbers and species carried; and

viii. emergency response procedures.

2. Preparation of animals for the journey

a. When animals are to be provided with a novel diet or unfamiliar methods of supplying of feed or water, they should be preconditioned.

b. There should be planning for water and feed availability during the *journey*. Feed should be of appropriate quality and composition for the species, age, condition of the animals, etc.

c. Extreme weather conditions are hazards for animals undergoing transport and require appropriate *vessel* design to minimise risks. Special precautions should be taken for animals that have not been acclimatised or which are unsuited to either hot or cold conditions. In some extreme conditions of heat or cold, animals should not be transported at all.

d. Animals more accustomed to contact with humans and with being handled are likely to be less fearful of being loaded and transported. Animals should be handled and loaded in a manner that reduces their fearfulness and improves their approachability.

e. Behaviour-modifying (such as tranquillisers) or other medication should not be used routinely during transport. Such medicines should only be administered when a problem exists in an individual animal, and should be administered by a *veterinarian* or other person who has been instructed in their use by a *veterinarian*. Treated animals should be placed in a dedicated area.

3. Control of disease

As animal transport is often a significant factor in the spread of infectious diseases, *journey* planning should take into account the following:

a. When possible and agreed by the *Veterinary Authority* of the *importing country*, animals should be vaccinated against diseases to which they are likely to be exposed at their destination.
b. Medications used prophylactically or therapeutically should only be administered by a *veterinarian* or other person who has been instructed in their use by a *veterinarian*.
c. Mixing of animals from different sources in a single consignment should be minimized.

4. Vessel and container design and maintenance

a. *Vessels* used for the sea transport of animals should be designed, constructed and fitted as appropriate to the species, size and weight of the animals to be transported. Special attention should be paid to the avoidance of injury to animals through the use of secure smooth fittings free from sharp protrusions and the provision of non-slippery flooring. The avoidance of injury to *animal handlers* while carrying out their responsibilities should be emphasised.
b. *Vessels* should be properly illuminated to allow animals to be observed and inspected.
c. *Vessels* should be designed to permit thorough cleaning and *disinfection*, and the management of faeces and urine.
d. *Vessels* and their fittings should be maintained in good mechanical and structural conditions.
e. *Vessels* should have adequate ventilation to meet variations in climate and the thermo-regulatory needs of the animal species being transported. The ventilation system should be effective when the *vessel* is stationary. An emergency power supply should be available to maintain ventilation in the case of primary machinery breakdown.
f. The feeding and watering system should be designed to permit adequate access to feed and water appropriate to the species, size and weight of the animals, and to minimise soiling of pens.
g. *Vessels* should be designed so that the faeces or urine from animals on upper levels do not soil animals on lower levels, or their feed or water.
h. Loading and stowage of feed and bedding should be carried out in such a way to ensure protection from fire hazards, the elements and sea water.
i. Where appropriate, suitable bedding, such as straw or sawdust, should be added to *vessel* floors to assist absorption of urine and faeces, provide better footing for animals and protect animals (especially young animals) from hard or rough flooring surfaces and adverse weather conditions.
j. The above principles apply also to *containers* used for the transport of animals.

5. Special provisions for transport in road vehicles on roll-on/ roll-off vessels or for containers

a. Road *vehicles* and *containers* should be equipped with a sufficient number of adequately designed, positioned and maintained securing points enabling them to be securely fastened to the *vessel*.
b. Road *vehicles* and *containers* should be secured to the ship before the start of the sea journey to prevent them being displaced by the motion of the *vessel*.
c. *Vessels* should have adequate ventilation to meet variations in climate and the thermo-regulatory needs of the animal species being transported, especially where the animals are transported in a secondary *vehicle*/*container* on enclosed decks.
d. Due to the risk of limited airflow on certain decks of a *vessel*, a road *vehicle* or *container* may require a forced ventilation system of greater capacity than that provided by natural ventilation.

6. Nature and duration of the journey

The maximum duration of a *journey* should be determined taking into account factors that determine the overall welfare of animals, such as:

a. the ability of the animals to cope with the stress of transport (such as very young, old, lactating or pregnant animals);
b. the previous transport experience of the animals;
c. the likely onset of fatigue;
d. the need for special attention;
e. the need for feed and water;
f. the increased susceptibility to injury and disease;
g. *space allowance* and *vessel* design;
h. weather conditions;
i. *vessel* type used, method of propulsion and risks associated with particular sea conditions.

7. Space allowance

a. The number of animals which should be transported on a *vessel* and their allocation to different pens on the *vessel* should be determined before *loading*.
b. The amount of space required, including headroom, depends on the species of animal and should allow the necessary thermoregulation. Each animal should be able to assume its natural position for transport (including during *loading* and *unloading*) without coming into contact with the roof or upper deck of the *vessel*. When animals lie down, there should be enough space for every animal to adopt a normal lying posture.
c. Calculations for the *space allowance* for each animal should be carried out in reference to a relevant national or international document. The size of pens will affect the number of animals in each.
d. The same principles apply when animals are transported in *containers*.

8. Ability to observe animals during the journey

Animals should be positioned to enable each animal to be observed regularly and clearly by an *animal handler* or other responsible person, during the *journey* to ensure their safety and good welfare.

9. Emergency response procedures

There should be an emergency management plan that identifies the important adverse events that may be encountered during the *journey*, the procedures for managing each event and the action to be taken in an emergency. For each important event, the plan should document the actions to be undertaken and the responsibilities of all parties involved, including communications and record keeping.

Article 3.7.2.6.

Documentation

1. Animals should not be loaded until the documentation required to that point is complete.

2. The documentation accompanying the consignment should include:
a. *journey* travel plan and emergency management plan;
b. time, date and place of *loading*;

c. the *journey* log – a daily record of inspection and important events which includes records of morbidity and mortality and actions taken, climatic conditions, food and water consumed, medication provided, mechanical defects;

d. expected time, date and place of arrival and *unloading*;

e. veterinary certification, when required;

f. *animal identification* to allow animal *traceability of animals* to the premises of departure, and, where possible, to the premises of origin;

g. details of any animals considered at particular risk of suffering poor welfare during transport (point 3e) of Article 3.7.2.7.);

h. number of *animal handlers* on board, and their competencies; and

i. *stocking density* estimate for each load in the consignment.

3. When veterinary certification is required to accompany consignments of animals, it should address:

a. when required, details of *disinfection* carried out;

b. fitness of the animals to travel;

c. *animal identification* (description, number, etc.); and

d. health status including any tests, treatments and vaccinations carried out.

Article 3.7.2.7.

Pre-journey period

1. General considerations

a. Before each *journey*, *vessels* should be thoroughly cleaned and, if necessary, treated for animal and public health purposes, using chemicals approved by the *Competent Authority*. When cleaning is necessary during a *journey*, this should be carried out with the minimum of stress to the animals.

b. In some circumstances, animals may require pre-*journey* assembly. In these circumstances, the following points should be considered:

　i. Pre-*journey* rest is necessary if the welfare of the animals has become poor during the collection period because of the physical environment or the social behaviour of the animals.

　ii. For animals such as pigs which are susceptible to motion sickness, and in order to reduce urine and faeces production during the *journey*, a species-specific short period of feed deprivation prior to *loading* is desirable.

　iii. When animals are to be provided with a novel diet or unfamiliar methods of supplying feed or water, they should be preconditioned.

c. Where an *animal handler* believes that there is a significant risk of disease among the animals to be loaded or significant doubt as to their fitness to travel, the animals should be examined by a *veterinarian*.

d. Pre-*journey* assembly/holding areas should be designed to:

　i. securely contain the animals;

　ii. maintain an environment safe from hazards, including predators and disease;

　iii. protect animals from exposure to adverse weather conditions;

　iv. allow for maintenance of social groups; and

　v. allow for rest, watering and feeding.

2. Selection of compatible groups

Compatible groups should be selected before transport to avoid adverse animal welfare consequences. The following guidelines should be applied when assembling groups of animals:

a. animals of different species should not be mixed unless they are judged to be compatible;
b. animals of the same species can be mixed unless there is a significant likelihood of aggression; aggressive individuals should be segregated (recommendations for specific species are described in detail in Article 3.7.2.12.). For some species, animals from different groups should not be mixed because poor welfare occurs unless they have established a social structure;
c. young or small animals may need to be separated from older or larger animals, with the exception of nursing mothers with young at foot;
d. animals with horns or antlers should not be mixed with animals lacking horns or antlers, unless judged to be compatible; and
e. animals reared together should be maintained as a group; animals with a strong social bond, such as a dam and offspring, should be transported together.

3. Fitness to travel

a. Animals should be inspected by a *veterinarian* or an *animal handler* to assess fitness to travel. If its fitness to travel is in doubt, it is the responsibility of a *veterinarian* to determine its ability to travel. Animals found unfit to travel should not be loaded onto a *vessel*.
b. Humane and effective arrangements should be made by the owner or agent for the handling and care of any animal rejected as unfit to travel.
c. Animals that are unfit to travel include, but may not be limited to:
   i. those that are sick, injured, weak, disabled or fatigued;
   ii. those that are unable to stand unaided or bear weight on each leg;
   iii. those that are blind in both eyes;
   iv. those that cannot be moved without causing them additional suffering;
   v. newborn with an unhealed navel;
   vi. females travelling without young which have given birth within the previous 48 hours;
   vii. pregnant animals which would be in the final 10% of their gestation period at the planned time of *unloading*;
   viii. animals with unhealed wounds from recent surgical procedures such as dehorning.
d. Risks during transport can be reduced by selecting animals best suited to the conditions of travel and those that are acclimatised to expected weather conditions.
e. Animals at particular risk of suffering poor welfare during transport and which require special conditions (such as in the design of facilities and *vehicles*, and the length of the *journey*) and additional attention during transport, may include:
   i. very large or obese individuals;
   ii. very young or old animals;
   iii. excitable or aggressive animals;
   iv. animals subject to motion sickness;
   v. animals which have had little contact with humans;
   vi. females in the last third of pregnancy or in heavy lactation.
f. Hair or wool length should be considered in relation to the weather conditions expected during transport.

Article 3.7.2.8.

Loading

1. Competent supervision
a. *Loading* should be carefully planned as it has the potential to be the cause of poor welfare in transported animals.

b. *Loading* should be supervised by the *Competent Authority* and conducted by *animal handler(s)*. *Animal handlers* should ensure that animals are loaded quietly and without unnecessary noise, harassment or force, and that untrained assistants or spectators do not impede the process.

2. Facilities

a. The facilities for *loading*, including the collecting area at the wharf, races and loading ramps should be designed and constructed to take into account the needs and abilities of the animals with regard to dimensions, slopes, surfaces, absence of sharp projections, flooring, sides, etc.

b. Ventilation during *loading* and the *journey* should provide for fresh air, and the removal of excessive heat, humidity and noxious fumes (such as ammonia and carbon monoxide). Under warm and hot conditions, ventilation should allow for the adequate convective cooling of each animal. In some instances, adequate ventilation can be achieved by increasing the *space allowance* for animals.

c. *Loading* facilities should be properly illuminated to allow the animals to be easily inspected by *animal handlers*, and to allow the ease of movement of animals at all times. Facilities should provide uniform light levels directly over approaches to sorting pens, chutes, loading ramps, with brighter light levels inside *vehicles/containers*, in order to minimise baulking. Dim light levels may be advantageous for the catching of some animals. Artificial lighting may be required.

3. Goads and other aids

When moving animals, their species specific behaviour should be used (see Article 3.7.2.12.). If goads and other aids are necessary, the following principles should apply:

a. Animals that have little or no room to move should not be subjected to physical force or goads and other aids which compel movement. Electric goads and prods should only be used in extreme cases and not on a routine basis to move animals. The use and the power output should be restricted to that necessary to assist movement of an animal and only when an animal has a clear path ahead to move. Goads and other aids should not be used repeatedly if the animal fails to respond or move. In such cases it should be investigated whether some physical or other impediment is preventing the animal from moving.

b. The use of such devices should be limited to battery-powered goads on the hindquarters of pigs and large ruminants, and never on sensitive areas such as the eyes, mouth, ears, anogenital region or belly. Such instruments should not be used on horses, sheep and goats of any age, or on calves or piglets.

c. Useful and permitted goads include panels, flags, plastic paddles, flappers (a length of cane with a short strap of leather or canvas attached), plastic bags and rattles; they should be used in a manner sufficient to encourage and direct movement of the animals without causing undue stress.

d. Painful procedures (including whipping, tail twisting, use of nose twitches, pressure on eyes, ears or external genitalia), or the use of goads or other aids which cause pain and suffering (including large sticks, sticks with sharp ends, lengths of metal piping, fencing wire or heavy leather belts), should not be used to move animals.

e. Excessive shouting at animals or making loud noises (e.g. through the cracking of whips) to encourage them to move should not occur as such actions may make the animals agitated, leading to crowding or falling.

f. The use of well trained dogs to help with the *loading* of some species may be acceptable.

g. Animals should be grasped or lifted in a manner which avoids pain or suffering and physical damage (e.g. bruising, fractures, dislocations). In the case of quadrupeds, manual lifting by a person should only be used in young animals or small species, and in a manner appropriate to the species; grasping or lifting animals only by their wool, hair, feathers, feet, neck, ears, tails, head, horns, limbs causing pain or suffering should not be permitted, except in an emergency where animal welfare or human safety may otherwise be compromised.

h. Conscious animals should not be thrown, dragged or dropped.

i. Performance standards should be established in which numerical scoring is used to evaluate the use of such instruments, and to measure the percentage of animals moved with an electric instrument and the percentage of animals slipping or falling as a result of their usage.

Article 3.7.2.9.

Travel

1. General considerations

a. *Animal handler(s)* should check the consignment immediately before departure to ensure that the animals have been loaded according to the load plan. Each consignment should be checked following any incident or situation likely to affect their welfare and in any case within 12 hours of departure.

b. If necessary and where possible adjustments should be made to the *stocking density* as appropriate during the *journey*.

c. Each pen of animals should be observed on a daily basis for normal behaviour, health and welfare, and the correct operation of ventilation, watering and feeding systems. There should also be a night patrol. Any necessary corrective action should be undertaken promptly.

d. Adequate access to suitable feed and water should be ensured for all animals in each pen.

e. Where cleaning or *disinfestation* is necessary during travel, it should be carried out with the minimum of stress to the animals.

2. Sick or injured animals

a. Sick or injured animals should be segregated.

b. Sick or injured animals should be appropriately treated or humanely killed, in accordance with a predetermined emergency response plan (Article 3.7.2.5.). Veterinary advice should be sought if necessary. All drugs and products should be used according to recommendations from a *veterinarian* and in accordance with the manufacturer's instructions.

c. A record of treatments carried out and their outcomes should be kept.

d. When humane killing is necessary, the *animal handler* must ensure that it is carried out humanely. Recommendations for specific species are described in Appendix 3.7.6. on killing of animals for disease control purposes. Veterinary advice regarding the appropriateness of a particular method of euthanasia should be sought as necessary.

Article 3.7.2.10.

Unloading and post-journey handling

1. General considerations

a. The required facilities and the principles of animal handling detailed in Article 3.7.2.8. apply equally to *unloading*, but consideration should be given to the likelihood that the animals will be fatigued.

b. *Unloading* should be carefully planned as it has the potential to be the cause of poor welfare in transported animals.

c. A livestock *vessel* should have priority attention when arriving in port and have priority access to a berth with suitable *unloading* facilities. As soon as possible after the *vessel*'s arrival at the port and acceptance of the consignment by the *Competent Authority*, animals should be unloaded into appropriate facilities.

d. The accompanying veterinary certificate and other documents should meet the requirements of the *importing country*. The veterinary inspection should be completed as quickly as possible.

e. *Unloading* should be supervised by the *Competent Authority* and conducted by *animal handler(s)*. The *animal handlers* should ensure that animals are unloaded as soon as possible after arrival but sufficient time should be allowed for *unloading* to proceed quietly and without unnecessary noise, harassment or force, and that untrained assistants or spectators do not impede the process.

2. Facilities

a. The facilities for *unloading* including the collecting area at the wharf, races and unloading ramps should be designed and constructed to take into account of the needs and abilities of the animals with regard to dimensions, slopes, surfaces, absence of sharp projections, flooring, sides, etc.

b. All *unloading* facilities should have sufficient lighting to allow the animals to be easily inspected by the *animal handlers*, and to allow ease of movement of animals at all times.

c. There should be facilities to provide animals with appropriate care and comfort, adequate space, access to quality feed and clean drinking water, and shelter from extreme weather conditions.

3. Sick or injured animals

a. An animal that has become sick, injured or disabled during a journey should be appropriately treated or humanely killed (see Appendix 3.7.6.). When necessary, veterinary advice should be sought in the care and treatment of these animals.

b. In some cases, where animals are non-ambulatory due to fatigue, injury or sickness, it may be in the best welfare interests of the animal to be treated or humanely killed aboard the *vessel*.

c. If *unloading* is in the best welfare interests of animals that are fatigued, injured or sick, there should be appropriate facilities and equipment for the humane *unloading* of such animals. These animals should be unloaded in a manner that causes the least amount of suffering. After *unloading*, separate pens and other appropriate facilities and treatments should be provided for sick or injured animals.

4. Cleaning and disinfection

a. *Vessels* and *containers* used to carry the animals should be cleaned before re-use through the physical removal of manure and bedding, by scraping, washing and flushing *vessels* and *containers* with water until visibly clean. This should be followed by *disinfection* when there are concerns about disease transmission.

b. Manure, litter and bedding should be disposed of in such a way as to prevent the transmission of disease and in compliance with all relevant health and environmental legislation.

Article 3.7.2.11.

Actions in the event of a refusal to allow the importation of a shipment

1. The welfare of the animals should be the first consideration in the event of a refusal to import.

2. When animals have been refused import, the *Competent Authority* of the *importing country* should make available suitable isolation facilities to allow the *unloading* of animals from a *vessel* and their secure holding, without posing a risk to the health of the national herd, pending resolution of the situation. In this situation, the priorities should be:

    a. The *Competent Authority* of the *importing country* should provide urgently in writing the reasons for the refusal.

    b. In the event of a refusal for animal health reasons, the *Competent Authority* of the *importing country* should provide urgent access to an OIE-appointed *veterinarian(s)* to assess the health status of the animals with regard to the concerns of the *importing country*, and the necessary facilities and approvals to expedite the required diagnostic testing.

    c. The *Competent Authority* of the *importing country* should provide access to allow continued assessment of the ongoing health and welfare situation.

    d. If the matter cannot be promptly resolved, the *Competent Authorities* of the *exporting* and *importing countries* should call on the OIE to mediate.

3. In the event that the animals are required to remain on the *vessel*, the priorities should be:

    a. The *Competent Authority* of the *importing country* should allow provisioning of the *vessel* with water and feed as necessary.

    b. The *Competent Authority* of the *importing country* should provide urgently in writing the reasons for the refusal.

    c. In the event of a refusal for animal health reasons, the *Competent Authority* of the *importing country* should provide urgent access to an OIE-appointed *veterinarian(s)* to assess the health status of the animals with regard to the concerns of the *importing country*, and the necessary facilities and approvals to expedite the required diagnostic testing.

    d. The *Competent Authority* of the *importing country* should provide access to allow continued assessment of the ongoing health and other aspects of the welfare of the animals, and the necessary actions to deal with any issues which arise.

    e. If the matter cannot be urgently resolved, the *Competent Authorities* of the *exporting* and *importing countries* should call on the OIE to mediate.

4. The OIE should utilise its dispute settlement mechanism to identify a mutually agreed solution which will address the animal health and welfare issues in a timely manner.

Article 3.7.2.12.

Species specific issues

Camelids of the new world in this context comprise llamas, alpacas, guanaco and vicuna. They have good eyesight and, like sheep, can negotiate steep slopes, though ramps should be as shallow as possible. They load most easily in a bunch as a single animal will strive to rejoin the others. Whilst they are usually docile, they have an unnerving habit of spitting in self-defence. During transport, they usually lie down. They frequently extend their front legs forward when lying, so gaps below partitions should be high enough so that their legs are not trapped when the animals rise.

Cattle are sociable animals and may become agitated if they are singled out. Social order is usually established at about two years of age. When groups are mixed, social order has to be re-established and aggression may occur until a new order is established. Crowding of cattle may also increase aggression as the animals try to maintain personal space. Social behaviour varies with age, breed and sex; *Bos indicus* and *B. indicus*-cross animals are usually more temperamental than European breeds. Young bulls, when moved in groups, show a degree of playfulness (pushing and shoving) but become more aggressive and territorial with age. Adult bulls have a minimum personal space of six square metres. Cows with young calves can be very protective, and handling calves in the presence of their mothers can be dangerous. Cattle tend to avoid "dead end" in passages.

Goats should be handled calmly and are more easily led or driven than if they are excited. When goats are moved, their gregarious tendencies should be exploited. Activities which frighten, injure or cause agitation to animals should be avoided. Bullying is particularly serious in goats. Housing strange goats together could result in fatalities, either through physical violence, or subordinate goats being refused access to food and water.

Horses in this context include all solipeds, donkeys, mules, hinnies and zebra. They have good eyesight and a very wide angle of vision. They may have a history of *loading* resulting in good or bad experiences. Good training should result in easier *loading*, but some horses can prove difficult, especially if they are inexperienced or have associated *loading* with poor transport conditions. In these circumstances, two experienced *animal handlers* can load an animal by linking arms or using a strop below its rump. Blindfolding may even be considered. Ramps should be as shallow as possible. Steps are not usually a problem when horses mount a ramp, but they tend to jump a step when descending, so steps should be as low as possible. Horses benefit from being individually stalled, but may be transported in compatible groups. When horses are to travel in groups, their shoes should be removed.

Pigs have poor eyesight, and may move reluctantly in unfamiliar surroundings. They benefit from well lit loading bays. Since they negotiate ramps with difficulty, these should be as level as possible and provided with secure footholds. Ideally, a hydraulic lift should be used for greater heights. Pigs also negotiate steps with difficulty. A good 'rule-of-thumb' is that no step should be higher than the pig's front knee. Serious aggression may result if unfamiliar animals are mixed. Pigs are highly susceptible to heat stress.

Sheep are sociable animals with good eyesight and tend to "flock together", especially when they are agitated. They should be handled calmly and their tendency to follow each other should be exploited when they are being moved. Sheep may become agitated if they are singled out for attention and will strive to rejoin the group. Activities which frighten, injure or cause agitation to sheep should be avoided. They can negotiate steep ramps.

Guidelines for the Transport of Animals by Land

PREAMBLE: These guidelines apply to the following live domesticated animals: cattle, buffaloes, camels, sheep, goats, pigs, poultry and equines. They will also be largely applicable to some other animals (e.g. deer, other camelids and ratites). Wild, feral and partly domesticated animals may need different conditions.

Article 3.7.3.1.

The amount of time animals spend on a *journey* should be kept to the minimum.

Article 3.7.3.2.

I. Animal behaviour

Animal handlers should be experienced and competent in handling and moving farm livestock and understand the behaviour patterns of animals and the underlying principles necessary to carry out their tasks.

The behaviour of individual animals or groups of animals will vary depending on their breed, sex, temperament and age and the way in which they have been reared and handled.

Despite these differences, the following behaviour patterns, which are always present to some degree in domestic animals, should be taken into consideration in handling and moving the animals.

Most domestic livestock are kept in herds and follow a leader by instinct.

Animals which are likely to harm each other in a group situation should not be mixed.

The desire of some animals to control their personal space should be taken into account in designing *loading* and *unloading* facilities, transport *vessels* and *containers*.

Domestic animals will try to escape if any person approaches closer than a certain distance. This critical distance, which defines the flight zone, varies among species and individuals of the same species, and depends upon previous contact with humans. Animals reared in close proximity to humans (i.e. tame) have a smaller flight zone, whereas those kept in free range or extensive systems may have flight zones which may vary from one metre to many metres. *Animal handlers* should avoid sudden penetration of the flight zone which may cause a panic reaction which could lead to aggression or attempted escape.

Animal handlers should use the point of balance at the animal's shoulder to move animals, adopting a position behind the point of balance to move an animal forward and in front of the point of balance to move it backward.

Domestic animals have a wide-angle vision but only have a limited forward binocular vision and poor perception of depth. This means that they can detect objects and movements beside and behind them, but can only judge distances directly ahead.

Although all domestic animals have a highly sensitive sense of smell, they may react differently to the smells encountered during travel. Smells which cause fear or other negative responses should be taken into consideration when managing animals.

Domestic animals can hear over a greater range of frequencies than humans and are more sensitive to higher frequencies. They tend to be alarmed by constant loud noises and by sudden noises, which may cause them to panic. Sensitivity to such noises should also be taken into account when handling animals.

An example of a flight zone (cattle)

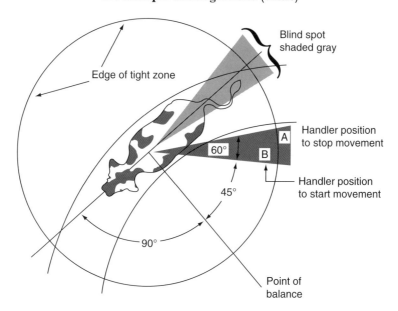

Handler movement pattern to move cattle forward

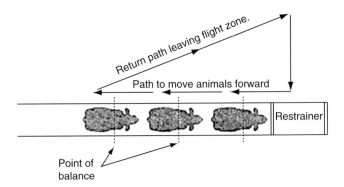

2. Distractions and their removal

Design of new *loading* and *unloading* facilities or modification of existing facilities should aim to minimise the potential for distractions that may cause approaching animals to stop, baulk or turn back. Below are examples of common distractions and methods for eliminating them:

a. reflections on shiny metal or wet floors – move a lamp or change lighting;
b. dark entrances – illuminate with indirect lighting which does not shine directly into the eyes of approaching animals;
c. animals seeing moving people or equipment up ahead – install solid sides on chutes and races or install shields;
d. dead ends-avoid if possible by curving the passage, or make an illusory passage;
e. chains or other loose objects hanging in chutes or on fences – remove them;
f. uneven floors or a sudden drop in floor levels – avoid uneven floor surfaces or install a solid false floor to provide an illusion of a solid and continuous walking surface;
g. sounds of air hissing from pneumatic equipment – install silencers or use hydraulic equipment or vent high pressure to the external environment using flexible hosing;
h. clanging and banging of metal objects – install rubber stops on gates and other devices to reduce metal to metal contact;
i. air currents from fans or air curtains blowing into the face of animals – redirect or reposition equipment.

Article 3.7.3.3.

Responsibilities

Once the decision to transport the animals has been made, the welfare of the animals during their *journey* is the paramount consideration and is the joint responsibility of all people involved. The individual responsibilities of persons involved will be described in more detail in this Article.

The roles of each of those responsible are defined below:

1. The owners and managers of the animals are responsible for:
    a. the general health, overall welfare and fitness of the animals for the *journey*;
    b. ensuring compliance with any required veterinary or other certification;

c. the presence of an *animal handler* competent for the species being transported during the *journey* with the authority to take prompt action; in case of transport by individual trucks, the truck driver may be the sole *animal handler* during the *journey*;

d. the presence of an adequate number of *animal handlers* during *loading* and *unloading*;

e. ensuring that equipment and veterinary assistance are provided as appropriate for the species and the *journey*.

2. Business agents or buying/selling agents are responsible for:

a. selection of animals that are fit to travel;

b. availability of suitable facilities at the start and at the end of the *journey* for the assembly; *loading*, transport, *unloading* and holding of animals, including for any stops at *resting points* during the *journey* and for emergencies.

3. *Animal handlers* are responsible for the humane handling and care of the animals, especially during *loading* and *unloading*, and for maintaining a *journey* log. To carry out their responsibilities, they should have the authority to take prompt action. In the absence of a separate *animal handler*, the driver is the *animal handler*.

4. Transport companies, *vehicle* owners and drivers are responsible for planning the *journey* to ensure the care of the animals; in particular they are responsible for:

a. choosing appropriate *vehicles* for the species transported and the *journey*;

b. ensuring that properly trained staff are available for *loading*/*unloading* of animals;

c. ensuring adequate competency of the driver in matters of animal welfare for the species being transported in case a separate *animal handler* is not assigned to the truck;

d. developing and and keeping up-to-date contingency plans to address emergencies (including adverse weather conditions) and minimise stress during transport;

e. producing a *journey* plan which includes a *loading* plan, *journey* duration, itinerary and location of resting places;

f. *loading* only those animals which are fit to travel, for their correct *loading* into the *vehicle* and their inspection during the *journey*, and for appropriate responses to problems arising; if its fitness to travel is in doubt, the animal should be examined by a *veterinarian* in accordance with point 3a) of Article 3.7.3.7.;

g. welfare of the animals during the actual transport.

5. Managers of facilities at the start and at the end of the *journey* and at *resting points* are responsible for:

a. providing suitable premises for *loading*, *unloading* and securely holding the animals, with water and feed when required, until further transport, sale or other use (including rearing or slaughter);

b. providing an adequate number of *animal handlers* to load, unload, drive and hold animals in a manner that causes minimum stress and injury; in the absence of a separate *animal handler*, the driver is the *animal handler*;

c. minimising the opportunities for disease transmission;

d. providing appropriate facilities, with water and feed when required;

e. providing appropriate facilities for emergencies;

f. providing facilities for washing and disinfecting *vehicles* after *unloading*;

g. providing facilities and competent staff to allow the humane *killing* of animals when required;

h. ensuring proper rest times and minimal delay during stops.

6. The responsibilities of *Competent Authorities* include:

a. establishing minimum standards for animal welfare, including requirements for inspection of animals before, during and after their travel, defining 'fitness to travel' and appropriate certification and record keeping;

b. setting standards for facilities, *containers* and *vehicles* for the transport of animals;

c. setting standards for the competence of *animal handlers*, drivers and managers of facilities in relevant issues in animal welfare;

d. ensuring appropriate awareness and training of *animal handlers*, drivers and managers of facilities in relevant issues in animal welfare;

e. implementation of the standards, including through accreditation of/interaction with other organisations;

f. monitoring and evaluating the effectiveness of standards of health;

g. monitoring and evaluating the use of veterinary medications;

h. giving animal consignments priority at frontiers in order to allow them to pass without unnecessary delay.

7. All individuals, including _veterinarians_, involved in transporting animals and the associated handling procedures should receive appropriate training and be competent to meet their responsibilities.

8. The receiving _Competent Authority_ should report back to the sending _Competent Authority_ on significant animal welfare problems which occurred during the _journey_.

Article 3.7.3.4.

Competence

1. All people responsible for animals during _journeys_, should be competent according to their responsibilities listed in Article 3.7.3.3. Competence may be gained through formal training and/or practical experience.

2. The assessment of the competence of _animal handlers_ should at a minimum address knowledge, and ability to apply that knowledge, in the following areas:

a. planning a _journey_, including appropriate _space allowance_, and feed, water and ventilation requirements;

b. responsibilities for animals during the _journey_, including _loading_ and _unloading_;

c. sources of advice and assistance;

d. animal behaviour, general signs of disease, and indicators of poor animal welfare such as stress, pain and fatigue, and their alleviation;

e. assessment of fitness to travel; if fitness to travel is in doubt, the animal should be examined by a _veterinarian_;

f. relevant authorities and applicable transport regulations, and associated documentation requirements;

g. general disease prevention procedures, including cleaning and _disinfection_;

h. appropriate methods of animal handling during transport and associated activities such as assembling, _loading_ and _unloading_;

i. methods of inspecting animals, managing situations frequently encountered during transport such as adverse weather conditions, and dealing with emergencies, including humane _killing_;

j. species-specific aspects and age-specific aspects of animal handling and care, including feeding, watering and inspection; and

k. maintaining a _journey_ log and other records.

Article 3.7.3.5.

Planning the journey

1. General considerations

a. Adequate planning is a key factor affecting the welfare of animals during a _journey_.

b. Before the _journey_ starts, plans should be made in relation to:

     i. preparation of animals for the *journey*;
     ii. choice of road, rail, roll-on roll-off vessels or *containers*;
     iii. nature and duration of the *journey*;
     iv. *vehicle* design and maintenance, including roll-on roll-off vessels;
     v. required documentation;
     vi. *space allowance*;
     vii. rest, water and feed;
     viii. observation of animals en route;
     ix. control of disease;
     x. emergency response procedures;
     xi. forecast weather conditions (e.g. conditions being too hot or too cold to travel during certain periods of the day);
     xii. transfer time when changing mode of transport, and
     xiii. waiting time at frontiers and inspection points.

c. Regulations concerning drivers (for example, maximum driving periods) should take into account animal welfare whenever is possible.

2. Preparation of animals for the journey

a. When animals are to be provided with a novel diet or method of water provision during transport, an adequate period of adaptation should be planned. Species-specific short period of feed deprivation prior to *loading* may be desirable.

b. Animals more accustomed to contact with humans and with being handled are likely to be less fearful of being loaded and transported. *Animal handlers* should handle and load animals in a manner that reduces their fearfulness and improves their approachability.

c. Behaviour-modifying compounds (such as tranquillisers) or other medication should not be used routinely during transport. Such compounds should only be administered when a problem exists in an individual animal, and should be administered by a *veterinarian* or other person who has been instructed in their use by a *veterinarian*.

3. Nature and duration of the journey

The maximum duration of a *journey* should be determined according to factors such as:

a. the ability of the animals to cope with the stress of transport (such as very young, old, lactating or pregnant animals);
b. the previous transport experience of the animals;
c. the likely onset of fatigue;
d. the need for special attention;
e. the need for feed and water;
f. the increased susceptibility to injury and disease;
g. *space allowance*, *vehicle* design, road conditions and driving quality;
h. weather conditions;
i. *vehicle* type used, terrain to be traversed, road surfaces and quality, skill and experience of the driver.

4. Vehicle and container design and maintenance

a. *Vehicles* and *containers* used for the transport of animals should be designed, constructed and fitted as appropriate for the species, size and weight of the animals to be transported. Special attention should be paid to avoid injury to animals through the use of secure smooth fittings free from sharp protrusions. The avoidance of injury to drivers and *animal handlers* while carrying out their responsibilities should be emphasised.

b. *Vehicles* and *containers* should be designed with the structures necessary to provide protection from adverse weather conditions and to minimise the opportunity for animals to escape.

c. In order to minimise the likelihood of the spread of infectious disease during transport, *vehicles* and *containers* should be designed to permit thorough cleaning and *disinfection*, and the containment of faeces and urine during a *journey*.

d. *Vehicles* and *containers* should be maintained in good mechanical and structural condition.

e. *Vehicles* and *containers* should have adequate ventilation to meet variations in climate and the thermo-regulatory needs of the animal species being transported; the ventilation system (natural or mechanical) should be effective when the *vehicle* is stationary, and the airflow should be adjustable.

f. *Vehicles* should be designed so that the faeces or urine from animals on upper levels do not soil animals on lower levels, nor their feed and water.

g. When *vehicles* are carried on board ferries, facilities for adequately securing them should be available.

h. If feeding or watering while the *vehicle* is moving is required, adequate facilities on the *vehicle* should be available.

i. When appropriate, suitable bedding should be added to *vehicle* floors to assist absorption of urine and faeces, to minimise slipping by animals, and protect animals (especially young animals) from hard flooring surfaces and adverse weather conditions.

5. Special provisions for transport in vehicles (road and rail) on roll-on/roll-off vessels or for containers

a. *Vehicles* and *containers* should be equipped with a sufficient number of adequately designed, positioned and maintained securing points enabling them to be securely fastened to the *vessel*.

b. *Vehicles* and *containers* should be secured to the *vessel* before the start of the sea *journey* to prevent them being displaced by the motion of the *vessel*.

c. Roll-on/roll-off *vessels* should have adequate ventilation to meet variations in climate and the thermo-regulatory needs of the animal species being transported, especially where the animals are transported in a secondary *vehicle/container* on enclosed decks.

6. Space allowance

a. The number of animals which should be transported on a *vehicle* or in a *container* and their allocation to compartments should be determined before *loading*.

b. The space required on a *vehicle* or in a *container* depends upon whether or not the animals need to lie down (for example, pigs, camels and poultry), or to stand (horses). Animals which will need to lie down often stand when first loaded or when the *vehicle* is driven with too much lateral movement or sudden braking.

c. When animals lie down, they should all be able to adopt a normal lying posture which allows necessary thermoregulation.

d. When animals are standing, they should have sufficient space to adopt a balanced position as appropriate to the climate and species transported.

e. The amount of headroom necessary depends on the species of animal. Each animal should be able to assume its natural position for transport (including during *loading* and *unloading*) without coming into contact with the roof or upper deck of the *vehicle*, and there should be sufficient headroom to allow adequate airflow over the animals.

f. Calculations for the *space allowance* for each animal should be carried out using the figures given in a relevant national or international document. The number and size of pens on the *vehicle* should be varied to where possible accommodate already established groups of animals while avoiding group sizes which are too large.

g. Other factors which may influence *space allowance* include:

    i. *vehicle*/ *container* design;
    ii. length of *journey*;
    iii. need to provide feed and water on the *vehicle*;
    iv. quality of roads;
    v. expected weather conditions;
    vi. category and sex of the animals.

7. Rest, water and feed

a. Suitable water and feed should be available as appropriate and needed for the species, age, and condition of the animals, as well as the duration of the *journey*, climatic conditions, etc.

b. Animals should be allowed to rest at *resting points* at appropriate intervals during the *journey*. The type of transport, the age and species of the animals being transported, and climatic conditions should determine the frequency of rest stops and whether the animals should be unloaded. Water and feed should be available during rest stops.

8. Ability to observe animals during the journey

a. Animals should be positioned to enable each animal to be observed regularly during the *journey* to ensure their safety and good welfare.

b. If the animals are in crates or on multi-tiered *vehicles* which do not allow free access for observation, for example where the roof of the tier is too low, animals cannot be inspected adequately, and serious injury or disease could go undetected. In these circumstances, a shorter *journey* duration should be allowed, and the maximum duration will vary according to the rate at which problems arise in the species and under the conditions of transport.

9. Control of disease

As animal transport is often a significant factor in the spread of infectious diseases, *journey* planning should take the following into account:

a. mixing of animals from different sources in a single consignment should be minimised;

b. contact at *resting points* between animals from different sources should be avoided;

c. when possible, animals should be vaccinated against diseases to which they are likely to be exposed at their destination;

d. medications used prophylactically or therapeutically should be approved by the *Veterinary Authority* of the *importing country* and should only be administered by a *veterinarian* or other person who has been instructed in their use by a *veterinarian*.

10. Emergency response procedures

There should be an emergency management plan that identifies the important adverse events that may be encountered during the *journey*, the procedures for managing each event and the action to be taken in an emergency. For each important event, the plan should document the actions to be undertaken and the responsibilities of all parties involved, including communications and record keeping.

11. Other considerations

a. Extreme weather conditions are hazardous for animals undergoing transport and require appropriate *vehicle* design to minimise risks. Special precautions should be taken for animals that have not been acclimatised or which are unsuited to either hot or cold conditions. In some extreme conditions of heat or cold, animals should not be transported at all.

b. In some circumstances, transportation during the night may reduce thermal stress or the adverse effects of other external stimuli.

Article 3.7.3.6.

Documentation

1. Animals should not be loaded until the documentation required to that point is complete.
2. The documentation accompanying the consignment should include:
   a. *journey* travel plan and emergency management plan;
   b. date, time and place of *loading* and *unloading*;
   c. veterinary certification, when required;
   d. animal welfare competencies of the driver (under study);
   e. *animal identification* to allow *animal traceability* to the premises of departure and, where possible, to the premises of origin;
   f. details of any animals considered at particular risk of suffering poor welfare during transport (point 3e) of Article 3.7.3.7.;
   g. documentation of the period of rest, and access to feed and water, prior to the *journey*;
   h. *stocking density* estimate for each load in the consignment;
   i. the *journey* log – daily record of inspection and important events, including records of morbidity and mortality and actions taken, climatic conditions, rest stops, travel time and distance, feed and water offered and estimates of consumption, medication provided, and mechanical defects.
3. When veterinary certification is required to accompany consignments of animals, it should address:
   a. fitness of animals to travel;
   b. *animal identification* (description, number, etc.);
   c. health status including any tests, treatments and vaccinations carried out;
   d. when required, details of *disinfection* carried out.
   At the time of certification, the *veterinarian* should notify the *animal handler* or the driver of any factors affecting the fitness of animals to travel for a particular *journey*.

Article 3.7.3.7.

Pre-journey period

1. General considerations

a. Pre-*journey* rest is necessary if the welfare of animals has become poor during the collection period because of the physical environment or the social behaviour of the animals. The need for rest should be judged by a *veterinarian* or other competent person.
b. Pre-*journey* assembly/holding areas should be designed to:
   i. securely hold the animals;
   ii. maintain a safe environment from hazards, including predators and disease;
   iii. protect animals from exposure to severe weather conditions;
   iv. allow for maintenance of social groups;
   v. allow for rest, and appropriate water and feed.
c. Consideration should be given to the previous transport experience, training and conditioning of the animals, if known, as these may reduce fear and stress in animals.

d. Feed and water should be provided pre-*journey* if the *journey* duration is greater than the normal inter-feeding and drinking interval for the animal. Recommendations for specific species are described in detail in Article 3.7.3.12.

e. When animals are to be provided with a novel diet or method of feed or water provision during the *journey*, an adequate period of adaptation should be allowed.

f. Before each *journey*, *vehicles* and *containers* should be thoroughly cleaned and, if necessary, treated for animal health and public health purposes, using methods approved by the *Competent Authority*. When cleaning is necessary during a *journey*, this should be carried out with the minimum of stress to the animals.

g. Where an *animal handler* believes that there is a significant risk of disease among the animals to be loaded or significant doubt as to their fitness to travel, the animals should be examined by a *veterinarian*.

2. Selection of compatible groups

Compatible groups should be selected before transport to avoid adverse animal welfare consequences. The following guidelines should be applied when assembling groups of animals:

a. Animals reared together should be maintained as a group; animals with a strong social bond, such as a dam and offspring, should be transported together.

b. Animals of the same species can be mixed unless there is a significant likelihood of aggression; aggressive individuals should be segregated (recommendations for specific species are described in detail in Article 3.7.3.12.). For some species, animals from different groups should not be mixed because poor welfare occurs unless they have established a social structure.

c. Young or small animals should be separated from older or larger animals, with the exception of nursing mothers with young at foot.

d. Animals with horns or antlers should not be mixed with animals lacking horns or antlers unless judged to be compatible.

e. Animals of different species should not be mixed unless they are judged to be compatible.

3. Fitness to travel

a. Each animal should be inspected by a *veterinarian* or an *animal handler* to assess fitness to travel. If its fitness to travel is in doubt, the animal should be examined by a *veterinarian*. Animals found unfit to travel should not be loaded onto a *vehicle*, except for transport to receive veterinary treatment.

b. Humane and effective arrangements should be made by the owner and the agent for the handling and care of any animal rejected as unfit to travel.

c. Animals that are unfit to travel include, but may not be limited to:

    i. those that are sick, injured, weak, disabled or fatigued;

    ii. those that are unable to stand unaided and bear weight on each leg;

    iii. those that are blind in both eyes;

    iv. those that cannot be moved without causing them additional suffering;

    v. newborn with an unhealed navel;

    vi. pregnant animals which would be in the final 10% of their gestation period at the planned time of *unloading*;

    vii. females travelling without young which have given birth within the previous 48 hours;

    viii. those whose body condition would result in poor welfare because of the expected climatic conditions.

d. Risks during transport can be reduced by selecting animals best suited to the conditions of travel and those that are acclimatised to expected weather conditions.

e. Animals at particular risk of suffering poor welfare during transport and which require special conditions (such as in the design of facilities and *vehicles*, and the length of the *journey*) and additional attention during transport, may include:

    i. large or obese individuals;

    ii. very young or old animals;

    iii. excitable or aggressive animals;

    iv. animals which have had little contact with humans;

    v. animal subject to motion sickness;

    vi. females in late pregnancy or heavy lactation, dam and offspring;

    vii. animals with a history of exposure to stressors or pathogenic agents prior to transport;

    viii. animals with unhealed wounds from recent surgical procedures such as dehorning.

4. Specific species requirements

Transport procedures should be able to take account of variations in the behaviour of the species. Flight zones, social interactions and other behaviour vary significantly among species and even within species. Facilities and handling procedures that are successful with one species are often ineffective or dangerous with another.

    Recommendations for specific species are described in detail in Article 3.7.3.12.

Article 3.7.3.8.

Loading

1. Competent supervision

a. *Loading* should be carefully planned as it has the potential to be the cause of poor welfare in transported animals.

b. *Loading* should be supervised and/or conducted by *animal handlers*. The animals are to be loaded quietly and without unnecessary noise, harassment or force. Untrained assistants or spectators should not impede the process.

c. When *containers* are loaded onto a *vehicle*, this should be carried out in such a way to avoid poor animal welfare.

2. Facilities

a. The facilities for *loading* including the collecting area, races and loading ramps should be designed and constructed to take into account the needs and abilities of the animals with regard to dimensions, slopes, surfaces, absence of sharp projections, flooring, etc.

b. *Loading* facilities should be properly illuminated to allow the animals to be observed by *animal handler(s)*, and to allow the ease of movement of the animals at all times. Facilities should provide uniform light levels directly over approaches to sorting pens, chutes, loading ramps, with brighter light levels inside *vehicles/containers*, in order to minimise baulking. Dim light levels may be advantageous for the catching of poultry and some other animals. Artificial lighting may be required.

c. Ventilation during *loading* and the *journey* should provide for fresh air, the removal of excessive heat, humidity and noxious fumes (such as ammonia and carbon monoxide), and the prevention of accumulations of ammonia and carbon dioxide. Under warm and hot conditions, ventilation should allow for the adequate convective cooling of each animal. In some instances, adequate ventilation can be achieved by increasing the *space allowance* for animals.

3. Goads and other aids

When moving animals, their species specific behaviour should be used (see Article 3.7.3.12.). If goads and other aids are necessary, the following principles should apply:

a. Animals that have little or no room to move should not be subjected to physical force or goads and other aids which compel movement. Electric goads and prods should only be used in extreme cases and not on a routine basis to move animals. The use and the power output should be restricted to that necessary to assist movement of an animal and only when an animal has a clear path ahead to move. Goads and other aids should not be used repeatedly if the animal fails to respond or move. In such cases it should be investigated whether some physical or other impediment is preventing the animal from moving.

b. The use of such devices should be limited to battery-powered goads on the hindquarters of pigs and large ruminants, and never on sensitive areas such as the eyes, mouth, ears, anogenital region or belly. Such instruments should not be used on horses, sheep and goats of any age, or on calves or piglets.

c. Useful and permitted goads include panels, flags, plastic paddles, flappers (a length of cane with a short strap of leather or canvas attached), plastic bags and rattles; they should be used in a manner sufficient to encourage and direct movement of the animals without causing undue stress.

d. Painful procedures (including whipping, tail twisting, use of nose twitches, pressure on eyes, ears or external genitalia), or the use of goads or other aids which cause pain and suffering (including large sticks, sticks with sharp ends, lengths of metal piping, fencing wire or heavy leather belts), should not be used to move animals.

e. Excessive shouting at animals or making loud noises (e.g. through the cracking of whips) to encourage them to move should not occur, as such actions may make the animals agitated, leading to crowding or falling.

f. The use of well trained dogs to help with the *loading* of some species may be acceptable.

g. Animals should be grasped or lifted in a manner which avoids pain or suffering and physical damage (e.g. bruising, fractures, dislocations). In the case of quadrupeds, manual lifting by a person should only be used in young animals or small species, and in a manner appropriate to the species; grasping or lifting animals only by their wool, hair, feathers, feet, neck, ears, tails, head, horns, limbs causing pain or suffering should not be permitted, except in an emergency where animal welfare or human safety may otherwise be compromised.

h. Conscious animals should not be thrown, dragged or dropped.

i. Performance standards should be established in which numerical scoring is used to evaluate the use of such instruments, and to measure the percentage of animals moved with an electric instrument and the percentage of animals slipping or falling as a result of their usage.

Article 3.7.3.9.

Travel

1. General considerations

a. Drivers and *animal handlers* should check the load immediately before departure to ensure that the animals have been properly loaded. Each load should be checked again early in the trip and adjustments made as appropriate. Periodic checks should be made throughout the trip, especially at rest or refuelling stops or during meal breaks when the *vehicle* is stationary.

b. Drivers should utilise smooth, defensive driving techniques, without sudden turns or stops, to minimise uncontrolled movements of the animals.

2. Methods of restraining or containing animals

a. Methods of restraining animals should be appropriate to the species and age of animals involved and the training of the individual animal.

b. Recommendations for specific species are described in detail in Article 3.7.3.12.

3. Regulating the environment within vehicles or containers

a. Animals should be protected against harm from hot or cold conditions during travel. Effective ventilation procedures for maintaining the environment within *vehicles* or *containers* will vary according to whether conditions are cold, hot and dry or hot and humid, but in all conditions a build-up of noxious gases should be prevented.

b. The environment within *vehicles* or *containers* in hot and warm weather can be regulated by the flow of air produced by the movement of the *vehicle*. In warm and hot weather, the duration of *journey* stops should be minimised and *vehicles* should be parked under shade, with adequate and appropriate ventilation.

c. To minimise slipping and soiling, and maintain a healthy environment, urine and faeces should be removed from floors when necessary and disposed of in such a way as to prevent the transmission of disease and in compliance with all relevant health and environmental legislation.

4. Sick, injured or dead animals

a. A driver or an *animal handler* finding sick, injured or dead animals should act according to a predetermined emergency response plan.

b. Sick or injured animals should be segregated.

c. Ferries (roll-on roll-off) should have procedures to treat sick or injured animals during the *journey*.

d. In order to reduce the likelihood that animal transport will increase the spread of infectious disease, contact between transported animals, or the waste products of the transported animals, and other farm animals should be minimised.

e. During the *journey*, when disposal of a dead animal becomes necessary, this should be carried out in such a way as to prevent the transmission of disease and in compliance with all relevant health and environmental legislation.

f. When *killing* is necessary, it should be carried out as quickly as possible and assistance should be sought from a *veterinarian* or other person(s) competent in humane *killing* procedures. Recommendations for specific species are described in Appendix 3.7.6. on killing of animals for disease control purposes.

5. Water and feed requirements

a. If *journey* duration is such that feeding or watering is required or if the species requires feed or water throughout, access to suitable feed and water for all the animals (appropriate for their species and age) carried in the *vehicle* should be provided. There should be adequate space for all animals to move to the feed and water sources and due account taken of likely competition for feed.

b. Recommendations for specific species are described in detail in Article 3.7.3.12.

6. Rest periods and conditions including hygiene

a. Animals that are being transported should be rested at appropriate intervals during the *journey* and offered feed and water, either on the *vehicle* or, if necessary, unloaded into suitable facilities.

b. Suitable facilities should be used en route, when resting requires the *unloading* of the animals. These facilities should meet the needs of the particular animal species and should allow access of all animals to feed and water.

7. In-transit observations

a. Animals being transported by road should be observed soon after a *journey* is commenced and whenever the driver has a rest stop. After meal breaks and refuelling stops, the animals should be observed immediately prior to departure.

b. Animals being transported by rail should be observed at each scheduled stop. The responsible rail transporter should monitor the progress of trains carrying animals and take all appropriate action to minimise delays.

c. During stops, it should be ensured that the animals continue to be properly confined, have appropriate feed and water, and their physical condition is satisfactory.

Article 3.7.3.10.

Unloading and post-journey handling

1. General considerations

a. The required facilities and the principles of animal handling detailed in Article 3.7.3.8. apply equally to *unloading*, but consideration should be given to the likelihood that the animals will be fatigued.

b. *Unloading* should be supervised and/or conducted by an *animal handler* with knowledge and experience of the behavioural and physical characteristics of the species being unloaded. Animals should be unloaded from the *vehicle* into appropriate facilities as soon as possible after arrival at the destination but sufficient time should be allowed for *unloading* to proceed quietly and without unnecessary noise, harassment or force.

c. Facilities should provide all animals with appropriate care and comfort, adequate space and ventilation, access to feed (if appropriate) and water, and shelter from extreme weather conditions.

d. For details regarding the *unloading* of animals at a *slaughterhouse*, see Appendix 3.7.5. on slaughter of animals for human consumption.

2. Sick or injured animals

a. An animal that has become sick, injured or disabled during a *journey* should be appropriately treated or humanely killed (see Appendix 3.7.6. on killing of animals for disease control purposes). If necessary, veterinary advice should be sought in the care and treatment of these animals. In some cases, where animals are non-ambulatory due to fatigue, injury or sickness, it may be in the best welfare interests of the animal to be treated or killed aboard the *vehicle*. Assistance should be sought from a *veterinarian* or other person(s) competent in humane *killing* procedures.

b. At the destination, the *animal handler* or the driver during transit should ensure that responsibility for the welfare of sick, injured or disabled animals is transferred to a *veterinarian* or other suitable person.

c. If treatment or humane *killing* is not possible aboard the *vehicle*, there should be appropriate facilities and equipment for the humane *unloading* of animals that are non-ambulatory due to fatigue, injury or sickness. These animals should be unloaded in a manner that causes the least amount of suffering. After *unloading*, separate pens and other appropriate facilities should be available for sick or injured animals.

d. Feed, if appropriate, and water should be available for each sick or injured animal.

3. Addressing disease risks

The following should be taken into account in addressing the greater risk of disease due to animal transport and the possible need for segregation of transported animals at the destination:

a. increased contact among animals, including those from different sources and with different disease histories;

b. increased shedding of pathogens and increased susceptibility to infection related to stress and impaired defences against disease, including immunosuppression;

c. exposure of animals to pathogens which may contaminate *vehicles*, *resting points*, *markets*, etc.

4. Cleaning and disinfection

a. *Vehicles*, crates, *containers*, etc. used to carry the animals should be cleaned before re-use through the physical removal of manure and bedding by scraping, washing and flushing with water and detergent. This should be followed by *disinfection* when there are concerns about disease transmission.

b. Manure, litter, bedding and the bodies of any animals which die during the *journey* should be disposed of in such a way as to prevent the transmission of disease and in compliance with all relevant health and environmental legislation.

c. Establishments like livestock *markets*, *slaughterhouses*, resting sites, railway stations, etc. where animals are unloaded should be provided with appropriate areas for the cleaning and *disinfection* of *vehicles*.

Article 3.7.3.11.

Actions in the event of a refusal to allow the completion of the journey

1. The welfare of the animals should be the first consideration in the event of a refusal to allow the completion of the *journey*.

2. When the animals have been refused import, the *Competent Authority* of the *importing country* should make available suitable isolation facilities to allow the *unloading* of animals from a *vehicle* and their secure holding, without posing a risk to the health of national herd or flock, pending resolution of the situation. In this situation, the priorities should be:

a. the *Competent Authority* of the *importing country* should provide urgently in writing the reasons for the refusal;

b. in the event of a refusal for animal health reasons, the *Competent Authority* of the *importing country* should provide urgent access to a *veterinarian*, where possible an OIE *veterinarian(s)* appointed by the Director General, to assess the health status of the animals with regard to the concerns of the *importing country*, and the necessary facilities and approvals to expedite the required diagnostic testing;

c. the *Competent Authority* of the *importing country* should provide access to allow continued assessment of the health and other aspects of the welfare of the animals;

d. if the matter cannot be promptly resolved, the *Competent Authorities* of the *exporting* and *importing countries* should call on the OIE to mediate.

3. In the event that a *Competent Authority* requires the animals to remain on the *vehicle*, the priorities should be:

a. to allow provisioning of the *vehicle* with water and feed as necessary;

b. to provide urgently in writing the reasons for the refusal;

c. to provide urgent access to an independent *veterinarian(s)* to assess the health status of the animals, and the necessary facilities and approvals to expedite the required diagnostic testing in the event of a refusal for animal health reasons;

d. to provide access to allow continued assessment of the health and other aspects of the welfare of the animals, and the necessary actions to deal with any animal issues which arise.

4. The OIE should utilise its dispute settlement mechanism to identify a mutually agreed solution which will address animal health and any other welfare issues in a timely manner.

Article 3.7.3.12.

Species specific issues

Camelids of the new world in this context comprise llamas, alpacas, guanaco and vicuna. They have good eyesight and, like sheep, can negotiate steep slopes, though ramps should be as shallow as possible. They load most easily in a bunch as a single animal will strive to rejoin the others. Whilst they are usually docile, they have an unnerving habit of spitting in self-defence. During transport, they usually lie down. They frequently extend their front legs forward when lying, so gaps below partitions should be high enough so that their legs are not trapped when the animals rise.

Cattle are sociable animals and may become agitated if they are singled out. Social order is usually established at about two years of age. When groups are mixed, social order has to be re-established and aggression may occur until a new order is established. Crowding of cattle may also increase aggression as the animals try to maintain personal space. Social behaviour varies with age, breed and sex; *Bos indicus* and *B. indicus*-cross animals are usually more temperamental than European breeds. Young bulls, when moved in groups, show a degree of playfulness (pushing and shoving) but become more aggressive and territorial with age. Adult bulls have a minimum personal space of six square metres. Cows with young calves can be very protective, and handling calves in the presence of their mothers can be dangerous. Cattle tend to avoid "dead end" in passages.

Goats should be handled calmly and are more easily led or driven than if they are excited. When goats are moved, their gregarious tendencies should be exploited. Activities which frighten, injure or cause agitation to animals should be avoided. Bullying is particularly serious in goats and can reflect demands for personal space. Housing strange goats together could result in fatalities, either through physical violence, or subordinate goats being refused access to food and water.

Horses in this context include donkeys, mules and hinnies. They have good eyesight and a very wide angle of vision. They may have a history of *loading* resulting in good or bad experiences. Good training should result in easier *loading*, but some horses can prove difficult, especially if they are inexperienced or have associated *loading* with poor transport conditions. In these circumstances, two experienced *animal handlers* can load an animal by linking arms or using a strop below its rump. Blindfolding may even be considered. Ramps should be as shallow as possible. Steps are not usually a problem when horses mount a ramp, but they tend to jump a step when descending, so steps should be as low as possible. Horses benefit from being individually stalled, but may be transported in compatible groups. When horses are to travel in groups, their shoes should be removed. Horses are prone to respiratory disease if they are restricted by period by tethers that prevent the lowering and lifting of their heads.

Pigs have poor eyesight, and may move reluctantly in unfamiliar surroundings. They benefit from well lit loading bays. Since they negotiate ramps with difficulty, these should be as level as possible and provided with secure footholds. Ideally, a hydraulic lift should be used for greater heights. Pigs also negotiate steps with difficulty. A good 'rule-of-thumb' is that no step should be higher than the pig's front knee. Serious aggression may result if unfamiliar animals are mixed. Pigs are highly susceptible to heat stress.

Sheep are sociable animals with good eyesight, a relatively subtle and undemonstrative behaviour and a tendency to "flock together", especially when they are agitated. They should be handled calmly and their tendency to follow each other should be exploited when they are being moved. Crowding of sheep may lead to damaging aggressive and submissive behaviours as animals try to maintain personal space. Sheep may become agitated if they are singled out for attention, or kept alone, and will strive to rejoin the group. Activities which frighten, injure or cause agitation to sheep should be avoided. They can negotiate steep ramps.

Index